Lecture Notes in Artificial Intelligence 1257

Subseries of Lecture Notes in Comp
Edited by J. G. Carbonell and J. Siek

Lecture Notes in Computer Science

Edited by G. Goos, J. Hartmanis and J. van Leeuwen

Springer
Berlin
Heidelberg
New York
Barcelona
Budapest
Hong Kong
London
Milan
Paris
Santa Clara
Singapore
Tokyo

Dickson Lukose   Harry Delugach
Mary Keeler   Leroy Searle   John Sowa (Eds.)

Conceptual Structures: Fulfilling Peirce's Dream

Fifth International Conference
on Conceptual Structures, ICCS'97
Seattle, Washington, USA, August 3-8, 1997
Proceedings

Springer

Volume Editors

Dickson Lukose
The University of Calgary, Department of Computer Science
2500 University Drive N.W., Calgary, Alberta, Canada T2N 1N4
E-mail: lukose@cpsc.ucalgary.ca

Harry Delugach
University of Alabama in Huntsville, Department of Computer Science
Huntsville, AL 35899, USA
E-mail: delugach@cs.uah.edu

Mary Keeler
Leroy Searle
University of Washington, Center for the Humanities
Box 353910, Seattle, WA 98195, USA
E-mail: mkeeler@u.washington.edu
          LSEARLE@humanities.badm.washington.edu

John Sowa
SUNY Binghamton
Binghamton, USA
E-mail: sowa@west.poly.edu
          sowa@watson.ibm.com

Cataloging-in-Publication Data applied for

Die Deutsche Bibliothek - CIP-Einheitsaufnahme

Conceptual structures : fulfilling Peirce's dream ; proceedings / 5th
International Conference on Conceptual Structures, ICCS '97, Seattle,
Washington, USA, August 3 - 8, 1997. Dickson Lukose ... (ed.). -
Berlin ; Heidelberg ; New York ; Barcelona ; Budapest ; Hong Kong
; London ; Milan ; Paris ; Santa Clara ; Singapore ; Tokyo : Springer,
1997
    (Lecture notes in computer science ; Vol. 1257 : Lecture notes in
    artificial intelligence)
    ISBN 3-540-63308-1

CR Subject Classification (1991): I.2, H.2.1, G.2.2

ISBN 3-540-63308-1 Springer-Verlag Berlin Heidelberg New York

© Springer-Verlag Berlin Heidelberg 1997
Printed in Germany

Typesetting: Camera ready by author
SPIN 10549789   06/3142 – 5 4 3 2 1 0   Printed on acid-free paper

Preface

The International Conference on Conceptual Structures (ICCS) is the annual conference and principal research forum in the practice and theory of conceptual structures. Conceptual graphs are a form of conceptual structures that constitute a logic-based formalism for knowledge representation. Conceptual graphs are based on the existential graphs of Charles S. Peirce and semantic networks. The Fifth International Conference on Conceptual Structures (ICCS'97) marks the twelfth international gathering of researchers focusing on conceptual graphs and conceptual structures. Since 1986, there have been seven workshops and four conferences. Participation in conceptual structures research has continued to grow; at present, there are hundreds of researchers in over 30 countries. There are a dozen countries represented at the present conference. Efforts have spanned a number of research areas, such as natural language processing, knowledge acquisition, information retrieval, and the philosophical foundations of logic.

In organizing this conference, we have made a deliberate effort to include supporting theories (e.g., formal concept analysis) and practical applications (e.g., the CGTOOLS sessions) in order to enhance and expand the value of conceptual graphs. These proceedings, first of all, cover a very wide range of topics, seen from different points of view. For the first time, we have also included a section in the proceedings for extended abstracts of the "Conceptual Graphs Tools" sessions that have traditionally concluded ICCS meetings, to provide a formal record of presentation and to disseminate information about tools and applications. This conference marks the first one in its series in which the submission of papers, distribution of papers to reviewers, and submission of reviews was performed almost entirely using the Internet (principally the World Wide Web). We believe Peirce would be pleased at this appropriate use of technology in making information available to people.

The chairs of this conference take this opportunity to thank the members of the Programme Committee and the CGTOOLS Committee who have reviewed the papers we received. Additional reviewers include Diana Cukierman, Stephane Lapalut, and Stephen Callaghan. The chairs extend a very special thanks to the members of the Editorial Board who had the very difficult task of making recommendation on all papers that attracted large differences in reviewers comments and ratings. This conference would not have been possible without the hard work of these editorial board members. We also thank University of New England (Australia), University of Calgary (Canada), University of Alabama in Huntsville (USA), and University of Washington (USA) for their support. We are also very grateful to AAAI (USA), CSCSI (Canada), ACS-AIES (Australia) and The Boeing Company (USA) for their generous support.

Seattle, August 1997

Dickson Lukose
Harry Delugach, Mary Keeler
Leroy Searle and John Sowa

Programme Committee

Table of Contents

KNOWLEDGE MODELING

FORMAL CONCEPT ANALYSIS

FORMAL REASONING

APPLICATIONS OF CONCEPTUAL GRAPHS

CONCEPTUAL GRAPH TOOLS

Fulfilling Peirce's Dream: Conceptual Structures and Communities of Inquiry

Leroy Searle[1], Mary Keeler[1], John Sowa[2], Harry Deluagch[3], and Dickson Lukose[4]

Center for the Humanities, University of Washington, USA[1]
Philosophy and Computers and Cognitive Science, SUNY Binghamton, USA[2]
Department of Computer Science, University of Alabama in Huntsville, USA[3]
Department of Mathematics, Statistics, and Computer Science, University of New England, Australia[4]

Abstract. C. S. Peirce, as a philosopher, logician, and scientist, argued that we should "adopt our logic as our metaphysics," and that a system of graphical logic could provide an experimental means to "illustrate the general course of thought." In particular, Peirce hoped that his Existential Graphs would develop into "the logic of the future." This paper provides an introduction to the proceedings of the International Conference on Conceptual Structures '97, and explores these two propositions as the basis for what may be the first articulate model of information and knowledge processing, and the direct antecedant of Conceptual Graphs, first developed in the work of John Sowa [19] [20]. The kind of "experiments" Peirce foresaw are exact (but still intuitively available) representations of problems and conceptual relations that permit the development of research groups and communities, such as the Conceptual Graphs and Formal Concept Analysis communities today. The paper outlines the development of Conceptual Graphs from Peirce's Existential Graphs and more recent sources in logic, computational linguistics, semantic networks, and artificial intelligence. A brief account of the planning and the program for ICCS'97 follows, detailing the intention of the Program Committee and the Editorial Board to broaden participation in this conference, and in the growing community of inquiry in many fields where a graphical representation of conceptual structures is proving fruitful.

1 Introduction

In November, 1866, Charles Sanders Peirce began the last of his lectures at the Lowell Institute with the observation that the only successful mode of conducting metaphysical research "is that of adopting our logic as our metaphysics" [13]. Though barely twenty five, Peirce had already made a profound and penetrating study of Kant, had developed incisive critiques of Boole and Whewell, Hamilton and John Stuart Mill, and had established a formidable local reputation as a logician, a mathematician, and a scientific experimentalist. In context, however, there is a striking contrast of language between this high minded, even lofty lecture on the question, "What is Man?" (which first introduces one of Peirce's favorite themes, "Man's Glassy Essence"), and all the lectures that had come before. The Lowell lectures, though perhaps less technical than his Harvard lectures of 1865, "The Logic of Science," show Peirce as a formal logician in the modern mode he was even then helping to create, not very much inclined to protect a general audience from the rigors of exact formal reasoning. To link logic to metaphysics, therefore, might just have passed as one more among many hard things to think about.

Forty years later, in 1906, we find another striking remark which opens Peirce's "Prolegomena to an Apology for Pragmaticism," inviting the reader to join him to "construct a diagram to illustrate the general course of thought; I mean a System of diagrammatization by means of which any course of thought can be represented with exactitude." Peirce was referring, of course, to his system of Existential Graphs which he had developed to a reasonable state of exactitude in the preceding decade, as a means for making "exact experiments upon uniform diagrams." Peirce meant this quite literally, asserting that "operations upon diagrams, whether external or imaginary, take the place of experiments upon real things that one performs in chemical and physical research" [12]. There is little doubt that this idea was virtually incomprehensible to Peirce's contemporaries, even his friends and supporters. While Peirce hoped that his existential graphs would be "the logic of the future," the fact is that at least until early work by Jay Zeeman in 1964 [23] and Don Roberts's pivotal work in 1973, no one had "agreed with [Peirce] regarding the value of his graphs" [16].

A century after Peirce's remarkable invention, it is much easier to see that his idea of representing "the general course of thought" by "uniform diagrams" is arguably the first articulate model of knowledge and information processing; and as the essays in this volume indicate even more clearly, there is a growing and diverse community that does indeed see the value of Peirce's graphs. Roberts's demonstration that the Alpha and Beta parts of Peirce's Existential Graphs are consistent and complete, vis a vis first order predicate calculus [16] (a point on which Peirce had already satisfied himself), showed at least that Peirce was partly justified about his "logic of the future." John Sowa's path-breaking book on conceptual graphs [20], however, provided a formal implementation of a Peirce-based graphical logic that was also computationally tractable. Conceptual graphs, from that point of departure, have provided a rich and fruitful instrument for research and the development of tools for knowledge processing and the construction of knowledge bases, independently verified by Eileen Way in her keynote address at ICCS'94 [22].

As the papers in this volume show, there has been significant progress toward the future that Peirce imagined, but they also indicate that there is still serious and significant disagreement about how to interpret or specify the value of a graphical approach to logic and knowledge representation. In the community of researchers who have concentrated primarily on conceptual graphs, following Sowa, fundamental theoretical problems continue to emerge, while the development of working applications and research tools has opened up an array of possible paths for future work. But other approaches, such as Formal Concept

Analysis, as exemplified in the work of Rudolph Wille and others, appearing also in this volume, make clear that the field of knowledge representation, acquisition, and processing is carrying out, in diverse ways, Peirce's dream of making "exact experiments upon uniform diagrams" into an essential part of the "logic of the future."

In retrospect, however, it is the relation between Peirce's claim that one's logic is one's metaphysics and his conception of logic as a primary form of experimentation that indicates a broader scope to Peirce's dream. It would be too narrow a view to interpret Peirce's commitment to diagrammatic thinking as a valuable strategic tool, or to take the fact that his existential graphs can be demonstrated to be consistent, complete, and transformable to the algebraic standard of contemporary logical theory as a final and sufficient assimilation of Peirce. The harder (and certainly more controversial) point is that Peirce's approach to logic does involve a persistent metaphyisical claim, that "experiments upon uniform diagrams" reveal, over time, something fundamental about the structure of reality. Peirce understood as well as any of his philosophical ancestors, that it is never easy to be a metaphysical realist, particularly if one embraces, as Peirce did, the necessity of experimentation and the modes of the possible, the probable, and chance as themselves constituents of reality.

Whether in the context of contemporary studies in logic or in intellectual history, Peirce remains a problem and an anomaly for just this reason: he does not quite fit our most familiar intellectual paradigms. Indeed, we might say that the whole cloth of Peirce's dream was the creation of an intellectual community of inquiry in which the linkage between logic and metaphysics was taken as axiomatic, and the experimental logic such inquiry requires would be available as a well exemplified and widely disseminated system of "uniform diagrams." Peirce perhaps more than any other thinker would have understood that such a result is not attained by fiat, but can emerge only through the actual inquiries of actual communities.

Thus it is no surprise that Peirce has been "picked up," by many disparate groups, in formal logic and philosophy, computer science, artificial intelligence, anthropology and sociology, psychology, intellectual history, and literary studies. But neither is it surprising that this has resulted in so many partial and sometimes incommensurable portraits of Peirce or versions of his philosophical thought that one is tempted to ask, "Will the Real Charles Sanders Peirce please step forward?" It is a notorious problem that Peirce's papers and published editions of his works have been vexed and remain incomplete or inaccessible; but one might suggest that this is as much an effect of the problem under consideration here as it is a cause. We are, collectively, still learning how to recognize themes, trends, and connections as essential to the structure of Peirce's thought, and not merely coincidentally interesting or useful–a problem that affects Peirce's contemporary editors every bit as much as it affects researchers who are actively trying to use Peirce in some specific research context.

Especially in his later writings, Peirce recognized the difficulty of his thought, even as it had grown transparent to him, as a function of its radical philosophical simplicity. It is not just interesting thoughts about familiar things: it is a different way of thinking. Thus, the tensions that appear between these papers may sometimes reflect fundamental disagreements over methodology, ontology, theory construction, or practical and computational utility, in an on-going (and expanding) dialogue about how to integrate or assimilate or "translate" Peirce, in order to make use of his enormously suggestive work. It could be argued, for example, that current ideas about how to establish rigorous proofs of consistency, formally expressed in standard set theoretic and algebraic terms, particularly in the interest of creating specific computer applications or strategies of research in Artificial Intelligence are subtly at odds with Peirce's distinctive version of philosophical realism, or his view of the heuristic and speculative dimension of scientific research.

Clearly, such questions will not soon be settled, and no one should look here for broad and general answers. What is more important, surely in Peirce's view as in our own, is that there is evidence here of an emerging and growing community of research, in which the heuristic and explanatory potential that Peirce saw in his graphical logic goes hand in hand with his particular version of Pragmatism. It is not the sort of "anything-that-works" instrumentalism that led Max Horkheimer and Theodor Adorno to dismiss pragmatism (really, Dewey and James) with barely a second thought [14], nor is it the breezy relativism which Richard Rorty has sometimes claimed as the point of pragmatism [17]. It is, rather, the view that the meaning of any proposition is the sum of the consequences that follow from accepting it (to abbreviate Peirce's famous pragmatist maxim, [[12], 5.9, 5.438]); and it is a mode of research that remains radically open to the possibilities that emerge when a community of inquiry focuses collective attention on a set of concretely representable problems. The relation between logic and metaphysics in such a context is a kind of experimental realism that demands formal precision, willing and able to reflect upon its own conditions of existence– but also willing and able to seize intellectual opportunities as they appear.

In the sections of this introduction that follow, we offer an overview of Conceptual Graphs and the ICCS '97 conference, starting with the modern convergence of early approaches to graph representation, semantic networks, and Peirce's pioneering work that led to the development of Conceptual Graphs.

2 From Existential Graphs to Conceptual Graphs

Conceptual graphs (CGs) are a synthesis of Peirce's existential graphs (EGs) with the semantic networks that had been independently developed for artificial intelligence and computational linguistics. The French linguist Lucien Tesniere used graph representations for his system of dependency grammar, which was posthumously published in a textbook [21] that is still used in courses on linguistics. In the United States, David Hays [7] and Klein and Simmons [9] adopted dependency grammar as a basis for their work on machine translation (MT). They influenced Roger Schank, who shifted the emphasis from syntactic dependencies to conceptual dependencies [18]. The first implementations of semantic networks were developed for MT in the late 1950s and early 1960s. Margaret Masterman's system at Cambridge University [10] was the first one to be called a semantic network. Another early MT system was based on the correlational nets by Silvio Ceccato [1], who labeled the arcs with 56 different relation types. The linguists Charles Fillmore [3] and Jeffrey Gruber [6] later analyzed the relations in greater depth and classified them more systematically. Their case roles or thematic relations supplied the labels that are used on the arcs of the semantic networks. At a major conference on machine translation in 1961, Ceccato, Hays, and Masterman presented papers on their networks; and other early researchers, such as Ross Quillian, were on the attendance list. For his PhD dissertation, Quillian [15] implemented a version of semantic networks that had a strong influence on the AI community.

In 1968, John Sowa attended Marvin Minsky's AI course at MIT. Minsky presented a variety of topics taken from a book that he had just published [11], which included an abridged version of Quillian's dissertation. For his term paper in that course, Sowa combined Quillian's networks with the Tesniere-Hays dependency graphs to form a semantic representation for natural language. The box and circle notation was influenced by the plastic templates used for drawing computer flow charts. The first published version of conceptual graphs [19] used the same notation, but with a more systematic presentation of the formation rules and an application to database query. Some of the CG terms, such as "join" and "projection", were adapted Ted Codd's [2] relational database theory. In effect, conceptual graphs can be considered intensional descriptors of the extensional database relations.

During the 1960s and 1970s, at least a dozen different versions of semantic networks were developed [4], many of which included representations for the operators and quantifiers of first-order logic. Yet none of the AI researchers were aware of Peirce's existential graphs. The graphs of the 1960s, including the ones by Quillian, Hays, and Schank, had the same logical limitations as Peirce's relational graphs of 1882: they could easily represent conjunction and the existential quantifier, they could represent local negations and disjunctions, but they could not show the scope of quantifiers or the scope of interacting Boolean operators. To overcome those limitations, Peirce experimented with a large variety of notational conventions during the 1880s and early 1890s, but he didn't find any that had the flexibility and expressive power of the algebraic notation for logic that he invented in 1883. The AI journals of the 1960s published new versions of the approaches that Peirce had tried and rejected eighty years before. Some of the representations finally attained the expressive power of first-order logic, but their notations were more cumbersome than Peirce's existential graphs of 1897.

In 1978, Martin Gardner published an article about Peirce in the mathematical games column of the Scientific American. After discussing Peirce's categories, Gardner also mentioned Don Roberts' book on Peirce's existential graphs. John Sowa happened to see that article and ordered a copy of Roberts' book. Peirce's version of existential graphs provided the missing ingredients that were needed for a rich and flexible version of semantic networks: a simple and elegant notation together with graph-based rules of inference instead of the substitutional rules of the algebraic notation. When he discovered Peirce's graphs, Sowa had a half-finished version of the book Conceptual Structures, which Addison-Wesley had already included in their catalog. To accommodate the new foundation based on Peirce's logic, Sowa threw away half of that half and wrote the bulk of the book from 1980 to 1983.

When Conceptual Structures appeared at the end of 1983, it did not include recent work by Hans Kamp [8] on discourse representation theory. After spending some time translating English to logic, Kamp had realized that the mapping to predicate calculus involved convoluted distortions of the sentence structure of English and other natural languages. In order to simplify the mapping, Kamp designed his discourse representation structures (DRSs) and formulated rules for representing and resolving indexicals in the DRS notation. Although Kamp's motivation was very different from Peirce's and he had no awareness of Peirce's graphs, Kamp's DRS notation happened to be isomorphic to Peirce's EGs. Since CGs are based on EGs, they are also isomorphic to DRSs; therefore, Kamp's rules for resolving indexicals in the DRS notation can be applied directly to EGs and CGs. A century after Peirce's pioneering work on both graph logic and indexicals, the former proved to be ideally suited for representing the latter.

The differences between EGs and CGs result from differences in their origin and motivation. In formulating EGs, Peirce was trying to find the simplest possible primitives for representing logic and operating on logical statements. In formulating CGs, Sowa was trying to find the most direct mapping from natural languages to logic. Since ordinary language has a much richer variety of expressive forms than Peirce's primitives, the CG notation includes many notational extensions that are designed to represent the major semantic forms of natural languages. To balance expressive power with simplicity, CGs provide both a basic and an extended notation. The basic CG notation is essentially a typed version of Peirce's EGs with the same logical primitives: negation, conjunction, and the existential quantifier. The extended CG notation, however, provides mechanisms for defining an open-ended range of new forms of expression, each of which has a direct mapping to some natural language expression and a formally defined expansion into the basic primitives. Following are the ways in which the extended CGs differ from EGs:

In the basic CGs, Peirce's line of identity is represented by one or more concept nodes linked by dotted lines called coreference links. As an alternate notation, the dotted lines may be replaced by symbols, called coreference labels, such as *x and ?x. One label marked with * is called the defining node, and the others marked with ? are called bound nodes. The labels are sanctioned by one of Peirce's observations that any line of identity could be cut with the two ends labeled to show how they could be reconnected.

In the extended CGs, generalized quantifiers can be introduced, including the universal, plurals, and various kinds of indexicals. Each of the extended forms, however, is defined by a formal mapping to the basic notation.

Each concept includes a type label inside the concept box. That label corresponds to a monadic predicate that could be attached to an EG line of identity. In the extended notation, the type labels may be replaced by lambda expressions that define them. The term lambda expression is taken from Alonzo Church, but the idea was anticipated by Peirce, who used labeled hooks instead of the Greek letter lambda.

In CGs, Peirce's ovals for negation are represented by concept boxes with the type label PROPOSITION and an attached negation relation NEG (or its abbreviation by a ˆ symbol). Instead of Peirce's tinctures for representing modalities, other relations can be attached to the concept box, such as PSBL for possibility or NECS for necessity.

Besides a type label, a concept box includes a referent field that identifies or partially identifies the referent of the concept. Following Peirce, the three kinds of referents are icons (represented by a diagram or picture), indexes (represented by a name or other indexical), and symbols (represented by a nested conceptual graph that describes the entity).

The linguistic work on thematic roles or case relations has been adopted for the basic types of conceptual relations. New types of relations can be defined in terms of them by lambda expressions, which like the lambda expressions for concept types can be written in place of the relation labels.

Except for notation, the basic CGs are minimal variants of EGs. The extended CGs provide a richer notation that maps almost one-to-one to the semantic structures of natural languages. The formal mapping from the extended to the basic CGs helps to bridge the gap between language and logic.

3 International Conference on Conceptual Structures (ICCS'97)

The Fifth International Conference on Conceptual Structures marks the twelfth international gathering of researchers focusing on conceptual graphs and conceptual structures. Starting in 1986, there have been seven workshops and four previous conferences. Participation in conceptual structures research has continued to grow. At present, there are hundreds of researchers in over 30 countries. There are a dozen countries represented at the present conference. Efforts have spanned a number of research areas, such as natural language processing, knowledge acquisition, information retrieval, and the philosophical foundations of logic.

As the theory and practice of conceptual graphs have matured, it becomes more and more apparent that working systems based (in whole or in part) on conceptual graphs are feasible. In organizing this conference, we have made a deliberate effort to include supporting theories (e.g., formal concept analysis) and practical applications (e.g., the CGTOOLS sessions) in order to enhance and expand the value of conceptual graphs. We adopted this strategy for three reasons:

1. In a practical sense, the success of conceptual graphs in society depends upon gaining support from industry, which has important problems to solve and is willing to invest in effective solutions to them;
2. Peirce's philosophy of pragmatism dictates that knowledge is not useful in itself, but only in its relation to society and how it affects the human condition. It is therefore time for conceptual graphs to prove themselves by significantly augmenting current knowledge-based problem-solving strategies; and
3. Peirce's view of knowledge was that scientific "facts" must be tested by a community of researchers, whose evolving collective agreement is what makes scientific progress possible. By expanding our community, we afford ourselves the opportunity to expand the body of available knowledge.

As our society becomes ever more oriented toward electronic communication, we have clearly gained more opportunities for such collective enterprises. This conference marks the first one in which the submission of papers, distribution of papers to reviewers, and submission of reviews was performed almost entirely using the Internet (principally the World Wide Web). We believe Peirce would be pleased at this appropriate use of technology in making information available to people. We believe he would also be pleased that the evaluations and decision-making were performed by a widely dispersed human community.

While technology certainly makes a conference like this possible, we hope the experience of this conference serves a multitude of human needs. We still need face-to-face communication and interaction. We need it in improving our understanding of the ideas presented in the papers. We need it for getting new and provocative ideas from each other. We need it for developing mere ideas into organized theories and systems. If any of us would think electronic communication would ever suffice, this conference should serve as a counter-example: consider how much more progress is made in a single week by our collective partipation in a single-site. As ideas are developed through the papers and through the interaction of this conference, we hope to see results disseminated widely. Perhaps then we can begin to realize Peirce's dream.

4   Conference Programme

In view of the aims and objectives of this conference (i.e., to promote and nurture the growth of the conceptual structures communities around the world), we departed from prior practices at ICCS meetings in several respects. These proceedings, first of all, cover a very wide range of topics, seen from different points of view. We have accepted only papers that the reviewers and editorial board also accepted for publication, thereby eliminating the need for any supplementary proceedings. We have included not only papers on essential theoretical research topics, but papers that describe the problems of conceptual structures (embracing both conceptual graphs and formal concept analysis) in fields ranging from philosophy and textual studies to medicine and agriculture, as well as papers concerning diverse applications of current theories for industry and tool development.

For the first time, we have also included a section in the proceedings for extended abstracts of the "CG Tools" sessions that have traditionally concluded ICCS meetings, to provide a formal record of presentation and to disseminate information about tools and applications. In the same spirit, we have organized two evening panel discussions (as indicated below), with other less formal occasions for conversation and discussion. There are no concurrent technical sessions, so all participants can attend all formal presentations and keynote addresses.

In outline, the conference programme includes the following categories of events:

1. **Pre-Conference Tutorials**: the following four half-day pre-conference tutorials:
   (a) An Historical View of Peirce - by Leroy Searle
   (b) Formal Concept Analysis - by Selma Strahringer
   (c) Formal Reasoning with Conceptual Graphs - by John Sowa
   (d) Gamma Graphs - the Modal Part of Existential Graphs - by Peter Oehrstroem

2. **Keynote Speakers**: the following four international researchers have been invited to give keynote addresses during this conference:
   (a) Peirce's Graphs - by Jay Zeman
   (b) A Pragmatic Understanding of "Knowing That" and "Knowing How": The Pivotal Role of Conceptual Structures - by Daniel Rochowiak
   (c) A Peircean Foundation for the Theory of Contexts - by John Sowa
   (d) The CORALI Project: From Conceptual Graphs to Conceptual Graphs via Labelled Graphs - by Michel Chein

3. **Technical Papers**: set of 34 technical papers were selected by the members of the editorial board to be included in this conference programme, and published in the proceedings. These papers were selected from a set of 51 papers that were submitted from around the world.

4. **Conceptual Graphs Tools**: application developers have been invited to present their work at this conference. A short extended abstract of each of these tools are included in this volume.

5. **Panel Sessions**: the following two panel sessions have been organized to promote collaboration between the conceptual graphs research community and various other research communities and industry:
   (a) Existential Graphs
   (b) Document Management

In organizing this conference, we have been especially conscious of the many and varied contributions to the study of conceptual structures that have been made by researchers and scholars around the world. In making more intensive use of international networks at every stage–in planning and publicizing the conference, and in using the World Wide Web as a medium for practical professional exchange–we have also been aware of the need for long-term planning to provide continuity between ICCS conferences, and a higher degree of coordination and information exchange, not only for researchers themselves, but for future program committees and chairs. Accordingly, at the conclusion of this conference, we will produce a report for conference participants, but especially for the planners and organizers of ICCS '98.

5   Conclusion

In the workshops and conferences that have preceded ICCS '97, it has been clear that Peirce's hopes for Existential Graphs, as the "logic of the future," have actually been inflected and interpreted by a concrete research history. Wherever Peirce has been "picked up," he has been found more fruitful and prescient than any of his contemporaries or early successors would credit. At the same time, however, Peirce himself would have been gratified to see that his larger concern for the conduct of research itself, as always referrable to the community actually doing it, has continued his work, though in many ways, in directions and with results that he could not possibly have anticipated.

In broadening the scope of this conference, foregrounding possible relations between Conceptual Graphs and Formal Concept Analysis; including papers with both theoretical and industrial aims; and representing in this publication the on-going development of CG and related tools, we have taken a chance on opening up a discussion that may go in directions we could not possibly anticipate.

This may be only a recognition that when Peirce linked logic to metaphysics, and graphs to experimentation, he was, as Gallie first recognized in 1952, selecting "certain features of experimental science for generalization" that apparently "hold true of other forms of human knowledge; for example, mathematics, history, metaphysical speculation, and indeed our everyday practical judgments of fact" [5]. What is remarkable is Peirce's recognition that the first step is always to make something representable (all the more so when one has to invent some new means for representation), for that is the condition under which thought is made public, communicable, and subject to precise correction and expansion by a community of inquiry–to which these proceedings are addressed.

References

1. Ceccato, Silvio, Automatic Translation of Languages, *Information Storage and Retrieval* 2:3, 1964, 105-158.
2. Codd, E. F. A Relational Model of Data for Large Shared Data Banks, *Communications of the ACM* vol. 13, no. 6, 1970, 377-387.
3. Fillmore, Charles J., The Case for Case, in E. Bach and R. T. Harms (eds.), *Universals in Linguistic Theory*, Holt, Rinehart and Winston, New York, 1968, 1-88.
4. Findler, Nicholas V., *Associative Networks: Representation and Use of Knowledge by Computers*, Academic Press, New York, 1979.
5. Gallie, W. B. *Peirce and Pragmatism*, Penguin Books, Harmondsworth, 1952.

6. Gruber, Jeffrey S., "Studies in Lexical Relations," PhD dissertation, MIT, 1965. Revised version published as *Lexical Structures in Syntax and Semantics*, North Holland Publishing Co., Amsterdam, 1976.

7. Hays, David G., Dependency Theory: A Formalism and Some Observations, *Language* 40(4), 511-525 (1964).

8. Kamp, Hans, Events, Discourse Representations, and Temporal References, *Langages*, vol. 64, 1981, pp. 39-64.

9. Klein, Sheldon, and Robert F. Simmons, Syntactic Dependence and the Computer Generation of Coherent Discourse, *Mechanical Translation* 7, 1963.

10. Masterman, Margaret, Semantic Message Detection for Machine Translation Using an Interlingua, *Proceedings of the 1961 International Conference on Machine Translation*, 1961, 438-475.

11. Minsky, Marvin, (ed.), *Semantic Information Processing*, MIT Press, Cambridge, MA., 1968.

12. Peirce, C.S., *Collected Papers*, vols. 1-8, Harvard University Press, Cambridge, Mass., USA, 1931-1958.

13. Peirce, C.S., *Writings of Charles Sanders Peirce. A Chronological Edition.* University of Indiana Press, Bloomington, USA, 1982.

14. Posnock, Ross. Bourne, Dewey, Adorno: Reconciling Pragmatism and The Frankfurt School. Milwaukee, WI: *Center for Twentieth Century Studies, Working Papers no. 4.* Fall-Winter, 1989-90.

15. Quillian, M. Ross, "Semantic Memory," PhD dissertation, Carnegie-Mellon University, Pittsburgh. Abridged version in Minsky (1968) pp. 227-270.

16. Roberts, D.D., *The Existential Graphs of Charles Sanders Peirce*, Mouton, The Hague, Netherlands, 1973.

17. Rorty, R., *Consequences of Pragmatism*, University of Minnesota Press, Minneapolis, USA, 1982.

18. Schank, Roger C., and Larry G. Tesler, A Conceptual Parser for Natural Language, *Proceedings, IJCAI-69*, 1969, 569-578.

19. Sowa, John F., Conceptual Graphs for a Data Base Interface, *IBM Journal of Research and Development*, vol. 20, 1976, pp. 336-357.

20. Sowa, John F., *Conceptual Structures: Information Processing in Mind and Machine*, Addison-Wesley, Reading, MA., 1984.

21. Tesniere, Lucien, *Elements de Syntaxe Structurale*, Librairie C. Klincksieck, Paris, 1959.

22. Way, E. C., Conceptual Graphs - Past, Present, and Future, in *Proceedings of the Second International Conference on Conceptual Structures*, Maryland, August 1994, pub. Springer Verlag, Berlin, 1994.

23. Zeeman, Jay. "The Graphical Logic of C. S. Peirce." Ph.D. diss., University of Chicago, 1964.

Peirce's Graphs

Jay Zeman
Department of Philosophy
University of Florida

Over a decade ago, John Sowa [11] did the AI community the great service of introducing it to the Existential Graphs of Charles Sanders Peirce. EG is a formalism which lends itself well to the kinds of thing that Conceptual Graphs are aimed at. But it is far more; it is a central element in the mathematical, logical, and philosophical thought of Peirce; this thought is fruitful in ways that are seldom evident when we first encounter it. In one of his major works on Existential Graphs, Peirce remarks that

> one can make exact experiments upon uniform diagrams; and when one does so, one must keep a bright lookout for unintended and unexpected changes thereby brought about in the relations of different significant parts of the diagram to one another. Such operations upon diagrams, whether external or imaginary, take the place of the experiments upon real things that one performs in chemical and physical research. Chemists have ere now, I need not say, described experimentation as the putting of questions to Nature. Just so, experiments upon diagrams are questions put to the Nature of the relations concerned (4.530)[1].

That Peirce's Graphs are likely to be a fruitful formalism in the context of Conceptual Graphs is a given; I would like to discuss some of the aspects of EG and of the Mathematical, Semiotical, and Logical thought in which Peirce embedded them. This may enhance the fruitful application of Peirce's logic in CG, and indeed, may provide a perspective which is helpful in work with Conceptual Graphs in general.

Arguably by 1880 and certainly by 1885 Peirce had developed the algebra of logic to something we can recognize as equivalent to the Classical algebraic logic of today; he had independently discovered and given his own twist to much of what Frege [3] had developed across the Atlantic [18]. Yet he grew dissatisfied with the algebra of logic, and in the last decade of the last century, he worked out a mathematical formalism which differed greatly from the algebraic, but which was mathematically equivalent to it. Why'd he do that? Wasn't it enough to invent the standard logic, without going into this graphical stuff? Seeking the answer to this will take us in the direction of understanding the fruitfulness of his approach.

Peirce, who considered himself primarily a logician, had a very clear idea of what he meant by logic:

> The different aspects which the algebra of logic will assume for the [logician in contrast to the mathematician] is instructive .... The mathematician asks what value this algebra has as a calculus. Can it be

---

[1] References to Peirce, 1931-58 are as standard in Peirce scholarship, with volume number, point, paragraph number. Thus 4.530 is paragraph number 530 of volume 4.

applied to unravelling a complicated question? Will it, at one stroke, produce a remote consequence? The logician does not wish the algebra to have that character. On the contrary, the greater number of distinct logical steps, into which the algebra breaks up an inference, will for him constitute a superiority of it over another which moves more swiftly to its conclusions. He demands that the algebra shall analyze a reasoning into its last elementary steps. Thus, that which is a merit in a logical algebra for one of these students is a demerit in the eyes of the other. The one studies the science of drawing conclusions, the other the science which draws necessary conclusions (4.239).

For Peirce, deductive logic is the study of a *process*, the process of reasoning necessarily. It is an empirical science, but it is not to be confused with psychology; as he comments,

> Logic is not the science of how we do think; but, in such sense as it can be said to deal with thinking at all, it only determines how we ought to think; nor how we ought to think in conformity with usage, but how we ought to think in order to think what is true. That a premiss should be pertinent to such a conclusion, it is requisite that it should relate, not to how we think, but to the necessary connections of different sorts of fact (2.52).

Logic, then, is not just a science, but a *normative* science. Note that Peirce speaks of deductive logic as exploring the "necessary connections of different sorts of fact"; his approach here is not to be confused with that of many thinkers on the topic, who imagine that logical truth emerges, somehow, full-panoplied from the head of the thinker, bearing an automatic and unchallengeable normative relationship to fact; this kind of apriorism is strongly at odds with the scientific thought of Peirce. Note him as he comments further upon the relationship between logic and mathematics:

> It might, indeed, very easily be supposed that even pure mathematics itself would have need of one department of philosophy; that is to say, of logic. Yet a little reflection would show, what the history of science confirms, that that is not true. Logic will, indeed, like every other science, have its mathematical parts. There will be a mathematical logic just as there is a mathematical physics and a mathematical economics. If there is any part of logic of which mathematics stands in need—logic being a science of fact and mathematics only a science of the consequences of hypotheses—it can only be that very part of logic which consists merely in an application of mathematics, so that the appeal will be, not of mathematics to a prior science of logic, but of mathematics to mathematics (2.247).

I call your attention to two of the features of logic as Peirce discusses it in the above; first of all, that logic is "a science of fact," and secondly that "Logic will, indeed, like every other science, have its mathematical parts. There will be a mathematical logic just as there is a mathematical physics and a mathematical economics." This view of logic and of the relationship of logic to mathematics is at odds with what is the opinion of many, probably most, philosophers. The (philosophically) prevalent view is roughly that of Russell and Whitehead [12], as derived from Frege and Peano; on that view (as Frege put it)

> arithmetic would be only a further developed logic, every arithmetic theorem a logical law, albeit a more developed one. ([4], 107)

(Now, Philosophers may believe that, but very few mathematicians do!) Peirce sees the relationship between logic and math as analogous to that between physics and math—an adequate study of logic demands mathematics as a tool, no less than does an adequate study of physics.

And just as physics is a science of fact, so too is logic. Say What? How can that be? Just as an example, consider the relationship between logic and probability theory (e.g., [16], [1], [2], [5], 340-416). A sample-space is, essentially, a set of *evidence*, evidence which may be taken as confirming or refuting propositions (in standard logical terminology, we would speak of a confirmed proposition as *true*, and a refuted one as *false*). *a* entails *b*, then, would mean that the evidence for (confirming) *a* is included in the evidence for *b*. Empirical aspects of logic so examined emerge when we ask about the conditions of confirmation and refutation. Classically speaking, no matter how complicated an empirical situation may appear, we may always think of ourselves as employing one and only one physical operation to gather evidence; the Classical Probability Theory of Kolmogorov [6], in fact, effectively *identifies* this physical operation with the sample space. And it is precisely this postulation which forces the logic of Classical sample-spaces to be Boolean. But empirical (not a priori) developments around the turn of the century (19th to 20th, that is) forced us to recognize that the Classical view is not general enough—the empirical development was, of course, the advent of Quantum Mechanics, and its upshot for Probability Theory was that there are situations for we cannot make the classical assumption of the refinability of all physical operations for a sample space into one[2]. The result is that not only is the physics of the subatomic different—more general than!—the physics of large objects, but the logic as well is more general (being orthomodular rather than just Boolean).

The empirical aspects of logic might manifest in a number of ways; Peirce saw the empirical in logic emerging in the "experiments on diagrams [which] are questions put to the Nature of the relations concerned" (4.530). There is here an intricate and intimate interplay with mathematics (which, as we have noted, plays a vital role in Peircean deductive logic). Mathematics, the "Science which reasons necessarily," does its reasoning by diagrams. Creative mathematical work deals with these diagrams, and does so by a process of inquiry involving Abduction, Deduction, and Induction within the domain of these diagrams; indeed, the necessity of inquiring thus within this domain is the origin of creativity in the area of necessary reasoning. But by changing the slant of our inquiry in the domain of diagrams, we may make it as well the locus of creative work in Logic; the difference is as Peirce stated earlier between what he

---

[2] This classical assumption is equivalent to holding that observer-observed interaction can always be made as small as we wish.

would call a "Calculus" and what he would call a "Logic"; the Calculus is aimed at getting to a conclusion as rapidly (and of course accurately) as possible: "Can it be applied to unravelling a complicated question?" The "Logic," while just as interested in accuracy as the Calculus, is not so concerned with the conclusion as it is with how we got there: the Logic "shall analyze a reasoning into its last elementary steps." But both mathematics and logic operate within the field of diagrams (note, by the way, that "diagram" here is very general—it doesn't necessarily imply "graphics" or Cartesian coordinates; the formulas used in algebra are diagrams, for example, as are the very numerals which name numbers).

This opens up, by the way, another area of Peirce's thought which is integral to the matters we are discussing: the Semiotic[3]. Although the theory of signs has connections throughout Peirce's logic, we must here advert to some of his classifications of signs, in particular, to how the sign represents its object. The sign as so representing may be

> Icon, Index, [or] Symbol. The Icon has no dynamical connection with the object it represents; it simply happens that its qualities resemble those of that object, and excite analogous sensations in the mind for which it is a likeness. But it really stands unconnected with them. The index is physically connected with its object; they make an organic pair, but the interpreting mind has nothing to do with this connection, except remarking it, after it is established. The symbol is connected with its object by virtue of the idea of the symbol-using mind, without which no such connection would exist (2.299).

Most to the point here are signs considered as icons. The resemblance which constitutes an icon is very general. In fact, the best mathematical characterization of iconicity is in the notion of a *mapping*. And Peirce has something considerable in mind; note in what follows that he has broadened his terminology regarding the Graphs. A *Pheme* is a sentence, though conceived as including interrogative and imperative as well as indicative signs (4.538). Thus the "Phemic Sheet" is the Sheet of Assertion, but in a broader sense than in the simple Alpha-Beta Existential Graphs. The *Leaf* is a sign which might be thought of as a container for Phemic Sheets. Peirce here takes us beyond the simple two-valued logic in the same way that possible-worlds semantics does in the study of modal logic; Peirce's vision, however is even broader than that of the possible-worlds approach. He comments:

> The entire Phemic Sheet and indeed the whole Leaf is an image of the universal field of interconnected Thought (for, of course, all thoughts are interconnected). The field of Thought, in its turn, is in every thought, confessed to be a sign of that great external power, that Universe, the Truth (4.553 n2).

So Peirce is aiming here at a mapping, an icon, of some of the important features of "mind"; I believe that he was able to carry this project considerably further than has

---

[3] For a general discussion of Peirce's Semiotic with many references, see [17].

generally been recognized (the text in question is that from which we have been quoting, his *Prolegomena to an Apology for Pragmaticism,* 1906 (4.530-572)).

Peirce in his work on EG is trying to set up a mathematical logic which will enable the appropriate description and analysis of deductive reasoning; as I have indicated, he was endeavoring here to be quite comprehensive in this effort. Although he is not explicitly aiming his logical endeavors at a *technology* of reasoning, I must remind you that theoretical physics is not aimed explicitly at a technology to control and manipulate the physical world, either. Physics is an effort to understand that world. But that understanding has been most fruitful in the generation of such a technology. I would suggest that the theoretical study of EG may well have the same result in the development of a technology of mind. So it seems to me that we could do worse than to follow his efforts and attempt to understand logic as he did, and to take him as seriously in his efforts to construct Gamma graphs as you have in his work on Alpha and Beta. And by the way, I believe that a major part of this must involve the study of his broader thought; and Existential Graphs is only a part of that thought. This study would include, ideally, an examination of his Semiotic, his Phenomenology, and his Pragmatism as well as his mathematical logic.

Before I go into any technical matters concerning the graphs, I would like to address myself briefly to a question I raised earlier: why did Peirce, who had developed a successful algebraic logic as early as 1880 go through the additional effort of developing EG at all? He answers this question himself. In his discussions of EG, he introduces an alternative notation to his Lines of Identity; objects in the universe of discourse may be represented by what he calls *selectives* as well as by the distinctive Line of Identity. The Selective is more like a conventional variable in algebraic logic, though like the Line of Identity, its quantification is implicit [13] [14]; thus

X is red
X is round

says the same thing as does

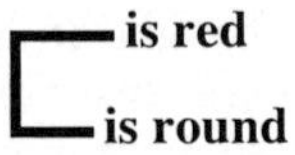

Figure 1

But Peirce preferred the latter Line-of-Identity notation. His reasons for doing so will also be the essential reasons why he preferred the Graphs to the Algebras as a notation for Logic, "the science of [that investigates] necessary reasoning."

> [The] purpose of the System of Existential Graphs ... [is] to afford a method (1) as simple as possible (that is to say, with as small a number of arbitrary conventions as possible), for representing propositions (2) as iconically, or diagrammatically and (3) as analytically as possible. ... These three essential aims of the system are, every one of them, missed by Selectives (4.563 n1).

In the present context, we can readily see Peirce's call for *simplicity* as fulfilled in EG as opposed to ordinary logical notation; this is, I believe one of the things that has made Existential Graphs so attractive in CG work. Iconicity and Analyticity of representation might be considered together; he notes that the analytic purpose of a logic "is infringed by selectives" (and so also by the variables of ordinary algebraic logic);

> Selectives are not as analytical as they might be, and therefore ought to be ... in representing identity. The identity of the two [X's in the red-round diagram] above is only symbolically expressed. . . . Iconically, they appear to be merely coexistent; but by the special convention they are interpreted as identical, though identity is not a matter of interpretation ... but is an assertion of unity of Object .... The two [X's] are instances of one symbol, and that of so peculiar a kind that they are interpreted as signifying, and not merely denoting, one individual. There is here no analysis of identity. The suggestion, at least, is, quite decidedly, that identity is a simple relation. But the line of identity which [is in the lower diagram] substituted for the selectives very explicitly represents Identity to belong to the genus Continuity and to the species Linear Continuity (*ibid.*).

Peirce goes on to comment on other aspects of iconicity and analyticalness exhibited by important signs of EG. It is very clear that he sees EG as serving the purpose of a Logic far better than does the standard algebraic logical calculus; it aids us to follow the process of reasoning with greater ease and acuity. I note that the aim of CG seems, in large part, to be movement toward the ideal of Artificial Intelligence; the employment of a deductive system fitting Peirce's norms for a Logic (as does EG) would seem to be far more appropriate for this than the narrower aim of a specific-purpose "Calculus." So again, it would seem that attention to Peirce's broader thought is likely to prove most fruitful in the development of Conceptual Graphs.

I now wish to look at a theme in Peirce's thought which—as do so many of his ideas—anticipates contemporary developments in mathematical logic, and which receives, in his presentations of Existential Graphs, a treatment and a twist which goes beyond what most contemporary logicians have done. Peirce had been concerned with what he called "hypothetical propositions" and their relationship to the *de inesse* (truth-functional) conditional (see [20]) since at least 1880; in 1885 we find him commenting that

> The question is what is the sense which is most usefully attached to the hypothetical proposition in logic? Now the peculiarity of the hypothetical proposition is that it goes out beyond the actual state of things and declares what *would* happen were things other than they are or may be. The utility of this is that it puts us in possession of a rule, say that "if *A* is true, *B* is true," such that should we hereafter learn something of which we are now ignorant, namely that *A* is true, then, by virtue of this rule, we shall find that we know something else, namely, that *B* is true. There can be no doubt that the Possible, in its primary meaning, is that which may be true for aught we know, that

whose falsity we do not know. The purpose is subserved, then, if throughout the whole range of possibility, in every state of things in which *A* is true, *B* is true too. The hypothetical proposition may therefore be falsified by a single state of things, but only by one in which *A* is true while *B* is false (3.374).

This is a theme which we see in Peirce's logical work for the rest of his life. Here it takes the form of a contrast between "If-then" as what he calls a "Hypothetical," and "If-then" as truth-functional. The truth-functional If-then is an *instantiation*, just one concrete instance of the If-then as Hypothetical; the Hypothetical is a general, a universal. While Peirce's Non-Relative algebraic logic, and later his Alpha Existential Graphs, gives an adequate treatment of the truth-functional *de inesse* conditional, Peirce the logician spent a great deal of his energy in the last part of the 19th Century seeking an adequate treatment of the *Hypothetical* If-then and related matters. A central theme here is that of a range of possible situations. Peirce finds at least a partial answer to the logic of the Hypothetical in such a framework in quantification as he develops it in his Logic of Relatives. In 1902 he comments that

> In a paper which I published in 1880, I gave an imperfect account of the algebra of the copula. I there expressly mentioned the necessity of quantifying the possible case to which a conditional or independential proposition refers. But having at that time no familiarity with the signs of quantification which I developed later, the bulk of the chapter treated of simple consequences *de inesse*. Professor Schröder accepts this first essay as a satisfactory treatment of hypotheticals; and assumes, quite contrary to *my* doctrine, that the possible cases considered in hypotheticals have no multitudinous universe. This takes away from hypotheticals their most characteristic feature (2.349),

which is that they are generals, that they represent a quantifiable range of situations of which each instantiation would be a *de inesse* conditional! Note that the domain with which we deal here isn't one of people, books, machines, and other such ordinary individuals, but of *situations*, of what Peirce elsewhere called "States of Information" (see, for example, 4.517). And this will suggest immediately the contemporary semantics of modal logic, which involves domains whose members are usually called "Possible Worlds."

And here we have, I think, yet another reason that Peirce preferred the Graphs as a notation for logic. His algebras of logic did not offer a medium for the simple, iconic, and analytic presentation of modality, of the realms of possibility and necessity. The Graphs, on the other hand, had some features which made them most suitable to this purpose. Although he experiments with graphical analogs of modal operators (note his "broken cut" (4.515 ff.) which means "possibly not"), his emphasis is on possible worlds; before introducing the broken cut, he writes of what amounts to a "universe of universes" of possibilities and of fact, and

> in order to represent to our minds the relation between the universe of possibilities and the universe of actual existent facts, if we are going to think of the latter as a surface, we must think of the former as

three-dimensional space in which any surface would represent all the facts that might exist in one existential universe (4.514).

Clearly, a representation of such a universe might be found in, say, a *book* of Sheets of Assertion. Peirce did indeed explore such representations. And he does this explicitly, stating that for the Gamma Graphs,

> in place of a sheet of assertion, we have a book of separate sheets, tacked together at points, if not otherwise connected. For our alpha sheet, as a whole, represents simply a universe of existent individuals, and the different parts of the sheet represent facts or true assertions made concerning that universe. At the cuts we pass into other areas, areas of conceived propositions which are not realized. In these areas there may be cuts where we pass into worlds which, in the imaginary worlds of the outer cuts, are themselves represented to be imaginary and false, but which may, for all that, be true (4.512).

Again, we have material suggestive of present-day possible-world semantics for modal logic (a source is [15]) which then must be considered a rediscovery of something that Peirce did a century ago. Peirce worked with the "Book of Sheets" model and variations thereof in a number of locations. As was his wont, however, he also tried out other ways of representing this material. One of the most interesting, and I think, one with a great deal of applicability to Conceptual Graphs, is in the work I quoted at the start of this paper, his 1906 "Prolegomena to an Apology for Pragmaticism." In this paper, he develops what he calls "Tinctured Existential Graphs." "Tincture" in this sense is a technical term of heraldry. The designers of coats-of-arms needed a way of representing the appearance of their product in a day when the exact picturing of colors, metals, and furs on paper was impossible, or at least difficult and expensive. Thus the "Tinctures" used in the graphical description of coats-of-arms. Peirce employs the tinctures for similar reasons (lest we think of how benighted those times were, let us reflect on the fact that the use of color is restricted in many publications even today!); the Tinctures were a way of indicating in black and white what could far better be done in color. In fact, we may for present think of the Tinctures as Colors; "Color" was one of the "Modes of Tincture" which Peirce wished to employ; this mode was the mode in which Possibility and Necessity would be dealt with, and this is our prime interest today.

As we have noted, from very early on, Peirce saw "states of information" as values of quantified variables; note that the "quantified subject of a hypothetical proposition" he refers to below might just as well be a "possible world" in the sense of contemporary logic:

> the quantified subject of a hypothetical proposition is a *possibility*, or *possible case*, or *possible state of things*. In its primitive state, that which is *possible* is a hypothesis which in a given state of information is not known, and cannot certainly be inferred, to be false. The assumed state of information may be the actual state of the speaker, or it may be a state of greater or less information. Thus arise various kinds of possibility (2.347).

Let us examine very quickly the mechanisms of Possible-Worlds semantics; the familiar treatments of this go back to the well-known work of Kripke (Beginning with Kripke [7]), as well as to that of Prior (Beginning with Prior [9]). The earliest and best-known such systems are what Segerberg [10] later called "relational" modal logics. Relational modal semantics works with a pair <W,R>; W is a set of what may be called "possible worlds"; intuitively, if x ∈ W, then it makes sense to speak of a proposition, say *p*, being "true at x" or "true relative to x" or some such; if (as is one common interpretation) x, y ∈ W are "instants of time," then we can see how *p* might be true at x, say, but false at y; what "possible world" means intuitively will vary tremendously, depending on the features of the semantics in question, and the correlative concept of modality (possibility and necessity) that goes with it. Now note that in the "instants of time" example, features of the semantics are dependent on how these instants are *related*. The relation in question is the second member of the pair <W,R>. R is such that Rxy makes sense for x, y ∈ W; in the temporal example, Rxy is most often the relation "y is in the future of x (or is the same as x)." We note that R so construed will have definite properties; it will be reflexive and transitive, anyway. R is commonly called "The accessibility relation"; note that temporal access as construed above will differ from, say, spatial access (which we would probably think of as being symmetrical as well as reflexive and transitive). We then can associate modality with possible worlds by the expressions:

(1) Possibly *p*, i.e., *Mp*: $\quad$ *Mp* holds at world *x* iff $\exists y(Rxy$ & *p* holds at *y*)

(2) Necessarily *p*, i.e., *Lp*: $\quad$ *Lp* holds at world *x* iff $\forall y(Rxy \supset p$ holds at *y*)

A central feature of Modal Logic as interpreted in Relational Semantics is that the meaning of modality—of possibility and necessity—is intimately and precisely linked to the properties of the accessibility relation. Thus, reflexive and transitive access give us a semantics appropriate to the well-known modal logic S4, while adding symmetry gives us the modal logic S5. The modal logic S5 is a limiting case of this type of modality; it is the system in which Necessarily *p* can be taken to mean simply that *p* holds in *all* possible worlds.

Peirce had had the notion of "possible world" well before he got into EG; his development of Gamma Graphs, however, supplied him with the basis of a mechanism for handling the relations between these possible worlds, and so of a treatment of modality which could be integrated with his logic as a whole. In the Beta Graphs, quantification is handled, remarkably, without explicit quantifiers (this in spite of the fact that Peirce was co-inventor of the quantifier). Peirce's preferred method of handling quantified variables is by the Line of Identity (more generally, by the Ligature). This is not, as we have already noted, because of mathematical deficiencies in the alternative representation—selectives—but because of reasons relating to the representation *vis-a-vis* Logic: the Line of Identity does a better job of showing us what's going on with quantified variables. The Line of Identity will have an analog in the realm of possible worlds within the Graphs, but it may be best to start off with a concept of *Selective* here.

Peirce had experimented with representations of possible worlds as definite individuals—Peircean seconds. We see this in his notion that

> in the gamma part of [Existential Graphs] all the old kinds of signs take new forms. ... Thus in place of a sheet of assertion, we have a book of separate sheets, tacked together at points, if not otherwise connected. For our alpha sheet, as a whole, represents simply a universe of existent individuals, and the different parts of the sheet represent facts or true assertions made concerning that universe. At the cuts we pass into other areas, areas of conceived propositions which are not realized. In these areas there may be cuts where we pass into worlds which, in the imaginary worlds of the outer cuts, are themselves represented to be imaginary and false, but which may, for all that, be true (5.512)

But as far back as 1880, as we have noted, he was aware that an adequate treatment of the topic required not just examination of definite individual worlds, but of a quantifiable range of possible states of affairs (see 2.349); and he had experimented with representations for such; in discussing one of his graphical modal operators (the "broken cut") he remarks that

> You thus perceive that we should fall into inextricable confusion in dealing with the broken cut if we did not attach to it a sign to distinguish the particular state of information to which it refers. And a similar sign has then to be attached to the simple *g*, which refers to the state of information at the time of learning that graph to be true. I use for this purpose cross marks below, thus:

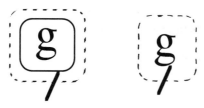

Figure 2

> These selectives are very peculiar in that they refer to states of information as if they were individual objects. They have, besides, the additional peculiarity of having a definite order of succession, and we have the rule that from Figure 3 we can infer Figure 4.
> These signs are of great use in cleaning up the confused doctrine of *modal propositions* as well as the subject of logical breadth and depth (4.518).

Figure 3   Figure 4

My suggestion (first made in [19]) is that the Tinctures of 4.530 ff. may be regarded as selectives rather than as representations of definite individual possible worlds. A line-of-identity representation is only a short hop away; in fact, we find Peirce making this hop in his continuation of the above:

> Now suppose we wish to assert that there is a conceivable state of information in which the knower would know *g* to be true and yet would not know another graph *h* to be true. We shall naturally express this by Figure 5.

Figure 5

> Here we have a new kind of ligature, which will follow all the rules of ligatures. We have here a most important addition to the system of graphs. There will be some peculiar and interesting little rules, owing to the fact that what one knows, one has the means of knowing what one knows—which is sometimes incorrectly stated in the form that whatever one knows, one knows that one knows, which is manifestly false (4.521).

And he develops this even further:

> The truth is that it is necessary to have a graph to signify that one state of information follows after another. If we scribe

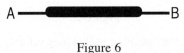

Figure 6

> to express that the state of information **B** follows after the state of information **A**, we shall have

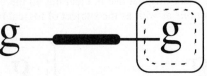

Figure 7

This last is a version—employing lines of identity ("ligatures") for states of information—of the rule of necessitation which is a feature of the most commonly studied modal logics.

So we see that Peirce did do explicit graphical work with concepts that are familiar to the contemporary modal logician. The Tinctured Existential Graphs present a medium for the extension of this work. For simplicity, let us think of Tinctures only in terms of colors for now; two areas of the "Phemic sheet" which are the same color are thought of as continuous with each other. We can even picture them as "cross-sections"of a special Line of Identity (just as the "tic-mark" selectives Peirce uses above are like cross-sections of a Ligature for States of Information); the sameness of color of a given tincture ties in with the continuity of that LI—which has to be embedded in dimensions beyond our usual three. This line of identity will then be a quantified variable for possible worlds. The rules for such Tinctures can be worked out if we understand the Beta rules; I do this explicitly in Zeman [19]. A Tincture as bearer of modality is *structured*: it actually involves two *colors*; although they are closely related, as we shall note. From this perspective, a Sheet of Assertion (which, of course, we associate with a "world," with a locus where propositions can be true or false) has two "sides": a *recto*, or true side and a *verso*, or false side; the verso is "seen" through the Cuts (which Peirce often describes as actually-cut-through-the-paper-and-turned-over). If the recto is a color in the R-G-B model, the verso will be its color complement—a Tincture whose recto is Red <255,0,0> will have a verso of Cyan <0,255,255>. But the pictured Sheet of Assertion will not represent a specific member of a domain of possible worlds; rather, it is a "cross-section" of a continuum (which, of course, would require an extra dimension or more for its representation). The continuum is a Line of Identity (more generally, a Ligature) and the pictured SA is a *Selective* associated with that line—thus a *variable* for possible worlds rather than a constant name for a possible world.

And the Tinctures as Ligatures/Selectives operate by the same rules for implicit quantification laid out by Peirce in the Beta Graphs (see Zeman [19] for the specifics of this). There is more; as we have noted, we must be able to deal with an "accessibility relation" between possible worlds to get the well-known contemporary treatments of possible-worlds semantics. The ability to do this is provided by the colors involved in the Tinctures. In the notation we introduced earlier, we interpret Rxy as meaning that x has access to y; with the tinctures, this would hold iff the recto color for x is $<a,b,c>$ (in the R-G-B model), the recto color for y is $<d,e,f>$, and all of the following hold: $a \leq d, b \leq e$, and $c \leq f$. This gives an accessibility relation which is essentially a partial order (and so would have the Lewis-Modal S_4 as its basic system), but which is open to many different variations for specific purposes (of course, White would be the Supremum in this p.o., but recall that each Tincture involves the complement of its recto color, making the basic p.o. of Tinctures a tree (a semilattice).

It seems to me that the Gamma graphs of Peirce as we have been examining them present us with great opportunities for the enrichment of the study of CG. The precise directions that this enrichment will take is dependent on the ingenuity of researchers in the area.

References

1. D.J. Foulis, and C.H. Randall, Operational Statistics I, *Journal of Mathematical Physics* 13, 1972, 1667-75.
2. D.J. Foulis, and C.H. Randall, Operational Statistics II, *Journal of Mathematical Physics* 14, 1973, 1472-80.
3. G. Ferge, Begriffschrift, *From Frege to Gödel: a source book in mathematical logic 1879-1931*, Jean van Heijenoort (ed.), Cambridge: Harvard, 1967, 1-82.
4. G. Ferge, The Concept of Number, *Philosophy of Mathematics*, ed. Paul Benacerraf and Hilary Putnam, Englewood Cliffs, NJ: Prentice-Hall, 1964.
5. M. Jammer, *The Philosophy of Quantum Mechanics*, New York: Wiley, 1974.
6. A.N., Kolmogorov, *Foundations of the Theory of Probability*, New York: Chelsea, 1950.
7. S. Kripke, A Completeness Theorem in Modal Logic," *Journal of Symbolic Logic* 24, 1959, 1-14.
8. C.S. Peirce,*The Collected Papers of Charles Sanders Peirce* Vol. 1-6, ed. Charles Hartshorne & Paul Weiss (1931-6); Vol. 7-8, Ed. Arthur Burks (1958), Cambridge: Harvard.
9. A. Prior,*Time and Modality*, Oxford: The Clarendon Press, 1957.
10. K. Segerberg, *An Essay in Classical Modal Logic : FILOSOFISKA STUDIER utgivna av Filosopfiska Föreningen och Filosofiska Institutionen vid Uppsala Universitet, nr 13*, Uppsala, 1971, Vol. 1-3.
11. J. Sowa,*Conceptual Structures: Information Processing in Mind and Machine*, Reading, Mass: Addison-Wesley, 1984.
12. A.N. Whitehead, and B. Russell, *Principia Mathematica*, Vol. 1-3, Cambridge, 1910.
13. J.J. Zeman, *The Graphical Logic of C. S. Peirce*, Doctoral Dissertation, The University of Chicago, 1964.
14. J.J. Zeman, A System of Implicit Quantification, *Journal of Symbolic Logic* 32, 1967, 480-504.
15. J.J. Zeman, *Modal Logic*, Oxford: The Clarendon Press, 1973.
16. J.J. Zeman, Generalized Normal Logic, *Journal of Philosophical Logic* 7, 1978, 225-43.
17. J.J. Zeman, Peirce's Theory of Signs, *A Perfusion of Signs*, ed. T. Sebeok, Bloomington: Indiana U. Press, 1978, 22-39.
18. J.J. Zeman, Peirce's philosophy of logic, *Transactions of the C. S. Peirce Society* 22 (1986), 1-22.
19. J.J. Zeman, The Tinctures and Implicit Quantification Over Worlds, *The Rule of Reason: The Philosophy of Charles Sanders Peirce*, ed. Jacqueline Brunning and Paul Forster, Toronto: University of Toronto Press, 1997, 96-119.
20. J.J. Zeman, Peirce and Philo, *Studies in the Logic of C. S. Peirce*, ed. N. Houser, Don Roberts, and James Van Evra, Bloomington: Indiana U Press, 1997, to appear.

A Pragmatic Understanding of "Knowing That" and "Knowing How": The Pivotal Role of Conceptual Structures

Daniel Rochowiak

The University of Alabama in Huntsville

Abstract : What is the difference between knowing that a cake is baked and knowing how to bake a cake? In each, the core concepts are the same, "cake" and "baking," yet there seems to be a significant difference. The classical distinction between "knowing that" and "knowing how" points to the pivotal role of conceptual structures in both reasoning about and using knowledge. Peirce's recognition of this pivotal role is most clearly seen in the pragmatic maxim that links theoretical and practical maxims. By extending Peirce's pragmatism with the notion of a general argument pattern, the relation between conceptual structures and these ways of knowing can be understood in terms of the "filling instructions" for concepts. Since a robust account of conceptual structures must be able to handle both the context of "knowing that" and "knowing how," it would seem reasonable to think that there will be multiple representations for the "filling instructions." This in turn suggests that a methodological principle of tolerance between those approaches that stress the theoretical understanding of concepts appropriate to "knowing that" and those that stress the proceduralist understanding of concepts appropriate to "knowing how" is desirable.

1 Knowing How and Knowing That

What is the difference between knowing that a cake is baked and knowing how to bake a cake? In each, the core concepts are the same, "cake" and "baking," yet there seems to be a significant difference.

Classically philosophers and other have recognized a distinction between declarative knowledge and procedural knowledge. This distinction recognizes that there are at least two different ways of knowing. For example, I may know that a cake is something that is backed without knowing how to bake a cake. The converse is a bit more contentious, but it might be the case that I can know how to bake a cake without knowing that a cake is something that is baked. In either case, it seems clear that there is at least a conceptual unity in the two kinds of knowledge claims since they both use the core concepts of "cake" and "baking". It is important to distinguish the ways in which the two kinds of knowledge claims are different and similar.

2 Classical Theory of Knowledge

Using a classical understanding of knowledge there are three dimensions along which knowledge claims can vary: justification, truth, and belief.

Declarative and procedural knowledge differ in terms of the justifications that can be offered for the knowledge claim. In the case of declarative knowledge, the justification flows from either deductive or empirical-evidential reasoning. In the former case, some claims are taken as known and deductions are performed to justify the claim in question. In the later case, other evidential techniques derived from statistical analysis and empirical induction are used to establish the claim in question as knowledge. In both of these cases, there are formal reasoning techniques that are to be followed to establish that the evidence correctly supports the knowledge claim. In the case of procedural knowledge, there is no such formal reasoning techniques. Indeed the support for the procedural knowledge claim, the "knowing how" claim, is the actual performance of the action. In this sense one know how to bake a cake just in case one does bake a cake.

Different understandings of truth are appropriate to declarative and procedural knowledge. The proposition imbedded in declarative knowledge in some way either corresponds to the way the world is or is coherent with a body of claims already accepted as true. Neither of these seems to be appropriate to procedural knowledge. Rather, the truth of the imbedded claim would be demonstrated by the performance of the specified action. Thus, the notion of truth that is relevant to "Harry knows that a cake is baked," is the notion that "A cake is baked" either corresponds to the world or that is coherent with all claims accepted as true. The notion of truth that is relevant to "Harry knows how to bake a cake," is a record of either past or current performance of cake-baking actions. Thus, to say that "Harry knows that a cake is baked" is to say that is to say that the proposition "A cake is baked" either corresponds to the world or is coherent with accepted propositions. On the other hand to say that "Harry knows how to bake a cake" is to say that Harry could perform a series of actions that would be correctly described as baking a cake.

The belief dimensions of declarative and procedural knowledge offer yet another point of contrast. In the declarative case Harry's belief amounts to the claim that Harry would assert the truth of the proposition that "A cake is baked." In the procedural case the belief is a habit. The belief is the disposition in certain circumstances to engage in a series of actions that would be described as baking a cake.

The classical distinction between "knowing that" and "knowing how" points to the pivotal role of conceptual structures in both reasoning about and using knowledge. The distinctions between declarative and procedural knowledge illustrate the ways in which conceptual structures unify a domain of discourse across declarative and procedural knowledge. The concepts of 'cake' and 'baking' hold the domain together. Both occur in the declarative and procedural knowledge. If we assume that there is but one domain - the domain of cake-baking - then the declarative and procedural knowledge claims represent different facets of knowledge. The different facets are related and independent. They are related in that they are about the same domain and they are independent in that the conditions of justification, truth, and belief are distinct. It is in this way that it is reasonable to think that a person could know that a cake is baked without knowing how to bake a cake. Focusing on these three dimensions of knowledge it is also quit possible that another person knows how to bake a cake without knowing that a cake is baked. For that person the conditions of knowledge for procedural knowledge could be satisfied without the conditions of knowledge for declarative knowledge being satisfied. Thus, that person might have the capacity to perform the cake baking actions and actually have done so without having applied the rules of evidence to establish the truth of the claim that "A cake is baked." Ordinarily, one would like to have a higher-order coincidence of declarative and procedural knowledge, yet it does not seem to be reasonable to claim that procedural knowledge claims are correct if and only if the corresponding declarative knowledge claims are correct.

- The declarative and procedural knowledge about a domain can be distinguished in terms of justification, truth and belief.
- Concepts unify a domain.
- Declarative and procedural knowledge about a domain share common concepts.

3 Knowledge Promotion

The traditional account of knowledge treats knowledge claims as atemporal in the sense that they lack a natural history. An alternative is to consider knowledge as have a natural history in which there is a process of knowledge promotion. Rather than considering knowledge as one sort of thing (justified true beliefs, for example), it might be better to consider knowledge, especially scientific knowledge, as being comprised of different items at different levels. This emphasis on scientific knowledge should not be thought of as a severe limitation. Following Peirce there is a broad sense in which the scientific method is a general method of "fixing belief" and is the best possible method for doing so.

Modifying Collins work, it seems reasonable to think that scientific knowledge incorporates unarticulated - but acquirable - skills, heuristics, and 'context-free' representations. [1] At the skill level, cultural, manual, and perceptual skills are found. These skills are acquired through apprenticeship, but are often not articulated, and may, as a whole, be impossible to articulate. These skills in a particular domain are the backdrop for scientific activity and furnish a ground upon which more well articulated pronouncements can be made. However, since these are not, in general, the sorts of things that can be taught, they are also not the sorts of things that can be directly embodied in a declarative form of representation. Further, these are the things that most resist knowledge acquisition processes, since even the expert has not learned them in a discursive way, and may never have attempted to represent them explicitly. In the cake-domain it is quite possible that the expert baker has formed very little in the way of discursive representations, and those representations that have been formed may use concepts the are not, from a logical point of view, clearly defined.

Some skills, however, are vital to the continuation of the research efforts of either individual scientists or groups of scientists. Those that are sufficiently vital are formed into heuristics that can be expressed, although not in a manner that is either sufficiently explicit, detailed, or precise to become 'context-free.' When knowledge is promoted to the heuristic level, the scientist is able to indicate 'rules of thumb' that are useful in problem solving or filling in the currently unfilled parts of a scientific argument even though they do not guarantee success. In the cake-baking domain this is the point at which cake baking can be expressed in a recipe that in a broad measure indicates ingredients and procedures. This procedure even if followed does not guarantee success and relies upon the skills built in the process of baking cakes.

Finally, the process of doing science produces 'context-free' knowledge claims. These 'context-free' claims are not, in any absolute sense, context-free. Rather, they are claims within a particular domain that incorporate in an explicit, articulate way accepted knowledge that is presumed stable across the introduction of new theories and hypotheses in that domain. An example, might help. At the level of skill, a chemist adds a chemical to a solution until she 'feels' the right amount has been added. At the heuristic level, a rule might be that the chemical should be added until the solution just begins to turn cloudy. At the 'context-free' stage, the articulated knowledge might be expressed as "the chemical should be added until the recording meter indicates a positive value greater then 0.1, which will be accompanied by a change in the clarity of the solution," or as "compound xyz is formed by adding A to B." Not all of the claims in science have a natural history of this sort. Once a claim is promoted to the 'context-free' level apprenticeship is no longer required and heuristics are largely replaced by measuring devices, computational techniques, and explicit representation. Indeed, a more fully 'context-free' representation would be in the form of a machine that would, upon instruction, determine if a state can be achieved given the current representations of conditions and domain knowledge, perform the appropriate operation and stop when the intended result had been achieved. Returning to the cake-baking domain, the level of context-free knowledge would be achieved if all of the requisite claims and tests were made explicit, ambiguity was removed from concepts, logical connections among the representations determined, and specific test conditions established. At this stage, skill and heuristics have context-free representations and the whole process of cake baking can be automated. Further, explicit reasoning about cake baking can be formulated and it can be determined whether, given the explicit declarative principles of cake baking, a particular cake can be baked.

Interestingly, the heuristic level is the level at which much of the 'expert system' enterprise has been focused. The difficulties of knowledge acquisition and representation are sever in this arena because the knowledge that is the target of the system is heuristic knowledge and this knowledge has been promoted from the skill level which is not generally articulable, but provides the grounds on which heuristic knowledge is interpreted. At this level, the expert's knowledge is said to be compiled. However, it is misleading to think that it is compiled in the sense that a discursive part of knowledge is thrown away and the behavior kept. While this is part of what happens, a more important part is that the skills have become so important that an effort is made to make them discursively available.

When heuristic knowledge is promoted to 'context-free' knowledge the knowledge is further removed from the skill level. However, the interpretation of 'context-free' knowledge is dependent on the heuristic knowledge that lies below it. What is important to notice is that at the level of 'context-free' knowledge, knowledge is more easily acquired. Knowledge is representable in a discursive way, and can be considered to be a library from which all can draw. In this sense, much of the particular skill aspect has been removed and the context that would ordinarily provide the means of interpretation of the knowledge is largely, but not wholly, built into the 'context-free' knowledge. At this level the knowledge does not need to be decompiled. The knowledge is already in a discursive and largely explicit form. The end of the promotion of scientific knowledge is a readily usable supply of publicly accessible knowledge representations.

- There is a natural history to knowledge that can be understood in terms of a process of knowledge promotion.
- The natural history of knowledge involves both declarative and procedural knowledge.
- Part of the interpretation of context-free knowledge is provided by skills and heuristics.

4 Definition, Family Resemblance, and Concepts

If it is difficult to determine the way in which knowledge is communicated and arguments are structured, it is tempting to think that the reason for this is that the analysis has proceeded at too high a level. When one is tempted in this way, the seductive strategy is to reduce the level of the problem and go to a more basic level. The reduction strategy in this case of knowledge might be the following. Knowledge is manifest in statements and these statements are linked together in argument patterns. Thus, statements are more primitive than either knowledge or argument patterns. Since this is so, the components of statements should be examined. If one can get clear here, then these components can be used to build statements. From these statements, knowledge and arguments can then be built. This would be all well and good if it were clear what a concept is and how concepts function in building statements, knowledge, and arguments.

There are two general accounts about the meaning of concepts. The definition account, following Aristotle, holds that the meaning of a concept is a specification of the necessary and sufficient conditions for either the correct application of a concept, or the correct replacement of a concept symbol. The family resemblance account, following Wittgenstein, holds that the meaning of a concept is a specification of the typical characteristics of a prototypical instance that people generally use to decide if an object is an instance of a particular concept.

According to the definition theory, the information content of a concept is a definition that gives the necessary and sufficient conditions that an object must meet to fall under the concept. If this account were accepted as correct, then one would need to adopt only (formal) logical techniques and practices. The symbols for the concepts would be linked to other symbols by equivalence relations, and the formal manipulation of the symbols would be sufficient for all reasoning about concepts and their relations. [2, 3, 4]

Following the family resemblance theory, however, the information content of a concept is given in terms of a collection of features that are neither necessary nor sufficient, but characterize typical instances. If one allows that concepts should be understood in terms of family resemblance, then logical techniques will not be sufficient and other reasoning procedures would be needed. Since the symbols for the concepts would not be linked to other symbols only through the equivalence relations, a formal logic would not be able to describe the whole information content of the concept. [5, 6]

If it were simply a choice of which account to accept, then for computational purposes the definitional account would be preferred. However, the empirical evidence indicates that this is not the way that people think and there is support for the family resemblance theory. Family resemblance accounts for the high typicality factors for subordinates to superordinates, and family resemblance is a good predictor of a person's immediate intuitions of typical and atypical instances. [5] Further, there is a crossing of definitional and typicality effects such that even when the definition of concepts are known there are contexts in which typicality effects appear. For example, people treat some rectangles as being more typical than others. This would not be at all reasonable on a strictly definitional view. On that view all instances are equally typical. A possibility is that for some concepts there are core (definitional) features and characteristic (family-resemblance) features for identification and guessing. Also, it should be noted that the family resemblance effect can swamp or override definitional reasoning in traditionally definitional areas. [7]

A few other points that should be kept in mind regarding the way in which humans behave in their thinking and cognition.

If a subject is asked to decide if an item is an instance of a concept, then the more typical the instance the quicker the decision. [8]

If a subject is asked to generate a list of instances of a concept, the more typical instances are generated before the less typical ones. [9]

In learning a concept, one is able to name or identify the typical instances before the atypical ones. [9]

If a person is asked to determine the validity of an incomplete deductive argument, the missing premise is produced faster when it is typical than when it is atypical. Thus, a person performing a logical reasoning task with incomplete information, will try to 'fill in the blanks' with typical information. [10]

In inductive arguments, when told that an initial instance of a concept has a novel property and asked to estimate the likelihood that another instance has that property, the subject's estimates are higher the more typical the initial instance. This might mean that the person's anticipation of what will happen next or what should be done next will be guided by typicality and not by logic. [11]

The similarity of an instance to a prototype can be computed using a contrast rule where the similarity of two concepts is the weighted measure of their common properties minus their distinctive ones. [12, 13]

In broader terms, concepts can be considered to be prototypes. A prototype is a schematic representation of an exemplary instance. The prototype is an instance that is highly typical and relatively free from distracting features. The properties of a prototype are identified by asking people to list them. Thus, it will be necessary to ask a person questions in order to determine the meaning of a concept. [14] Note that the asking-people account of typicality has several biases. For example, people are likely to present only the easily codable properties, and are likely to present only the properties that discriminate among current instances.[7, 15]

Returning to the cake-baking domain, it is clear that even the apparently simple declarative assertion that a cake is baked can become complex. Either or both of the key notions of 'cake' and 'baking' may have family resemblance characterizations and lack definition in terms of necessary and sufficient conditions, Or again even though a person might know that a cake is baked and know the definition of 'cake', their concept of cake might include only those instance close to the cake prototype and their reasoning might be focused on these.

- The content of concepts can be specified in terms of definitions and family resemblance.
- There is empirical support for the claim that human thought uses concepts given in terms of family resemblance.
- In actual reasoning the definition and family resemblance aspects of a concept are mixed.

5 The Complexity of Concepts

The complexities of knowledge and concepts motivates two distinct views of human cognition. The first view is that the world of human cognition is simply a mess! Much of normal cognitive life lacks clarity and precision and succeeds only in a general way. What is needed is a regimentation of ordinary cognition so that it come to be aligned with a normative account of how reasoning and conceptualization ought to occur. The second view is that the world of cognition is highly successful in operating under constraints of time, space, and imperfect information. While formalizations are useful at times, they do not match actual human productive thought. In short, formalized declarative reasoning about 'context free' knowledge with explicitly defined concepts may be able to avoid errors but at the cost of missing solutions, engaging in rigid actions, and, perhaps, missing the truth. These are the extreme cases.

A reconstruction of the preceding discussions of knowledge and concepts suggests that the two extreme views may be better thought of as complementary accounts. Consider the stage or maturity of the domain in question. If the domain is sufficiently mature, then the content of the domain can be represented as declarative knowledge in a 'context free' manner where the concepts can be given clear definitions. Alternatively, if a domain is not sufficiently, mature then the content of the domain may be represented as procedural knowledge based on heuristics and skills where the concepts pick out family resemblance relations. However, it might be better to think of all of knowledge as being somewhere between birth and complete maturity. If this is so, then in any given domain and in any given reasoning about the domain, one would expect to find a mixture of features. This sort of mixture is congenial to Peirce's pragmatism.

6 Peirce and Pragmatism

Peirce's recognition of the pivotal role of concepts in linking declarative and procedural knowledge role is clearly seen in the pragmatic maxim . Peirce expressed the pragmatic maxim as: [16]

> Consider what effects, that might conceivably have practical bearings, we conceive the object of our conception to have. Then, our conception of these effects is the whole of our conception of the object. 5.402

Or in somewhat different form as:

> Pragmatism is the principle that every theoretical judgment expressible in a sentence in the indicative mood is a confused form of thought whose only meaning, if it has any, lies in its tendency to enforce a corresponding practical maxim expressible as a conditional having its apodosis in the imperative mood. 5.18

A key to understanding Peirce's pragmatism is the way in which practical issues, issues of action and procedure, are intimately mingled with theoretical constructs and logical argumentation in the pragmatic maxim. Taking a very liberal stance, the pragmatic conception of meaning is that every concept fuses together formal or theoretical elements and practical elements. The practical dimension is always needed to attain clarity in our conceptions. Thus, in terms of the cake-baking domain, it would appear that the two key concepts of "cake" and "baking" should be represented as a fusion of theoretical and practical maxims. In the case of such concepts, there is a unification of the theoretical and practical dimensions of the concept such that these dimensions can be retrieved at some later point when there is a need to generate declarative or procedural knowledge claims. Deviating only slightly from Peirce's position, the theoretical dimensions of the concept can be understood in terms of hierarchies, formal definition, and the regime of formal logic. The practical dimension of the concept can be understood in terms of associations, family resemblance, and procedures.

The merging of the declarative and procedural aspects of knowledge can also be seen in Peirce's account of inquiry. Rejecting Cartesian accounts of knowledge, Peirce developed a view in which the beliefs would stand so long as there was no real doubt about the belief. Once there was a doubt, however, a process of inquiry would commence in an effort to fix a new belief.

> That the settlement of opinion is the sole end of inquiry is a very important proposition. It sweeps away, at once, various vague and erroneous conceptions of proof. 5.375

Although there are several methods for fixing belief, the most desirable is the scientific method and this method has an essentially social aspect.

> To satisfy our doubts, therefore, it is necessary that a method should be found by which our beliefs may be determined by nothing human, but by some external permanency - by something upon which our thinking has no effect ... It must be something

which affects, or might affect, every man. And, though these affections are necessarily as various as are the individual conditions, yet the method must be such that the ultimate conclusion of every man shall be the same. Such is the method of science. 5.384

The ideas that the point of inquiry is to fix opinion, that an external permanence is needed, and that science is social, leads Pierce to the ultimate end of inquiry.

... All the followers of science are animated by a cheerful hope that the process of investigation, if only pushed far enough, will give one certain solution to each question to which they apply it. ... This great hope is embodied in the conception of truth and reality. The opinion which is fated to be ultimately agreed to by all who investigate, is what we mean by truth, and the object represented in this opinion is real. 5.407

Assuming that scientific method would lead to knowledge having a natural history and that in making our concepts clear habits of mind and behavior are needed, it would appear that a Peircean account of knowledge and concepts would require both a procedural and declarative aspect. Further, given the sort of natural history of knowledge that Peirce would expect the scientific method to generate, one could rightly assume that at any point prior to the completion of all inquiry these aspects would be mixed.

If a Peircean sort of account of knowledge and conceptualization makes sense, then it would seem that a framework for argumentation that was broader than, but included, that needed for argumentation with declarative knowledge is needed. In examining the goodness of arguments, Peirce recognized that there are three kinds of reasoning.

These three kinds of reasoning are Abduction, Induction, and Deduction. Deduction is the only necessary reasoning. It is the reasoning of mathematics. It starts from a hypothesis, the truth or falsity of which has nothing to do with the reasoning; and of course its conclusions are equally ideal. ... The only thing that induction accomplishes is to determine the value of a quantity. It sets out with a theory and it measures the degree of concordance of that theory with a fact. ... All the ideas of science come to it by the way of Abduction.. Abduction consists in studying facts and devising a theory to explain them. 5.145

What is needed is not only a framework that will allow for these three kinds of reasoning but will also allow for the notions of declarative knowledge and definition as well as procedural knowledge and family resemblance.

• Pragmatism combines the procedural and declarative aspects of knowledge and reasoning.
• Pragmatism emphasizes the action aspect of knowledge and meaning.
• Pragmatism allows for a natural history of knowledge.
• Pragmatism allows for different kinds of reasoning.

7 General Argument Patterns

By extending Peirce's pragmatism with the notion of a general argument pattern, the relation between conceptual structures and the different ways of knowing can be understood in terms of the "filling instructions" for concepts.

In advancing the notion of explanatory unification, Kitcher developed the notion of a general argument pattern. [17] A general argument pattern consists of:
      1) a schematic pattern,
      2) a set of filling instructions for the schematic elements in the pattern,
      3) a classification describing the inferential characteristics of the pattern.
This general notion becomes the logical notion of an argument pattern in those cases where the conditions on the classification are as stringent as possible and the filling instructions are as relaxed as possible. The general argument pattern becomes the logical argument pattern when there are no restrictions on the filling instructions and the classification is determined exclusively by the logical connectives in the schematic pattern. The notion of a general argument pattern does not preclude the production of logical argument patterns, but these are produced only under special conditions. As actual conditions become more stringent general argument patterns more closely resemble the micro-theories of open systems and these general argument patterns will contain semantic elements that preclude the use of wholly syntactic (declarative) techniques.

The filling instructions do two things. First, they indicate semantically the kinds of things that the schematic variables represent. Thus, the filling instructions indicate the types of substitutions that are legitimate in a domain. In Newtonian physics, for example, these first sort of filling instruction would restrict the sorts of things that can be substituted for 'F,' 'm,' and 'a.' While we typically think that 'F=ma' means that "Force is equal to mass times acceleration" this is not obvious from the schema as it is presented. 'F,' 'm,' and 'a' could represent other things. Second, the filling instructions indicate how to replace the schematic terms. For example, a filling instruction may specify that 'a' is to replaced by a function of an object's coordinates and its time derivatives. Or again, the filling instruction might simply allow that 'a' should be replaced by the reading of some instrument or that a certain instrument should be read, and a specific computation should be made on the basis of the value that is read.

Different domains can have different filling instructions. Further, the filling instructions may change as knowledge is promoted in a domain. In a typical sequence of events, a filling instruction for a schematic term might be based on the intuition of a particular individual skilled at a particular operation. The intuition might be taken as a starting value, or a reasonable value for a certain sort of phenomenon in a particular domain. In a sense, the filling instruction at this point is "Ask person X for the value." At a slightly later stage, when heuristic knowledge becomes available the filling instruction might change. At this stage, the filling instruction might be "Try a value though to be close to V; if the result is unacceptably high then alter V by a reasonable increment i in direction d and if unacceptably low, alter V by i in direction not d." At the last stage of knowledge promotion the filling instruction has become a bit of 'context free' knowledge. The filling instruction in this case might simply be "Calculate V using equation X." This sort of knowledge is now 'context free' since all researchers in the domain know how to calculate V. The social dimension of knowledge promotion is clearly evident. At the skill level, it is important to ask a skilled person. At the heuristic level, there are more individuals who can fill in the blanks in the schematic patterns, but these individuals still require experience in the domain and their number is limited. At the final level of 'context free' knowledge, the level of skill has become very small. Filling in the value of the schematic term can be accomplished by those who are relatively new to the domain, since the filling instruction has been promoted to a calculation. It is but a small step from this notion to the idea that the whole of the filling can be done automatically.

This notion of a general argument pattern is sufficiently broad to handle both declarative and procedural knowledge as well as concepts characterized in definitions and family resemblance. Beginning with general argument patterns where the classification is as stringent as possible, the filling instructions are as relaxed as possible, and all concepts have definitions, and progressing through general argument patterns where the classification is relaxed, the filling instructions are restricted and concepts are given by family resemblance, the broad range of argumentation and conceptualization can be represented. Further, the natural history of the argument can be traced in the changes to the filling instructions. Moreover, the filling instructions themselves will need to contain representations that are appropriate for both declarative and procedural knowledge. Finally depending on what is known in the general argument pattern and what is to be determined, deductive, inductive, and abductive reasoning can be covered.

- General argument patterns can represent strict logical arguments.
- General argument patterns allow for the use of both declarative and procedural knowledge
- The filling instruction component can represent the definition and family resemblance aspects of concepts.
- The classification component can be used to represent a broad range of reasoning patterns.

8 Conceptual Structures

Sowa attempted to construct a framework which would be a compromise between Aristotle's defnitional account of concepts and Wittgenstien's family resemblance account. [18, p.17] However, in recognizing that many topics in artificial intelligence would have to come to grips with the problems of declarative and procedural knowledge, he seems to have assumed that procedural knowledge was declarative knowledge in formation.

> Issue in the procedural-declarative controversy appear throughout this book: the representation of schemata in human memory; semantics defined in model theory or computation; parsing techniques for natural language; and the form of inference rules for expert systems. In all these areas, procedures allow a "quick and dirty" solution when theoretical issues are still unsolved. As theory progresses, more of the knowledge can be removed from procedures and put in declarative form. [18, p.24]

Sowa proposes to build on Peirce's work and construct a system of conceptual graphs. Conceptual graphs provide a mechanism for representing the formal, logical aspects of reasoning and the definitional meaning for concepts. Further, the graphical nature of the approach provides for a clearer human understanding of the logical relations and arguments, and the basic framework can be extended beyond strict first-order logic. Finally, by using the conceptual graph approach, the virtues of logical systems in terms of valid and reliable rules of inference can be preserved. However, several assumption must be made to even begin the conceptual graphs project. These include:
- Concepts are discrete units.
- Combinations of concepts are not diffuse mixtures, but ordered structures.
- Only discrete relationships are recorded in concepts [18, p.73].

Further, he claims that perception provides the working model that represents and interprets sensory input and that the more abstract part of the working model is a conceptual graph. [18, p.69]

While this represents part of a Peircean approach to knowledge, conceptualization, and reasoning, it leaves much of Peirce's pragmatism to the side. There are two points to be considered in this regard. The first is the action orientation of pragmatism and the second is the kind of ordered structure that a concept exhibits.

Peirce's pragmatism is based on interaction with the world. The pragmatic maxim indicates the way in which practical effects determine the meaning of a concept. While a pragmatic account does not deny the utility of a definitional account of concepts it does require that practical difference be drawn if the meaning of a concept is to be clear. For many concepts this will mean that some procedure will be specified whereby some perception will signal a difference in concept. If no such procedure can be specified for a concept, then that concept would have no meaning. This also accords with Collins view of the natural history of knowledge. Even at the point at which knowledge becomes 'context free,' skills and heuristics provide the background for interpretation. Thus, it is both perception and action that play a role in conceptualization and knowledge.

The ordered structure of a concept need not be resolved into a collection of logical relations among concepts. Between logical structure and diffuse mixture is a large ground. As noted above even when there are explicit definitions of concepts the typicality effects of family resemblance may play a role in reasoning. Further, the specification of a conceptual structure can make certain sorts of reasoning either difficult or impossible. Consider the case of abduction. In this case there is a collection of propositions that seems to have a unity, but has not yet been unified. Using the notion of a general argument pattern this can be represented as a template in which a collection of propositions fills a schematic pattern, the filling instructions are restricted, and the classification is not strict. In other words the actual propositions fill in the schematic variables that lead from them to a unifying proposition that stands as the conclusion of the argument. The filling instructions are restricted to those appropriate to the domain in question and reasoning principles are those for abductive reasoning.

- Conceptual graph theory focuses on the declarative and definitional aspects of knowledge and reasoning.
- Conceptual graph theory focus on deductive reasoning.
- Conceptual graph theory advances part of the Peircean program, but does not address the pragmatic aspects of the program

9 Conclusion

A robust account of knowledge and conceptualization should be able to account for knowing that a cake is baked and knowing how to bake a cake. It must be able to account for both declarative and procedural knowledge. If Peirce is right, it is only in the theoretical limit that all of knowledge would have a declarative form. If so, and if knowledge has a natural history, then knowledge about a given domain will continue to have both a declarative and procedural aspect. Likewise, the concepts that unify a domain will continue to have both definitional and family resemblance aspects. Finally given the constraints under which human reasoning takes place it is reasonable to think that there will be multiple reasoning patterns.

The notion of a general argument pattern can be used to represent the diverse elements in the actual circumstances of reasoning and conceptual graphs and be used to represent one of the modes of reasoning and conceptualization. In addition to Peirce's notions of inductive and abductive reasoning can be added the studies on minimal rationality [19], belief revision [20], modeling and judgment [21], cognitive labor [22], case-based reasoning [23], analogy [24], and more. These modes of reasoning while not possessing the virtues of a formal logic, are not wholly unprincipled. Thought they may draw their principles from the cognitive sciences, they are still principled.

Since a robust account of conceptual structures must be able to handle both the context of "knowing that" and "knowing how," as well as variability in the meaning of concepts and reasoning, it would seem reasonable to think that there will be multiple representations for knowledge and concepts. This in turn suggests that a methodological principle of tolerance between those approaches that stress the notions of declarative knowledge, definition, and logic appropriate to "knowing that" and those that stress procedural knowledge, family resemblance, and the varieties of reasoning appropriate to "knowing how" is desirable. Further since the Peircean convergent community of inquirers does not appear on the horizon, it will be a practical necessity to use all of these. Pragmatically our concepts unify domains and allow us to both make statements about them and to interact with them. To modify the old Kantian saying, procedural knowledge without declarative knowledge is blind, and declarative knowledge without procedural knowledge is empty. Peirce's pragmatism was neither blind nor empty.

References

1. Collins, H, Artificial Experts: Social Knowledge and Intelligent Machines, Cambridge, MA: MIT Press, 1990.
2. Smith, E. E. and Medin, D. L., Categories and Concepts, Cambridge, MA: Harvard University Press, 1981.
3. Schwartz, S. "Natural kind terms." Cognition 7 (1979): 301-315.
4. Fodor, J., Garrett, M., Walker, E., and Parkes, C. M., "Against definitions," Cognition 8 (1980): 263-367.
5. Rosch, E. and Mervis, C. B., "Family resemblances: studies in the internal structures of categories," Cognitive Psychology 3 (1975): 382-429.
6. Wittgenstein, L., Philosophical Investigations. New York, NY: Macmillan, 1953.
7. Armstrong, S. L., Gleitman, L. R. and Gleitman, H. "What some concepts might not be," Cognition 13 (1983): 263-308.
8. Smith, E. E., Shoben, E. J., and Rips, L. J., "Structure and process in semantic memory," Psychological Review 81 (1974): 214-352.
9. Rosch. E., "Principles of categorization," in E. Rosch and B. B. Lloyd, eds., Cognition and Categorization, Hillside, NJ: Erlbaum, 1978.
10. Cherniak, C., "Prototypicality and deductive reasoning," Journal of Verbal Learning and Verbal Behavior 23 (1984): 625-642.
11. Rips, L. J., "Inductive judgments about natural categories," Journal of Verbal Learning and Verbal Behavior 14 (1975): 665-681.

12. Tversky, A., "Features of similarity," Psychological Review 84 (1977): 327-352.
13. Tversky, A. and Hutchinson, J. W., "Nearest neighbor analysis of psychological spaces," Psychological Review 93 (1986): 3-22.
14. Malt, B. C. and Smith E. E., "Correlated properties in natural categories," Journal of Verbal Learning and Verbal Behavior 23 (1984): 250-269.
15. Shepard, R. N., "Representation of structure in similarity data," Psychometrika 39 (1974): 373-421.
16. Hartshorne, C. and Weiss, P. Collected Papers of Charles Sanders Peirce, Volume 5 and Volume 6, Harvard University Press, 1965.
17. Kitcher, P., "Explanatory Unification," Philosophy of Science 48 (1981): 507-531.
18. Sowa, J. Conceptual Structures: Information Processing in Mind and Machine, Reading, MA: Addison-Wesley, 1984.
19. Cherniak, C., Minimal Rationality, Cambridge, MA: MIT Press, 1986.
20. Harman, G., Change in View, Cambridge, MA: MIT Press, 1986.
21. Giere, R., Explaining Science, Chicago, University of Chicago Press, 1988.
22. Kitcher, P., The Advancement of Science, New York: Oxford University Press, 1993.
23. Kolodner, J., Case-Based Reasoning, San Mateo, CA: Morgan Kaufmann, 1993.
24. Holyoak, K and Thagard, P., ambridge, MA: MIT Press, 1995.

Peircean Foundations for a Theory of Context

John F. Sowa

Philosophy and Computers and Cognitive Science
Binghamton University

Abstract. Contexts have become a topic of major interest in linguistics, philosophy, and artificial intelligence, but no one has been able to give a definitive definition of context that is widely accepted. Yet the basic ideas for a theory of context were all present in Peirce's writings in the late nineteenth and early twentieth century. This paper shows how Peirce's work provides a foundation that unifies ideas on context that are scattered in modern research on logic, philosophy, linguistics, and artificial intelligence.

1 Search for a Theory of Contexts

The word *context* has been used with a variety of conflicting meanings. Some of the confusion results from an ambiguity in the informal senses of the word *context*. Dictionaries list two major meanings:

- The basic meaning is a section of linguistic text or discourse that surrounds some word or phrase of interest.

- The derived meaning is a nonlinguistic situation, environment, domain, setting, background, or milieu that includes some entity, subject, or topic of interest.

The word *context* may refer to the text, to the information contained in the text, to the thing that the information is about, or to the possible uses of the text, the information, or the thing itself. The ambiguity about contexts results from which of these aspects happens to be the central focus. These informal senses of the word suggest criteria for distinguishing the formal functions:

- *Syntax.* The syntactic function of context is to group, delimit, or package a section of text.

- *Semantics.* The quoted text of a context may describe or refer to some real or hypothetical situation. That nonlinguistic referent, which may also be called the context, constitutes the derived meaning of the term.

- *Pragmatics.* The word *interest*, which occurs in both senses of the English definition, suggests some reason or purpose for distinguishing the section of linguistic text or nonlinguistic situation. That purpose is the pragmatics or the reason why the text is being quoted.

The three-way classification of syntax, semantics, and pragmatics is based on Peirce's categories of Firstness, Secondness, and Thirdness.

Peirce's friend and fellow pragmatist William James (1897) gave an example of the importance of purpose in determining what should be included in a context:

> Can we realize for an instant what a cross-section of all existence at a definite point of time would be? While I talk and the flies buzz, a sea gull

catches a fish at the mouth of the Amazon, a tree falls in the Adirondack wilderness, a man sneezes in Germany, a horse dies in Tartary, and twins are born in France. What does that mean? Does the contemporaneity of these events with one another, and with a million others as disjointed, form a rational bond between them, and unite them into anything that means for us a world? Yet just such a collateral contemporaneity, and nothing else, is the real order of the world. It is an order with which we have nothing to do but to get away from it as fast as possible. As I said, we break it: we break it into histories, and we break it into arts, and we break it into sciences; and then we begin to feel at home. We make ten thousand separate serial orders of it, and on any one of these we react as though the others did not exist.

Peirce's categories, his emphasis on purpose, his graph logic with its notation for contexts, his work on modal and intentional logics, his context-dependent references or *indexicals*, and his notion of *laws* as governing Thirdness constitute a tightly integrated theory that can help to unify the disconnected research of more recent logicians.

2 Syntax of Contexts

In 1883, Peirce invented the algebraic notation for predicate calculus. A dozen years later, he developed a graphical notation that more clearly distinguished contexts. Figure 1 shows his graph notation for delimiting the context of the proposition under discussion. In explaining that graph, Peirce (1898) said "When we wish to assert something about a proposition without asserting the proposition itself, we will enclose it in a lightly drawn oval." The line attached to the oval links it to a relation that makes a *metalevel* assertion about the nested proposition.

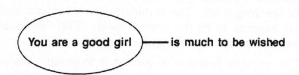

Figure 1. One of Peirce's graphs for talking about a proposition

The oval supports the basic syntactic function of grouping related information in a package. But besides notation, Peirce also developed a theory of the semantics and pragmatics of contexts and the rules of inference for importing and exporting information into and out of the contexts. To support first-order logic, the only necessary metalevel relation is *negation*. By combining negation with the existential-conjunctive subset of logic, Peirce developed his *existential graphs* (EGs), which are based on three primitives:

1. Existential quantifier: A bar or linked structure of bars, called a *line of identity*, represents ∃.

2. Conjunction: The *juxtaposition* of two graphs in the same context represents ∧.

3. Negation: An *oval enclosure* with no lines attached to it represents ~ or the denial of the enclosed proposition.

When combined in all possible ways, these three primitives can represent full first-order logic. When used to state propositions about nested contexts, they form a metalanguage that can be used to define modal and higher-order logic.

To illustrate the use of negative contexts for representing FOL, Figure 2 shows an existential graph and a conceptual graph for the sentence *If a farmer owns a donkey, then he beats it.* The EG on the left has two ovals with no attached lines; by default, they represent negations. It also has two lines of identity, represented as linked bars: one line, which connects *farmer* to the left side of *owns* and *beats*, represents an existentially quantified variable ($\exists x$); the other line, which connects *donkey* to the right side of *owns* and *beats* represents another variable ($\exists y$).

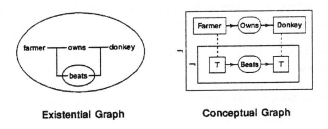

    Existential Graph   **Conceptual Graph**

Figure 2. EG and CG for "If a farmer owns a donkey, then he beats it."

In CGs, a context is defined as a concept whose referent field contains nested conceptual graphs. Since every context is also a concept, it can have a type label, coreference links, and attached conceptual relations. Syntactically, Peirce's ovals are squared off to form boxes, and the negation is explicitly marked by a ¬ symbol in front of the box. The primary difference between EGs and CGs is in the treatment of lines of identity. In EGs, the lines serve two different purposes: they represent existential quantifiers, and they show how the arguments are connected to the relations. In CGs, those two functions are split: the concepts [Farmer] and [Donkey] represent typed quantifiers ($\exists x$:Farmer) and ($\exists y$:Donkey), and arcs marked with arrows show the connections of relations to their arguments. In the inner context, the two concepts represented as [T] are connected by coreference links to concepts in the outer context.

When the EG of Figure 2 is translated to predicate calculus, *farmer* and *donkey* map to monadic predicates; *owns* and *beats* map to dyadic predicates. The implicit conjunctions are represented with the ∧ symbol. The result is an untyped formula:

$\sim(\exists x)(\exists y)(\text{farmer}(x) \wedge \text{donkey}(y) \wedge \text{owns}(x,y) \wedge \sim\text{beats}(x,y))$.

The CG maps to the equivalent typed formula:

$\sim(\exists x{:}\text{Farmer})(\exists y{:}\text{Donkey})(\text{owns}(x,y) \wedge \sim\text{beats}(x,y))$.

A nest of two ovals, as in Figure 2, is what Peirce called a *scroll*. It represents implication, since ~(p∧~q) is equivalent to p⊃q. Using the ⊃ symbol, the two formulas may be rewritten

(∀x)(∀y)((farmer(x) ∧ donkey(y) ∧ owns(x,y)) ⊃ beats(x,y)).

(∀x:Farmer)(∀y:Donkey)(owns(x,y) ⊃ beats(x,y)).

The algebraic formulas with the ⊃ symbol illustrate a peculiar feature of predicate calculus: in order to keep the variables x and y within the scope of the quantifiers, the existential quantifiers in the phrases *a farmer* and *a donkey* must be moved to the front of the formula and be translated to universal quantifiers. This puzzling feature of logic has posed a problem for linguists and logicians since the middle ages. The *donkey sentence* represented in Figure 2 is one of a series of sentences that the scholastics used to illustrate the problems of mapping language to logic.

Besides attaching a relation to an oval, Peirce also used colors or *tinctures* to distinguish contexts other than negation. Figure 3 shows one of his examples, but with shading instead of color. The graph contains four ovals: the outer two form a scroll for *if-then*; the inner two represent possibility (shading) and impossibility (shading inside a negation). The outer oval may be read *If there exist a person, a horse, and water*; the next oval may be read *then it is possible for the person to lead the horse to the water and not possible for the person to make the horse drink the water.*

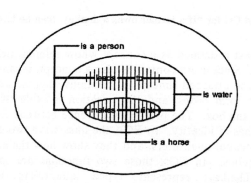

Figure 3. EG for "You can lead a horse to water, but you can't make him drink."

The notation -leads-to- represents a triadic predicate leadsTo(x,y,z), and -makes-drink- represents makesDrink(x,y,z). In the algebraic notation with ◇ for possibility, Figure 3 maps to the following formula:

~(∃x)(∃y)(∃z)(person(x) ∧ horse(y) ∧ water(z) ∧
   ~(◇leadsTo(x,y,z) ∧ ~◇makesDrink(x,y,z))).

With the symbol ⊃ for implication, this formula becomes

(∀x)(∀y)(∀z)((person(x) ∧ horse(y) ∧ water(z)) ⊃
   (◇leadsTo(x,y,z) ∧ ~◇makesDrink(x,y,z))).

This version may be read *For all x, y, and z, if x is a person, y is a horse, and z is water, then it is possible for x to lead y to z, and not possible for x to make y drink z*. These readings, although logically explicit, are not as succinct as the proverb *You can lead a horse to water, but you can't make him drink.*

Discourse representation theory. The logician Hans Kamp once spent a summer translating English sentences from a scientific article to predicate calculus. During the course of his work, he was troubled by the same kinds of irregularities that puzzled the medieval scholastics. In order to simplify the mapping from language to logic, Kamp (1981) developed *discourse representation structures* (DRSs) with an explicit notation for contexts. In terms of those structures, Kamp defined the rules of *discourse representation theory* for mapping quantifiers, determiners, and pronouns from language to logic (Kamp & Reyle 1993).

Although Kamp had not been aware of Peirce's existential graphs, his DRSs were structurally equivalent to Peirce's EGs. The diagram on the left of Figure 4 is a DRS for the donkey sentence, *If there exist a farmer x and a donkey y and x owns y, then x beats y*. The two boxes connected by an arrow represent an implication where the antecedent includes the consequent within its scope.

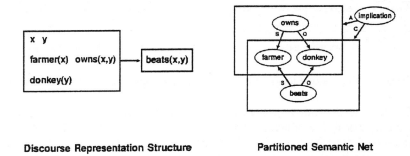

Discourse Representation Structure   **Partitioned Semantic Net**

Figure 4. DRS and PSN for "If a farmer owns a donkey, then he beats it."

The DRS and EG notations look quite different, but they are exactly isomorphic: they have the same primitives, the same scoping rules for variables or lines of identity, and the same translation to predicate calculus. Therefore, the DRS in Figure 4 maps to the same formula as the EG in Figure 2:

$$\sim(\exists x)(\exists y)(\mathtt{farmer}(x) \wedge \mathtt{donkey}(y) \wedge$$
$$\mathtt{owns}(x,y) \wedge \sim\mathtt{beats}(x,y)).$$

Not all context notations have the same structure. On the right of Figure 2 is another notation that uses boxes: the *partitioned semantic network* (PSN) designed by Gary Hendrix (1979). Instead of nested contexts, Hendrix adopted overlapping contexts. Like EGs, CGs, and DRSs, the PSN notation supports complete first-order logic, but the PSN overlapping contexts do not have the same scope as the EG, CG, and DRS forms.

Peirce's motivation for the EG contexts was to simplify the logical structure and rules of inference. Kamp's motivation for the DRS contexts was to simplify the mapping from language to logic. Remarkably, they converged on context representations that are isomorphic. Therefore, Peirce's rules of inference and Kamp's discourse rules apply equally well to contexts in the EG, CG, or DRS notations. For notations with a different structure, such as PSN or predicate calculus, those rules cannot be applied without major modifications.

Resolving indexicals. Besides inventing a logical notation for contexts, Peirce coined the term *indexical* for context-dependent references, such as pronouns and words like *here*, *there*, and *now*. In CGs, the symbol # represents the general indexical, which is usually expressed by the definite article *the*. More specific indexicals are marked by a qualifier after the # symbol, as in #here, #now, #he, #she, or #it. Figure 5 shows two conceptual graphs for the sentence *If a farmer owns a donkey, then he beats it*. The CG on the left represents the original pronouns with indexicals, and the one on the right replaces the indexicals with the coreference labels ?x and ?y.

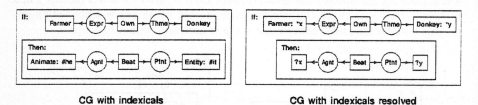

Figure 5. Two conceptual graphs for "If a farmer owns a donkey, then he beats it."

In the concept [Animate: #he], the label Animate indicates the semantic type, and the indexical #he indicates that the referent must be found by a search for some type of Animate entity for which the masculine gender is applicable. In the concept [Entity: #it], the label Entity is synonymous with т, which may represent anything, and the indexical #it indicates that the referent has neuter gender. The search for referents starts in the inner context and proceeds outward to find concepts of an appropriate type and gender. The CG on the right of Figure 5 shows the result of resolving the indexicals: the concept for *he* has been replaced by [?x] to show a coreference to the farmer, and the concept for *it* has been replaced by [?y] to show a coreference to the donkey.

Predicate calculus does not have a notation for indexicals, and its syntax does not show the context structure explicitly. Therefore, the CG on the left of Figure 5 cannot be translated directly to predicate calculus. After the indexicals have been resolved, the CG on the right can be translated to the following formula:

$(\forall x{:}\text{Farmer})(\forall y\text{Donkey})(\forall z{:}\text{Own})$
$\quad(\text{expr}(z,x) \wedge \text{thme}(z,y)) \supset$
$\quad\quad(\exists w{:}\text{Beat})(\text{agnt}(w,x) \wedge \text{ptnt}(w,y))).$

Note that this formula and the graph it was derived from are more complex than the CG in Figure 2. In order to compare the EG and CG directly, that diagram represented the verbs by relations Owns and Beats, which do not explicitly show the linguistic roles. In Figure 5, the concept Own represents a state with an experiencer (Expr) and a theme (Thme). The concept Beat, however, represents an action with an agent (Agnt) and a patient (Ptnt). In general, the patient of an action is more deeply affected or transformed than a theme. See Appendix B.4 for a summary of these linguistic relations.

In analyzing the donkey sentences, the scholastics defined transformations or conversion rules from one logical form to another. As an example, a sentence with the word *every* can be converted to an equivalent sentence with an implication. The sentence *Every farmer who owns a donkey beats it* is equivalent to the one represented in Figures 2, 4, and 5. In CGs, the word *every* maps to a universal quantifier in the referent of some concept:

    [[Farmer: λ]←(Expr)←[Own]→(Thme)→[Donkey]: ∀]-
      (Agnt)←[Beat]→[Entity: #it].

In this graph, the quantifier ∀ does not range over the type Farmer, but over the subtype defined by the nested lambda express: just those farmers who own a donkey. The quantifier ∀ is an example of a *defined quantifier*, which is not one of the primitives in the basic CG notation. It is defined by a macro, whose expansion produces the following CG in the if-then form:

    [If: [Farmer: *x]←(Expr)←[Own]→(Thme)→[Donkey]
    [Then: [?x]←(Agnt)←[Beat]→[Entity: #it]]].

This graph, which may be read *If a farmer x owns a donkey, then x beats it*, is halfway between the two graphs in Figure 5. The indexical that relates the nested agent of beating to the farmer has already been resolved to the coreference pair *x-?x by the macro expansion. The second indexical for the pronoun *it* remains to be resolved to the donkey. This example shows how two sentences that have different surface structures may be mapped to different semantic forms, which are then related by a separate inference step.

The expansion of a universal quantifier to an implication has been known since medieval times. But the complete catalog of all the rules for resolving indexicals is still an active area of research in linguistics and logic. For the sentence *You can lead a horse to water, but you can't make him drink*, many more conversions must be performed to generate the equivalent of Peirce's EG in Figure 3. The first step would be the generation of a logical form with indexicals, such as the CG in Figure 6, which may be read literally *It is possible* (Psbl) *for you to lead a horse to water, but it is not possible* (¬ Psbl) *for you to cause him to drink the liquid*. The relation ¬Psbl is defined as a combination of ¬ and Psbl:

    relation ¬Psbl(*p) **is**
      ¬[Proposition: (Psbl)→[Possible: ?p]].

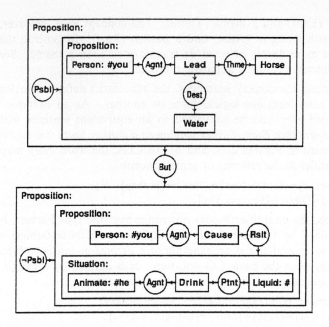

Figure 6. CG for "You can lead a horse to water, but you can't make him drink."

A parser and semantic interpreter that did a purely local or *context-free* analysis of the English sentence could generate the four concepts marked as indexicals by # symbols in Figure 6:

- The two occurrences of *you* would map to the two concepts of the form [Person: #you].

- The pronoun *him* represents a masculine animate indexical in the objective case, whose concept is [Animate: #he].

- The missing object of the verb *drink* would be supplied by a *canonical graph* for the type Drink, which would show an expected patient of type Liquid. Since a context-free parser could not determine the correct referent, the concept [Liquid: #] is marked with an indexical.

The indexicals would have to be resolved by a context-dependent search, proceeding outward from the context in which each indexical is nested.

Conversational implicatures. Sometimes no suitable referent for an indexical can be found. In such a case, the person who hears or reads the sentence must make some further assumptions about implicit referents. The philosopher Paul Grice (1975) observed that such assumptions, called *conversational implicatures*, are often necessary to make sense out of the sentences in ordinary language. They are justified by the charitable assumption that the speaker or writer was trying to make a meaningful statement, but for the sake of brevity, happened to leave some background information unspoken. To resolve the indexicals in Figure 6, the listener would have to make the following kinds of assumptions to fill in the missing information:

1. The two concepts of the form [Person: #you] would normally be resolved to the listener or reader of the sentence. Since no one is explicitly mentioned in any containing context, some such person must be assumed. That assumption corresponds to drawing an if-then nest of contexts with a hypothetical person x coreferent with *you*:

```
[If: [Person: *x]- - -[Person: #you]
   [Then: ...]].
```

The entire graph in Figure 6 would be inserted in place of the three dots in the then part; then every occurrence of #you could be replaced by ?x. The resulting graph could be read *If there exists a person x, then x can lead a horse to water, but x can't make him drink the liquid.*

2. The concept [Animate: #he] might be resolved to either a human or a beast. Since the reader is referred to as *you*, the most likely referent is the horse. But in both CGs and DRSs, coreference links can only be drawn between concepts under one of the following conditions:

   - The antecedent concept [Horse] and the indexical [Animate: #he] both occur in the same context.

   - The antecedent occurs in a context that includes the context of the indexical.

   In Figure 6, neither of these conditions holds. To make the second condition true, the antecedent [Horse] can be *exported* or *lifted* to some containing context, such as the context of the hypothetical reader x. This assumption has the effect of treating the horse as hypothetical as the person x. After a coreference label is assigned to the concept [Horse: *y], the indexical #he could be replaced by ?y.

3. The liquid, which had to be assumed to make sense of the verb *drink*, might be coreferent with the water. But in order to draw a coreference link, another assumption must be made to lift the antecedent concept [Water: *z] to the same hypothetical context as the reader and the horse. Then the concept [Liquid: #] would become [Liquid: ?z].

The result would be the following CG with all indexicals resolved:

```
[If: [Person: *x] [Horse: *y] [Water: *z]
  [Then:
    [Proposition:
      (Psbl)→[Proposition: [Person: ?x]←[Lead]-
        (Thme)→[Horse: ?y]
        (Dest)→[Water: ?z] ]]→(But)-
    [Proposition:
      (¬Psbl)→[Proposition: [Person: ?x]←(Agnt)←[Cause]-
        (Rslt)→[Situation:
          [Animate: ?y]←(Agnt)←[Drink]→(Ptnt)→[Liquid: ?z]] ]]]].
```

This CG may be read *If there exist a person x, a horse y, and water z, then the person x can lead the horse y to water z, but the person x can't make the animate being y drink the liquid z.* This graph is more detailed than the EG in Figure 3, because it explicitly shows the conjunction *but* and the linguistic roles Agnt,

Thme, Ptnt, Dest, and Rslt. Before the indexicals are resolved, the type labels are needed to match the indexicals to their antecedents. Afterwards, the *bound concepts* [Person: ?x], [Horse: ?y], [Animate: ?y], [Water: ?z], and [Liquid: ?z] could be simplified to just [?x], [?y], or [?z].

As this example illustrates, indexicals frequently occur in the intermediate stages of translating language to logic, but their correct resolution may require nontrivial assumptions. Many programs in AI and computational linguistics are able to follow the rules of discourse representation theory to resolve indexicals. The problem of making the correct assumptions about conversational implicatures is more difficult. The kinds of assumptions needed to understand ordinary conversation are similar to the assumptions that are made in nonmonotonic reasoning. Both of them depend partly on context-independent rules of logic and partly on context-dependent background knowledge.

3 Semantics of Contexts

As William James observed, an arbitrary region of space-time has no intrinsic meaning. The best way to deal with the bewildering confusion of events in some region of space and time is to "break it.... We make ten thousand separate serial orders of it, and on any one of these we react as though the others did not exist." A context is a package of information about one of those separated chunks of the world. Semantics determines how those packages relate to those chunks.

Situations and Propositions. Logicians such as Rudolf Carnap (1947), Saul Kripke (1963a,b), and Richard Montague (1974) developed theories of semantics based on models of possible worlds. Each model represents an unbounded region of space-time with all the heterogeneous complexity of William James's example. To avoid such large, open-ended models, Jon Barwise and John Perry (1983) developed *situation semantics* as a theory that related the meaning of sentences to smaller, more manageable chunks called *situations*. Each situation is a configuration of some aspect of the world in a bounded region of space and time. It may be a static configuration that remains unchanged for some period of time, or it may include processes and events that are causing changes. It may include people and things with their actions and speech; it may be real or imaginary; and its time may be present, past, or future.

Situation semantics is a theory about the flow of information: from situations in the world, to speakers who perceive and talk about those situations, and then to listeners who interpret the speech by thinking about and acting upon the situations.

- Speaker's information flow: Situation ⇒ Perception ⇒ Statement.
- Listener's information flow: Statement ⇒ Interpretation ⇒ Action ⇒ Modified situation.

This flow relates the abstract symbols of language to the physical situations that people live in and talk about. Without the physical situations at both ends, the symbols would be *ungrounded*. Symbols acquire meaning by the process of *symbol grounding*, which as Peirce insisted depends on triadic relationships: the speaker expresses a *concept* of an *object* by a *symbol*, which the listener interprets by an equivalent concept "or perhaps a more developed one."

Figure 7 shows a concept of type Situation, which is linked by two image relations (Imag) to two different kinds of images of that situation: a picture and the associated sound. The description relation (Dscr) links the situation to a proposition that describes some aspect of it. That proposition is linked by three statement relations (Stmt) to statements of the proposition in three different languages: an English sentence, a conceptual graph, and a formula in the Knowledge Representation Language (KIF).

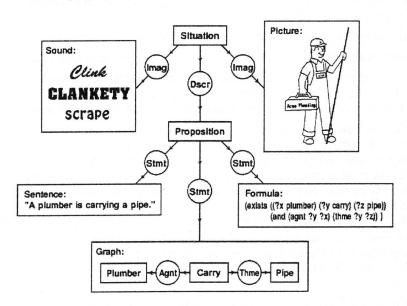

Figure 7. A CG representing a situation of a plumber carrying a pipe

The Imag relation links an entity to an icon that shows what it looks like or sounds like. The Dscr relation or the corresponding predicate dscr(*x,p*) links an entity *x* to a proposition *p* that describes some aspect of *x*. In the metatheory about logic, the symbol $\models$, called the *double turnstile*, is used to say that some proposition *p* is *entailed by* some entity *x*. Semantic entailment $x \models p$ means that the proposition *p* makes a true assertion about some entity *x*; an alternate terminology is to say that the entity *x* *satisfies p*. Semantic entailment is equivalent to the description predicate dscr(*x,p*):

$$(\forall x{:}\text{Entity})(\forall p{:}\text{Proposition})(\text{dscr}(x,p) \equiv x \models p).$$

Literally, for every entity *x* and proposition *p*, *x* has a description *p* if and only if *x* semantically entails *p*. Informally, the terms *semantic entailment*, *description*, and *satisfaction* have been used by different philosophers with different intuitions, but formally, they are synonymous.

As Figure 7 illustrates, the proposition expressed in any of the three languages represents a tiny fraction of the total information available. Both the sound image and the picture image capture information that is not in the sentence, but even they are only partial representations. A picture may be worth a thousand words, but a situation can be worth a thousand pictures. Yet the less detailed sentences have the advantage of being easier to think about and talk about.

McCarthy's contexts. John McCarthy is one of the founding fathers of AI, whose collected work (McCarthy 1990) has frequently inspired and sometimes revolutionized the application of logic to knowledge representation. In his "Notes on Formalizing Context," McCarthy (1993) introduced the predicate ist($\mathscr{C},p$), which may be read "the proposition p is true in context $\mathscr{C}$." As examples, he gave the following assertions:

> ist(contextOf("Sherlock Holmes stories"), "Holmes is a detective").

> ist(contextOf("U.S. legal history"), "Holmes is a Supreme Court Justice").

In these examples, the context *disambiguates* the referent of the name *Holmes* either to the fictional character Sherlock Holmes or to Oliver Wendell Holmes, Jr., the first appointee to the Supreme Court by President Theodore Roosevelt.

One of McCarthy's reasons for developing a theory of context was his uneasiness with the proliferation of new logics for every kind of modal, temporal, epistemic, and nonmonotonic reasoning. The ever-growing number of modes presented in AI journals and conferences is a throwback to the scholastic logicians who went beyond Aristotle's two modes *necessary* and *possible* to *permissible, obligatory, doubtful, clear, generally known, heretical, said by the ancients,* or *written in Holy Scriptures.* The medieval logicians spent so much time talking about modes that they were nicknamed the *modistae.* The modern logicians have axiomatized their modes and developed semantic models to support them, but each theory includes only one or two of the many modes. McCarthy (1977) observed

> For AI purposes, we would need all the above modal operators in the same system. This would make the semantic discussion of the resulting modal logic extremely complex.

Instead of an open-ended number of modes, McCarthy hoped to develop a simple, but universal mechanism that would replace all the modal logics with first-order logic supplemented with metalanguage about contexts. His student R. V. Guha (1991) implemented contexts in the Cyc system and showed that FOL supplemented with metalanguage about contexts could support versions of modal, temporal, default, and higher-order reasoning.

McCarthy and his students have shown that the ist-predicate can be a powerful tool for building knowledge bases, but they have not clearly distinguished the syntax of contexts from their semantics in terms of some subject matter. In fact, the ist-predicate (is-true-in) mixes the syntactic notion of containment (is-in) with the semantic notion of truth (is-true-of). One way to resolve the semantic status of McCarthy's contexts is to assume a mapping from contexts to Barwise and Perry's situations:

$$(\forall \mathscr{C}{:}\text{Context})(\exists s{:}\text{Situation})(\forall p{:}\text{Proposition})(\text{ist}(\mathscr{C},p) \equiv s \models p).$$

This formula says that for every context $\mathscr{C}$ in McCarthy's sense, there exists a situation s in Barwise and Perry's sense; furthermore, for every proposition p, p is true in the context $\mathscr{C}$ if and only if the situation s semantically entails p. For Figure 7, the context $\mathscr{C}$ would include the proposition that a plumber is carrying a pipe, but it would also include all the propositions that are entailed by the sound and picture images: the plumber is carrying a toolbox, he works for Acme Plumbing Co., he is dragging the pipe with a clankety noise, he is wearing a cap with the letter A on it, and so forth. The predicate $ist(\mathscr{C},p)$ would then mean that p is one of those propositions.

At a workshop on contexts at the IJCAI'95 conference, McCarthy was asked whether a context $\mathscr{C}$ could be defined as the set of all propositions entailed by a situation s:

$$\mathscr{C} = \{p \mid s \models p\}.$$

McCarthy, however, objected that this proposed definition did not fully capture his intuitions. He gave an example where one person John believes everything that another person Mary believes. Then both of their belief systems would contain exactly the same propositions. Yet McCarthy claimed that the context of John's beliefs is not the same as the context of Mary's beliefs. That claim would imply that the situation s_1 of Mary's belief is not identical to the situation s_2 of John's belief, even though s_1 and s_2 have the same descriptions. The conceptual graph in Figure 8 represents that implication.

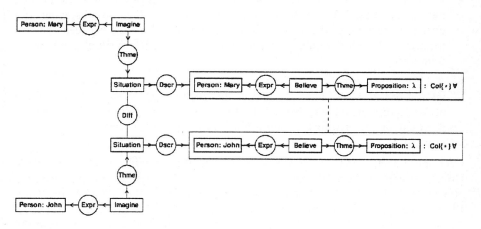

Figure 8. Distinguishing a situation from its description

Figure 8 shows that Mary imagines a situation and John imagines a situation, but the two are different (by the relation Diff, which is defined as not coreferent). Mary's situation is described by the set of all propositions that Mary believes, and John's situation is described by the set of all propositions that John believes. The two sets of propositions are coreferent, even though the two situations are not. The referent $\{*\}\forall$, which may be read *all*, combines the universal quantifier $\forall$ with the generic set $\{*\}$. The prefix Col indicates that the whole set collectively describes the situation.

The distinction between an entity and its description provides a way of interpreting McCarthy's claim about John's beliefs and Mary's beliefs. In ordinary speech, it is common to say that John bought "the same car" that Mary bought when he actually bought a different instance of the same model. The cars or the situations may be distinct, even though they have the same descriptions. For the theory of contexts, Barwise and Perry's term *situation* can be used to refer to the entities that are described by the propositions contained in McCarthy's collections. The mapping from situations to contexts helps to relate ideas in one theory to ideas in the other theory, but further analysis is necessary to clarify the meaning of both theories.

Meaningful situations. In their 1983 book, Barwise and Perry identified a situation with a bounded region of space-time. But as William James observed, an arbitrary region of space and time contains "disjointed events" with no "rational bond between them." A meaningful situation is far from arbitrary, as the following examples illustrate:

- A college lecture could be considered a situation bounded by a 50-minute time period in a spatial region enclosed by the walls of a classroom. But if the time were moved forward by 30 minutes, the region would include the ending of one lecture and the beginning of another. That time shift would create an unnatural "situation."

- If the space were shifted left by half the width of a classroom, it would include part of one class listening to one teacher, part of another class listening to a different teacher speaking on a different topic, and a wall between the two lectures. That shift would create an even more unnatural situation than the time shift.

- Another transformation might fix the coordinate system relative to the sun instead of the earth. Then the region that included the class at the beginning of the lecture would stay behind as the earth moved. Within a few minutes, it would be in deep space, containing nothing but an occasional hydrogen atom.

Even more complex situations would be needed for the referents of the Sherlock Holmes stories or the U.S. legal history. The first is fictional, and the second is intertwined with all the major events that happened in the United States from 1776 to the present. The space-time region for a fictional situation does not exist, and the space-time region for the U.S. legal history cannot be separated from the region of its political history, economic history, or cultural history.

In discussing the development of situation theory, Keith Devlin (1991) observed that the definitions were modified to the point where situations "include, but are not equal to any of simply connected regions of space-time, highly disconnected space-time regions, contexts of utterance (whatever that turns out to mean in precise terms), collections of background conditions for a constraint, and so on." After further discussion, Devlin admitted that they cannot be defined: "Situations are just that: situations. They are abstract objects introduced so that we can handle issues of context, background, and so on."

For Devlin, situations are undefinable objects whose purpose is to simplify the problems of reasoning about contexts. For McCarthy, contexts are undefinable objects whose purpose is to simplify the problems of reasoning about situations. Peirce and James, the two founders of *pragmatism*, focused on what is common in these two circular definitions: the notion of purpose. Space-time coordinates are not sufficient to distinguish a meaningful situation from a disjointed collection of events. Some agent for some purpose must pick and choose what is relevant.

Tinctured existential graphs. Peirce (1906) introduced colors to distinguish contexts of different types. Conveniently, the *heraldic tinctures*, which were used to paint coats of arms in the middle ages, were grouped in three classes: *metal*, *color*, and *fur*. Peirce adopted them for his three-way distinction of *actual*, *modal*, and *intentional* contexts:

1. Actuality is Firstness, because it is what it is, independent of anything else. Peirce used the metallic tincture *argent* (white background) for "the actual or true in a general or ordinary sense," and three other metals (*or, fer,* and *plomb*) for "the actual or true in some special sense."

2. Modality is Secondness, because it distinguishes the mode of a situation relative to what is actual: whenever the actual world changes, the possibilities must also change. Peirce used four heraldic colors to distinguish modalities: *azure* for logical possibility (dark blue) and subjective possibility (light blue); *gules* (red) for objective possibility; *vert* (green) for "what is in the interrogative mood"; and *purpure* (purple) for "freedom or ability."

3. Intentionality is Thirdness, because it depends on how some agent (first) distinguishes a situation (second) from what is actual (third). Peirce used four heraldic furs for intentionality: *sable* (gray) for "the metaphysically, or rationally, or secondarily necessitated"; *ermine* (yellow) for "purpose or intention"; *vair* (brown) for "the commanded"; and *potent* (orange) for "the compelled." Bertrand Russell called these modes *propositional attitudes* because they represent some agent's attitude towards a proposition.

Throughout his analyses, Peirce distinguished the logical operators, such as ∧, ~, and ∃, from the tinctures, which, he said, do not represent

> ...differences of the *predicates*, or *significations* of the graphs, but of the predetermined objects to which the graphs are intended to refer. Consequently, the Iconic idea of the System requires that they should be represented, not by differentiations of the Graphs themselves but by appropriate visible characters of the surfaces upon which the Graphs are marked.

In effect, Peirce did not consider the tinctures to be part of logic itself, but of the metalanguage for describing how logic applies to the universe of discourse:

> The nature of the universe or universes of discourse (for several may be referred to in a single assertion) in the rather unusual cases in which such precision is required, is denoted either by using modifications of the heraldic tinctures, marked in something like the usual manner in pale ink upon the surface, or by scribing the graphs in colored inks.

Peirce's later writings are fragmentary, incomplete, and mostly unpublished, but they are no more fragmentary and incomplete than most modern publications about contexts. In fact, Peirce was more consistent in distinguishing the syntax (oval enclosures), the semantics ("the universe or universes of discourse"), and the pragmatics (the tinctures that "denote" the "nature" of those universes).

Classifying contexts. The first step towards a theory of context is a classification of the types of contexts and their relationships to one another. Any of the tinctured contexts may be nested inside or outside of the ovals representing negation. When combined with negation in all possible ways, each tincture can represent a family of related modalities:

1. The first metallic tincture, argent, corresponds to the white background that Peirce used for his original existential graphs. When combined with existence and conjunction, negations on a white background support classical first-order logic about what is actually true or false "in an ordinary sense." Negations on the other metallic backgrounds support FOL for what is "actual in some special sense." A statement about the physical world, for example, would be actual in an ordinary sense. But Peirce also considered mathematical abstractions, such as Cantor's hierarchy of infinite sets, to be actual, but not in the same sense as ordinary physical entities.

2. In the algebraic notation, $\Diamond p$ means that p is possible. Then necessity $\Box p$ is defined as $\sim\Diamond\sim p$. Impossibility is represented as $\sim\Diamond p$ or equivalently $\Box\sim p$. Instead of the single symbol $\Diamond$, Peirce's five colors represent different versions of possibility; for each of them, there is a corresponding interpretation of necessity, impossibility, and contingency:

   - *Logical possibility.* A dark blue context, Peirce's equivalent of $\Diamond p$, would mean that p is consistent or not provably false. His version of $\Box p$, represented as dark blue between two negations, would therefore mean that p is provable. Impossible $\sim\Diamond p$ would mean inconsistent or provably false.

   - *Subjective possibility.* In light blue, $\Diamond p$ would mean that p is believable or not known to be false. $\Box p$ would mean that p is known or not believably false. This interpretation of $\Diamond$ and $\Box$ is called *epistemic logic.*

   - *Objective possibility.* In red, $\Diamond p$ would mean that p is physically possible. As an example, Peirce noted that it was physically possible for him to raise his arm, even when he was not at the moment doing so. $\Box p$ would mean physical necessity according to the laws of nature.

   - *Interrogative mood.* In green, $\Diamond p$ would mean that p is questioned, and $\Box p$ would mean that p is not questionably false. This interpretation of $\Diamond p$ corresponds to a proposition p in a Prolog goal or the where-clause of an SQL query.

   - *Freedom.* In purple, $\Diamond p$ would mean that p is free or permissible; $\Box p$ would mean that p is obligatory or not permissibly false; $\sim\Diamond p$ would mean that p is not permissible or illegal; and $\Diamond\sim p$ would mean that p is permissibly false or optional. This interpretation of $\Diamond$ and $\Box$ is called *deontic logic.*

3. The heraldic furs represent various kinds of intentions, but Peirce did not explore the detailed interactions of the furs with negations or with each other. Don Roberts (1973) suggested some combinations, such as negation with the tinctures gules and potent to represent *The quality of mercy is not strained.*

Although Peirce's three-way classification of contexts is useful, he did not work out their implications in detail. He wrote that the complete classification of "all the conceptions of logic" was "a labor for generations of analysts, not for one."

In the current generation, theories like situation semantics address the problems of classifying and reasoning about intentional contexts, but the number of intentional combinations is very large. Many intentional verbs come in pairs like *hope* and *fear*, *know* and *believe*, or *seek* and *avoid*. When combined with negations, these verbs generate more complex patterns than the relationships between the two basic modes $\lozenge$ and $\square$. As an example, the predicate hope(a,p) could mean that the agent a hopes that the proposition p will become true. Then the predicate fear(a,p) would mean that a hopes that p will not become true:

fear(a,p) $\equiv$ hope($a,\sim p$).

An agent a is *indifferent* to p if a neither hopes nor fears that p:

indifferent(a,p) $\equiv$ $\sim$hope(a,p) $\wedge$ $\sim$fear(a,p).

An agent a is *ambivalent* about p if a both hopes that p and fears that p:

ambivalent(a,p) $\equiv$ hope(a,p) $\wedge$ fear(a,p).

Given the definition of fear in terms of hope, ambivalence would mean that a hopes that p will come true and that p will not come true:

ambivalent(a,p) $\equiv$ hope(a,p) $\wedge$ hope($a,\sim p$).

If conjunction is assumed to commute with hope, this formula would imply that a hopes for a contradiction $p \wedge \sim p$. Such states occur in science-fiction movies when someone like Captain Kirk presents a computer with an unresolvable dilemma.

4 Possible Worlds, Situations, and Contexts

Leibniz introduced the notion of *possible worlds* as the foundation for modal semantics: a proposition p is necessarily true if it is true in every possible world, and p is possible if it is true in some possible world. In the algebraic notation, Peirce followed Leibniz by representing necessity with a universal quantifier Π_ω where the variable ω ranges over all "states of affairs." For the graphic notation, he suggested a pad of paper instead of a single "sheet of assertion." Graphs that are necessarily true would be copied on every sheet; those that are possibly true would be drawn on some, but not all sheets. The top sheet would represent what is asserted about the real world, and the other sheets would contain descriptions of other possible worlds.

Kripke's semantics. Besides having a notation, a theory of modality must relate the modal axioms (which were presented in Section 1.4) to formal models of the possible worlds or states of affairs. Saul Kripke (1963a,b) developed such a theory with *model structures* having three components:

- *Possible worlds.* A set K of entities called *possible worlds*, of which one *privileged world* w_0 represents the real world.
- *Accessibility relation.* A relation $R(u,v)$ defined over K, which says that world v is *accessible* from world u.
- *Evaluation function.* A function $\Phi(p,w)$, which maps a proposition p and a possible world w to one of the two truth values $\{T,F\}$. The world w semantically entails p if $\Phi(p,w)$ has the value T, and w entails $\sim p$ if $\Phi(p,w)$ has the value F:

$$w \models p \equiv \Phi(p,w) = T.$$
$$w \models \sim p \equiv \Phi(p,w) = F.$$

For the real world w_0, the evaluation function Φ determines whether p is contingently true or false. To determine whether p is necessary or possible, Kripke defined $\Diamond p$ and $\Box p$ by considering the truth of p in the worlds that are accessible from w_0:

- *Possibility.* p is possible in the real world w_0 if p evaluates to T for some world w accessible from w_0:

$$\Diamond p \equiv (\exists w{:}\text{World})(R(w_0,w) \wedge \Phi(p,w) = T).$$

- *Necessity.* p is necessary in w_0 if p evaluates to T for every world w accessible from w_0:

$$\Box p \equiv (\forall w{:}\text{World})(R(w_0,w) \supset \Phi(p,w) = T).$$

These definitions are a formal statement of Leibniz's intuition. Kripke's contribution was to show how the modal axioms determine constraints on R:

- *System T.* Two basic axioms of System T are $\Box p \supset p$ (Necessity implies truth) and $p \supset \Diamond p$ (Truth implies possibility). They require every world to be accessible from itself; hence, R must be reflexive:

$$\text{reflexive}(R) \equiv (\forall w{:}\text{World})R(w,w).$$

- *System S4.* System T with Lewis's axiom S4, $\Box p \supset \Box\Box p$, requires R to be transitive:

$$\text{transitive}(R) \equiv (\forall u,v,w{:}\text{World})((R(u,v) \wedge R(v,w)) \supset R(u,w)).$$

- *System S5.* System S4 with axiom S5, $\Diamond p \supset \Box \Diamond p$, requires R to be symmetric:

$$\text{symmetric}(R) \equiv (\forall u,v{:}\text{World})(R(u,v) \supset R(v,u)).$$

For System S5, the properties of reflexivity, transitivity, and symmetry make R an *equivalence relation.* Those properties cause the collection of all possible worlds to be partitioned in disjoint equivalence classes. Within each class, all the worlds are accessible from one another, but no world in one class is accessible from any world in another class.

Hintikka's model sets. Instead of assuming possible worlds, Jaakko Hintikka (1961, 1963) independently developed an equivalent semantics for modal logic based on collections of propositions, which he called *model sets.* He also assumed an *alternativity relation* between model sets, which serves the same purpose as Kripke's accessibility relation between worlds. As collections of propositions, Hintikka's model sets describe Kripke's possible worlds in the same way that

McCarthy's contexts describe Barwise and Perry's situations. The assumption about McCarthy's ist-predicate can be used to relate a Hintikka model set $\mathcal{M}$ to a Kripke world w:

$$(\forall \mathcal{M}:\text{ModelSet})(\exists w:\text{World})(\forall p:\text{Proposition})(\text{ist}(\mathcal{M},p) \equiv w \models p).$$

This formula says that for any model set $\mathcal{M}$, there exists a possible world w for which a proposition p is true in $\mathcal{M}$ if and only if w semantically entails p.

The primary difference between model sets and contexts is size: Hintikka defined model sets as *maximally consistent* sets of propositions that could describe everything in the real world or any possible world. But as William James observed, a collection of information about the entire world is far too large and disjointed to be comprehended and manipulated in any meaningful way. Although contexts might be extendible to infinite model sets, Peirce and McCarthy used finite contexts to describe a limited part of the world from a particular point of view. A context is an excerpt from a model set in the same sense that a situation is an excerpt from a possible world.

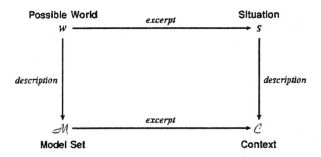

Figure 9. Ways of mapping possible worlds to contexts

Figure 9 shows mappings from a Kripke possible world w to a description of w as a Hintikka model set $\mathcal{M}$ or a finite excerpt from w as a Barwise and Perry situation s. Then $\mathcal{M}$ and s may be mapped to a McCarthy context $\mathcal{C}$. This is an example of a *commutative diagram*, which shows a family of mappings that lead to the same result by multiple routes. From a possible world w, the mapping to the right extracts an excerpt as a situation s, which may be described by the propositions in a context $\mathcal{C}$. From the same world w, the downward mapping leads to a description of w as a model set $\mathcal{M}$, from which an equivalent excerpt would produce the same context $\mathcal{C}$.

The combined mappings in Figure 9 replace the mysterious possible worlds with finite, computable contexts. Hintikka's model sets support operations on familiar symbols instead of possible worlds, but they may still be infinite. Situations are finite, but they consist of physical or fictitious objects that are not computable. The contexts in the lower right of Figure 9 are the only things that can be represented and manipulated in a digital computer. Any theory of semantics that is stated in terms of possible worlds, model sets, or situations must ultimately be mapped to a theory of contexts in order to be computable.

Dunn's laws and facts. As long as the accessibility relation R is assumed as an *a priori* primitive, it cannot explain modality in terms of anything more fundamental. To make accessibility a derived relation, Michael Dunn (1973) added more structure to Hintikka's propositions. For any model set $\mathcal{M}$, he called the propositions true in $\mathcal{M}$ the *facts* of $\mathcal{M}$; then he distinguished a subset $\mathcal{L}$ of facts, which he called the *laws* of $\mathcal{M}$. Finally, Dunn showed how accessibility from one model set to another is determined by the laws and facts they contain. As a result, accessibility is no longer primitive, and modality depends on which propositions are chosen as laws. The ultimate responsibility for modality devolves upon some agent, called the *lawgiver*, who makes that choice.

Philosophers since Aristotle have recognized that modality is related to laws. Dunn's innovation lay in making the relationships explicit. Let $\mathcal{M}_1$ be a model set with laws $\mathcal{L}_1$ that describes a possible world w_1, and let $\mathcal{M}_2$ be another model set with laws $\mathcal{L}_2$ that describes a world w_2. Dunn defined accessibility from the world w_1 to the world w_2 to mean that the laws $\mathcal{L}_1$ are a subset of the facts in $\mathcal{M}_2$:

$$R(w_1, w_2) \equiv \mathcal{L}_1 \subset \mathcal{M}_2.$$

According to this definition, the laws of the first world w_1 remain true in the second world w_2, but they may be demoted from the status of laws to just ordinary facts. Dunn then restated the definitions of possibility and necessity in terms of laws and facts. In Kripke's version, possibility $\Diamond p$ means that p is true of some world w accessible from the real world w_0:

$$\Diamond p \equiv (\exists w \text{World})(R(w_0, w) \wedge w \models p).$$

By substituting the laws and facts for the possible worlds, Dunn derived an equivalent definition:

$$\Diamond p \equiv (\exists \mathcal{M}\text{:ModelSet})(\text{laws}(\mathcal{M}) \subset \mathcal{M}_0 \wedge p \in \mathcal{M}).$$

Now possibility $\Diamond p$ means that there exists a model set $\mathcal{M}$ whose laws are a subset of the facts of the real world $\mathcal{M}_0$ and p is a fact in $\mathcal{M}$. By the same substitutions, the definition of necessity becomes

$$\Box p \equiv (\forall \mathcal{M}\text{:ModelSet})(\text{laws}(\mathcal{M}) \subset \mathcal{M}_0 \supset p \in \mathcal{M}).$$

Necessity $\Box p$ means that in every model set $\mathcal{M}$ whose laws are a subset of the facts of the real world $\mathcal{M}_0$, p is also a fact in $\mathcal{M}$.

Dunn performed the same substitutions in Kripke's constraints on the accessiblity relation. The result is a restatement of the constraints in terms of the laws and facts:

- *System T.* The two axioms that $\Box p \supset p$ and $p \supset \Diamond p$ require every world to be accessible from itself. That property follows from Dunn's definition because the laws $\mathcal{L}$ of any world are a subset of the facts: $\mathcal{L} \subset \mathcal{M}$.

- *System S4.* System T with Lewis's axiom S4, $\Box p \supset \Box \Box p$, requires that R must also be transitive. It imposes the tighter constraint that the laws of the first world must be a subset of the laws of the second world: $\mathcal{L}_1 \subset \mathcal{L}_2$.

- *System S5.* System S4 with axiom S5, $\Diamond p \supset \Box \Diamond p$, requires that R must also be symmetric. It constrains both worlds to have exactly the same laws: $\mathcal{L}_1 = \mathcal{L}_2$.

In Dunn's version of the definitions and constraints, all references to possible worlds have been eliminated: $\Box p$, $\Diamond p$, and the modal axioms depend only on the laws and facts. In the model sets themselves, all the formulas are purely first order, and the symbols $\Box$ and $\Diamond$ never appear in any of them.

Completing the pushout. In the commutative diagram of Figure 9, the downward arrow on the left corresponds to Dunn's mapping of possible worlds to model sets, and the rightward arrow at the top corresponds to Barwise and Perry's mapping of possible worlds to situations. The branch of mathematics called *category theory* has methods of completing such diagrams by deriving the other mappings. Given the two arrows at the left and the top, the technique called a *pushout* defines the two arrows on the bottom and the right:

- *Left.* The downward mapping on the left side of the diagram replaces each possible world w with a model set $\mathcal{M}$ of facts that describe w. Every proposition p in $\mathcal{M}$ must be true of w: $w \models p$. The laws $\mathcal{L}$ of $\mathcal{M}$ must be chosen to preserve the constraints on the accessibility relation R and the evaluation function Φ.

- *Top.* The rightward mapping at the top replaces w with a finite region s selected from w called a situation. This selection is made by some agent who decides how much of the world is relevant and what level of granularity is appropriate for its description.

- *Bottom.* The rightward mapping on the bottom is determined by the selection of s at the top. Every proposition p in $\mathcal{M}$ that is true of s ($s \models p$) is copied to the context $\mathcal{C}$. The laws of $\mathcal{C}$ are those laws in $\mathcal{M}$ that happened to be copied: $\mathcal{C} \cap \mathcal{L}$.

- *Right.* The downward mapping at the right replaces the situation s with a context $\mathcal{C}$ that describes s. It produces the same result as the mapping on the bottom, because it must satisfy the same constraints: every proposition p in $\mathcal{M}$ that is true of s must be in $\mathcal{C}$; the laws of $\mathcal{C}$ are the laws in $\mathcal{M}$ that are included in $\mathcal{C} \cap \mathcal{L}$.

Since each of the four mappings in Figure 9 applies to some noncomputable structure, the mappings themselves are not computable. Their purpose is not to support computation, but to determine how the theories that apply to the non-computable possible worlds and situations are carried over to the computable contexts. After the theories have been transferred to contexts, the possible worlds and situations are unnecessary for further reasoning and computation.

Situations as pullbacks. The inverse of a pushout, called a *pullback*, is an operation of category theory that "pulls" some structure or family of structures backwards along an arrow of a commutative diagram. For the diagram in Figure 9, the model set $\mathcal{M}$ and the context $\mathcal{C}$ are symbolic structures that have been studied in logic for many years. The situation s, as Devlin observed, is not as clearly defined. One way to define a situation is to assume the notion of context as more basic and to say that a situation s is whatever is described by a context $\mathcal{C}$. In terms of the diagram of Figure 9, the pullback would start with the two mappings from w to $\mathcal{M}$ and from $\mathcal{M}$ to $\mathcal{C}$. Then the situation s in the upper right and the two arrows $w\rightarrow s$ and $s\rightarrow\mathscr{C}$ would be derived by a pullback from the starting arrows $w\rightarrow\mathscr{M}$ and $\mathscr{M}\rightarrow\mathscr{C}$.

The definition of situations in terms of contexts may be congenial to logicians for whom collections of propositions are familiar notions. For people who like to think in terms of concrete objects, the notion of a situation as a chunk of the real world may seem more familiar. The commutative diagram provides a way of reconciling the two views: starting with a situation, the pushout determines the propositions in the context; starting with a context, the pullback defines the situation. The two complementary views are useful for different purposes: for a mapmaker, the context is derived as a description of some part of the world; for an architect, the concrete situation is derived by some builder who follows an abstract description.

Specifying laws. The description relation dscr(x,p) is a dyadic relation that links an entity x to a proposition p that describes x. A triadic relation lex(a,p,x) could be used to relate an agent a who specifies a proposition p as a law for some entity x. The following formula says that Tom specified some proposition as a law for a lottery game:

($\exists p$:Proposition)($\exists x$:LotteryGame)(person(Tom) $\wedge$ lex(Tom,p,x)).

By Dunn's convention, the laws $\mathscr{L}$ of any entity x must be a subset of the facts $\mathscr{M}$ that describe x. That condition may be stated as an axiom:

($\forall a$:Agent)($\forall p$:Proposition)($\forall x$:Entity)(lex(a,p,x) $\supset$ dscr(x,p)).

This formula says that for every agent a, proposition p, and entity x, if a specified p as a law of x, then x has a description p. Like the triadic purpose relation in Section 5.1, the triadic lex relation is an example of Mediation or Peirce's category of Thirdness.

As the multiple axioms for modal logic indicate, there is no single version that applies to all problems. The complexities increase when different interpretations of modality are mixed, as in Peirce's five versions of possibility, which could be represented by colors or by subscripts, such as $\lozenge_1$, $\lozenge_2$, ..., $\lozenge_5$. Each of those modalities is derived from a different set of laws, which interact in various ways with the other laws:

- The combination $\square_3\lozenge_1p$, for example, would mean that it is subjectively necessary that p is logically possible.

- According to the definition of $\square_3$, someone must know that $\lozenge_1p$.

- Since what is known must be true, the following theorem would hold for that combination of modalities:

$$\square_3\lozenge_1p \supset \lozenge_1p.$$

Similar analysis would be required to derive the axioms and theorems for all possible combinations of the five kinds of possibility with the five kinds of necessity. Since subjective possibility depends on the subject, the number of combinations increases further when multiple agents interact.

Multimodal reasoning. By introducing contexts, McCarthy hoped to reduce the proliferation of modalities to a single mechanism of metalevel reasoning about the propositions in a context. By supporting a more detailed representation than the operators $\diamondsuit$ and $\square$, the relations lex and dscr enable metalevel reasoning about the implications of the laws and facts. Following are some implications of Peirce's five kinds of possibility:

- *Logical possibility.* The only statements that are logically necessary are tautologies: those statements that are provable from the empty set. No special lawgiver is needed for the empty set; alternatively, every agent may be assumed to specify the empty set:

    $$\{\} = \{p:\text{Proposition} \mid (\forall a:\text{Agent})(\forall x:\text{Entity})\text{lex}(a,p,x)\}.$$

    The empty set is the set of all propositions p where every agent a holds p as a law of every entity x.

- *Subjective possibility.* A proposition p is subjectively possible for an agent a if a does not know p to be false. The subjective laws for any agent a are all the propositions that a knows:

    $$\text{SubjectiveLaws}(a) = \{p:\text{Proposition} \mid \text{know}(a,p)\}.$$

    For any entity x, lex(a,p,x) if a knows p and p refers to x.

- *Objective possibility.* The laws of nature define what is physically possible. The symbol *God* may be used as a place holder for the lawgiver:

    $$\text{LawsOfNature} = \{p:\text{Proposition} \mid (\forall x:\text{Entity})\text{lex}(\text{God},p,x)\}.$$

    If God is assumed to be omniscient, this set is the same as everything God knows or SubjectiveLaws(God). What is subjective for God is objective for everyone else.

- *Interrogative mood.* A proposition is not questioned if it is part of the common knowledge of the parties to a conversation. For two agents a and b, common knowledge can be defined as the intersection of their subjective knowledge or laws:

    $$\text{CommonKnowledge}(a,b) = \text{SubjectiveLaws}(a) \cap \text{SubjectiveLaws}(b).$$

- *Freedom.* Whatever is free or permissible is consistent with the laws, rules, regulations, ordnances, or policies of some lawgiver who has *jurisdiction* or the right to say what is the law. This interpretation, which defines deontic logic, makes it a weak version of modal logic since consistency is weaker than truth. The usual modal axioms $\square p \supset p$ and $p \supset \diamondsuit p$ do not hold for deontic logic, since people can violate the laws.

Reasoning at the metalevel of laws and facts is common practice in courts. In the United States, the Constitution is the supreme law df the land; any law or regulation of the U.S. government or any state, county, or city in the U.S. must be consistent with the U.S. Constitution. But the tautologies and laws of nature are established by an even higher authority. No one can be forced to obey a law that is logically or physically impossible.

References.

Barwise, Jon, & John Perry (1983) *Situations and Attitudes*, MIT Press, Cambridge, MA.

Carnap, Rudolf (1947) *Meaning and Necessity*, University of Chicago Press, Chicago.

Devlin, Keith (1991) "Situations as mathematical abstractions," in J. Barwise, J. M. Gawron, G. Plotkin, & S. Tutiya, eds., *Situation Theory and its Applications*, CSLI, Stanford, CA, pp. 25-39.

Dunn, J. Michael (1973) "A truth value semantics for modal logic," in H. Leblanc, ed., *Truth, Syntax and Modality*, North-Holland, Amsterdam, pp. 87-100.

Grice, H. Paul (1975) "Logic and conversation," in P. Cole & J. Morgan, eds., *Syntax and Semantics 3: Speech Acts*, Academic Press, New York, pp. 41-58.

Hintikka, Jaakko (1961) "Modality and quantification," *Theoria* **27**, pp. 110-128.

Hintikka, Jaakko (1963) "The modes of modality," *Acta Philosophica Fennica, Modal and Many-valued Logics*, pp. 65-81.

James, William (1897) *The Will to Believe and Other Essays*, Longmans, Green & Co.

Kamp, Hans (1981a) "Events, discourse representations, and temporal references," *Langages* **64**, 39-64.

Kamp, Hans (1981b) "A theory of truth and semantic representation," in *Formal Methods in the Study of Language*, ed. by J. A. G. Groenendijk, T. M. V. Janssen, & M. B. J. Stokhof, Mathematical Centre Tracts, Amsterdam, 277-322.

Kamp, Hans, & Uwe Reyle (1993) *From Discourse to Logic*, Kluwer, Dordrecht.

Kripke, Saul A. (1963a) "Semantical considerations on modal logic," *Acta Philosophica Fennica, Modal and Many-valued Logics*, pp. 65-81.

Kripke, Saul A. (1963b) "Semantical analysis of modal logic I," *Zeitschrift für mathematische Logik und Grundlagen der Mathematik* **9**, 67-96.

McCarthy, John (1990) *Formalizing Common Sense*, Ablex, Norwood, NJ.

McCarthy, John (1993) "Notes on formalizing context," *Proc. IJCAI-93*, Chambéry, France, pp. 555-560.

Montague, Richard (1974) *Formal Philosophy*, Yale University Press, New Haven.

Peirce, Charles Sanders (1905) "What Pragmatism Is," *The Monist*, reprinted in CP 5.411-36.

Peirce, Charles Sanders (1906) "Prolegomena to an apology for pragmaticism," *The Monist*, vol. 16, pp. 492-497.

Peirce, Charles Sanders (CP) *Collected Papers of C. S. Peirce*, ed. by C. Hartshorne, P. Weiss, & A. Burks, 8 vols., Harvard University Press, Cambridge, MA, 1931-1958.

Sowa, John F. (1984) *Conceptual Structures: Information Processing in Mind and Machine*, Addison-Wesley, Reading, MA.

The CORALI Project:
From Conceptual Graphs to Conceptual Graphs via Labelled Graphs

Michel Chein

LIRMM (CNRS & University Montpellier 2)
Email: chein@lirmm.fr

Abstract. The scientific objectives of the "COnceptual gRAphs at LIrmm" project and the method used to reach these objectives are presented.

1 Introduction

Sowa proposed the conceptual graphs model, in the data base context, in [4], and then developed it with a very large scope in his 1984 seminal book [5]. Sowa's multiple interests (e.g., linguistics, logics, computer science, artificial intelligence, philosophy) led to a multifaceted and highly diversified CG community. A look at the impressive bibliography compiled by Wermelinger clearly shows this, and therefore we will not present our views on all aspects of the CG model.

This paper is simply aimed at presenting the scientific objectives of the "COnceptual gRAphs at LIrmm" project, and the method used to reach these objectives. Some of our results were presented during ICCS, others were presented elsewhere, or only published in French. I hope that this presentation of the overall picture will fill some holes.

Prior to the CORALI project there was a system, called DILEM which was developed in the late 80s by Dicky, Cogis, and myself (see for instance [1]). DILEM is a reflex programming system, based on labelled trees, for representing reactive systems. It was used to construct a Therapy Adviser in Oncology (TAO ESPRIT project 1592/86). Sowa's book showed me that his way of considering labelled bipartite graphs -and not only trees- opened interesting perspectives, that differ from those usually proposed in semantic networks literature. From its very beginning in 1991, the CORALI project, based on [16], has been scientifically managed by Mugnier and myself. However, many other people have also been involved. Boksenbaum and Libourel brought their experience in database theory, Cogis with his skill in graph algorithms, Carbonneill, Guinaldo, Haemmerlé, and Leclère, who spent three years preparing their PhD theses on the CORALI project (they are now involved in different laboratories and companies). Next year, the only ones left will be Salvat who is finishing his PhD thesis and Genest who will begin his thesis.

However, different ideas have come from researchers (e.g., Levinson, Ellis, Lehmann, Willems, and others) who, in the CG community, put labelled graphs and

orders at the core of their work, at some time or another. Our fundamental scientific objective can be put forward by the following question : *How far it is possible to go, in knowledge representation, by exclusively using labelled graphs to represent knowledge and labelled graph operations to do reasonings?* To study this question, we use the classical four-stroke experimental methodology:
- build a theoretical formal model and algorithms for solving problems in this model;
- construct software tools implementing this theory;
- use the preceding two points to build applications useful for real problems; and
- evaluate the systems built, and loop through this 4-cycle process until satisfactory results have been obtained.

This approach is outlined in Figure 1. Let us begin our look at this figure with one comment. The objects used at the different levels (user interface, formal, physical) are always labelled bipartite graphs. We believe that it is an essential feature for building intelligent cooperative systems, i.e. for knowledge programming. Indeed, there must be an "isomorphism" between what is seen by the user and the formal model, to enable faithful modeling of the actual data and problems, and to correctly interpret the results and the computations. There must be an "isomorphism" between what is seen by the user and how objects and operations are implemented, in order to understand why and how results have been obtained. The most secure way to do this is to use a homogeneous model: the same kind of objects occur at each fundamental level. For similar reasons, links I, II, III, IV which represent intuitive semantics should be in correspondence.

2 The Simple Conceptual Graphs (SCG) model

2.1 Main Simplification

Conceptual graphs are introduced in the first paragraph of chapter 3 [5]. In the second paragraph, entitled "Semantic Network", Sowa states: "a conceptual graph has no meaning in isolation. Only through the semantic network are its concepts and relations linked to context, language, emotion, and perception." Figure 3.5 [5], highlights the idea that a CG has a meaning only if it is embedded in a large semantic network. Otherwise, it is simply a labelled bipartite graph. A canon is the first part of such a semantic network proposed by Sowa. However, as he notes, a canon only contains "very little knowledge of the world, and more information has to be packed into other structures..." [5; p. 96].

Generally, a formal system is inductively defined by a primitive set of objects (the basis) and by a set of rules for building new objects from existing ones (the inductive step). We used this classical way to define data of the SCG model, i.e. simple conceptual graphs.

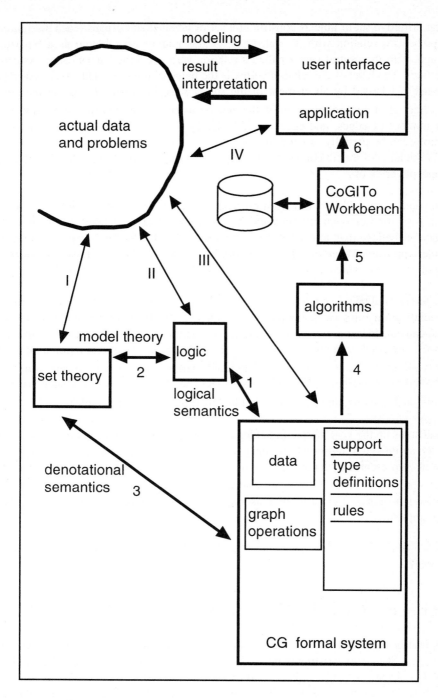

Fig. 1 . Classical Four-Stroke Experimental Methodology

The primitive set of objects consists of a set of labelled star graphs (each star graph represents a relation with its signature). A star graph is considered as a bipartite graph, with the central vertex called a relation vertex and the leaves called concept vertices. The label of the relation vertex is the name of the star graph, and a concept vertex is labelled by an element of the set T_CxMx$\{*\}$, where T_C is the partially ordered set of concept types and * the generic marker. The arity of a star graph r is its number of edges, and edges are labelled by 1, 2, ..., arity(r). This set of star graphs can be partially ordered, in this case if $r \leq r'$ then arity(r) = arity(r') and for all i the label of the ith concept vertex of r is less than or equal to the label of the ith concept vertex of r'.

(In some papers, Robert Levinson proposed directed hypergraphs. We think that in doing so the homogeneity between the models used at each level might be lost, since we believe that, at least at the user interface level, a labelled graph has a much wider applicable range than a hypergraph. Note also that the first notion is that of a labelled vertex, which is similar to a term occurrence in any linear formula.)

Different equivalent construction rule sets have been proposed. Sowa called them specialization rules (with their inverses called generalization rules). For instance, one can consider the set: {simplification (s), relation or concept label restriction (dr or dc), external or internal join (j_1 or j_2)}. With concept label restriction a new set M is introduced, called the set of individual markers, and then a concept vertex is labelled by an element of the partially ordered set T_CxMx$\{*\}$ (* is greater than all individual markers, and any two individual markers are uncomparable).

That sums up the very simple (and elegant) SCG formal system. This formal system is too simplistic to represent concepts and relations between them, indeed a concept is only represented by its position in the hierarchy T_C, and by star graphs in which it appears. In this case, the semantic network is so poor that we propose to simply called it a *support*. Furthermore, reasonings based solely on these notions might also be simplistic.

Hence, the obvious first question to study is: What can be done with the SCG model?

2.2 First Results on the SCG Formal System

As briefly mentioned above, a composition of elementary rules enables derivation of a new SCG from existing ones. To ensure that a derivation sequence of a SCG G solely consists of SCGs actually used to build G, we define a *derivation* as follows [35]. A derivation of a SCG G is a directed graph, without directed cycles, whose vertices are labelled by conceptual graphs, such that:
1. there is a unique sink labelled by G,
2. let x be any vertex with label H; then,

(a) x is a source, and H is a SCG, or

(b) x has one predecessor labelled K, and $H = s(K)$ or $dr(K)$ or $dc(K)$ or $j_1(K)$ or

(c) x has two predecessors, labelled K_1 and K_2, and $H = j_2(K_1, K_2)$

This allows to give a specific definition of the *specialization* relation [5; p.97] between SCGs: G is a specialization of H iff H appears in a derivation of G. This relation is denoted $G \leq H$. It is straightforward to dualize these notions by considering generalization operations and to prove that G is a specialization of H iff H is a generalization of G.

There is a problem in theorem 3.5.2 [5] because the specialization relation is not antisymmetric, it is simply a preorder relation not an order relation. We thus studied the equivalence classes of SCG in [16], and proved, for instance, that in any class there is a unique irredundant SCG, which is the SCG of the class with the minimal cardinality. Concerning the FOL semantics of SCG discussed later, note that if $\leq$ is a partial order then, in the FOL fragment associated to SCG, two equivalent formulas are always obtained by a one-to-one correspondance between variables (this is also the case when projection is restricted to injective projection).

Studying the relation $\leq$ through derivations is not very simple. However, it is possible to characterize the specialization relation by a unique operation that Sowa called projection operator, and which is usually called *projection*. A projection is a *morphism* on SCG considered as labelled bipartite graphs, i.e. preserving the two classes of vertices, preserving the edge labels, and not increasing the vertex labels. In his theorem 3.5.4, Sowa proved that if $G \leq H$ then there is a projection from H to G, we proved the opposite in [16]. Based on these results a derivation can be replaced by a single operation. Furthermore, as for any class of mathematical objects, this operation is the fundamental notion for SCG, since it is the notion of a morphism between SCGs.

With this result, the specialization relation can be called structural *subsumption* and the main kind of reasonings on SCGs involve subsumption computing [6]. There is an obvious analogy with descriptive logics, which blossomed from KL-ONE (see for instance [7]), and the following fundamental question can be put forward for a knowledge representation model: Is the SCG model "equivalent" to some fragment of FOL? More specifically, is it possible to identify a fragment of FOL which is a sound and complete formal semantics for the SCG formal system SCG? The function Φ introduced by Sowa is a natural candidate for such semantics since theorem 3.5.3 [5], where a set of wff $\Phi(S)$ associated with a support S is added, proved that: If G and H are two SCGs such that G is subsumed by H then $\Phi(G)$, $\Phi(S) \to \Phi(H)$. The reciprocal was proven by Mugnier [32], and published in [16]. This requires $\Phi(S)$ and one additional notion. The SCGs can not have two concept vertices with the same individual marker (they are said to be in *normal form*). Then one gets:

let $\Phi(S)$ be the set of formulas associated with a support S. Let g and h be formulas associated with two SCGs G and H in normal form. If h is a logical consequence of $\Phi(S)$ and g, then H subsumes G.

Identifying a computable (the existence problem of a projection between SCGs is trivially computable) fragment of FOL (link 1 in Figure 1, and then with the model theory in FOL link 2) was the first step in the study and development of the CG model as a knowledge representation model "autonomous" from logic, i.e. a formal system having, at least for some parts, a sound and complete logical semantics. The Figure 2 details link 1 of Figure 1.

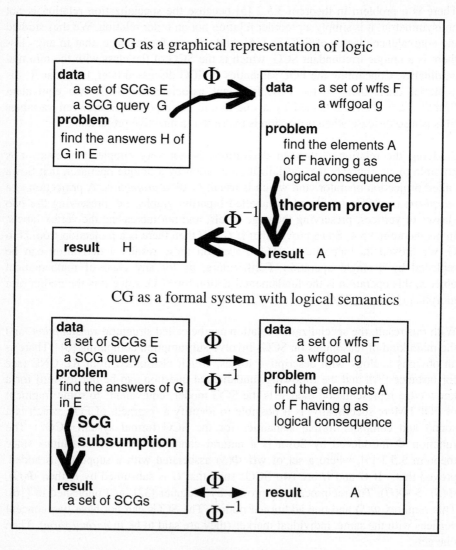

Fig. 2. CG as Graphical Representation of Logic and Formal System with Logical Semantics

The next step is to study algorithmic problems. More precisely, labelled bipartite graph morphism algorithms must be developed. First, we proved that - as for any interesting knowledge representation model - all basic algorithmic problems in the SCG model are NP-hard ([16], [17]). Secondly, bactracking algorithms for projection computing between any SCGs have been constructed, as well as polynomial algorithms for some special cases. These algorithms are the basis of most algorithms we have developed ([31], [335], and link 4 in figure 1).

These first results provide a solid kernel for a CG model. Sowa's chapter 3 [5] is essential: the SCGs formal system is based on three theorems which specify and complete three of the first four theorems of chapter 3 [5] ! (The last one (theorem 3.2.6) deals with the denotation operator. Following the same approach we used it to introduce a set theoretic formal semantics [36] (link 3 in Figure 1).)

Extensions for enriching the semantic network (in Sowa's meaning), and which has until now only contained a support, can be considered. From here on, we were guided by theoretical considerations and also by actual problems. Software tools were required to tackle actual problems, and then Haemmerlé ([26], [28]) developed the CoGITo workbench.

2.3 The CoGITo Workbench

Different objectives led to the main choices concerning CoGITo. CoGITo is the basis on which different applications are built, it must be extendible, portable, easy to maintain, and must be implemented in a widespread environment. The object oriented programming paradigm was then chosen, and the core of CoGITo is a set of classes in the GNU freeware version of C++ under UNIX. With the choice of object oriented programming made, the classes have to be chosen.

In human/machine interactions, graphical representations of graphs allow the construction of useful interfaces. In particular, convivial graph editors are easy to build (this is the important - for applications - visual language aspect of CGs). However, a graph is a mathematical object having "natural" implementations by cells and pointers, which are analogous to the represented graphs. Explanation modules may be built by combining this property with graphical representations, allowing the user to follow computations on a faithful and "readable" image of the model. For algorithmic considerations and interface purposes, the principal classes thus correspond to objects of the theoretical model (e.g., support, hierarchy of concept types, relation signatures, concept vertex, relation vertex, SCG). These classes are equiped with methods corresponding to the operations of the model: mainly specialization operations and projection. The implemented algorithms were those which were developed in our group. Based on the same ideas (at all levels, a representation of a SCG has to be provided) we proposed BCGCT, a file format whose fields correspond to "natural" objects: e.g. support, graph, concept vertex, and relation vertex. This external format is used for saving in permanent memory and exchanging knowledge bases.

2.4 First Applications

The two first applications we considered were: a question/answering system and database modeling. These topics are the basis of Sowa's first paper, but we did not know it at that time. Having a SCG database F, where each SCG represents an asserted fact, the first problem is to query this base. More specifically, if Q is a query SCG, what is the subset $Q(F)$ of F which is the set of specializations of Q? This is the classical subsumption reasoning applied to SCGs, and it can be performed with projections between SCGs.

To show what novelties SCGs can bring for this classical problem, Carbonneill developed his ROCK system, considering both exact reasoning and approximate reasonings ([13], [10]). An important idea was that the database is not changed and only the query is modified. Different relaxations of the query have been used: mainly, splitting of the query, and replacing some concept types appearing in the query by neighbors in T_C. A notion of partial projection has also been introduced and implemented. Note here that it is quite simple to propose approximate reasonings on SCGs (and more generally on any graph-based model). However, logical interpretations of these kinds of approximate reasonings can be difficult to find, and logical interpretations for the approximate reasonings implemented in ROCK have not yet been studied.

ROCK also processes SCG concept type definitions (see III.1). One difficulty appears because concept type levels in the query and in the database may differ. Once again, the database is not changed and the query alone is transformed by type contraction and expansion.

Several conceptual models were introduced for the relational database model. Boksenbaum, Libourel, Carbonneill, and Haemmerlé, [9] encorporated Sowa's ideas and proposed specific definitions, in the CG formalism extended with built-in relations, for relational database concepts such as schemas, data, queries, and views. In particular, they show how updates and SQL queries can be transformed into graph operations and queries. They also show how this modelization can be used to complete uncomplete queries [11] [12].

Two important extensions of the simple model emerged during these investigations: first, an extension of the semantic network in order to consider both primitive and defined types. Secondly, extension of the simple graph model to nested graphs.

2.5 A Management System for SCG Sets

The need for a SCG database management system soon arose. Indeed, we planned to develop applications with large CG sets for dictionary and chemical database projects. Guinaldo built such a system, which is now integrated in CoGITo [22] [23] [24] [25]. In his system, the basic operations of any database management system (i.e., adding, deleting, and modifying a SCG), which can be searched by its name and also its content, have been implemented. He thus developed an SCG isomorphism algorithm which uses an original filter. The system can also answer a SCG query by giving the closest specializations or generalizations to the query. According to Levinson, Ellis and Lehmann, management techniques are then based on classification (with subsumption) and ordered set management. An interesting combination of classification with hashing techniques is also used.

3 Extensions of the Semantic Network

3.1 Canonical Graphs and Type Definitions

Sowa showed how some constraints can be modeled by a canonical basis, which is a set of SCGs. In [34] we proposed a study of this notion. First, beginning with the notion of well-formed SCGs, we give several equivalent definitions of a canonical SCG, each corresponding to a particular viewpoint. We then show that the correspondence between projection and specialization remains true for well-formed SCGs and for canonical SCGs (we thus introduce a notion of a linear canonical derivation). Finally, we propose an algorithm for canonical SCG recognition whose complexity is polynomially related to the complexity of computing a projection between two SCGs. For instance, the problem is polynomial when the canonical basis is a tree set, and we develop an algorithm for this particular case.

Michel Leclère ([15], [30] [31]) considered concept and relation type definitions as defined by Sowa. Introducing this notion in the semantic network leads to some complications, mainly: where in T_C should a new defined type be inserted? how can one reason when facts and queries may contain defined types? These difficulties arise even if a defined type has only one definition, and if type definitions are considered as (mathematical) definitions, i.e. a defined type is equivalent to its definition. Type contraction - substitution of a defined type to its definition - and type expansion - a defined type is replaced by its definition - must preserve equivalence. As Sowa states [5; p. 107]: "Type contraction deletes a complete subgraph and incorporates the equivalent information in the type label of a single concept."

The first problem is the fundamental problem in descriptive logics: i.e. computation of the subsumption relation between defined terms, when the subsumption relation between primitive terms is known. The second problem arises because types may occur with different descriptive levels in a query and in some SCG facts. For instance, it is possible that there is no projection from Q to G, even if G is a specialization of Q. In [31], a simple and specific framework is stated in which these problems can be solved using only projection and graph operations. Michel Leclère also gave a logical interpretation which is sound and complete for SCG formal system. The formula $\Phi(def(t))$ associated with the type definition $def(t)$ of t : $t(x_1,..., x_k) = \lambda x_1..x_k G$, is equal to the universal closure of $t(x_1,..., x_k)$ <-> $\Phi'(\lambda x_1..x_k G)$, where Φ of a lamda abstraction is the usual function Φ applied to G except that it leaves free the variables $x_1,..., x_k$. Michel Leclère also proposed a methodology for building a support extended by defined types.

3.2 Rules

Production rule sets is one of the oldest models for knowledge representation which has been used in many applications, and therefore we studied SCG rules [38]. A rule $r: G_1 \to G_2$ is a couple of lambda-abstractions $(\lambda x_1..x_k G_1, \lambda x_1...x_k G_2)$, where the x_i are co-referent. A rule $r : G_1 \to G_2$, applies to a graph G if there is a projection, Π, from G_1 to G. The resulting graph, denoted by $r(G)$, is built from G and G_2 by merging each x_i of G_2 with $\Pi(x_i)$, the image of x_i of G_1 by Π. If one has a support, S, a SCG set (on S), F, a SCG rule set (on S), R, then the closure of F by R is classically defined as the SCG set, $R(F)$, which can be obtained from F by a finite number of rule applications. To provide a logical semantics to rule application and closure, a formula $\Phi(r)$ must be associated to a rule r. $\Phi(\lambda x_1...x_k G_1, \lambda x_1...x_k G_2)$ is the universal closure of $\Phi(\lambda x_1...x_k G_1) \to \Phi'(\lambda x_1...x_k G_2)$. It can now be proven that the forward chaining mechanism is sound and complete. More precisely, if a SCG G is in the closure of F then $\Phi(G)$ is a logical consequence of $\Phi(S)$, $\Phi(F)$, $\Phi(R)$. Reciprocally, if F and R are in normal form, and g is a formula associated with a normal SCG G, and g is a logical consequence of $\Phi(S)$, $\Phi(F)$, $\Phi(R)$, then there is a SCG H in the closure of F which is subsumed by G. Backward chaining has also been studied, and was also found to be sound and complete. Furthermore, a backward chaining algorithm using particular graph notions was constructed.

An important point to note is that formulas associated with SCG rules are not (Horn) clauses because, in a SCG rule, the variables which are exclusive to the conclusion are existentially quantified (and not universally quantified as in clauses). (Horn) clauses can be seen as a special case of our rules only if they have no functions and all variables of the conclusion appear in the hypothesis. In these conditions, this becomes a very particular case of a SCG rule, since G_2 would be composed of only one relation whose neighbors all have an individual marker or a co-reference marker.

Introducing and studying SCG rules from our viewpoint led to new results which should be useful in logical programming.

4 Positive Nested Graphs : An Extension of the SCG Model

In this part, we no longer investigate extensions of the semantic network, but an extension of the data. We will not detail this topic since it is presented during this conference [19]. Let us just mention that nested graphs were encountered in several applications we were involved in: in the Menelas project [8], in the modeling of databases, in the GRAFIA project [2], and more recently in document retrieval (see [21] this conference), and in a knowledge acquisition problem (see [Bos & Botella] this conference).

We consider several nested CG models which form knowledge representation models for reasoning on level-structured knowledge. They all generalize the SCG model since they involve objects which are generalizations of SCGs (essentially rooted trees of SCGs), and reasoning on these objects is based on a projection which generalizes SCG projection. During this study, we introduced a general categorical framework (graphs of graphs, and trees of graphs, morphisms between these objects, and equivalence classes of such objects) for representing hierarchic structures based on graphs. We think that this could be useful in other contexts, for instance for the study of equivalence problems between graph formalisms provided with subsumption relations.

Two logical sound and complete semantics have been given for positive nested graphs. First, Anne Preller [37] proposes a specific logical language where occurrences of terms are explicitly differentiated (coloured formulas). She proposes a Gentzen system which is sound and complete for projection between NCGs. Geneviève Simonet [39] then proposes a FOL sound and complete semantics extending that of SCGs. Briefly said, a new argument is added to each predicate, which represents the context (i.e. the concept vertex) in which the predicate appears.

5 Conclusion

In order to validate a hypothesis and methodology, it is essential to be faithful to this hypothesis and methodology. We started with conceptual graphs and maximally simplified the knowledge around these graphs, obtaining labelled bipartite graphs also called SCGs. We provided these objects with subsumption, and studied formal semantics. As the powerfulness of the formalism is interesting (e.g., basic asserted facts are richer than in the relational data model), we developed algorithms and a workbench for constructing applications, using SCGs at each level (user interface, formal, physical). In our opinion, this latter property is mandatory if one wants to develop intelligent cooperative systems (which can also be called knowledge programming systems). Any knowledge representation model and its associated reasonings require a logical semantics, nevertheless this does not imply that this model must be restricted to an intermediary language between a user and logic. In different actual problems, it was soon necessary to leave labelled graphs and return to notions closer to Sowa's initial notion of conceptual graphs. We then enriched the semantic network by defined types and rules, and provided a FOL semantics for the introduced notions and operations. Finally, once again forced by reality, we considered nested CGs.

We are now studying the extension of type definitions and of rules for positive nested CGs with co-reference links, and also extension of CoGITo. A kind of negation is already present in type definitions and rules (since in these cases some variables are universally quantified), but this is not sufficient. We are also working on some forms of negation which can considered with our tools (it is possible to do interesting things without negation in facts, but allowing some negations in queries, while assuming the closed world assumption).

As for the SCG and positive NCG models, our viewpoint might lead to identification of fragments of FOL computable by graph specific techniques. We will also study, for document retrieval purposes, non-exact reasonings with CGs (e.g., plausible reasonings with some maximal join operations, relaxation reasonings with partial projections, possible reasonings with weighted partial projections) without looking, at least at the first step, for logical semantics. We hope that it will possible to actually evaluate systems and ideas on two current major actual problems: acquisition and simulation of human behaviors, and document retrieval.

Since they emerged, the death of semantic network models is regularly announced, by part of the AI community. We firmly believe that regardless of the tribulations, Sowa's conceptual graph model will remain of major importance, especially if the bricks are solid, and the building is carefully constructed.

Bibliography

We only give few references. Works of the quoted people from the CG community can be found in the bibliography compiled by Wermelinger.

1. M. Chein, O. Cogis. A Simple Knowledge Representation Scheme with Precise Formal Properties. In *Intern. Journal of Systems Research and Information Science*, vol.2, 1988, p.215-229.
2. M. Chein, J. Bouaud, J.P. Chevallet, R. Dieng, B. Levrat, G. Sabah. Graphes concceptuels. In *Actes des 5-ihmes Journies nationales du PRC-GDR Intelligence Artificielle*. Hermès, 1995, p. 179-212.
3. F. Lehmann (edt). *Semantic Networks in Artificial Intelligence*. Pergamon, 1992.
4. J.F. Sowa. Conceptual Graphs for a Data Base Interface. *IBM Journal of Research and Development*, vol. 20, 4, 1976, p. 336-357.
5. J.F. Sowa. *Conceptual Structures. Information Processing in Mind and Machine*. Addison-Wesley, 1984.

6. W.A. Woods. Understanding Subsumption and Taxinomy: A Framework for Progress. In *Principles of Semantic Networks*, J.F. Sowa (edt), Morgan Kaufmann, 1991, p. 45-94.
7. W.A. Woods, J.G. Schmolze. The KL-ONE Family. In [Lehmann, 1992], p.133-177.
8. P. Zweigenbaum, B. Bachimont, J. Bouaud, J. Charlet, J.F. Boisvieux. Issues in the Structuration and Acquisition of an Ontology for medical language undrstanding. In Natural Language and Medicazl Concept Representation, C. Safran, C. Chute, J.R. Scherrer (edts), Vevey, 1994.

Bibliography of CORALI

9. C. Boksenbaum, B. Carbonneill, O. Haemmerlé, T. Libourel. Conceptual Graphs for Relational Databases. In *Proceedings of the 1st International Conference on Conceptual Structures*, ICCS'93, Quebec City, Canada, August 1993, LNAI #699, Springer Verlag, p. 142-161.
10. B. Carbonneill, O. Haemmerlé. *Proceedings of the 2nd International Workshop on PEIRCE*, Quebec City, Canada, August 1993, p. 29-32
11. B. Carbonneill, O. Haemmerlé. Standardizing and Interfacing Relational Databases using Conceptual Graphs. In *Proceedings of the 2nd International Conference on Conceptual Structures*, ICCS'94, College Park MD, USA, August 94, LNAI #699, Springer Verlag, p. 311-330.
12. B. Carbonneill, O. Haemmerlé. Conceptual Graphs for Relational Databases : Implementation and Perspectives. In *Proceedings of the 3rd International Workshop on PEIRCE*, College Park MD, USA, August 94, p. 54-66.
13. B. Carbonneill. *Vers un système de représentation de connaissances et de raisonnement fondé sur les graphes conceptuels* . Ph. D. thesis, University Montpellier 2, 1996.
14. B. Carbonneill, M. Chein, O. Cogis, O. Guinaldo, O. Haemmerlé, E. Salvat, M.L. Mugnier. COnceptualgRaphs At LIrmm. In *Proceedings of the 1st CGTOOLS Workshop*, Sydney, Australia, August 1996, p. 5-8.
15. M. Chein, M. Leclère. A cooperative program for the construction of a concept type lattice. In *Supplement Proceedings of the 2nd International Conference on Conceptual Structures* , ICCS'94, Washington, August 94, p.16-30.
16. M. Chein, M.L. Mugnier. Conceptual Graphs : Fundamental notions. *Revue d'intelligence artificielle*, 6, 4, 1992, 365-406.
17. M. Chein, M.L. Mugnier. Specialization: where do the Difficulties Occur ? In "Conceptual Structures: Theory and Implementation", *Lecture Notes in Artificial Intelligence*, #754, pp 19-28, H.D. Pfeiffer and T.E. (eds), Springer-Verlag. Collected papers from the 7th Annual Workshop on Conceptual Structures.
18. M. Chein, M.L. Mugnier. Conceptual Graphs are also Graphs (1995). *Actes des. 4èmes journées du LIPN*, Villetaneuse, 18-19 septembre 1995, 81-97, and Research Report LIRMM 95-004.

19. M. Chein, M.L. Mugnier. Positive Nested Conceptual Graphs. *Proceedings of the 5th International Conference on Conceptual Structures*, ICCS'97 (these proceedings).

20. O. Cogis, O. Guinaldo. Linear Descriptor for Conceptual Graphs and a Class for Polynomial Isomorphism Test. In *Proceedings of the 3rd International Conference on Conceptual Structures*, ICCS'95, Santa Cruz, CA, USA, August 1995. Lecture Notes in AI #954, Springer-Verlag, p. 263-277.

21. D. Genest, M. Chein. An Experiment in Document Retrieval Using Conceptual Graphs *Proceedings of the 5th International Conference on Conceptual Structures*, ICCS'97 (these proceedings).

22. O. Guinaldo. *Étude d'un gestionnaire d'ensembles de graphes conceptuels*. Ph. D. thesis, University Montpellier 2, 1996.

23. O. Guinaldo. Filtering and Hashing Techniques for the Search for Isomorphic Conceptual Graphs. In *Proceedings of the 4th International Workshop on PEIRCE*, Santa Cruz, CA, USA, August 1995, p. 58-69.

24. O. Guinaldo. Techniques d'indexation pour aider la classification dans le modèle des graphes conceptuels.In *Actes du 2ème colloque Langages et Modèles à Objets,* Nancy, Octobre 1995, p. 53-66.

25. O. Guinaldo. Conceptual Graphs Isomorphism - Algorithm and Use. In *Proceedings of the 4th International Conference on Conceptual Structures*, ICCS'96, Sydney, Australia, August 1996. Lecture Notes in AI #1115, Springer-Verlag, p. 160-174.

26. O. Haemmerlé. *CoGITo : Une plate-forme de développement de logiciels sur les graphes conceptuels* . Ph. D. thesis, University Montpellier2, 1995.

27. O. Haemmerlé, B. Carbonneill. Interfacing a Relational Database Using Conceptual Graphs. In *Proceedings of the 7th International Workshop on Database and Expert Systems Applications*, DEXA'96, Zurich, Switzerland,September 1996, p. 499-505.

28. O. Haemmerlé. Implementation of Multi-Agent Systems using Conceptual Graphs for Knowledge and Message Representation : the CoGITo Platform. In *Supplement Proceedings of the 3rd International Conference on Conceptual Structures*, ICCS'95, Santa Cruz CA, USA, August 95, p. 13-24.

29. G. Kerdiles et E. Salvat. A sound and complete proof procedure for conceptual graphs combining projections with analytic tableaux.In *Proceedings of the 5th International Conference on Conceptual Structures*, ICCS'97, (these proceedings).

30. M. Leclère. Reasoning with type definitions. In *Proceedings of the 5th International Conference on Conceptual Structures*, ICCS'97 (these proceedings)

31. M. Leclère. C-CHiC : Construction coopérative de hiérarchies de catégories. Revue d'Intelligence Artificielle, vol.10, n°1, 1996, p. 57-100.

32. M.L. Mugnier. Contributions algorithmiques pour les graphes d'héritage et les graphes conceptuels. Ph. D. thesis, University Montpellier 2, 1992.

33. M.L. Mugnier, On Specialization/Generalization for Conceptual Graphs. *Journal of Experimental and Theoretical Artificial Intelligence*, vol.7, 3, 1995, p. 325-344.

34. M.L. Mugnier, M. Chein, Polynomial algorithms for projection and matching. In *Conceptual Structures: Theory and Implementation, Lecture Notes in Artificial Intelligence,* #754, 1992, pp 49-58, H.D. Pfeiffer and T.E. Nagle(eds), Springer-Verlag. Collected papers from the 7th Annual Workshop on Conceptual Graphs.
35. M.L. Mugnier, M. Chein, Characterization and Algorithmic Recognition of Canonical Conceptual Graphs. In *Proceedings of the First International Conference on Conceptual Structures,* ICCS'93, *Lecture Notes in Artificial Intelligence,* #699, Springer Verlag, 1993, 294-311.
36. M.L. Mugnier, M. Chein. Représenter des connaissances et raisonner avec des graphes. *Revue d'Intelligence Artificielle* , vol.10, 1, 1996, 7-56.
37. A. Preller, M.L. Mugnier, M. Chein. A Logic for Nested Graphs . Research Report LIRMM 95-038, Juin 95, to be published in *Computational Intelligence* (ref. CI 95-02-558).
38. E. Salvat et M.L. Mugnier. Sound and complete forward and backward chainings of graph rules.In *Proceedings of the 4th International Conference on Conceptual Structures,* ICCS'96, Sydney, Australia, August 1996. Lecture Notes in AI #1115, Springer-Verlag, p. 248-262.
39. G. Simonet. Une sémantique logique pour les graphes emboités. Research Report, LIRMM 96047, 1996

Contexts: A Formal Definition of Worlds of Assertions

Guy W. Mineau
Department of Computer Science
Université Laval
Quebec City, Canada
G1K 7P4
tel.: (418) 656-5189
fax: (418) 656-2324
email: mineau@ift.ulaval.ca

Olivier Gerbé
Groupe DMR Inc.
1200 McGill College
Montreal, Canada
H3B 4G7
tel.: (514) 877-3301
fax: (514) 866-0423
email: olivier.gerbe@dmr.ca

Abstract. For many years now on-going discussions, not to say endless discussions, about the intrinsic definition of contexts have been at the center stage of every single meeting of the conceptual graph community. It is our opinion that this lack of consensus about contexts is in direct relation to its lack of a formal definition. As a matter of fact, no formal definition of contexts was given up to this moment, not even in John Sowa's original book. Being a vital issue in conceptual graph theory, this paper addresses the problem of providing formal semantics to the definition of contexts, when used for information packaging. It proposes a definitional framework for contexts, based on formal concept analysis [17], bridging these two research areas. It also presents how querying a knowledge base structured as a lattice of contexts can be done.

Keywords: contexts, conceptual graph theory, semantics, concept formation, knowledge representation

1 Introduction

Contexts are a vital notion of the conceptual graph theory. Inferences are based on the notion of contexts; contexts are useful for packaging information. Over the years, they were extensively used in many areas of conceptual graph related research, particularly in knowledge definition [2, 3, 4], in knowledge structuring [13], in natural language processing [1, 9, 12, 11], in the representation of modalities, intentional verbs and temporal relations [9], in multi-agent systems [10] and in object systems [16]. Despite this obvious need for contexts, there is still no consensus on their definition and usage in the community. This is mainly due to the fact that they are ill-defined. As a matter of fact, as surprising as this may be, there exists no formal definition of contexts so far, not even in John Sowa's original book. The only intuitive definition given in Sowa's book is that a context exists whenever some proposition is asserted: a proposition is said to be true in a particular context. No way of defining this context is provided, resulting in different uses for different types of applications. To our opinion, this lack of semantics when defining contexts is its source of problems.

For inferential purposes, contexts are useful to delimit the scope of negation, pretty much like parentheses in Peano-Russel's notation. They introduce a notion of subsumed (dominated) contexts that is not formally defined in the cg literature so far, but which is useful for the mechanics of the inference process. One realizes that this scope delimitation mechanism could be represented otherwise. However, the information packaging capabilities that these contexts provide proved to be useful for other application areas such as modal logic, linguistics and multi-agent applications. The implied notion of subsumption between contexts is useful, though it needs to be well defined so that different *worlds of assertion* may exist in relation to one another, as is required with these types of applications. When the truth value of a set of assertions is conditional to the same set of conditions, then we say that a *world of assertion* is created, and that it is defined by both the assertions which describe it and the inherent conditions which allow its existence.

This paper presents a formal definition of contexts (when used for packaging information[1]), based on the notion of *intention* and *extension*. We define a context in terms of what propositions it *allows* and *represents*. The propositions that it represents (a set of conceptual graphs) is said to be its extension. These graphs are conjunctively true (asserted) in that context. Its intention will be the conditions upon which the context exists. These conditions are the premises under which the extension of the context is said to be true. These conditions are expressed as a set of conceptual graphs conjunctively true in that context[2]. Together, the intention and extension of a context provide a formal definition which explicitly states what is (the extension) and what can be (according to the intention) represented in this context. They are each represented as a set of conceptual graphs. The association of an intention set and an extension set provides a formal definition for a context since the graphs of the extension set have to abide by the conditions of those of the intention set; and the graphs of the intention set are the conditions that must be enforced to allow the extension set to exist. As will be seen below, not all such associations are possible; not all such associations are necessary. A minimal representation of these associations is sufficient to represent the whole knowledge base. This representation is a *concept lattice*.

Concept lattices were introduced in [17]; this paper adapts these definitions to the conceptual graph terminology and presents how they can be used to properly structure a knowledge base defined in terms of a set of contexts. This structure, which we named *context lattice*, can be used to optimize query mechanisms which can address their queries to particular contexts, or which can now inquire about the links between intention and extension sets, i.e., about the structure of the knowledge base itself, which represents the relations among different worlds of assertions.

Section 2 below presents a brief literature survey on contexts in the conceptual graph theory. Section 3 states our proposition. Section 4 presents different

---

[1] This paper will not present contexts for inferential purposes, though we feel that the work presented here will eventually permit such extensions to be defined; it only presents its information packaging capability, used to create a lattice of subsumed worlds of assertions. However, section 6 presents how certain characteristics of contexts used in an inference engine for existential graphs, are also present with the definitional framework that we propose in this paper.

[2] In this paper, the intention of a context will be a unique conceptual graph, called *intention graph*, stating under what condition the context exists. Section 5 shows how this graph is produced. Other constraints on the extension of a context could be added to intention of a context, as will be done in a forthcoming paper.

functions which can be used to query a knowledge base structured as a context lattice. Section 5 describes the algorithm that produces the intention graphs used in section 3 to build the context lattice. Section 6 shows how some characteristics of contexts used for inferential purposes are present within the framework that we propose in this paper. Section 7 concludes.

2 Literature Survey

Contexts and concepts have been largely studied in the theory of conceptual graphs [2, 4, 14, 15]. It has been proven that contexts are fundamental to existential graphs and that the structuring of a knowledge base may benefit from using contexts for packaging information. Even though all authors agree on C.S. Peirce's preliminary definition of contexts as sheets of assertions, there is no common understanding of their underlying semantics.

[14] introduced the notion of context in the conceptual graph theory through the formal definition of propositions. He defined a proposition p as a concept where its concept type is PROPOSITION, and its referent, a set of conceptual graphs. These graphs are said to be true in the context of p. In [15], Sowa analyzed the semantic foundations of contexts through four theories. In this analysis, Sowa distinguished three syntactic aspects of contexts: a mechanism for packaging information, the contents of the package (a set of assertions), and the permissible operations on the package, including importing and exporting information to and from a package. In this study, Sowa himself points out that much of the controversy about the notion of context results from a lack of definitional semantics.

Traditionally, conceptual graphs specialists have used contexts to represent and partition (structure) the universe of discourse, as illustrated by [10, 11] and [1]. In the first case, Bernard Moulin gives a pragmatic interpretation of contexts. He defines the notion of discourse space that extends the notion of contexts by adding information grouped in an envelope describing a context. By analyzing the cognitive operations that human agents use when producing or understanding a discourse, Moulin identifies different kinds of discourse spaces. Among them, temporal discourse spaces support the representation of temporal structures, including situations; definitional discourse spaces support the definition of concept and extends the notion of white box proposed by Esch [2]).

In the second case, Judith Dick uses contexts to represent texts, especially legal texts, and notes that there is little clarity about the nature of contexts and how they ought to be used to help represent knowledge effectively. Contexts are definitely needed to represent natural language sentences and to package information. In order to fulfill all the requirements to successfully support this usage, it is obvious that contexts need to be semantically well defined.

In this paper, we propose to contribute to the semantic definition of contexts by adding a semantic dimension to the three syntactic aspects identified in [15]. This semantic dimension completely specifies the *raison d'être* of contexts when information packaging is concerned, thus bridging the gap between different points of view concerning the theory of contexts in the conceptual graph community.

3 Contexts: a Formal Definition

When a graph g represents a true statement, it is said to be an assertion. Except for universal truths, the truth value of a graph often depends on other assertions; it is conditional to these assertions which define the proper conditions under which it applies. These other assertions define situations where the graph g can be considered an assertion. The set of all situations in which g is an assertion is called its *scope*. Consequently, no assertion can be made without reference to the situations where it applies, i.e., to its scope; graphs are always asserted in relation to a situation. By default, in a conceptual graph system, all assertions are done on the *universal sheet of assertion* which represents universal truths, i.e., graphs whose truth value does not depend on some conditions, which in itself, is a particular situation. Consequently, since assertions are made in relation to some situations, there is an obvious need to define a situation where graphs which comply to the same conditions may be asserted. Such a situation will be called a *context*.

A context is a sheet of assertion whose existence depends on some graphs which describe the premises of its existence. All graphs in a context comply to the same premises. For example, modalities may be represented using such contexts. The assertion: "Mary thinks that Peter loves her" is universally true[3]: "Mary does think that Peter loves her", but does Peter really love Mary or not? We could not assert anything about that, except that "Peter loves Mary" in Mary's thoughts. So Mary's thoughts form a context under which the statement: "Peter loves Mary" becomes true. The graph of Figure 1 shows how this assertion would be represented as a conceptual graph[4].

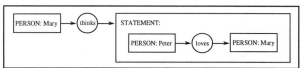

Fig. 1. The statement: "Mary thinks that Peter loves her".

As another example, we could have the following statement: "Mary thinks that Peter is handsome" (see Figure 2). Also, we could have embedded contexts as shown in Figure 3, where the following statement is represented: "Mary thinks that Peter thinks that he is handsome". As a shorter notation, Figure 4 may be used to represent all previous statements. In this figure, the set of statements instantiating the concept STATEMENT represents a sheet of assertion created by Mary's thoughts.

Analyzing the graphs reveals that the three statements of Figure 4 are all related to what Mary thinks. So the following context (see Figure 5) exists de facto. A context will be defined by both the conditions from which it is created, and the set of graphs asserted under it. Similarly with embedded contexts, the context of Figure 6 is implied by the graph of Figure 3; it represents what Mary thinks that Peter thinks. This latter

---

[3] Provided that the vocabulary is well defined and the instances Peter and Mary do exist.

[4] In this article, proper names such as Mary and Peter are used as unique identifiers. Consequently, they can be used where referents normally appear.

context is obviously part of the former one, as represented by the embodiment structure of the graph of Figure 3.

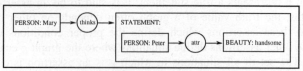

Fig. 2. The statement: "Mary thinks that Peter is handsome".

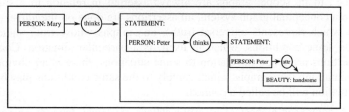

Fig. 3. The statement: "Mary thinks that Peter thinks that he is handsome".

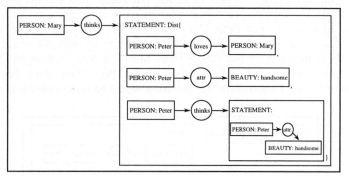

Fig. 4. A shorter notation for representing sets of statements in the same context.

As illustrated in Figures 5 and 6 below, a context is defined as two parts: an *intention*, a set of conceptual graphs which describe the conditions which make the asserted graphs true, and an *extension*, which is composed of all the graphs true under these conditions. Of course, a graph can appear in the extension of more than one context. For example, if Peter thinks that he loves Mary, this would imply the existence of the context of Figure 7, where the graph in the extension of this context already appears in some other context (in the extension set of the context shown in Figure 5).

Formally, a context C_i can be described as a tuple of two sets of conceptual graphs, T_i and G_i. T_i defines the conditions under which C_i exists, represented for now by a single intention graph; G_i is the set of conceptual graphs true in that context. So, for a context C_i, $C_i = <T_i, G_i> = <I(C_i), E(C_i)>$, where $I(C_i)$ is a single conceptual graph, the intention graph of C_i, and $E(C_i)$, the set of graphs conjunctively true in C_i, called the extension graphs. At this point, the reader should notice that a graph g may be true in $E(C_i)$ without having been explicitly asserted in that context. For instance, if g is asserted in some other context but is a generalization of a graph asserted in C_i, than g is considered to be part of $E(C_i)$. We define $E(C_i)$ as the transitive closure of the asserted graphs of Ci under the subsumption relation that exists among all graphs of the knowledge base.

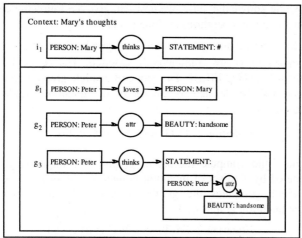

Fig. 5. The context of what Mary thinks.

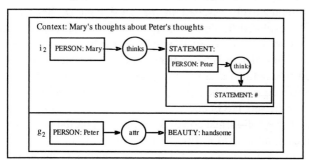

Fig. 6. The context of what Mary thinks that Peter thinks.

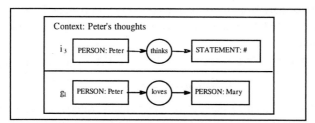

Fig. 7. The context of what Peter thinks.

Also, since a graph may appear in more than one context, if two contexts C_i and C_j are such that their respective intention graphs are related in a subsumption relation such that $I(C_j) < I(C_i)$, then the graphs true in C_j, i.e., the graphs of $E(C_j)$, are true in C_i as well, that is, $E(C_j) \subseteq E(C_i)$. So, we will compute $E(C_i)$ as the union of the graphs originally asserted in C_i and in C_j for all j such that $I(C_j) < I(C_i)$. After $E(C_i)$ is computed for all contexts, we say that g is *asserted* in C_i if g belongs to $E(C_i)$, no matter if it was originally asserted in that context or not. Knowing that $E(C_i)$ is computed using the subsumption relation that exists between the intentions graphs of all contexts, and using the subsumption relation that exists between the asserted graphs of the whole knowledge base, we now can give the following definitions using the intersection and union operations as done in [17].

First, let us define the *scope* S of a set of graphs G, i.e., the set of all contexts where the elements of G are conjunctively asserted. The scope of G is formally defined as: $S(G) = \{C_i \mid G \subseteq E(C_i)\}$. The scope of a single graph g could be computed as $S(\{g\})$. $I^*(G)$ will be defined as the set of the intention graphs of $S(G)$ (see equation 1 below).

$$\text{Equation 1: } I^*(G) = \cup_i I(C_i) \mid G \subseteq E(C_i)$$

Conversely, one could compute the set of graphs G conjunctively asserted in all contexts designated by T, a set of intention graphs. Equation 2 below introduces the function E^* which computes the graphs conjunctively asserted in all contexts identified by the intention graphs of T.

$$\text{Equation 2: } E^*(T) = \cap_i E(C_i) \mid I(C_i) \subseteq T$$

Queries about graphs asserted in different contexts may be as interesting as queries about graphs conjunctively asserted in the same context; links between different contexts may be explored through these queries. Consequently, it is useful to relate sets of contexts to sets of asserted graphs, providing a structure for the whole knowledge base. In order to achieve that, we now introduce the notion of a *formal context*, named C_i^*, defined as a tuple $<T_i, G_i>$, where $G_i = E^*(T_i)$ and $T_i = I^*(G_i)$ at the same time (see equation 3 below). Because two partial orders of inclusion defined respectively over the sets $I^*(G_i)$ and $E^*(T_i)$ exist, the set of all formal contexts form a lattice structure L, i.e., all formal contexts are part of a partial order $\leq$. In effect, $C_2^* < C_1^*$ iff $C_2^* \neq C_1^*$ and $E(C_2^*) \subset E(C_1^*)$, or conversely, iff $C_2^* \neq C_1^*$ and $I(C_1^*) \subset I(C_2^*)$. Assuming the completion of $E(C_i)$ for all contexts C_i under both subsumption relations (of the intention graph and of the extension graphs), we can adapt the notions presented here from those presented in [17]. Equations 3 and 4 below remind the reader of essential definitions used to develop the query mechanism presented in the next section.

$$\text{Equation 3: } C_i^* = <T_i, G_i> \text{ where } G_i = E^*(T_i) = E(C_i^*) \text{ and } T_i = I^*(G_i) = I(C_i^*)$$

$$\text{Equation 4: } L = < \{ C_i^* \}, \leq >$$

The context lattice L can be computed automatically without needing the knowledge engineer to intervene, providing an explanation and access structure to the knowledge base, and relating different worlds of assertions to one another. As explained earlier, all $E(C_i)$ sets are automatically computed from the sets of graphs originally asserted in each C_i. As will be shown in section 5, the intention graphs of each context C_i, i.e, $I(C_i)$, can also be automatically extracted from the graphs asserted in the system. Since all of this is automatic, the context lattice is built, i.e., the knowledge base is structured according to the semantics of the graphs it contains, without involving the knowledge engineer. With our example, the context lattice L of Figure 8 below would be produced.

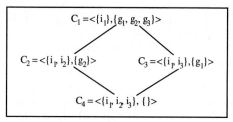

Fig. 8. The Hasse diagram of the context lattice produced from our example[5].

4 Querying a Knowledge Domain Structured as a Context Lattice

As mentioned earlier, queries about the truth value of any graph must always be computed according to a designated sheet of assertion, a context. The query mechanism then must include the information about the scope of its query graph. For that purpose, we now define a query mechanism which is based on formal contexts. Queries will be required to specify two parameters, one designating the appropriate context C^* of the query, the other being the query graph itself. Since contexts are built from two types of information: an intention and an extension, two queries mechanisms will be devised. The first one will query the extension of a context; while the second one will query the intention of a context. The corresponding functions are introduced as Equations 5 and 6.

$$\text{Equation 5: } \delta_{E*}(C^*,q) = \{g \in E(C^*) \mid g \leq q\}$$

$$\text{Equation 6: } \delta_{I*}(C^*,q) = \{g \in I(C^*) \mid g \leq q\}$$

Because queries can be embedded, i.e., the result of a query being the object of the next query, and because we may want to use the structure of L to navigate, explore, and query the structure of the knowledge base, we now introduce two functions which identify a context based on a set of graphs G representing a subset of some extension set (Equation 7), or based on a set of graphs T representing a subset of some intention set (Equation 8).

$$\text{Equation 7: } C_E(G) = \langle I^*(G), E^*(I^*(G)) \rangle$$

$$\text{Equation 8: } C_I(T) = \langle I^*(E^*(T)), E^*(T) \rangle$$

Combined together, the functions of Equations 5 and 7, and of Equations 6 and 8, are useful to change contexts based on the result of some previous query. When querying a knowledge base, going from one context to the next (navigating) can be done as illustrated in Figure 9, where we show how a change of context can be done based on the result of query q_1, prior to the evaluation of query q_2 (sent to a potentially different context)[6].

---

[5] Please notice that the graph identifiers in this figure, are those found in Figures 5, 6 and 7.

[6] Embedded queries can involve intention sets as well. We only show one simple example here to shorten the presentation.

$$\delta_{E*}(\ C_E(\delta_{E*}(C^*,q_1))\ ,\ q_2)$$

Fig. 9. Embedded queries.

As a shorter notation, we could define two functions ctx_{E*} and ctx_{I*}, shown in Equations 9 and 10, which produce a new context based on a previous context and a query. The appropriate subscript indicates to which part of the context the query q is sent to. Please notice that since $ctx_{E*}(C^*,q)$ and $ctx_{I*}(C^*,q)$ each identify a single formal context, because the knowledge base is structured as a lattice, one could use either one of them as parameters for the functions I and E in order to extract the intention set or the extension set of the new context, respectively.

$$\text{Equation 9: } ctx_{E*}(C^*,q) = C_E(\delta_{E*}(C^*,q))$$

$$\text{Equation 10: } ctx_{I*}(C^*,q) = C_I(\delta_{I*}(C^*,q))$$

Also, it is often the case that the result Q of a query q sent to the extension set of a context $E(C^*)$ is a subset of the extension set of the new context, i.e., $Q \subset E(ctx_{E*}(C^*,q)) \subseteq E(C^*)$. In that case, subsequent queries may need to consider only this subset Q and not the whole extension set of the new context, for obvious efficiency reasons. To accommodate that need, we now introduce variations on previous functions, where the set to which a query is sent may be smaller than the one to which the query would normally be sent to.

$$\text{Equation 11: } \delta_E(C^*,G,q) = \{g \in G \mid G \subseteq E(C^*) \text{ and } g \leq q\}$$

$$\text{Equation 12: } \delta_I(C^*,T,q) = \{g \in T \mid T \subseteq I(C^*) \text{ and } g \leq q\}$$

These definitions are equivalent to their previous counter-parts if $G = E(C^*)$ and $T = I(C^*)$ (see Equations 13 and 14). The user now has the possibility of restraining the set to which the query is sent to, even though the context may change as the result of the query, by using either δ_E or δ_{E*}, or similarly, either δ_I or δ_{I*}.

$$\text{Equation 13: } \delta_E(C^*,E(C^*),q) = \delta_{E*}(C^*,q)$$

$$\text{Equation 14: } \delta_I(C^*,I(C^*),q) = \delta_{I*}(C^*,q)$$

Embedded queries could then be expressed using variables representing sets of graphs obtained by previous queries. This way, long algebraic formulas can be broken down into shorter and more comprehensible ones. As such, these queries are much easier to write and read. An example is given in Figure 10. With such a framework, at any moment i in time, the set of graphs retrieved so far is s_i, and the current context is given by $C_E(s_i)$ (or by $C_I(s_i)$).

$$s_1 = \delta_{E*}(C^*, q_1)$$
$$s_2 = \delta_E(\ C_E(s_1)\ ,\ s_1\ ,\ q_2)$$
$$s_3 = \delta_E(\ C_E(s_2)\ ,\ s_2\ ,\ q_3)$$
$$s_4 = \delta_{I*}(C_E(s_3),\ q_4)$$

Fig. 10. An example of embedded and consecutive queries.

Given the context lattice of Figure 8, the graph of Figure 1 as the first query q_1 and the graphs of Figures 11 and 12 as subsequent queries q_2 and q_3, the embedded formulation of Figure 13 (explained below) would result in the navigation illustrated in Figure 14. This embedded query aims at determining if Mary thinks that Peter is aware of his love for her.

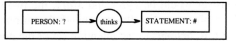

Fig. 11. Who thinks (the same thing)? (q_2)

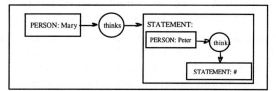

Fig. 12. Does Mary think that Peter shares her thought that he loves her? (q_3)

$$s_1 = \delta_{E*}(C_1, q_1)$$
$$s_2 = \delta_{I*}(C_E(s_1), q_2)$$
$$q_3 \in I(C_I(s_2))?$$

Fig. 13. The three consecutive queries q_1, q_2 and q_3.

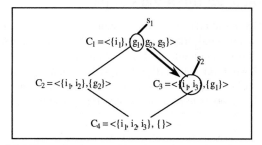

Fig. 14. The navigation resulting from the queries of Figure 13.

The first query, the graph of Figure 1 (q_1), could be sent as such to the universal sheet of assertion[7], or the context to which the most embedded assertion is made can be automatically determined (in our example, it is context C_1) before a simple graph (the most embedded graph in q_1, being the graph g_1 in Figure 5) can be sent for evaluation to the extension set of context C_1. This simplified query graph matches graph g_1 in the extension of C_1. So the answer to this query graph would be *true*, the set $s_1 = \{g_1\}$ being non-empty: Mary does think that Peter loves her. That gives us a starting point (a current context) in L, to navigate from. The computation of the most specific

---

[7] The universal sheet of assertion has not been modeled in our paper in order to simplify our example. However, it could be described using an intention and an extension set, as any other context, and could be part of the context lattice representing the knowledge base. However, it would introduce redundancy for graphs containing STATEMENT concepts, as the asserted information is then represented in some other context.

context containing the elements of s_1 as part of their extension set results in a downward move of the current context from C_1 to C_3 (when $C_E(s_1)$ is evaluated, see Figure 14). The second query q_2 is then sent to the intention set of C_3: "who shares this thought?". The answer would be *Peter* and *Mary*, the two referents instantiating the ? symbol in both graphs of s_2. Finally, we update the current context using the C_I function before testing if q_3 belongs to the intention set of the potentially new context. In our example, the current content does not change and the final answer is *no*, since i_2 is not part of the intention set of the current context. So, under a closed world assumption, Mary does not think that Peter is aware of his love for her.

This example should emphasize two advantages of our approach: 1) queries can be simplified (in some cases even flattened) by identifying a context to which a simpler query can be sent to, 2) the structure of the knowledge base itself can be queried by allowing queries to be sent to either the intention or extension set of a context. The process used to automatically identify a context when a graph is acquired (see section 5 below) is used once more when a query graph is sent to the knowledge base: the context where the query graph should be sent to is identified; the query is thus simplified; its evaluation is much simpler than it would normally be if sent to the universal sheet of assertion where a graph matching procedure must be applied recursively on embedded parts (which are graphs).

5 The Extraction of an Intension Graph

The intension graphs associated with the existence of contexts can be automatically extracted from the conceptual graphs acquired by the system. In fact, the extraction procedure given below is adaptable to different applications; the only requirement is that it must be used when the knowledge base is both acquired and queried. What is important is that intension graphs must be produced in a uniform manner. The example of Figure 15 below is used to show how this procedure works.

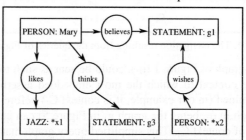

Fig. 15. An asserted graph on the universal sheet of assertion.

Step 1: Make sure that each generic concept is identified by a unique referent variable with regard to any other concept in the knowledge base, unless the concept should be in coreference with an already existing concept. In that case, it should be identified using the referent variable of this existing concept[8]. With our example, we suppose

---

[8] In this paper, we assume *global* coreference. As a matter of fact, all cg system developers know that global coreference is mandatory to distinguish concepts belonging to different graphs and to allow the coreference of concepts belonging to

that this step was already carried out, and that g1 and g3 refer to the graphs of Figure 5.

Step 2: Duplicate (copy) every STATEMENT concept so that they are used with only one relation, either as input or output parameter. Notice that a set of graphs may result from this step. With our example, we get the two graphs shown in Figure 16.

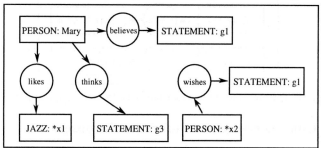

Fig. 16. The graph of Figure 15 after step 2.

Step 3: From the set of graphs produced by step 2, extract the relations which use a STATEMENT concept, either as input or output parameter. With our example, we would get the graphs of Figure 17.

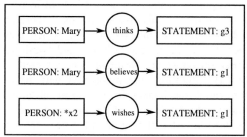

Fig. 17. The relations extracted by step 3.

Step 4: For each graph g produced by step 3, replace the referent r of g by the # symbol. With our example, we would get the graphs of Figure 18.

Step 5: For each graph produced by step 4, called g', if the corresponding graph in step 3 g was asserted in a context, use the intention graph of this context and replace its # symbol by g', producing the intension graph designating the context where the referent of g, r, should be asserted. Since the graph of Figure 15 was asserted on the universal sheet of assertion, this step does not apply to the first iteration of our example, and the intention graphs produced with our example are the ones shown in Figure 18. However, this step would apply to the second iteration of the procedure.

Step 6: Select the graphs of step 3 where the referent of the STATEMENT concept is an individual referent, that is, a conceptual graph, where a STATEMENT concept appears. Reapply steps 1 thru 6 to these graphs. With our example, only the first graph

graphs asserted independently. This may seem to be a philosophical difference from what is currently done in any cg system, but in fact, considering the iteration rule, it is not (see section 6 for more details).

of Figure 17 would be kept. From it, the procedure would produce the intention graph of the context shown in Figure 6.

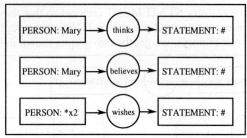

Fig. 18. The three intention graphs produced from the graph of Figure 15.

6 Using a Context Lattice for Inference Purposes

This paper proposes a way to automatically structure a knowledge base using a *context lattice*. Each context can be seen as a world of assertion. The extension of this work will lead to the definition of inference capabilities within each context. For that purpose, contexts used to delimit the scope of negation with regard to sets of propositions, called *basic contexts*, could be used within each of these worlds. In that case, their use is almost purely syntactical, and it is hard to find a mapping between them and the contexts described in this article. However, there exist similarities in the functionnalities that both types of contexts provide, which may lead to a mapping from one to the other. For instance, the "lifting" and "pushing" rules allowing the propagation of information between subsumed contexts could probably be defined using basic contexts. The iteration rule associated with basic contexts would be represented by a propagation mechanism based on the subsumption relation between contexts. The deiteration rule would be just the inverse process; it would imply that inheritance would be used in order to conduct inferences and avoid redundancy. The double negation rule would be an assertion rule with regard to some context. The insertion and erasure rules, however, need negation to be represented within these contexts in order to have their counter-part in what we propose in this article. Our work remains to be extended in that direction. However, one may think that an information propagation mechanism between contexts, coupled with inferential capabilities, could erase the need for basic contexts as used so far in the cg literature. Of course, that remains to be seen, as the burden of proof is still ours.

7 Conclusion and Future Developments

This paper proposes a definitional framework for defining contexts formally; we hope that with regard to the information packaging facilities that it offers, that it makes a consensus among the CG community. We explained how contexts could be automatically abstracted from a set of graphs; we presented query mechanisms useful to explore not only the asserted graphs, but also the structure of the contexts themselves, i.e., the way they relate to one another. The framework that we propose in this paper is based on formal concept analysis [17]; it links the intention and extension of a knowledge base in a single unique representation paradigm since both the intention and extension of a context are described using conceptual graphs. The knowledge base, structured as a *context lattice*, represents how the different sets of assertions are partitioned and shared among different agents who believe them. To our opinion, this structuring will prove to be useful for many application areas, particularly for natural language processing, modal logic and multi-agent systems. However, even though section 6 identified some trails as to how a context lattice may be used to fully implement an inferential mechanism based on these contexts, more research in that area is definitely needed. Still, we believe that the contexts used in an inference system for existential graphs should be defined with sufficient semantics to allow their full mapping to some other packaging constructs whose existence could be formally defined. We hope that this paper starts a discussion in that direction.

Implementation issues concerning lattices and lattice-based data structures have been on our research agenda for many years now. We have developed very efficient batch and incremental algorithms that create and maintain a lattice structure [6, 8]. We have been particularly concerned with the feasibility of our approach for large knowledge bases. As a result, we proposed in [5] a wide spectrum of lattice-based data structures useful for organizing and structuring large knowledge bases. We will soon address the problem of computing the $E(C_i)$ sets effectively. As mentioned before, there are a lot of interesting work about computing subsumption relations at the lowest cost possible, that could be adapted for our needs. All algorithmic and implementation aspects of that task will be a major concern of ours at least for the next year.

Meanwhile, we are currently extending the work presented in this paper to allow extension sets to be infinite. That way, contexts will not only represent sheets of assertions, but complete worlds of inferences. Agents in multi-agent systems need the capability to infer more than what they were originally told. It is our opinion that this possibility of distributing and partitioning inference capabilities will allow the conceptual graph formalism to be considered for distributed applications. Furthermore, we feel that there are application domains for which this distribution of inference capabilities will result in a gain of efficiency for the whole system, as mentioned in [7]. At this point in time, this remains to be seen; and we certainly plan to explore in that direction in the near future.

8 Bibliography

1. Dick, J. P. Using Contexts to Represent Text. In W.M. Tepfenhart, J.P. Dick & J.F. Sowa (Eds.), *Conceptual Structures: Current Practices,* (pp. 196-213). Springer-Verlag, 1994.

2. Esch, J. Contexts and Concepts, Abstraction Duals. In W.M. Tepfenhart, J.P. Dick & J.F. Sowa (Eds.), *Conceptual Structures: Current Practices,* (pp. 175-184). Springer-Verlag, 1994.

3. Esch, J. Contexts, Canons and Coreferent Types. In W.M. Tepfenhart, J.P. Dick & J.F. Sowa (Eds.), *Conceptual Structures: Current Practices,* (pp. 185-195). Springer-Verlag, 1994.

4. Esch, J. W. Contexts as white box concepts. In *Proceedings of the 1st Int. Conf. on Conceptual Structures*, G.W. Mineau, B. Moulin & J.F. Sowa (Ed.), Université Laval, pp. 17-29, 1993.

5. Godin, R., Mineau, G., Missaoui, R. & Mili, H. Méthodes de classification conceptuelle basées sur le treillis de Galois et applications. *Revue d'Intelligence Artificielle*, **9**(2), 105-137, 1995.

6. Godin, R., Mineau, G. W. & Missaoui, R. Incremental Structuring of Knowledge Bases. In *Proceedings of the 1st Int. Symposium on Knowledge Retrieval, Use, and Storage for Efficiency (KRUSE)*, G. Ellis, R. Levinson, A. Fall & V. Dahl (Ed.), Santa Cruz, CA.: Dept. of Computer Science, UCSC, CA., pp. 179-193, 1995.

7. Lenat, D. B. & Guha, R. V. *Building Large Knowledge-Based Systems, Representation and Inference in the Cyc Project*. Addison-Wesley, 1990.

8. Mineau, G. W. & Godin, R. Automatic Structuring of Knowledge Bases by Conceptual Clustering. *IEEE Transaction on Knowledge and Data Engineering*, **7**(5), 824-829, 1995.

9. Moulin, B. The representation of linguistic information in an approach used for modeling temporal knowledge in discourses. In G.W. Mineau, B. Moulin & J.F. Sowa (Eds.), *Conceptual Graphs for Knowledge Representation*, (pp. 182-204). Springer-Verlag, 1993.

10. Moulin, B. Discourse Spaces: A Pragmatic Interpretation of Contexts. In G. Ellis, R. Levinson, W. Rich & J.F. Sowa (Eds.), *Conceptual Structures: Applications, Implementation and Theory*, (pp. 89-104). Springer-Verlag, 1995.

11. Moulin, B. A Pragmatic Representational Approach of Context and Reference in Discourses. In G. Ellis, R. Levinson, W. Rich & J.F. Sowa (Eds.), *Conceptual Structures: Applications, Implementation and Theory*, (pp. 105-114). Springer-Verlag, 1995.

12. Moulin, B. & Dumas, S. The Temporal Structure of a Discourse and Verb Tense Determination. In W. M. Tepfenhart, J. P. Dick, & J. F. Sowa (Eds.), *Conceptual Structures: Current Practices*, (pp. 45-68). Springer-Verlag, 1994.

13. Moulin, B. & Mineau, G. W. Factoring Knowledge into Worlds. In *Proceedings of the 7th Ann. Workshop on Conceptual Graphs*, H. Pfeiffer (Ed.), New Mexico State University, Las Cruces, New Mexico: pp. 119-128, 1992.

14. Sowa, J. F. *Conceptual Structures: Information Processing in Mind and Machine*. Addison-Wesley, 1984.

15. Sowa, J. F. Syntax, Semantics and Pragmatics of Contexts. In G. Ellis, R. Levinson, W. Rich & J.F. Sowa (Eds.), *Conceptual Structures: Applications, Implementation and Theory*, (pp. 1-15). Springer-Verlag, 1995.

16. Sowa, J. F. Processes and Participants. In P.W. Eklund, G. Ellis & G. Mann (Eds.), *Conceptual Structures: Knowledge Representation as Interlingua*, (pp. 1-22). Springer-Verlag, 1996.

17. Wille, R. Restructuring Lattice Theory: an Approach Based on Hierarchies of Concepts. In I. Rival (Eds.), *Ordered Sets*, (pp. 445-470). Dordrecht-Boston: Reidel, 1982.

Positive Nested Conceptual Graphs

Michel Chein &[1] **Marie-Laure Mugnier**
LIRMM (CNRS & University of Montpellier)
email: chein@lirmm.fr, mugnier@lirmm.fr

Abstract. This paper deals with positive (i.e. without negation) nested conceptual graphs (NCGs). We first give a general framework - graphs of graphs provided with morphism - for defining classes of NCGs. Then we define a new class of NCGs - typed NCGs - and we show that known kinds of NCGs can be described very simply as classes of the general framework. All NCG models considered generalize the simple CG model in the sense that they involve objects which are generalizations of simple CGs and reasonings on these objects are based on a graph operation (projection) which is a generalization of that used for simple CGs. Furthermore, the general framework introduced allows one to consider all these models as slight variations of a unique notion. This study has been initiated by applications we are currently involved in.

1. Introduction

The work reported in this paper concerns conceptual graphs (CGs) for representing and reasoning on hierarchically structured knowledge. In these CGs, concept nodes may express complex information represented by CGs. We may call them *positive* nested conceptual graphs, in order to differentiate them from CGs with nested negative contexts. However, for the sake of brevity, we simply call them *nested CGs* (NCGs).

Let us first recall that we develop the CG model as a *graphical* knowledge representation model, where "graphical" is used in the sense of [1], that is a model that "uses graph-theoretic notions in an essential and nontrivial way". Basic objects are *simple CGs* (SCGs), which are connected labelled bipartite graphs (they correspond to the CGs introduced in [2], chap. 3). The fundamental operation for reasonings is *projection*, which is a morphism between these objects. Since it is sound [2] and complete (when graphs have a normal form, [3]) with respect to deduction in first order logic, reasonings on simple CGs can be based on projection computing. Furthermore, projection induces a structural *subsumption* relation on simple CGs. All NCG models considered in this paper generalize this simple CG model, while preserving its properties.

The simple CG model has been implemented on CoGITo [4] [5], a software platform for developing knowledge-based applications for which every piece of knowledge is represented by CGs. But in several applications this model proved to be inconvenient for expressing the knowledge involved. Let us give two examples of applications we are currently involved in. First, a simulation application of the required behavior of human agents in case of a catastrophe in a factory [6] [7]. This behavior is described by CGs which represent situations, pre-conditions and post-conditions of actions to be performed, each of these elements being themselves described by CGs. Second, a study of document retrieval for university libraries [8].

---

[1]& is commutative

The content of a document is partially represented by a CG. At first level, the title, the author and the subject are encoded. At second and deeper levels, the subject is described in further detail. In both cases, knowledge can be naturally structured by hierarchical levels and reasoning must respect these levels. A convenient way of encoding these levels is to put simple CGs within simple CGs. Descending from level to level can be seen as a "zooming" operation [9][10].

This led us to a first NCG model. In these NCGs, concept nodes are labelled with three fields: to the type and referent is added a *description*, which may be either generic (i.e. empty) or a set of NCGs. Projection is recursively defined according to the recursive structure of NCGs. Let us specify that in [11][12] a model close to NCGs is defined (but projection in this latter model is not recursive).

It appeared in applications that a mean of specifying the semantics of nestings was desirable. This led us to the *typed* NCG model. Briefly, besides the posets of concept and relation types we now have a poset of *nesting types*. To each graph of a concept description is associated a nesting type, which specifies the link between this graph and the including concept node. The typed NCG model is presented in this paper.

Several kinds of NCGs have been used by other researchers. Most of them follow more or less Sowa's formalism in [9][10]. Although Sowa does not consider graph operations for reasoning, a projection associated to these graphs can be defined. One obtains a new NCG model. Then a natural question arises: how can we compare these models with respect to their expressive power? More specifically, which model should we choose for extending the CoGiTo platform? In order to answer these questions, we propose a general framework in which various NCG models can be defined and compared. Fundamentally, we deal with objects (categories in the mathematical sense) which are graphs within graphs provided with morphism. We give some examples of known kinds of NCGs and prove that they are specific cases of such objects. A generalization relation over classes can be defined. Briefly said, a class C is more general than a class C' if there is a transformation from C' to C which preserves morphisms (similarly to the notion of problem reduction in complexity theory). This allows us to show that classes of NCGs considered in this paper are equivalent (each one is more general than the others). From a practical viewpoint, this means that any of these models may be chosen for implementation purposes. Several models may be used at an interface level, the choice depending on the application or the taste of the user. Solving a problem involving projection computing can be done by translating this problem into the implemented model, via one of the given transformations, and translating back the answer into the interface model.

The paper is organized as follows. In next part we define the general framework for defining and comparing models of graphs within graphs, with respect to morphism (or subsumption) computing. Part 3 is devoted to several NCG models. Their equivalence is proven in 3.4. For clarity reasons, these models are first presented without coreference links. Then part 3.5 shows how coreference links can be introduced. In conclusion, we discuss logical semantics of previous NCG models.

2. Graph of Graphs Framework

2.1. Graph of Graphs

In this part a combinatorial structure, provided with a morphism notion, is introduced. This structure, called a *graph of C-graphs*, is defined from any class C of graphs for which a morphism is defined (directed or undirected graphs, multigraphs, labelled graphs, SCGs,...). Let us recall that a *morphism* from a graph G to a graph H is a mapping, say f, from the node set of G to the node set of H, which preserves adjacency, i.e. for each edge xy of G, $f(x)f(y)$ is an edge of H. When G and H are labelled graphs, there may be additional constraints on labels. For instance, projection on simple CGs is a morphism, which keeps the ordering of edges incident on relation nodes and may decrease node labels (i.e. specialize types and, for concepts, replace the generic marker with an individual one).

Definition 1. *Let C be a class of graphs. A **graph of C-graphs** (Figure 1) is a couple $G = (\{G_1, ..., G_k\}, W)$ where :*
(1) any G_i belongs to C, and for all $i \neq j$, G_i and G_j are disjoint
(2) W is a set of triples (x, G_i, G_j), where x is a node of G_i and $i \neq j$.

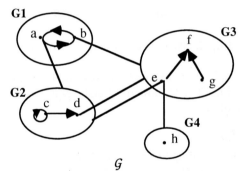

$$\mathcal{G}$$
$$W=\{(a, G_1, G_2), (b, G_1, G_3), (d, G_2, G_3), (e, G_3, G_2), (e, G_3, G_4)\}$$

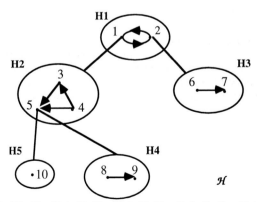

$$\mathcal{H}$$
$$W=\{(1, H_1, H_2), (2, H_1, H_3), (5, H_2, H_5), (5, H_2, H_4)\}$$
Fig. 1. Graphs of graphs, G and H.

When the class C is implicit we use the term *graph of graphs*. A triple (x, G_i, G_j) can be interpreted in the following way: x is a *complex node* of G_i, which is described by G_j. The *skeleton* of a graph of graph G is a graph which explicits the underlying structure of G, i.e. relationships among the G_i, and forgets the content of each G_i (Figure 2).

Definition 2. *The **skeleton**, $S(G)$, of $G = (\{G_1, ..., G_k\}, W)$, is a directed labelled graph, whose node set is in bijection with $\{G_1, ..., G_k\}$ and edge set is in bijection with W. There is a directed edge from G_i to G_j labelled by x for each triple (x, G_i, G_j) of W.*

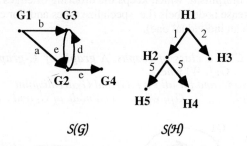

$S(G)$ $\qquad$ $S(\mathcal{H})$

Fig. 2. Skeletons of G and $\mathcal{H}$ (from Fig. 1).

A morphism from a graph of graphs G to a graph of graphs $\mathcal{H}$ combines graph morphisms from graphs of G to those of $\mathcal{H}$, and a graph morphism from the skeleton of G to the skeleton of $\mathcal{H}$.

Definition 3. *Let $G = (\{G_1, ..., G_k\}, W)$ and $\mathcal{H} = (\{H_1, ..., H_p\}, Y)$ be two graphs of C-graphs. A **morphism from** G **to** $\mathcal{H}$ is a set $\Pi = \{\Pi_1, ..., \Pi_k\}$ of mappings, where for all i, Π_i is a (C-graph) morphism from G_i to $H_{s(i)}$, s being a mapping from $\{1, ..., k\}$ to $\{1, ..., p\}$, which satisfies: if $(x, G_i, G_j) \in W$ then $(\Pi_i(x), H_{s(i)}, H_{s(j)}) \in Y$.*

Figure 3 shows a (graph of graphs) morphism $\Pi = \{\Pi_1, \Pi_2, \Pi_3, \Pi_4, \Pi_5\}$ between G and $\mathcal{H}$ of Figure 1. Graph morphism induces a preorder (reflexive and transitive relation) over graphs, defined as follows: a graph G subsumes a graph H ($G \geq H$) if there exists a morphism from G to H. Similarly, a graph of graphs morphism induces a subsumption relation over graph of graphs, which is also a preorder.

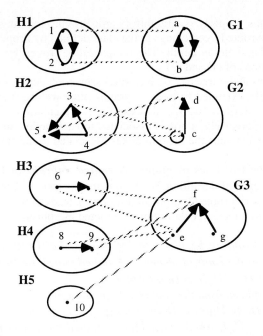

The 5 graph morphisms are represented with hachured lines.

Fig. 3. A morphism from $\mathcal{H}$ to $\mathcal{G}$ (from Fig. 1).

2.2. Tree of Graphs

Nested CGs are "graphs of simple CGs" with a tree underlying structure, as defined below.

Definition 4. *A graph of graphs is a **tree of graphs** if its skeleton is a rooted tree.*

For instance, the graph of graphs $\mathcal{H}$ of Figure 1 is a tree of graphs. It is useful to give a property which can be seen as a direct definition, by structural induction, of a tree of graphs.

Property 1. *A tree of graphs is a graph of graphs defined as follows:*
(1) a graph G with an empty set of triples, i.e. ({G}, ∅), is a tree of graphs with root G
(2) let A_1, ..., A_k be pairwise disjoint trees of graphs and $(x_1, ..., x_k)$ be a k-tuple of nodes of a graph G ($x_1...x_k$ are not necessarily distinct nodes). Let N be the set of graphs appearing in the A_i, added with G. Let W be the set of triples appearing in the A_i, added with triples $(x_i, G, root\ of\ A_i)$ for each i. Then (N, W) is a tree of graphs with root G.

The proof of this property is straightforward. Proving that any graph of graphs whose skeleton is a rooted tree belongs to the class defined by Property 1 can be done by induction on the skeleton depth. The reciprocal is proven by induction on the number of applications of (1) and (2).

Classical notions about trees — such as successor, root, depth, subtree induced by a node — are easily extended to trees of graphs, thus we use them with the same names and without definitions. We also use the representation of a tree by boxes within boxes. More precisely, a box is associated to each node c of the root of a tree of graphs. This box contains the disjoint boxes associated to the subtrees induced by the successors of c.

Let $\mathcal{A}$ and $\mathcal{A}'$ be two trees of graphs. A **rooted morphism** from $\mathcal{A}$ to $\mathcal{A}'$ is a morphism (in the sense of definition 3) which maps the root of $\mathcal{A}$ onto the root of $\mathcal{A}'$. Projection on NCGs is a rooted morphism. Property 1 leads to the following property which can be seen as a recursive definition of a rooted morphism.

Property 2. *Let $\mathcal{A}$ and $\mathcal{A}'$ be two trees of graphs. Let $D(\mathcal{A})$ and $D(\mathcal{A}')$ be the complex node sets of the roots of $\mathcal{A}$ and $\mathcal{A}'$. A rooted morphism from $\mathcal{A}$ to $\mathcal{A}'$ is a set of mappings $\{\Pi_0\} \cup \{\cup \{\Pi_c\}_c \in D(\mathcal{A})\}$ where:*

(1) Π_0 is a morphism from the root of $\mathcal{A}$ to the root of $\mathcal{A}'$

(2) for all c in $D(\mathcal{A})$, $\Pi_0(c) \in D(\mathcal{A}')$. Let $G_1, ..., G_k$ be the successors of c in $\mathcal{A}$. Then $\Pi_c = \{\Pi_1, ..., \Pi_k\}$, where each Π_i is a rooted morphism from the subtree induced by G_i to a subtree induced by a successor of $\Pi_0(c)$.

This property can be used for building an algorithm which computes a morphism (rooted or not) between two trees of graphs, whose complexity is polynomially related to the complexity of morphism computation on the class C [13].

2.3. Generalization Relation

Let us first recall that a *category*, say C, is a class of objects, called C-objects, together with a class of morphisms, called C-morphisms, which are related in the following way (see for instance [14]).

(1) with each couple of objects a and b is associated a set of morphisms, C-morph(a, b), such that each morphism of the C-morphism class belongs to just one C-morph(a, b)

(2) if $P \in C$-morph(a, b) and $Q \in C$-morph(b, c), there is a unique element of C-morph(a, c), denoted by PQ, called the composition of P and Q

(3) given $P \in C$-morph(a, b), $Q \in C$-morph(b, c) and $R \in C$-morph(c, d), $(PQ)R = P(QR)$

(4) to each object corresponds a morphism id_a of C-morph(a, a) called the identity morphism such that, for any $P \in C$-morph(a, b) and $Q \in C$-morph(b, a), $id_aP = P$ and $Qid_a = Q$.

It is straightforward to check that ordinary graphs provided with morphism, simple CGs provided with projection, graphs of graphs provided with morphism, nested CGs provided with projection (see part 3) are categories. In order to compare two models of nested CGs, or more generally two categories, we use the problem reduction notion of complexity theory (e.g. [15]). The following definition is obtained by considering the problem of the existence of a morphism between two objects, which is also the problem of subsumption checking.

Definition 5. *Let* C *and* $\mathcal{D}$ *be two categories.* $\mathcal{D}$ *is **more general than** C if there is a mapping* α *from* C*-objects to* $\mathcal{D}$*-objects, such that for any couple* (a, b) *of* C*-objects one has:* C*-morph* $(a, b) \neq \varnothing$ *if and only if* $\mathcal{D}$*-morph* $(\alpha(a), \alpha(b)) \neq \varnothing$.

In other words, $\mathcal{D}$ is more general than C when for all C-objects a and b, a subsumes b if and only if $\alpha(a)$ subsumes $\alpha(b)$. If $\mathcal{D}$ is more general than C, an algorithm for subsumption checking in C can be built using an algorithm for subsumption checking in $\mathcal{D}$ and an algorithm for computing α. Furthermore, if α is polynomially computable (i.e. for any object c of C, $\alpha(c)$ can be computed with a worst-case complexity polynomial with respect to the size of c) and if there exists a polynomial algorithm for subsumption checking on $\mathcal{D}$, then a polynomial algorithm can be built for subsumption checking on C. The generalization relation between categories is a preorder. Thus, it leads to an equivalence relation.

Definition 6. *Two categories are **equivalent** if each one is more general than the other one.*

As far as we know, the generalization relation between categories has not been studied yet, neither the equivalence relation which is weaker than the classical notion of isomorphism between categories.

3. Nested Conceptual Graphs

All kinds of (positive) nested CGs we found in papers can be defined as trees of C-graphs where C is a class of simple CGs (e.g. [10], [16], [17], [12]; more generally, see ICCS proceedings). This is not surprising since trees precisely are the underlying structure of boxes within boxes. As the case may be, trees of SCGs may have labelled edges and some constraints on their structure. Projection between NCGs can be defined as a rooted morphism between trees of C-graphs, where C is a class of simple CGs. A subsumption relation on NCGs is induced by projection: a NCG G subsumes a NCG H if and only if there exists a projection from G to H. Using results on trees of graphs, it can be proven that subsumption checking on NCGs (without explicit coreference links, see part 3.5) is not more difficult than subsumption checking on simple CGs, in the following meaning: given an algorithm A for projection checking (or computing) on the class C of simple CGs, an algorithm for projection checking (or computing) on the corresponding class of NCGs can be built, whose complexity is polynomially related to the complexity of A. Furthermore, the generalization relation over categories (definition 5) provides a mean of comparing NCG classes. Thus, the notions introduced in the preceding part define a unified framework for formalizing and comparing NCG classes provided with a subsumption relation. In what follows, we give three examples of NCG classes.

3.1. Untyped NCGs

A class of NCGs has been introduced in [18]. In order to distinguish this class from the others, we call it *untyped NCG* in this paper. Labels of concept nodes are triples *(t, m, d)*. *t* and *m* respectively represent the type and the referent of the concept node, as in simple CGs. The third field represents a description of the concept node. Its value may be **, called the generic description, or a set of untyped NCGs. Therefore, an intuitive semantics of *(t, m, d)* is *"m* denotes an object with type *t* and *partial internal description d"*. For instance, the untyped NCG of Figure 4 may encode the knowledge "Peter is thinking about the painting A, which represents a bucolic scene: a couple in a boat on a lake, one person being fishing, the other one being sleeping".

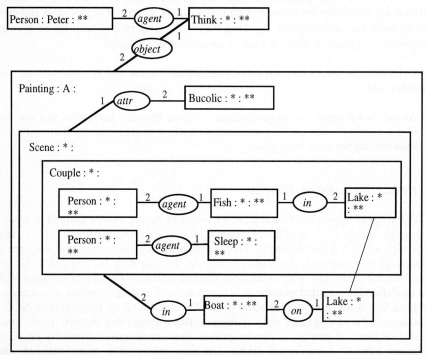

The coreference link between the "Lake" nodes is ignored in this part. See part 3.5

Fig. 4. An untyped NCG

The graph inside the node A is not supposed to fully characterize the painting, that is why it is called a "partial" description. This description is also called "internal" because it corresponds to a "zoom" (following Sowa's idea in [10]) on the node A, which behave like a *glass box*. In contrast, an information such as "the painting A is from Renoir" could rather be considered as an external information. It would be added to the NCG outside the node A, which in this case becomes a *black box*.

Let us recall that a *support* [3] encodes basic ontological knowledge. Among other elements, it possesses a concept type set T_C, a relation type set T_R, and a set of individual markers I. T_C and T_R are partially ordered by an A-Kind-Of relation, denoted by $\leq$. T_C has a greatest element, denoted by *Universal*. T_R is partitioned into sets of relation types with the same arity.

Definition 7. *Untyped NCG can be defined as follows.*

(1) *A basic untyped NCG is obtained from a SCG by adding to the label of each concept node c, a third field, denoted Desc(c), equal to* ** *(** can be considered as the empty description, and a trivial bijection exists between basic untyped NCGs and SCGs).*

(2) *Let G be a basic NCG, $(c_1, c_2, ..., c_k)$ a k-tuple of nodes of G, not necessarily distinct, and $G_1, G_2, ..., G_k$ NCGs. The graph obtained by substituting G_i to the description ** of c_i for $i = 1, ..., k$ is an untyped NCG.*

An untyped NCG G can be denoted by $G = (R, C, E, l)$ where R, C and E are respectively relation, concept, and edge sets of G, l is a labelling function of R and C such that for all $r \in R$, $l(r)=type(r) \in T_R$ and for all $c \in C$, $l(c) = (type(c), ref(c),$ $Desc(c))$ with $type(c) \in T_C$, $ref(c) \in I \cup \{*\}$, $Desc(c) = **$ or is equal to a set of untyped NCGs (in the latter case, c is said to be a complex node). Now, projection can be defined as follows.

Definition 8. Let $G=(R_G, C_G, E_G, l_G)$ and $H=(R_H, C_H, E_H, l_H)$ be two NCGs, with complex node sets D_G and D_H. A **projection** from G to H is a set of mappings $\Pi=\{\Pi_0\} \cup \{\cup \{\Pi_c\}_{c \in D_G}\}$, which satisfies:
(1) for all edge rc of E_G, $\Pi_0(r)\Pi_0(c)$ is an edge of E_H, and if c is the *ith* neighbor of r, then $\Pi_0(c)$ is the *ith* neighbor of $\Pi_0(r)$;
(2) for all node r of R_G, $type(r) \geq type(\Pi_0(r))$;
for all node c of C_G, with $l(c)=(t,m,d)$, let $l(\Pi_0(c))=(t',m',d')$, then: $t \geq t'$, $m \geq m'$ (i.e. a generic marker may be specialized into an individual one), and if $c \in D_G$, say $d=\{G_1, ..., G_k\}$, then $\Pi_0(c) \in D_H$, and $\Pi_c =\{\Pi_1, ..., \Pi_k\}$, where each Π_i is a projection from the NCG G_i to a NCG of d'.

It is straightforward to prove that untyped NCGs are trees of SCGs. The third field of a concept node c, $Desc(c)$ the description of c, is the set of subtrees induced by the successors of c. Figure 5 shows the NCG of Figure 4 represented as a tree of SCGs.

Property 3
An untyped NCG can be defined in an equivalent way as a tree of SCGs. Projection between untyped NCGs is then a rooted morphism between trees of SCGs.

3.2 Typed NCGs

In the untyped NCG model, a nesting does not have a semantics more specific than "the graph inside the concept node represents a partial internal description of the entity represented by the node". In Figure 4 for instance, the same notion is used for representing the nesting of the two persons who *constitute* a couple, the nesting of the couple in the *description* of the scene, and the nesting of the scene *represented* on the painting. It appeared in applications that a mean of specifying the nesting semantics was desirable. A way of solving this problem consists in adding nesting types to the untyped NCGs. A new type set T_N, called the nesting type set, is added to the support. T_N is partially ordered by an A-Kind-Of relation, as the concept and relation type sets, and possesses a greatest element called *Description*.

Definition 9. *Typed NCGs can be defined as follows.*
(1) *A basic typed NCG is a basic untyped NCG.*
(2) *Let G be a basic NCG and $(c_1, c_2, ..., c_k)$ be a k-tuple of non necessarily distinct nodes of G. Instead of substituting a graph to the third field of each c_i, one substitutes a couple (n_i, G_i), where n_i is a nesting type and G_i is a typed NCG. The graph obtained by substituting (n_i, G_i) to the description ** of c_i for $i = 1, ..., k$ is a typed NCG.*

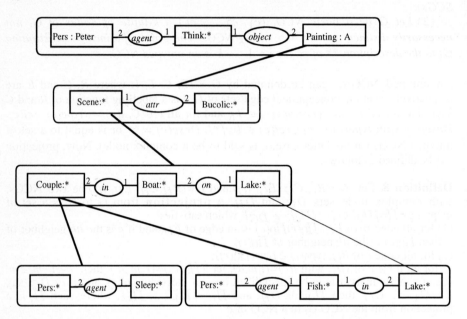

Fig. 5. An untyped NCG represented as a tree of graphs

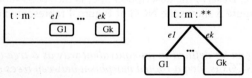

On the left: a concept node with label $(t, m, d = \{(e_1, G_1), (e_2, G_2), ..., (e_k, G_k)\})$
On the right: the tree induced by this node

Fig. 6. Representation of a typed nesting

Property 4. *A typed NCG is a tree of SCGs, where each edge is labelled by a nesting type.*

Figure 7 shows a typed NCG represented as a tree of SCGs with labelled edges. It represents the "painting A" example (see Figure 4 for a comparison with untyped NCGs). *Representation*, *Description* and *Component* are nesting types, with *Representation* ≤ *Description* and *Component* ≤ *Description*.

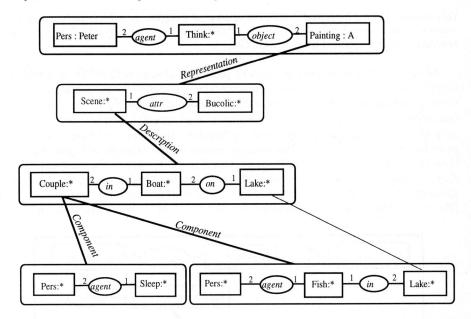

Fig. 7. A Typed NCG represented as a tree of graphs

Definition 10. *A **projection** between typed NCGs is a projection between the underlying untyped NCGs which may decrease the labels of the tree edges.*

3.3. 2-NCGs

In this paragraph a minor variant of the most used model of NCGs (e.g. [10]) is translated in our framework. In [10] two kinds of concept types are considered, *complex* and *simple* types. Nodes with complex type ("complex nodes") can have a graph for referent, and nodes with simple type ("simple nodes") can only have an atomic marker for referent. In the following formalization, we keep this distinction but any concept node must have a marker, and graphs are not referents but descriptions. Thus, in this model three pieces of information are associated to a complex node: its type, its referent which is a marker, and its description which is a set of graphs, possibly empty (note that this notion of a complex node does not exactly correspond to the notion of a complex node in previous NCGs, which assumes a non-empty description). A simple node only has a type and a referent.
Let us first define 2-SCGs, from which trees of 2-SCGs will be built. A *2-SCG* is a SCG built on a support, called a 2-support, which is more structured than the usual ones. In a 2-support the concept type set T_C, is composed of two disjoint sets, a simple concept type set, denoted T_1, and a complex concept type set, denoted T_2. T_1 has a greatest element called *Entity*, and T_2 has a greatest element called *Graph*. *Entity* and *Graph* are covered by *Universal*, the greatest element of T_C. No elements of T_1 are comparable with elements of T_2. The set of individual markers I is then

bipartitioned into I_1 and I_2, which are marker sets for simple and complex concept nodes respectively. Two generic markers are considered, * for a simple node and ** for a complex node. These two generic markers are incomparable , * is greater than any element of I_1, and ** is greater than any element of I_2.

Definition 11. *An **untyped** 2-NCG is a tree of 2-SCGs in which only complex concept nodes may have a description. Projection between untyped 2-NCGs is a rooted morphism between trees of 2-SCGs.*

NCGs of [10] can be seen as trees of particular classes of 2-SCGs, in which a support contains a particular binary relation type, say *descr*, with signature *(Entity, Graph)*, such that any relation type with an argument of type T_2 is less than *descr* (and therefore it is binary and its first argument is of type T_1). The second argument may be seen as a partial description of the first one. Note that introduction of *descr* provides another way of specifying nesting semantics. The example of Figure 7 could then be represented as partially shown in Figure 8, where the concept type *Representation* is a specialization of *Graph*. Or, in a dual way, *descr* could be specialized into a relation type *repr* . Then [Painting: A] -1-*(descr)*-2-[Representation: G] would be translated into [Painting: A] -1-*(repr)*-2-[Graph: G].

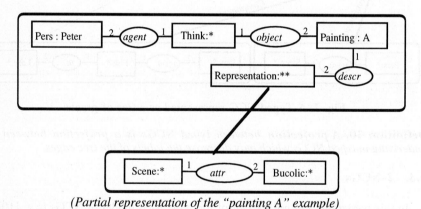

(Partial representation of the "painting A" example)

Fig. 8. A 2-NCG represented as a tree of graphs

3.4. Equivalence between classes of NCGs

Other classes of NCGs may be built in the same way as in previous part. For instance, *typed 2-NCGs* can be introduced, by adding nesting types to 2-NCGs or, equivalently, labels on the edges of trees. In [13] other classes are considered. An important point is that all these classes can be defined as trees of C-graphs for some class C of simple CGs. It is then easy to provide them with a projection which is a rooted morphism between these trees. The generalization relation introduced in part 2 of this paper allows to show that all these notions are only slight variations of a unique notion.

Theorem. *Typed or untyped NCGs, typed or untyped 2-NCGs, are equivalent classes.*

Proof. Let *UNCG(S), TNCG(S),* 2-*NCG(S)* be respectively classes of *untyped NCGs, typed NCGs* and *untyped* 2-*NCGs* on a support *S*.

Trivially, a graph *G* of 2-*NCG(S)* can be represented as a graph $\gamma(G)$ of *UNCG($\gamma(S)$)*, with $\gamma(S)$ is *S* where the distinction between *T*1 and *T*2 is forgotten. $\gamma(G)$ is obtained from *G* by replacing labels *(t, m)* of simple concept nodes with *(t,m,**)*, i.e. by adding an empty description ****.

A graph *G* of *UNCG(S)* can be represented by a graph $\alpha(G)$ of *TNCG($\alpha(S)$)*, where $\alpha(S)$ is obtained from *S* by adding a nesting type set restricted to a unique type, *Description*. $\alpha(G)$ is obtained from *G* by simply labelling all edges of the tree with *Description*.

A graph *G* of *TNCG(S)* can be represented by a graph $\beta(G)$ of *UNCG($\beta(S)$)*, where $\beta(S)$ is obtained from *S* by adding to the concept type set T_C of *S* a subset *T* in bijection with T_N and a new maximal type, and by deleting T_N. $\beta(G)$ is obtained from *G* by transforming labelled edges of the tree into concept nodes with types in *T*. More specifically, any edge in the tree of *G*, say from a node *c* to a graph *H* and labelled by a nesting type *n*, is replaced with the following construction: a new graph *K* with a unique concept node *c'* of type *n* is created and two edges are added to the tree, one from *c* to *K* and one from *c'* to *H*.

A graph *G* of *TNCG(S)*, with *S = (T_R, T_C, T_N, I)*, can be represented by a graph $\mu(G)$ of 2-*NCG($\mu(S)$)*, where $\mu(S)$ is built as follows: the simple concept type set T_1 is in bijection with T_C; the complex type set T_2 is restricted to *{Graph}*; the relation type set is in bijection with the union of T_R and T_N, where the *Description* nesting type corresponds to the *descr* binary relation ; the marker set for simple nodes is in bijection with *I* and there are no individual markers for complex nodes. $\mu(G)$ is obtained from *G* by transforming each concept node *c = [t : m : d]* into a simple concept node *c' = [t : m]*, whose neighbors are the images of *c* neighbors; and, if *d* is not the empty description, say *d = {(e_1, G_1), ..., (e_k, G_k)}*, then, for each *i* a complex concept node $c'_i = [Graph : \text{**} : \mu(G_i)]$ is created and a new relation node of type e_i links *c'* to c'_i.

A transformation similar to α (resp. β) can be used for transforming an *untyped* 2-*NCG* into a *typed* 2-*NCG* (resp. a *typed* 2-*NCG* into an *untyped* 2-*NCG*).

It is straightforward to check that transformations γ, α, β and μ preserve morphisms and rooted morphisms. $\qquad\Box$

3.5. Coreference links.

Previous NCG models can be extended with coreference links. Let us formally define coreference links, by way of an equivalence relation, say **coref**, over the set of concept nodes of all SCGs appearing in the tree associated with a NCG. Two distinct concept nodes are said to be coreferent if they belong to the same equivalence class. Concept nodes with the same individual marker necessarily belong to the same equivalence class (they are related with an implicit coreference link). And concept nodes with different individual markers necessarily belong to distinct classes. By definition, a node is coreferent with itself.

There are often constraints on labels of coreferent nodes ensuring that these nodes can be merged. In this case, coreference links internal to a simple CG do not present a fundamental interest, since an equivalent simple CG ("equivalent" in an intuitive meaning, and also according to logical semantics) without coreference link can be built in merging all coreferent concept nodes. But, when a coreference link relates two concept nodes, say c and c', from *distinct* simple CGs composing a NCG G, which do not belong to the description of the same concept node of G, the merging of c and c' is never possible.

Projection on simple CGs provided with coreference links, say Π from G to H, must preserve the *coref* relation, i.e. for all concepts nodes c and c' of G, if $coref(c,c')$ then $coref(\Pi(c),\Pi(c'))$. In other words, coreferent concept nodes either have the same image or distinct coreferent images. And the definition of a projection between NCGs provided with coreference links, say Π from G to H, is added with the same constraint: for all concepts nodes c and c' appearing in G, if $coref(c,c')$ then $coref(\Pi(c),\Pi(c'))$.

Note that the coref relation could be introduced in the general framework, as an equivalence relation on the set of nodes appearing in a graph of graphs.

4. Conclusion

All nested CG models considered in this paper generalize the simple CG model, in the sense that they involve objects which are generalizations of simple CGs and reasoning on these objets is based on a graph operation (projection) which generalizes that used for simple CGs. An important point, which we did not discuss, is that of the logical semantics of nested CGs. And the corollary question is whether projection is sound and complete with respect to this semantics.

This point has been studied for untyped NCGs provided with coreference links, with two approaches. In [19] a "non-classical" logic is defined, with a sequent calculus working on these formulas. Projection is proven to be sound and complete with respect to this calculus, i.e. there exists a projection from a NCG G to a NCG H if and only if the sequent leading from the formula associated to G to the formula associated to H is derivable. [20] defines a semantics in first-order logic. And projection (from G to H) is proven to be sound and complete with respect to implication (when H has a normal form). Equivalence of NCG classes allows transfer of these semantics to other NCG classes. For 2-NCGs, this transfer is "natural" since the transformation from a 2-NCG to an untyped NCG is trivial. On the other hand, for typed NCGs a new study seems to be more appropriate.

There are possible links between NCGs and contextual reasoning [21], although contextual reasoning is not our goal here. But both logical semantics mentioned above do not seem to be close to one of the logical formalizations of contextual reasoning (e.g. [22], [23]). However, a more in-depth comparison has to be done.

Acknowledgments

We would like to thank G. Kerdiles and E. Salvat for their comments on a previous version of this paper, and specially G. simonet for her suggestions that greatly improved clarity of this work.

References

1. L. K. Schubert, Semantic Networks are in the eye of the Beholder, in *Principles of Semantic Networks*, J. F. Sowa (ed), Morgan Kaufmann, 95-108, 1991.
2. J. F. Sowa, *Conceptual Structures — Information Processing in Mind and Machine*, Addison-Wesley, 1984.
3. M. Chein, and M.-L. Mugnier, Conceptual Graphs : fundamental notions, *Revue d'intelligence artificielle*, vol. 6, n°4, 365-406, 1992.

4. O. Haemmerlé, *La plate-forme CoGITo : manuel d'utilisation*, RR LIRMM 95012, 1995, 52 pages.
5. O. Guinaldo, *CoGITo v*3.3 *: module SGBD de graphes conceptuels*. RR LIRMM 96022, 1996, 10 pages.
6. C.Bos, and B.Botella, P.Vanheeghe, *Modelling and Simulating Human Behaviours with Conceptual Graphs*, R. R. Isen, 1996.
7. C.Bos, and B.Botella, Modelling Stereotyped Behaviours in Human Organizations, in *proceedings 7th Workshop on Knowledge Engineering: Methods and Languages*, Milton Keynes, United Kingdom, 1996.
8. D. Genest, *Une utilisation des graphes conceptuels pour la recherche documentaire*. Mémoire de DEA d'informatique, LIRMM & Université Montpellier II, Juin 1996.
9. J. F. Sowa, Conceptual Graphs as a universal knowledge representation, in *Semantic Networks in AI*, F. Lehmann (ed), Pergamon Press, 75-94, 1992.
10. J. F. Sowa, Logical foundations for representing object-oriented systems, *Journal of Experimental and Theoretical Artificial Intelligence*, vol. 5, 1994.
11. B.C. Ghosh, *Conceptual Graph Language*, Ph.D. Thesis, Asian Institute of Technology, Bangkok, 1996.
12. B.C. Ghosh, and V. Wuvongse, Computational Situation Theory in the Conceptual Graph Language, in *proceedings of ICCS'96*, 188-202, Lecture Notes in AI, 1115, Springer Verlag, Berlin, 1996
13. M. Chein, and M.-L. Mugnier, *Quelques classes de graphes emboîtés équivalentes*, R.R. LIRMM 96-063, 1996, 64 pages.
14. P. M. Cohn, *Universal Algebra*, Harper & Row, New-york.
15. M.R. Garey, and D.S. Johnson, *Computers and intractability — A guide to the Theory of NP-Completeness*, W.H. Freeman and Co, 1979.
16. A. Nazarenko, Representing natural language causality in conceptual graphs: the higher order conceptual relation problem, in *proceedings of ICCS'93*, 205-222, Lecture Notes in AI, 699, Springer Verlag, Berlin, 1993.
17. J.P. Dick, Using Contexts to Represent Text, in *proceedings of ICCS'94*, 196-213, Lecture Notes in AI, 835, Springer Verlag, Berlin, 1994.
18. M.-L. Mugnier, and M. Chein, Représenter des connaissances et raisonner avec des graphes, *Revue d'intelligence artificielle*, numéro spécial "graphes conceptuels", vol. 10, n°1, 7-56, 1996.
19. A. Preller, M.L. Mugnier, and M. Chein, A Logic for Nested Graphs, *Computational Intelligence Journal*, (to be published, reference number CI 95-02-558).
20. G. Simonet, *Une sémantique logique pour les graphes emboîtés*, R.R. LIRMM 96047, 1996.
21. J. McCarthy, Notes on Formalizing Context, in *Proceedings of IJCAI'93*, 555-560, 1993.
22. R. V. Guha, *Contexts: a formalization and some applications*, PhD Thesis, Stanford University, 1991.
23. S. Buvac, Quantificational Logic of Context, in *Proceedings AAAI'96*, 600-606,1996.

A Different Perspective on Canonicity

Michel Wermelinger

Departamento de Informática, Universidade Nova de Lisboa
2825 Monte da Caparica, Portugal
E-mail: mw@di.fct.unl.pt

Abstract. One of the most interesting aspects of Conceptual Structures Theory is the notion of canonicity. It is also one of the most neglected: Sowa seems to have abandoned it in the new version of the theory, and most of what has been written on canonicity focuses on the generalization hierarchy of conceptual graphs induced by the canonical formation rules. Although there is a common intuition that a graph is canonical if it is "meaningful", the original theory is somewhat unclear about what that actually means, in particular how canonicity is related to logic.

This paper argues that canonicity should be kept a first-class notion of Conceptual Structures Theory, provides a detailed analysis of work done so far, and proposes new definitions of the conformity relation and the canonical formation rules that allow a clear separation between canonicity and truth.

Topics: Conceptual Graph Theory, Knowledge Representation, Ontologies

1   Introduction

The development of Conceptual Structures Theory (CST) has been driven to great extent by natural language and its meaningfulness levels [8, p. 94]: gibberish; ungrammatical sequence; violation of selectional constraints; logically inconsistent; possibly false; empirically true. Syntax distinguishes the first two levels from the other ones, canonicity handles level 3, and logic the rest. Thus canonicity provides the ontological level that draws the borderline between meaningless and meaningful expressions (which are graphs in CST).

One could argue that syntax and logic are enough because any conceptual graph that obeys the arity of relations can be translated into a syntactically well-formed first-order formula and as such can be given a truth value. We feel however that logic alone does not distinguish the different "degrees" of falsehood: "Pigs fly" and "Portugal is a monarchy" are both false statements but not in the same way. Any knowledge representation theory should provide a way to capture our intuitions about such statements. In this paper, that role will be played by the ontology. Besides its conceptual importance, it has practical advantages: it can be shared by many knowledge bases; the knowledge representation system becomes more flexible and robust regarding arbitrary user input; processing an expression becomes more efficient.

Ideally, the ontology should be accompanied by two mechanisms: one to derive all the meaningful expressions, and the other to check whether an expression is meaningful. In CST the ontology is called *canon*, the derivation mechanism is given by the *canonical formation rules*, the checking mechanism is *projection*, and the meaningful expressions are the *canonical graphs*. A canon is specified by its types, markers, *conformity relation* and *canonical basis*. The conformity relation indicates for each marker all the types to which it is compatible, while the canonical basis is the initial set of graphs to which the canonical formation rules are applied.

In spite of its conceptual and practical importance, canonicity has been seldom a central theme of investigation. Most of the time only two notions "derived" from the canonical formation rules have received attention: projection and the generalization hierarchy. Only seldom have the knowledge representation aspects of canonicity been investigated: what are meaningful graphs? how is canonicity related to logic? This paper analyzes the work known to us, which can be summarized as follows. The original theory regards conformant concepts (i.e. those that obey the conformity relation) as true ones, while the canonical formation rules preserve falsehood: hence the relationship to logic is unclear since there are both true canonical graphs and true non-canonical graphs (and the same for false graphs). Kocura [5] considers that truth implies canonicity and therefore all non-canonical graphs must be false. Wermelinger [13] and Sexton [7] consider a graph to be canonical if it obeys the relation signatures given in the canonical basis. Finally, Sowa's new version of the theory [9] has no notion of canonicity: the definitions of conformity relation, canon, and canonical graphs have disappeared, and the canonical formation rules are just an auxiliary definition used by the inference rules.

This paper presents a different perspective on canonicity which fits better into the three meaningfulness levels (syntax, ontology, logic). A level characterizes a set of graphs, each level being a superset of the next one: canonical graphs are conceptual graphs (i.e., syntactically well-formed graphs), and true and false graphs are canonical. Seeing it the other way round, it does not make sense to speak about the canonicity of a non-conceptual graph or about the truth value of a non-canonical graph. To adhere to this principle, the conformity relation and the canonical formation rules (and thus the definition of projection) will be changed.

The structure of the paper is as follows. The next section is dedicated to the conformity relation. In the original theory, if marker m conforms to type t then "m is a t", i.e. m belongs to t's denotation. As will be seen, this interpretation of conformity has several problems. To allow false graphs to be canonical (and therefore conformant), the conformity relation will be relaxed.

The third section deals with the canonical basis. The main issue is to define what kinds of graphs the knowledge engineer may put into the canonical basis in order to specify useful *selectional constraints*. There have been several proposals, ranging from simple relation signatures to an arbitrary collection of graphs. We analyze those proposals and conclude that Sowa's original definition is still the most satisfactory one.

Section 4 handles the canonical formation rules. The original ones [8] only specialize the graphs they are applied to, but in [9] they may also generalize them. We follow the same approach because due to our principle true graphs (handled in [11]) should be canonical. Therefore the inference rules should be a special case of the formation rules. However, the formal definition of the latter will be changed to allow them to be applied directly instead of through the inference rules (as in [9]).

The last section characterizes canonical graphs in the usual two ways: using the canonical formation rules or projection. The definition of the latter will be extended to cope with the new definition of the former. An algorithm that decides whether a graph is canonical or not will be given. It is almost identical to the one for the original theory [6] and has the same complexity.

We use the following notational conventions: t and t' are concept types, i is an individual marker, m is an individual marker or the generic marker $*$, $\leq$ is the subtype relation, and $t \wedge t'$ denotes the maximal common subtype of both types. All references to pages (p. X) and to the original theory (Assumption/Theorem/Definition $x.y.z$) are to be understood in the context of [8]. The definitions given in the paper will be rather informal, since the formal details depend on the exact formalization of the basic notions (marker, type, concept, relation, etc.). Due to lack of space we will only deal with CST in its simplest form, assuming however that relation types may form a hierarchy [13]. The complete formalization of canonicity [12] also deals with higher-order types, a marker hierarchy including the absurd marker, contexts and coreference links.

2   The conformity relation

Sowa introduced the conformity relation :: as a test to be done when changing a concept's type: "if #98077 is a cat then CAT :: #98077 is true; otherwise, it is false"; in the second case "$\boxed{\text{ANIMAL: }\#98077}$ could not be restricted to $\boxed{\text{CAT: }\#98077}$" (p. 87). The sentence quoted first makes it very clear that an individual conforms to a type if and only if it belongs to the set denoted by the type. This is further stressed by the formal definition (Assumption 3.3.3) which imposes the following conditions:

1. for any concept $\boxed{t:\, m}$, $t :: m$;
2. if $t \leq t'$ and $t :: i$, then $t' :: i$;
3. if $t :: i$ and $t' :: i$, then $t \wedge t' :: i$;
4. for any i, $\top :: i$, but not $\bot :: i$;
5. for any t, $t :: *$.

In fact, if we define the conformity relation as the expression of the denotation (formally, $t :: m \Leftrightarrow m = * \vee m \in \delta t$) then we easily get condition 5 from the definition, and conditions 2 and 4 from the properties of δ: $\delta\top$ is the universal set, $\delta\bot$ is the empty set, and $t \leq t'$ implies $\delta t \subseteq \delta t'$.

As noted in [2], an individual marker cannot conform to two incompatible types (i.e., $t \wedge t' = \bot$). Otherwise conditions 3 and 4 would contradict each other. But the problem roots deeper. In fact, condition 3 implies that $\delta(t \wedge t') =$

$\delta(t) \cap \delta(t')$ which is called the lattice-theoretic interpretation of the type hierarchy in [1]. This contradicts the order-theoretic interpretation of Theorem 3.2.6: $\delta(t \wedge t') \subseteq \delta t \cap \delta t'$. Also, the lattice-theoretic approach has conceptual and practical drawbacks: a maximal common subtype must be interpreted as the "implication" of its supertypes, and the intersection of each pair of compatible types must be represented by an explicit type, leading to an explosion of conceptually irrelevant types.

Example 1. Consider a knowledge base about people, containing types for jobs and family relationships. According to the lattice-theoretic interpretation, **UNCLE** = **SON** ∧ **BROTHER** means that every person which is a son and a brother is also an uncle, clearly an undesired meaning. If **TEACHER** :: **#Michael** and **FATHER** :: **#Michael** then the type **FATHER-TEACHER** must exist for condition 3 to be satisfied, even if there is no other teacher with children in the knowledge base.

Even if condition 3 is abandoned other problems remain. Condition 1 forces every concept to be conformant. We will not impose this constraint as it is too strong in our opinion. Furthermore, notice that $\perp :: *$ but not $\perp :: i$. Logically, both are false statements since they correspond to $\exists x \in \delta\perp$ and $i \in \delta\perp$, contradicting $\delta\perp = \emptyset$. There is thus no valid reason to allow one but not the other.

Furthermore, if the conformity relation is just the indication of the types each individual marker is an instance of, then it is theoretically useless, because the same effect can be obtained by axioms in the knowledge base: assert in the outer context $\boxed{\perp : *}$ and $\boxed{t : i}$ whenever $t :: i$. Applying the generalization inference rule one gets $\boxed{t' : i}$ for any $t' \geq t$ (including $t' = \top$) and $\boxed{t : *}$ for any t. Also, since the first-order rules of inference are consistent, the graph $\boxed{\perp : i}$ can never be obtained. Thus conditions 2, 4, and 5 are satisfied.

The real problem however is not of formal but of conceptual nature. If $t :: i$ means that "i is a t", then all false concepts will not be conformant and as such cannot be generated by the canonical formation rules. As Sexton [7] noted, this is not consistent with the statement "The formation rules enforce selectional constraints, but they make no guarantee of truth or falsity" (p. 94). As a concrete example, take the one on page 92: if **BEAGLE** :: **#Snoopy** then $\boxed{\text{DOG: \#Snoopy}}$ cannot be restricted to $\boxed{\text{COLLIE: \#Snoopy}}$ since Snoopy is a beagle, not a collie. Sexton remarks that the latter concept is a meaningful one and therefore should be allowed in a canonical graph, although the individual marker does not belong to the denotation of (i.e., conform to) the type.

Therefore, in this paper a new notion of conformity relation is proposed: a marker m will conform to a type t if $\boxed{t : i}$ should be part of a true or false graph. Thus, conformity (as part of the broader notion of canonicity) does not imply truth any longer. Conversely, it does not make sense to speak about the truth or falsehood of a graph with non-conformant concepts.

Before presenting the formal definition, some observations are in order. First, as $\boxed{\perp : m}$ is false for any m, it will be considered a conformant concept. Second, if $\boxed{t : i}$ is true, so is $\boxed{t' : i}$ for any supertype t'; and if it is false, it is for any subtype t', too. To put it simply,

$$\text{if } t :: i \text{ then } t' :: i \text{ for any } t' \leq t \text{ or } t' \geq t.$$

Formally, however, we cannot state it this way, because $t :: i$ would imply $\top :: i$ (and $\bot :: i$) and therefore $t' :: i$ for any type t'. In other words, any individual marker would conform to any type, thus making the conformity relation meaningless. Even if we impose the restriction $\bot < t < \top$ in the above rule, an individual marker would still conform to concept types that are "zig-zag"-related in the type hierarchy. To circumvent this, we split the conformity relation into two relations: the base relation is given by the knowledge engineer and states for each individual marker what are the most relevant types it should conform to; the actual conformity relation is basically just the closure of the base relation over subtypes and supertypes.

Assumption 1. Given a relation R between concept types and markers, the *conformity relation* $::$ is the smallest superset of R such that

- for any m, $\top :: m$ and $\bot :: m$;
- for any t, $t :: *$;
- for any t and m, if $t R m$ and $t \leq t'$ or $t' \leq t$, then $t' :: m$.

Relaxing the definition of conformity is not only theoretically more elegant, it has also practical advantages: a conceptual graph processor can be made more robust and it can indicate the source of errors precisely. Consider for example a natural language processor that has to join the concepts $\boxed{\text{MAN: \#Lou}}$ and $\boxed{\text{WOMAN: \#Lou}}$. According to the original definition, the resulting concept $\boxed{\bot: \text{\#Lou}}$ does not obey the conformity relation and as such the join would fail (i.e. the text would not be parsed). An implementation could provide some *ad-hoc* way to indicate the source of error to the user, but it is always better to have a clean theoretical framework, as is the case with the new definition: the concept is meaningful, although false, and the absurd type clearly shows where the parsing has produced an inconsistency.

3 The Canon

The conformity relation is only a small part of the overall definition of an ontology to be used by one or more knowledge bases. In Conceptual Structures Theory the ontology is called *canon* and contains the types, the markers, the conformity relation, and an initial set of well-formed graphs, the *canonical basis* (Assumption 3.4.5). By applying the *canonical formation rules* to those graphs one obtains all *canonical graphs*, i.e. all graphs that "are meaningful" (p. 91).

However, the Conceptual Catalog [8, Appendix B] assigns a canonical graph to each concept or relation type in order to specify the selectional constraints to be observed by each type. Besides not being part of the formal definition of canonical basis, this association lead the Conceptual Structures community to use the term "canonical graph" in two different senses: (1) a graph that is derivable from the canonical basis, and (2) *the* graph in the canonical basis that is associated to a given type. Of course, these two senses are not incompatible, since (2) implies (1). To make the distinction clear, the elements of the canonical basis will be called *base graphs*.

The existence or not of associations between types and base graphs influences greatly the notion of canonical graph, because in the former case the base graph of type t must project on any graph using t. As Willems pointed out[1], this leads to another dual view of the canonical basis: whether it represents selectional constraints on the links between relations and concepts, or mandatory "arguments" of types.

Example 2 (adapted from Willems). Consider these graphs, the first two being base graphs:

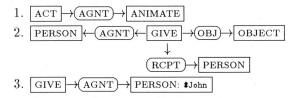

If base graphs are not assigned to types, then graph 3 is canonical because it can be derived from the first one. But if graph 2 is associated to concept type GIVE, then graph 3 is no longer canonical as it is missing two arguments of the verb.

Contrary to what Willems seems to imply, the problem is not the existence of associations *per se*, but the *kind* of associations done. As seen, assigning base graphs to concept types rules out meaningful graphs, that we would like to consider canonical, on the ground of having only partial information. This is not acceptable for a knowledge representation formalism. Moreover, we feel that the "arguments" view of the canonical basis is more appropriate of a lexicon [10].

It is however possible to retain the "selectional constraint" view as long as associations are restricted to relation types and if base graphs consist only of a single relation. Thus each base graph states the "signature" of the associated relation type, i.e. its arity and the maximal concept types of its arguments. This kind of base graph is called *star graph* and was introduced in [2]. Graph 1 of Example 2 could be the star graph of AGNT. This approach has been adopted by [13] and extended to handle relation type hierarchies: if $t \leq t'$ then the star graph of relation type t' must project into the star graph of t. Although not apparent at first sight, Sexton [7, Section VIII] also advocates the use of star graphs: "In order for a graph to be canonical, the type of each arc of each conceptual relation must be predicable [i.e., a supertype] of the type of the concept the arc points to". In other words, each relation must state the maximal type of each of its

---

[1] In a message sent to the CG mailing list on July 31, 1992.

arcs (i.e., arguments). To sum up, both [13, 7] consider a graph to be canonical if and only if all relations are used according to their signatures.

This is obviously a very weak notion of canonicity because the set of canonical graphs is too large. Star graphs are also too restrictive: by imposing the form and the number of base graphs (one for each relation type), the user can only specify very simple selectional constraints that take no contextual information into account. However, star graphs have computational advantages. As the recognition of a canonical graph is based on graph projection (see Section 5), if the elements of the canonical basis have a single relational vertex the complexity becomes polynomial.

The approach of Chein and Mugnier [2, 6] is better. They distinguish between the canonical basis and the *basis of the support*. The latter is the set of star graphs, and the graphs generated by the canonical formation rules from the star graphs are called *well-formed*. As in the original theory, the authors consider the canonical basis to be the generator set of the canonical graphs, but as expected they require the graphs of the canonical basis to be well-formed. This means that every canonical graph is well-formed. Therefore in this approach there is a new meaningfulness level between arbitrary conceptual graphs and canonical graphs. Notice that in the approaches mentioned above [13, 7] the canonical basis corresponds to the basis of the support and hence there is no distinction between well-formed and canonical graphs.

Chein and Mugnier's "mixed" approach is not as restrictive, but still there are conditions imposed on the elements of the canonical basis. This limits the knowledge engineer's flexibility to specify an ontology. Moreover, if the type of an argument of a relation depends on the type of another of its arguments, more than one star graph is necessary for that relation type. Also, it seems to us that the selectional constraints specified by star graphs are just a special case of the selectional constraints that base graphs are supposed to express. In fact, it is easy to provide a tool that checks whether the relations occurring in the canonical basis are used consistently.

To sum up, Sowa's original definition of a canonical basis as a set of conceptual graphs is still the most satisfactory one, as it provides all the flexibility required by a knowledge engineer, who is free to adhere to the specification discipline imposed by star graphs if he wishes. The formal definition of canon remains hence similar to Sowa's, but as expected it uses the new definition of conformity relation, which must be obeyed by every base graph. This is not explicitly stated in the original Assumption 3.4.5 since condition 1 of the original definition of conformity relation already required every concept to be conformant. As we have abandoned condition 1 in general, we must impose it for the base graphs.

Notice that the canonical basis may be redundant: it might be possible to derive exactly the same canonical graphs just from a proper subset of the canonical basis. A Conceptual Structures system can detect that case using Theorem 7 (Section 5) to verify for each base graph if it can be derived from the other ones.

4 The canonical formation rules

The rules proposed by Assumption 3.4.3 have several advantages: they are simple, they do not contain redundancies (i.e. they are independent from each other), and they are specialization rules. This means that their application (called a *canonical derivation*) establishes a relationship between the initial graphs and the resulting one that can be analyzed both from the logical (implication) as from the graph-theoretical viewpoint (projection).

The drawback of using just specialization rules is that not every true graph is a canonical one, clearly an undesirable state of affairs. Considering Example 2 again, if graph 3 is true and ACT $\leq$ EVENT then $\boxed{\text{EVENT}} \rightarrow (\text{AGNT}) \rightarrow \boxed{\text{PERSON: } \sharp \text{John}}$ is also true but as it cannot be obtained by specialization from the other three graphs, it is not a canonical graph. This is contrary to the idea that canonical graphs are "meaningful graphs that represent real or possible situations in the external world" (p. 91). In other words, the true graphs must be a subset of the canonical graphs. Hence, given a set of true graphs, the graphs derived from them using the inference rules must be canonical and as such should be obtained by applying the canonical formation rules to the same set of graphs. Put differently, the inference rules must be a particular case of the canonical formation rules.

This is the approach followed in the new version of the theory [9]. Sowa has made the canonical formation rules more general, and the inference rules limit the applicability of the formation rules. He made two kinds of changes. First, some rules do not apply to a single vertex or to a complete graph any more but to a subgraph. Second, rules have been divided into three groups: those that generate a logically equivalent graph, those that specialize the graph to which they are applied, and those that generalize it.

As the next example shows, the rules are no longer independent from each other. It is possible to get the same result from the same graph(s) applying different rules (to different subgraphs in some cases). Although [9] does not state what a subgraph is, from the rules we interpret it in the graph-theoretical way as a subset of vertices and edges. A subgraph therefore does not have to be a conceptual graph. In particular it may be just a single relation node. This allows the new rules to include the original simplification rule.

Example 3. From the graph $\boxed{\text{ACT}} \rightarrow (\text{AGNT}) \rightarrow \boxed{\text{ANIMATE}}$ one can derive

in two distinct ways. The first is to make a copy of the subgraph (AGNT), which shows that the two graphs are equivalent. The second one starts with a copy of the whole graph and then joins pairwise equal concepts.

The notion of canonicity does not exist in [9]. As such, the canonical formation rules are not used autonomously but by the inference rules. There are however good reasons to want to use the canonical formation rules directly:

- in many applications (like natural language understanding [10]) it is useful to process graphs whose truth value is unknown;
- it is desirable to have an "operational" characterization of canonical graphs;
- the canonical formation rules can be a starting point for the definition of other operations.

Since the new canonical formation rules may specialize part of a graph and generalize some other part, the notion of projection must be relaxed to allow the "declarative" characterization of the canonical graphs thus obtained.

Definition 2. Let g and g' be two conceptual graphs. A *semi-projection* $\pi : g \rightarrow g'$ is a function that maps g to a subgraph of g' such that

- for any vertex v of g, either $\pi(v) \leq v$ or $\pi(v) \geq v^2$;
- for any relation r of g, if its i-th arc a links r to concept c, then $\pi(a)$ is the i-th arc between $\pi(r)$ and $\pi(c)$.

If the function is a bijection then g' is called an *semi-instance* of g.

The new canonical formation rules can now be presented. They are similar to Sowa's. The changes made arose from the need to generate only ontologically meaningful graphs. Therefore some rules had to be restricted. Others had to be added to make sure that the first-order rules of inference are indeed a particular case of the canonical formation rules[3].

Assumption 3. Given a canon and zero or more conceptual graphs, the *canonical formation rules* generate new graphs. Some rules are defined in terms of subgraph duplication, removal and substitution. The definition of subgraph depends on the rule to be applied. In any case the operations also duplicate, remove, or substitute the arcs between subgraph vertices and external vertices. If the graphs to which the rules are applied obey the conformity relation then so must the resulting graph. In the following c is a context, either empty or containing the conceptual graphs g_1 and g_2 which might be the same one. Graphs g'_1 and g'_2 are subgraphs of g_1 and g_2, respectively.

- *Equivalence Rules.* In these rules, if a subgraph contains a concept, it also contains all relations linked to the concept.
  Copy Make a copy of g'_1.
  Simplify Remove g'_1 if g'_1 and g'_2 are identical and are linked to the same external vertices but have no vertices in common.
- *Specialization Rules.* In these rules, if a subgraph contains a relation, it also contains all concepts linked to the relation.
  Join Overlay g'_1 and g'_2 if they are identical.

---

[2] This is an extension of the partial order over types to concepts and relations.

[3] That does not happen in [9] and neither in this paper since we restrict ourselves to graphs without contexts or coreference links. The full version of the canonical formation rules [12] is however a generalization of the inference rules [11].

Restrict Substitute a vertex v of g_1 by a specialization if v has not been unrestricted before.

Insertion Insert a base graph in c.

- *Generalization Rules.* In these rules, if a subgraph contains a relation, it also contains all concepts linked to the relation.

  Detach Substitute g_1 by g'_1 if g'_1 is a semi-instance of some base graph.

  Unrestrict Substitute a vertex v of g_1 by a generalization if v has not been restricted previously.

  Remove Remove g_1.

The detailed explanation of the rules is given in [12]. The next subsections will just highlight the most important issues. First some general remarks. As in Sowa's original rules, the conformity relation must be checked before restricting or relaxing a concept. The formulation is also much more concise and simpler than those in [9, 4]. It is also clearer as it gives precise definitions of subgraphs. As Sowa's new rules, these are not independent from each other but they are so within each of the three groups.

4.1 Copy and Simplify

To see the reason for the given definition of subgraph, consider a concept c linked to a relation r such that c is part of the subgraph but r is not. Then the copy or simplification (i.e., removal) of c adds or removes the arc to r. In other words, the arity of r increments or decrements by one, and the resulting graph is not canonical. Therefore r must also be part of the subgraph as required by Assumption 3. Now, if r is duplicated or removed, its links to some external concept c' will be duplicated or removed, too, but that only changes the number of arcs attached to c', not the arity of r or its copy r' (see Example 3).

As for the simplify rule, two subgraphs g'_1 and g'_2 are duplicates only if they are connected to the same external vertices. There are two cases. If there are no such vertices then $g'_1 = g_1$ and $g'_2 = g_2$, which means that we are considering two copies of a complete graph. Hence one of them can be eliminated. In the second case, if the external vertices are the same, the two subgraphs must be part of the same graph: $g_1 = g_2$. In both cases the two subgraphs may not overlap. As for the first case that would amount to $g_1 = g_2$ and the simplify rule would become the remove rule. The problem in the second case is similar.

Example 4. Consider the relation **NTT** (not taller than) between persons. The subgraph $\boxed{(\text{NTT}) \rightarrow \boxed{\text{PERSON: }\sharp\text{John}} \rightarrow (\text{NTT})}$ occurs twice in

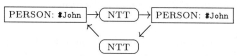

Eliminating one of the copies one gets just $\boxed{\text{PERSON: }\sharp\text{John}}$ which is not equivalent to the original.

Notice that the simplify rule stated in [9] does not impose any restriction on the duplicate subgraphs. As such, it is not an equivalence rule.

4.2 Restrict and Unrestrict

The restrict and unrestrict rules now allow to generalize *or* specialize *any* vertex (including relations), but they prevent the generalization *and* specialization of the same vertex. If that would be possible, any type t could be substituted by any other type t', even an incompatible one. This has been noted independently by [12, 5]. It means that the canonical formation rules could derive almost any non-canonical graph from the canonical basis, thus making the selectional constraints imposed by the base graphs useless.

Example 5. $[\text{ACT}]\!\rightarrow\!(\text{AGNT})\!\rightarrow\![\text{ANIMATE}]$ can be restricted to $[\text{ACT}]\!\rightarrow\!(\text{AGNT})\!\rightarrow\![\bot]$ and then unrestricted to $[\text{ACT}]\!\rightarrow\!(\text{AGNT})\!\rightarrow\![\text{IDEA}]$. The intermediate step could also be a generalization to $[\text{ACT}]\!\rightarrow\!(\text{AGNT})\!\rightarrow\![\top]$ followed by a specialization to the final graph.

In Sowa's approach, the canonical formation rules are only used by the inference rules. Since a graph cannot be simultaneously in an even context (where it can be generalized) and in an odd one (where it could be specialized), the restrict and unrestrict rules are never mixed. In our approach there were two possibilities: generalizations and specializations are forbidden for the same graph or just for the same vertex. The second option is more flexible and has been adopted.

The interplay between specialization and generalization is subtle. Certain vertices cannot be changed at all, namely those that were obtained by joining a vertex that has been generalized with one that was specialized.

Example 6. Let PERSON < ANIMAL < ANIMATE and let NTT be the relation of Example 4. From the conceptual graphs

$$[\text{ACT}]\!\rightarrow\!(\text{AGNT})\!\rightarrow\![\text{ANIMATE}] \qquad [\text{PERSON}]\!\rightarrow\!(\text{NTT})\!\rightarrow\![\text{PERSON}]$$

it is possible to obtain $[\text{ACT}]\!\rightarrow\!(\text{AGNT})\!\rightarrow\![\text{ANIMAL}]\!\rightarrow\!(\text{NTT})\!\rightarrow\![\text{PERSON}]$ by restricting $[\text{ANIMATE}]$ and unrestricting $[\text{PERSON}]$ followed by a join on both. The $[\text{ANIMAL}]$ concept can be no longer changed. Otherwise it would be possible to obtain e.g. $[\text{ACT}]\!\rightarrow\!(\text{AGNT})\!\rightarrow\![\text{DOG}]\!\rightarrow\!(\text{NTT})\!\rightarrow\![\text{PERSON}]$ through specialization and $[\text{ACT}]\!\rightarrow\!(\text{AGNT})\!\rightarrow\![\text{PHYSOBJ}]\!\rightarrow\!(\text{NTT})\!\rightarrow\![\text{PERSON}]$ through generalization, which violate NTT's and AGNT's selectional constraints, respectively.

To correctly implement these rules it is necessary to keep the history of each vertex. That can be done using two boolean variables, one indicating if the vertex has been generalized, the other whether it was specialized. An operation can be performed only if the variable corresponding to the *other* operation is set to false. When two vertices are joined, the variables of the resulting vertex are the conjunction of the corresponding variables of the original vertices.

4.3 Join and Detach

Both of Sowa's join rules [8, 9] only handle two (identical) concepts at a time. The rule of Assumption 3 allows one to join identical subgraphs. Notice that the rule allows them to belong to different graphs. That case is called an *external join* in [6]. As the join of two relations also involves the join of their arguments, the definition of subgraph is exactly the opposite of the one used by the equivalence rules.

One should also point out that the simplification can be simulated by an *internal join*, i.e. when the two subgraphs belong to the same graph. Indeed, overlaying two subgraphs is equivalent to eliminating one of them while keeping its arcs to the rest of the graph. As in the simplify rule both subgraphs have the same external links, the overlapping effect is obtained. Let us see an example using the graph of Example 4.

Example 7. Subgraph PERSON: ⋕John →(NTT)→ PERSON: ⋕John occurs twice in

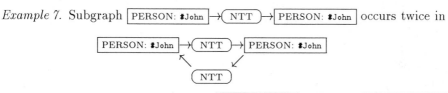

Overlaying the two of them one gets PERSON: ⋕John →(NTT)→ PERSON: ⋕John . The same result can be obtained by applying the simplify rule to the (NTT) subgraph.

Sowa's detach rule allows one to erase any subgraph. It is obvious that the remaining subgraph may not be canonical. Up to this part of the work, the only graphs that are guaranteed to be canonical are the base graphs. Therefore our rule must check that the remaining subgraph must be a base graph up to some generalizations or specializations. By repeated application of the copy and the detach rules it is possible to separate a graph into a cover of base components (compare with Theorem 7).

5 Canonical Graphs

Now that we have a generator set (the canonical basis) and the generation rules (the canonical formation rules) we can finally define the notion of canonical graph, which will be equivalent to the one of Assumption 3.4.5. Only the formulation differs. The original canonical formation rules (Assumption 3.4.3) use just the conformity relation and therefore the initial set of graphs to which the rules are applied must be explicitly stated. In our formulation that set (the canonical basis) is already part of the definition of the formation rules. Thus a canonical graph is a graph that can be generated from a canon and an "empty sheet".

Definition 4. A conceptual graph is called *canonical* regarding a given canon C if it is possible to derive it from the empty set of graphs through application of canonical formation rules using C.

In particular, applying the insertion rule one gets, as expected,

Proposition 5. *A base graph is canonical (regarding the canon it belongs to).*

Although the definition of canonical graph is the same as the original one, due to the differences in the definitions of the conformity relation and the canonical formation rules, given the same canonical basis both frameworks generate quite different sets of canonical graphs. The sets are incomparable (i.e. neither is a subset of the other) because Sowa's rules are not a subset of ours or vice-versa. However, the presented rules guarantee as wished that every true or false graph is canonical. Notice also that a graph can be considered canonical regarding a canon, and non-canonical regarding another one.

Besides forming new canonical graphs it is also convenient to be able to recognize them without explicitly constructing the derivation (the sequence of rules) that leads to their formation. The original canonical formation rules just specialize a graph. Hence the derivation process corresponds to a projection (Theorem 3.5.4). In addition, Mugnier and Chein [6] have shown that if there is a projection between two canonical graphs then there is a derivation. From this and other results they obtained the following characterization: a conceptual graph g is canonical if and only if there are projections of base graphs into g that cover the whole of g. As the new canonical formation rules also allow vertices to be generalized, projection is substituted by semi-projection (Definition 2) but the main idea remains.

Definition 6. A *cover* of a conceptual graph g is a finite set of conceptual graphs $\{g_1, \ldots, g_n\}$ such that each g_i is a subgraph of g and each vertex and arc of g occurs in at least one graph of the cover.

Theorem 7. *A conceptual graph is canonical regarding canon C if and only if it obeys the conformity relation and has a cover G such that each graph $g \in G$ is an semi-instance of a base graph of C.*

The importance of the theorem (proven in [12]) stems from the fact that it provides an algorithm to check whether a conceptual graph g (e.g. given by an user) is canonical or not. The method consists basically in finding base graphs whose semi-projections into g cover g completely. Since the semi-projection of a relation also includes the concepts it is linked to, the algorithm can be simply as follows. Choose a relation r of g and go through the canonical basis until finding a base graph whose semi-projection into g contains r. All relations covered by that semi-projection are marked and the process is repeated with an unmarked relation. This algorithm is identical to the one presented in [6] except that the projection is substituted by the semi-projection. Hence the algorithm still is polynomial in relationship to the complexity of the base operation (in this case semi-projection). Most of the time base graphs are trees. Therefore, in those cases semi-projection has polynomial complexity and recognizing a canonical graph takes polynomial time in the size of the graph and of the canonical basis.

6 Conclusions

This paper has argued that canonicity is fundamental to Conceptual Structures Theory since it corresponds to the intermediate ontological level between syntax and logic. Therefore it has both conceptual as practical advantages. We have analyzed some of the literature on canonicity and found out that the (implicit) meaning of canonicity is either too weak (as in the case of relation signatures) or its relationship to logic is vague or dubious. We have therefore explicitly adopted a very precise guideline: a graph should be canonical if we would like to make a definite statement about its truth or falsehood. Therefore canonical graphs are a superset of true and false graphs: it is meaningless to speak about the truth value of a non-canonical graph and, for a given knowledge base, the truth value of some canonical graphs may be unknown.

Based on this new perspective of canonicity, which provides a simple yet clear relationship between the ontological and logical levels, we have substantially changed the definitions of the conformity relation and the canonical formation rules and improved their formulation. To stress the relation between canonical and true graphs, the first-order rules of inference have become a special case of the canonical formation rules. Furthermore, the latter can now be used independently of the former without the risk of generating non-canonical graphs.

In order to keep a "declarative" definition of canonical graphs projection was generalized to semi-projection, but without computational impact since the recognition of canonical graphs still has the same complexity as for the original theory, in many cases being polynomial.

References

1. C. Beierle, U. Hedstück, U. Pletat, P. H. Schmitt, and J. Siekmann. An order-sorted logic for knowledge representation systems. Technical Report 113, IWBS, April 1990.
2. Michel Chein and Marie-Laure Mugnier. Conceptual graphs: fundamental notions. *Révue d'Intelligence Artificielle*, 6(4):365–406, 1992.
3. Gerard Ellis, Robert Levinson, William Rich, and John F. Sowa, editors. *Conceptual Structures: Applications, Implementation and Theory — Proceedings of the Third International Conference on Conceptual Structures*, volume 954 of *Lecture Notes in Artificial Intelligence*, Santa Cruz, USA, August 1995. Springer-Verlag.
4. John Esch and Robert Levinson. An implementation model for contexts and negation in conceptual graphs. In Ellis et al. [3], pages 247–262.
5. Pavel Kocura. Conceptual graph canonicity and semantic constraints. In Peter W. Eklund, Gerard Ellis, and Graham Mann, editors, *Conceptual Structures: Knowledge Representation as Interlingua — Auxiliary Proceedings of the Fourth International Conference on Conceptual Structures*, pages 133–145, Sydney, Australia, August 1996. University of New South Wales.
6. M. L. Mugnier and M. Chein. Characterization and algorithmic recognition of canonical conceptual graphs. In Guy W. Mineau, Bernard Moulin, and John F. Sowa, editors, *Conceptual Graphs for Knowledge Representation — Proceedings of the First International Conference on Conceptual Structures*, volume 699 of *Lecture*

Notes in Artificial Intelligence, pages 294–311, Québec City, Canada, August 1993. Springer-Verlag.

7. Clark A. Sexton. Types, type hierarchies, and canonical graphs. In Gerard Ellis, Robert A. Levinson, William Rich, and John F. Sowa, editors, *Conceptual Structures: Applications, Implementation and Theory — Supplementary Proceedings of the Third International Conference on Conceptual Structures,* pages 213–221, Santa Cruz, USA, August 1995.

8. John F. Sowa. *Conceptual Structures: Information Processing in Mind and Machine.* Addison-Wesley, 1984.

9. John F. Sowa. *Knowledge Representation: Logical, Philosophical, and Computational Foundations.* August 1994. Draft of a book to be published by PWS Publishing Company, Boston.

10. John F. Sowa and Eileen C. Way. Implementing a semantic interpreter using conceptual graphs. *IBM Journal of Research and Development,* 30(1):57–69, January 1986.

11. Michel Wermelinger. Conceptual graphs and first-order logic. In Ellis et al. [3], pages 323–337.

12. Michel Wermelinger. Teoria básica das estruturas conceptuais. Master's thesis, Universidade Nova de Lisboa, Lisbon, Portugal, January 1995.

13. Michel Wermelinger and José Gabriel Lopes. Basic conceptual structures theory. In William M. Tepfenhart, Judith P. Dick, and John F. Sowa, editors, *Conceptual Structures: Current Practices — Proceedings of the Second International Conference on Conceptual Structures,* volume 835 of *Lecture Notes in Artificial Intelligence,* pages 144–159, College Park, USA, August 1994. Springer-Verlag.

Aggregations in Conceptual Graphs

William M. Tepfenhart

AT&T
Middletown, NJ 07748
Email: william.tepfenhart@att.com

Abstract. In this paper issues associated with representing and reasoning about aggregations are examined. A hierarchy of aggregation types are described and the importance of individual types of aggregations in the reasoning endeavor is discussed. Some of the typical operations associated with aggregations are identified.

1. Introduction To Aggregations

Conceptual Graph Theory has limited itself to a model of aggregation that contains sets, bags, and sequences [2]. Sets are aggregations that can have duplicate elements. Bags are identical to sets with the exception that duplicate elements are not allowed. A sequence is an ordered bag. The linear form of Conceptual Graph Theory employs a set notation using the referent field of a concept as in:

[PERSON: {Liz, Kirby}]

This example demonstrates a set of two people, Liz and Kirby. This method of representing a set is appropriate for small sets in which the membership can explicitly be identified. In large sets, the membership is not specified. Instead a count is given:

[PERSON: { }@20]

This can be read as a set of 20 people. For even larger sets it is often not reasonable to explicitly identify individual members or even identify the number of members. Instead, membership is specified by a rule or property, as in:

[ANIMAL: {x | VERTEBRATE(x) & WARM-BLOODED(x) & HAS-HAIR(x) }]

This example can be read as the set of all x of type animal such that x is vertebrate, x is warm-blooded, and x has hair. Early on, Conceptual Graph Theory included a set type specifier of the form:

[PERSON: COLL{Jack, Jill}]

where COLL is a set type specifier. The theory allows for four set type specifiers: COLL, DISJ, DIST, or RESP. They identify collective, disjunctive, distributive, and respective sets respectively. In the initial linear form [2], a sequence was denoted by enclosing the elements in '<','>' as in:

[INTEGER: <1, 2, 3, 4>]

This example identifies a sequence of integers in the following order: 1, 2, 3, 4. However, it appears that the linear notation for a sequence has been dropped. This is clearly the case in recent publications associated with the linear form of Conceptual Graph Theory [1], [6] in which all mention of the use of '<' and '>' in the referent field has been dropped. This can be justified by recognizing that the use of RESP set type token conveys the same information.

In this paper, we address the concept of aggregation and various types of aggregations. We take a moderately different approach than the classical one represented by examples in linear notation given above. In those examples, the dominate semantics are derived from the type labels. The referent field, which can be interpreted as indicating individuals, is expanded to identify a set of individuals. In the approach taken here, an aggregation is a type separate from the type labels derived from the individual members. As a result of this difference, we can have hierarchy of aggregation types as opposed to a hierarchy of concept types for which aggregations can exist. For example, we can have:

[ANIMALS:*] >>
    [CATS: *] >>
        [LITTER_OF_CATS:*]
        [PRIDE:*]
    [DOGS: *] >>
        [LITTER_OF_DOGS: *]
        [PACK: *]

This allows us to capture and distinguish the semantics of sets of cats with arbitrary membership, sets of cats all born to a single mother, and a social group of cats. This kind of ontology captures the semantics of the set and establishes an identity for the set which can be independent of the membership. That is, the set [CATS: #1] has a separate identify from the set [CATS: #2]. For example, we can have two sets -- one which contains the cats I own and the other which contains cats currently in my house. The first set is (basically) static. The second set can grow and shrink based on where individual cats are currently located. However, there can be times in which the membership of both sets is identical - that is, all of my cats and no others are in my house. Distinguishing between the two sets is important in answering common questions such as, "Are all of our cats in the house?"

Linear Notation	Form
[CAT: {x I OWN([PERSON: Bill],x)}]	Intensional
[CAT: {fluffy, fuzzy, princess}]	Extensional

Table 1. Cats I Own

Linear Notation	Form
[CAT: {x I IN_HOUSE(h,x) & OWN([PERSON:Bill], h)}]	Intensional
[CAT: {fluffy, fuzzy, princess}]	Extensional

Table 2. Cats In My House

In the linear notation normally used to denote aggregations, the distinction of identity between the two sets can be lost. In conventional linear notation, we could denote the two sets in either of the two notations shown in the Table 1 and Table 2. When the predicate intensional form is employed, the distinction between the two sets is easily maintained. However, if the individuals are expressed extension in the referent field, then maintaining the distinction is not trivial. It is very easy to construct a case in which the four basic types of aggregations in Conceptual Graph Theory are insufficient. For example, a disjunctive set captures the concept of an exclusive or. This allows us to capture the semantics of a statement such as, "The ball is either red or blue." However, it does not allow us to deal with a non-exclusive or as exemplified in the statement, "Your pizza can be topped with any combination of olives, pepperoni, sausage, extra cheese, peppers, or anchovies." Using the set type specifier in dealing with new set types requires an extension to the linear notation for referents. This change would be in the form of a new set type specifier. One direct consequence of this is that any CG-based system for processing statements incorporating the new set types would have to be modified to accept and process the new token. One of the advantages of the approach employed in this paper is that it easily allows introduction of very different kinds of aggregation. It does so without requiring a modification of the notation employed in the referent field. Hence, we will not have to modify any CG-based system that processes statements containing such aggregations.

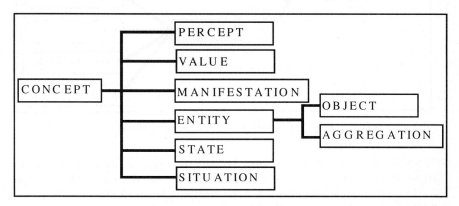

Fig.1. Top-Level Concept Hierarchy

2. Types Of Aggregations

In this paper, the term aggregation is used in the sense of a quantity of units taken as a whole. The individual units of an aggregation are the members. These members can be instances of any type of concept. This sense of aggregation does not imply any relationship among the members of an aggregation. This use is consistent with the concepts of aggregation used by Jackendoff, but differs from that of Sowa who views an aggregation as a tighter association than a set. Using the concepts as defined in the type hierarchy for concepts1 shown in Figure 1, the definition for an aggregation is:

type [AGGREGATION: *x] is
[ENTITY: *x] ->
    (hasMember) -> [CONCEPT: *]
    (numberMembers) -> [NUMBER: *]

The conceptual relation (hasMember) is defined as a subtype of an (entityConcept) conceptual relation as:

type (hasMember: *x) is
[AGGREGATION: *] -> (entityConcept:*x) -> [CONCEPT: *]

This definition for aggregation allows us to construct aggregations of all kinds of concepts, including percepts, values, manifestations, entities, states, and situations. The need for aggregations to include these types of concepts will be demonstrated in the next section. We treat all aggregations as abstract concepts, because they are

---

1 This is the top-level concept hierarchy as defined in the Situation Data Model [5].

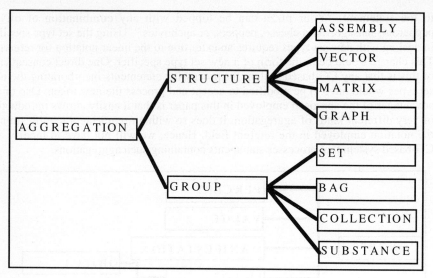

Fig. 2. Classifications Of Aggregations

observed by observing their members. Initially, this may not make sense for some aggregations particularly for some structures whose members are physical objects (e.g., an automobile). This position is defensible in terms of the concrete physical natures of such items. A car mechanic must necessarily view an automobile as an assembly. There are many types of aggregations and each is important for different reasons. However, there are a number of common characteristics among all kinds of aggregations. A classification scheme over the types of aggregation is shown in Figure 2. In this section we will discuss each of the major types of aggregations and demonstrate their importance with examples.

2.1. Structure

A structure is an aggregation in which the individuals within it are structured. This is an important category of aggregations and may, in fact, represent the aggregation with the largest extension. Mathematical entities such as arrays, matrices, spaces, and such are all structures because of the mathematical relationships among the members that are integral to their definition. A structure is an aggregation such that the members in it stand in some structural relationship with each other. The definition for a structure is:

     type [STRUCTURE: *x] is
     [AGGREGATION: *x] ->
          (hasMember: t) -> [CONCEPT: *y]
          (hasMember: t) -> [CONCEPT: *z]
     [CONCEPT: *y] -> (structuralRelation: t) -> [CONCEPT: *z]

This definition simply states that a structure has members of type concept and that the individual members stand in some relationship with each other. There are several important categories of structures: assembly, array, and graph.

2.1.1. Assembly

An assembly is a structure in which the elements form a structured whole. Access to an element in an assembly is achieved by dismantling the assembly. Addition of an element to an assembly is through the process of assembly. Examples of an assembly would include an automobile engine, an automobile, a television, and a radio to name a few. An assembly is defined as:

    type [ASSEMBLY: *x] is
    [STRUCTURE: *x] ->
            (hasMember: t) -> [CONCEPT: *]
            (numberElements: t) -> [INTEGER: *] -> (greater:t) -> [ONE: *]

This definition basically states that an assembly is a structure that has more than one part. Inherited within this definition is that the members stand in some structural relationship with each other. The relationship between an element within an assembly and the assembly is often referred to as a (partOf) relation. The inverse relation, (hasPart) is defined as a subtype of the (hasMember) relation as:

    type (hasPart: *x) is
    [ASSEMBLY: *] -> (hasMember: *x) -> [CONCEPT: *]

Of course, in this definition the target remains a concept of type [CONCEPT]. This allows us to deal with assemblies of features (e.g., parts of a face) and entities (e.g., dining room set or automobile). The assembly itself is often referred to as the 'whole.' Assemblies are cases in point where the whole is more than the sum of the parts. In particular, they have properties and capabilities that arise as a result of the elements that comprise them and the interactions among the elements. For example, none of the parts of an automobile directly give it the ability to go at 60mph. However, when the parts of an automobile are all functioning correctly cars can go at 60 mph. Assemblies are typically very stable - subject to infrequent change. While an assembly itself is stable, the parts that make up the assembly may not be. The substitution of a part does not affect the overall assembly. This explains 'grandfather's ax' which has had three handles and two heads. In terms of conceptual graph theory, grandfathers ax could be represented as:

    [AX: "grandfather's ax"] ->
            (hasPart:t) -> [AX_HEAD: *x]
            (hasPart:t) -> [AX_HANDLE: *y]
            (hasPart:t) -> [WEDGE: *z]
    [AX_HEAD: *x] -> (attachedTo) -> [AX_HANDLE: *y]
    [WEDGE: *z] -> (embeddedIn) -> [AX_HANDLE: *y]

This simplistic definition basically describes an ax as an assembly that has three parts: the head, the handle, and a wedge such that the head is attached to the handle and the wedge is embedded in the handle. The reason this is a simplistic definition is because it does not tell where the head is attached to the handle nor where the wedge is embedded on the handle. This prevents us from knowing that it is the wedge that keeps the head on the handle. The replacement of an ax handle is represented in terms of an actor <replaceAxHandle> which could be defined as:

    type < replaceAxHandle: *xx> is
    [[STATE:*a] ->
            (encompasses:t) ->[AX: *w] -> (hasPart:t) -> [AX_HANDLE:*x]
            (encompasses:t) -> [AX_HANDLE: *z]]
    -> <replaceMember: *xx> ->
    [[STATE:*a] ->

(encompasses:t) ->[AX: *w] -> (hasPart:t) -> [AX_HANDLE:*z]
(encompasses:t) -> [AX_HANDLE: *x]]

In short, the new ax handle is substituted for old ax handle while leaving the aggregation, namely the ax, untouched.

2.1.2. Array

An array is a structure in which the number of elements is fixed and access is achieved via indices into the structure. Arrays are typically described using mathematics with operations such as product, scaling, and normalization. However, the concept of an array is much more generalized than the mathematical definition. The example of a bit plane is one where the operations of product and normalization do not make sense. The representation of an array is one of the most difficult of the aggregations to represent in the sense that what are normally considered elements are not the only members of the array. There are also structured indices that identify an element's location within it. As a result, we use the concept of sites as holders of the actual elements in the array. This is illustrated for a two dimensional array containing alphabetic characters as elements in Figure 3.

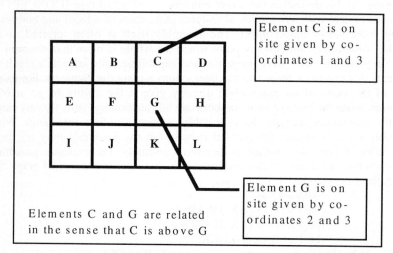

Fig. 3. Illustration of the relationship among array, site, and element

The definition that we will use for an array is as follows:

        type [ARRAY: *x] is
    [STRUCTURE: *x] ->
            (hasDimensionality) -> [DIMENSIONALITY: *a] ->
                    (hasValue) -> [INTEGER: *b] -> (greater:t) -> [ZERO: *]
            (hasDimensions) -> [SET: *] ->
                    (numberElements) -> [INTEGER: *b]
                    (hasMember) ->[DIMENSION: *c] ->
                            (hasMinValue:t) -> [INTEGER: *d]
                            (hasMaxValue) -> [INTEGER: *dd]
            (hasSite) -> [SITE: *e] ->
                    (hasIndices) -> [SET] ->
                            (numberElements) -> [INTEGER: *b]

```
                       (hasMember) -> [INDEX: *f] ->
                               (hasValue) -> [INTEGER: *g]
                               (forDimension) -> [DIMENSION: *c]
               (hasMember) -> [CONCEPT: *h]
       [CONCEPT: *a] -> (occupies) -> [SITE: *e]
       [INTEGER: *g] -> (>=) -> [INTEGER: *d]
       [INTEGER: *g] -> (<=) -> [INTEGER: *dd]
```

It is recognized that this definition is quite long and complex, but it does give all of the general properties of an array. First it has some integer dimensionality which is greater than zero. Second, it has a set of dimensions that has an equal number of members as the dimensionality of the array. Third, each dimension has some minimum and maximum integer value. Care has been take to assure that the minimum and maximum are not artificially bounded by zero or a finite number. Fourth, it has sites that are denoted by a set of indices for which each index is related to some dimension and has a value that is constrained to be between the minimum and maximum allowed values (inclusive) for that dimension. Finally, a site is occupied by a member of the aggregation. While it is recognized that an array is typically considered only within a mathematical context, there are additional uses of arrays. For example, many games are played upon boards where there are specific sites on which a game piece may reside. Some of these games, such as chess, checkers, and others, have a strong mathematical component to their play. However, operations such as scaling, cross-product, and addition do not apply for these games.

2.1.3. Vector

A vector is an array in which access to an element is achieved via a single index into the structure. Vectors can be employed to capture concepts such as [COORDINATE], [LOCATION], and [VELOCITY]. A vector is a restriction of an array and is defined as follows:

```
       type [VECTOR: *x] is
       [ARRAY: *x] ->
               (hasDimensionality) -> [DIMENSIONALITY: *a] ->
                       (hasValue) -> [ONE: *]
               (hasDimensions) -> [SET: *] ->
                       (numberElements) -> [ONE: *]
                       (hasMember) ->[DIMENSION: *b] ->
                               (hasMinValue:t) -> [INTEGER: *d]
                               (hasMaxValue) -> [INTEGER: *dd]
               (hasSite) -> [SITE: *e] ->
                       (hasIndices) -> [SET] ->
                               (numberElements) -> [ONE: *]
                               (hasMember) -> [INDEX: *f] ->
                                       (hasValue) -> [INTEGER: *g]
                                       (forDimension) -> [DIMENSION: *b]
               (hasMember) -> [CONCEPT: *h]
       [CONCEPT: *a] -> (occupies) -> [SITE: *e]
       [INTEGER: *g] -> (>=) -> [INTEGER: *d]
       [INTEGER: *g] -> (<=) -> [INTEGER: *dd]
```

As can be seen in this definition, a vector is limited to a single dimension. The number of entries in an array is determined by the difference in the maximum and minimum values allowed the index.

2.1.4. Matrix

A matrix is an array in which access to an element is achieved via two or more indices into the structure. A matrix corresponds to the general definition of an array with the constraint:

> type [MATRIX: *x] is
> [ARRAY: *x] ->
> > (hasDimensionality:t) -> [DIMENSIONALITY: *a] ->
> > > (hasValue:t) -> [INTEGER: *b] -> (greater:t) -> [ONE: *]

It remains a question as to whether the dimensionality of a matrix should be fixed at 2 or allowed to be greater. Fixing the index at 2 would require introducing another type that allows dimensionalities greater than 2.

2.1.5. Graph

A graph is a structure in which elements are organized according to a node-link scheme. Access to a particular node is achieved by traversing links. A generic definition of a graph is given below:

> type [GRAPH: *x] is
> [STRUCTURE: *x] ->
> > (hasNode) -> [CONCEPT: *a]
> > (hasNode) -> [CONCEPT: *b]
> [CONCEPT: *a] -> (hasLink) -> [CONCEPT: *b]

We shall restrict the type of link among the children concepts of graph for several types of graphs, namely lists, trees, and networks. An alternative definition for a graph could use sets for the nodes and arcs. However, we would still need to capture the relationships that exist between the elements within the node set and elements within arcs which are elements within the arc sets. There are many types of graphs. In this paper we concentrate on only three of the more important types - lists, trees, and networks. We can also have bi-partite graphs, directed graphs, and directed bi-partite graphs to name a few.

2.1.6. List

A list corresponds to the sequence of original Conceptual Graph Theory. A list is a graph in which the members are ordered according to a next scheme. The definition for a list is:

> type [LIST: *x] is
> [GRAPH: *x] ->
> > (hasNode) -> [CONCEPT: *a]
> > (hasNode) -> [CONCEPT: *b]
> [CONCEPT: *a] -> (next) -> [CONCEPT: *b]

Access to a member is achieved by traversing the list either from the beginning or the end. Adding an element requires adding a next link between the new element and first element in the list. Appending an element requires adding a next link between the last element of the list and the new element.

Standard List

This is a basic list as given by the definition for a list. It is included here so that we will have a leaf node in the classification scheme for basic lists.

Sorted List

This is a list in which the members are sorted according to some criteria. The definition for a sorted list is:

        type [SORTED_LIST: *x] is
        [LIST: *x] ->
                (hasNode) -> [CONCEPT: *a]
                (hasNode) -> [CONCEPT: *b]
        [CONCEPT: *a] -> (comparisonRelation) -> [CONCEPT: *b]
        [CONCEPT: *a] -> (next) -> [CONCEPT: *b]

This definition basically states that a sorted list is a list in which any two adjacent nodes also stand in some comparison relation. The comparison relation will depend on the types of concepts in the list.

2.1.7. Tree

A tree is a graph in which members are ordered according to a parent-child scheme. A member may only exist in a child of relation with one member, but may stand in a parent of relation with multiple members. Children of the same parent member exist in a sibling relationship with each other. The definition for a tree is:

        type [TREE: *x] is
        [GRAPH: *x] ->
                (head) -> [CONCEPT: *y]
                (hasMember) -> [CONCEPT: *z]
                (hasMember) -> [CONCEPT: *z1]
                (hasMember) -> [CONCEPT: *z2]
        [CONCEPT: *z] -> (parentOf) -> [CONCEPT: *z1]
        [CONCEPT: *z] -> (parentOf) -> [CONCEPT: *z2]
        [CONCEPT: *z1] -> (siblingOf) -> [CONCEPT: *z2]

A tree has a member which stands in a head relation with the aggregation. This member does not have a child relationship to any other member.

2.1.8. Network

A network is a graph in which an element is connected to other elements in a general fashion by links. For all intents and purposes, a network is a graph. The definition for a network is:

        type [NETWORK: *x] is
        [GRAPH: *x]

Networks are used to represent things like the long distance network in telecommunications or connections among a group of people (e.g., a network of friends). A network does not allow an element that is not connected in some fashion to any other element except in the case where it contains only a single element.

2.2. Group

Groups are aggregations of entities in which the elements contained within it are unstructured. Groups enable us to deal with quantities of things in an unstructured fashion. The definition of a group is identical to the definition of an aggregation. The definition for a group is:

> type [GROUP: *x] is
> [AGGREGATION: *]

This is an intermediate abstraction to distinguish its subtypes from structures. It also provides the semantics of the operations over membership that are common to all of the subtypes. These operations provide the ability to split the group into sub-groups.

2.2.1. Set

A set is an aggregation of objects in which membership is restricted in terms of criteria that must be met for an individual to be a member. Sets are unstructured. Membership is limited by a predicate or by quantity of elements. Consistent with Sowa's use of set, a set can have a duplicate elements. The definition for a set is:

> type [SET: *x] is
> [GROUP: *]

This definition has no major constraints that distinguish it from it's parent. However, it does add some additional operations which provide a further distinction from its siblings.

Predicate Set

A set in which membership is defined by some predicate. The predicate can be simple or complex. The predicate is expressed as a set of relationships which any member must have. The definition for a predicate set is:

> type [PREDICATE_SET: *X] is
> [SET:*x]->(anyMember)->[CONCEPT:*]->(conceptualRelation)->[CONCEPT: *]

A predicate set in this framework will still have extensional members.

Type Set

A set in which membership is determined solely by type. The extension of a type is an example of a type set. This corresponds to the sets in the original CG theory. A type set can be represented by:

> type [TYPE_SET: *X] is
> [SET: *x] -> (anyMember) -> [MEMBER: *] -> (hasType) -> [TYPE: *]

This definition basically says that a type set is a set in which any member has a specified type. An example would be:

> [CATS: *] ->
> (hasMember) -> [CAT: Fuzzy]
> (hasMember) -> [CAT: Fluffy]

One can consider the conventional linear notation for an extensional set,

> [CAT: {Fuzzy, Fluffy}]

as a short-hand for the representation employed in this paper.

Potential Member Set

A potential member set is an aggregation of individuals that can possibly be a member, but in which membership is not assured. There are two types of members that must be tracked, potential members and actual members. A definition for a potential member set is:

> type [POTENTIALMEMBERSET: *x] is

[SET: *x] -> (hasPotentialMember) -> [CONCEPT: *]

This type of set is introduced as an intermediate abstraction to distinguish the subtypes from basic sets.

Probability   Set

A probability set is a set in which some restricted set of potential members can exist, but in which the possibility of an individual being member is specified in terms of a probability. A probability set can have only a single actual member.

      type [PROBABILITYSET: *x] is
[POTENTIALMEMBERSET: *x] ->
          (number Members)->[ONE: *]
          (numberPotentialMembers)->[NUMBER: *] ->(greater)->[ONE: *]
          (hasPotentialMember)->[CONCEPT: *] ->
               (hasProbability)->[PROBABILITY: *]

This definition basically states that a potential member has some probability associated with it. Certain operations are defined for manipulating the probability of membership. This kind of set can deal with situations such as the odds of a horse winning a race, a coin landing heads up, or the color of a ball pulled from a collection of differently colored balls.

ExclusiveOr   Set

A exclusive or set is one in which there are several potential members, but only one of which can actually be an actual member. The basic definition is:

      type [EXCLUSIVEOR: *x] is
[POTENTIALMEMBERSET: *x] ->
          (number Members)->[ONE: *]
          (numberPotentialMembers)->[NUMBER: *]->(greater)->[ONE: *]
          (hasPotentialMember)->[CONCEPT: *]

The restriction that the number of potential members be greater than one is required. If the number of potential members is identical to one, than that member must be the one actual member of the set. This type of set corresponds to a set with the set type specifier of DISJ in conventional linear notation. This type of set can capture the semantics of statements such as a ball must be either blue or red. Whenever a potential member is proven not to be the actual member, it is removed as a potential member and the number of potential members decrements by one. In the event that the number of potential members becomes exactly equal to one, then the actual member is deductively determined.

NonExclusiveOr   Set

A non-exclusive or set is a set in which there are several potential members, any one of which can actually be a member. The basic definition is:

      type [NONEXCLUSIVEOR: *x] is
[POTENTIALMEMBERSET: *x] ->
          (number Members)->[NUMBER: *]->(greater)->[ZERO: *]
          (numberPotentialMembers)->[NUMBER: *]->(greater)->[ONE: *]
          (hasPotentialMember) -> [CONCEPT: *]

This definition allows us to have a set in which there are a number of potential members, any of which is an actual member. This type of set handles the pizza topping case described earlier.

2.2.2. Bag

A bag is much like a shopping bag or a cardboard box, you can put all kinds of things in it, in any order, and you can remove things from it in any order. Individuals are explicitly tracked in a container. The definition of a bag is identical to the definition of a group with the exception that every member of a container is unique. The definition for a bag is:

    type [BAG: *x] is
    [GROUP: *x]

This definition appears identical to the definition for a set. However, the operations over a bag precludes the introduction of a member twice. This difference in operations gives a bag slightly different semantics than a set.

2.2.3. Collection

A collection is much like a flock of birds, there are many members, but the individual members are not explicitly identified until removed from the collection. Membership in the collection is tracked by count.

2.2.4. Substance

A substance is an aggregation in which the number of members is so large that quantity is measured indirectly rather than counted. The derivation of substance has been previously described [4].

3. Summary

We did not approach this paper with the goal of providing a comprehensive model of aggregations for use in conceptual graph theory. Any paper of this length would be sure to fail in meeting that goal. The topic of aggregations is extremely broad and subtle. It is not one that can be addressed in fourteen pages. My goal was to demonstrate just how complex and large is the topic of aggregations. Even though we have only addressed a handful of the more important aggregations We have, at least, succeeded in demonstrating just that. Any one of the types of aggregations that we have described in this paper could take an entire paper just to partially describe the operations and representational uses of it.

This paper has only briefly touched upon some of the operations that can be performed over aggregations. All aggregations have a common set of operations that can be performed on them. We add, remove, substitute and count members to name a few. We can also treat an aggregation as a union of secondary aggregations. Depending on the type of aggregation, these secondary aggregations can be termed subsystems, subsets, or any other fitting term. As a result, there will also be various operations associated with decomposition of aggregations and unions of aggregations to form new ones.

In many cases, we have stated that the definition of two types of aggregations are essentially the same. This is not completely accurate. There is an actual, although not quite apparent, difference between such sets. It lies in the operations that can be performed over them. This is similar to the concept of abstraction in object-oriented systems where two subclasses differ in how instances behave rather than in the attributes associated with them. As we have stated, this paper only scratches the surface of this topic. There is much that remains to be explored. It will remain a fertile area of research for many years.

References

1. Esch, J, et. al, "LINEAR -- Linear Notation Interface," Proceedings of the Third International Workshop on PEIRCE: A Conceptual Graphs Workbench, University of Maryland, College Park, MA, August, 1994.
2. Sowa, J., *Conceptual Structures Information Processing in Mind and Machine*, Addison Wesley, Reading Massachusetts, 1984.
3. Tepfenhart, W.M., "Using the Situation Data Model to Construct a Conceptual Basis," *Conceptual Structures: Current Practice and Research*, (eds.) T.E. Nagle, J.A. Nagle, L.L. Gerholz, and P.W. Eklund, Ellis Horwood, August 1989.
4. Tepfenhart, W.M., "Representing Knowledge About Substances," *Conceptual Structures: Theory and Application,* (eds.) H.D. Pfieffer and T.E. Nagle, Springer-Verlag, August 1992.
5. Tepfenhart, W.M., *et al*, "Situation Data Model," Proceedings of the Third Annual Workshop on Conceptual Graphs," August 1988.
6. Wermelinger, Michel, "Conceptual Structures Linear Notation: A Proposal for PEIRCE," Proceedings of the Fourth International Workshop on PEIRCE, August, 1995.

The Representation of Semantic Constraints in Conceptual Graph Systems

Guy W. Mineau
Department of Computer Science
Université Laval
Quebec City, Canada
G1K 7P4
tel.: (418) 656-5189
fax: (418) 656-2324
email: mineau@ift.ulaval.ca

Rokia Missaoui
Dept. of Computer Science
Université du Québec à Montréal
C.P.8888, Succ. Centre-Ville
Montreal, Canada, H3C 3P8
tel.: (514) 987-3000 (ext. 7939)
fax: (514) 987-8477
email: missaoui.rokia@uqam.ca

Abstract. The conceptual graph formalism is both simple and expressive. It offers great potential as a modeling formalism for developing information systems. In fact, its potential was recognized by the ANSI X3H4.6 committee, which recommended its adoption as a standard for such modeling tasks [1]. However, it lacks the modeling capabilities required to represent a wide range of semantic constraints, even though this is a vital characteristic of any useful modeling formalism. In this article, we propose a representation based on generalization hierarchies as defined [15], which allows most semantic constraints found in database literature to be: 1) represented in a unified framework, 2) enforced at all times, 3) subject to a minimum of resources, and 4) compared with one another in terms of their scope. Also, this paper shows that no cg system which allows the generalization of concepts, such as with maximal type expansion, is sound without the explicit representation of certain semantic constraints.

Keywords: semantic constraints, conceptual graph theory, semantics

1 Introduction

Semantic integrity constraints represent a rich source of semantic knowledge of an application and must be taken into account during the design of an information system. The subsequent acquisition or inference of information must then be coherent with these constraints. A non-coherent system could produce false answers, erroneous advice or faulty behaviors. Consequently, maintaining coherence throughout the acquisition, handling, updating and inference of information is of the outmost importance for the viability of any such system.

The underlying philosophy of the conceptual graph theory with regard to the management of semantic constraints is to filter out any graph which is known to be false because it violates some constraint. For instance, no restriction of a concept can be achieved without the conformity relation being verified between the referent of the concept and its new concept type. Also, the derivation of any graph has to comply with the selectional constraints imposed on the cg operators by the canonical basis [15]. Therefore, no cg known to be false is allowed to be derived. The cg operators are said to be *falsity-preserving*; they only allow the derivation of *plausible* graphs.

The conceptual graph theory already provides for certain semantic constraints to be expressed. These are mainly *typing* and *selectional* constraints. Generally speaking, the theory needs to be extended to cover a wider range of constraints, for instance, those found in database literature. Of course, we propose in this paper, to represent semantic constraints in a declarative formalism related to conceptual graphs. This will provide uniformity and standardization of the content of a cg system, will allow the constraints to be part of the inference process, and will eventually permit second-order knowledge on the domain to take them into account.

There are two types of constraints: those restraining the description of a state of a system, called *state constraints*, and those restraining the transitions between subsequent states of a dynamic system, called *transition constraints*. This paper presents two complementary techniques for representing state constraints under the conceptual graph formalism. The first one proposes the enforcement of semantic constraints on a *set of conceptual graphs* through the explicit identification of invalid graphs. These constraints are called *topological constraints*, since validation is done based on a graph matching procedure between a newly acquired or inferred graph, and designated invalid graphs. The second modeling technique addresses the problem of representing semantic constraints on *sets of individuals* represented by concepts in a conceptual graph. These constraints are called *domain constraints*. Topological constraints are useful for restraining the set of valid graphs in the description language. However, the instantiation of variables (concepts) by some constants also obey certain constraints which are not easily covered by topological constraints. Consequently, we propose to enforce domain constraints using a very simple mechanism: the actor triggering capability. The novelty of our approach resides in the way we represent topological constraints, in a unified framework which meets the objectives cited above.

In brief, since the conceptual graph formalism offers typing mechanisms to determine the individuals to which a concept can be instantiated to, we propose to use these mechanisms as much as possible to enforce domain constraints (Section 3). As for topological constraints, *non-validity intervals* will be used (Section 2.1). Non-validity intervals represent subsets (subspaces) of the generalization hierarchy H^* (built with all graphs derivable from the vocabulary that describes the application domain), where invalid graphs (according to the semantic constraints of the domain) are identified. Together, these non-validity intervals define a *validity space* (Section 2.2). A validity space is a compact representation of H^*, which encompasses the information about the validity and non-validity of the graphs that it represents. A validity space is thus useful for the validation of any conceptual graph in the system, and therefore, can be used to validate the application of the cg operators. As an immediate motivation for the explicit representation of constraints, the section titled Disjointness presents the *join-everything-to-everything* problem which makes any cg system where concept generalization is possible, unsound, unless some constraints are explicitly represented. Section 4 presents a brief literature survey on how semantic constraints were dealt with previously in the literature. Section 5 introduces interesting future developments based on the definition framework described in this paper.

2 The Representation of Topological Constraints

The cg operators are not allowed to deliberately produce false graphs. If we could beforehand represent graphs which are known to be false according to the semantic constraints of the application domain, a new cg, either inferred or acquired, could be validated against these graphs. The identification of invalid graphs is thus relevant to the enforcement of semantic constraints. To represent them, we propose to use non-validity intervals (Section 2.1). Since these graphs are all part of H*, the subspaces of H* identified by the non-validity intervals are part of a partial order of generality/specificity. From these intervals and the partial order relation among them, we can built H, a compact representation of H*, which encompasses the information about the validity and non-validity of all graphs in H* (Section 2.2)[1].

2.1 Non-Validity Intervals

A non-validity interval is expressed as a pair of conceptual graphs u and v, such that u is more general than or equal to v in H*, written v $\leq$ u, and such that both graphs are represented within brackets (either closed or opened). For instance, the interval [u,v] means that any graph g which falls in this interval is to be considered invalid, that is, g is invalid if g $\leq$ u and g is not more specific than v, otherwise g is plausible according to this interval. Here is such an example [u_1,v_1] with the following constraint I_1: *no employee can be assigned to a project and manage it at the same time*, where u_1 and v_1 are shown in Figures 1 and 2 respectively. As with this example, when v is the absurd graph[2], this type of interval prevents overspecializations to be represented. Overspecializations could not be avoided in cg systems so far; what we propose here is therefore useful for that purpose.

This example supposes that u_1 is not plausible, as well as any graph between u_1 and the absurd graph (i.e., any specialization of u_1). If u_1 were acceptable (plausible) as a minimal specialization of some other graphs, then interval I_1 would be represented as]u_1,v_1]. In that case, the expression validating g would be: g is invalid if g < u_1 and g is not more specific than v_1, otherwise g is plausible according to I_1. If g falls outside this interval, then we know that the operation producing g does not violate I_1. Of course, to be plausible, g has to fall outside all non-validity intervals.

On the other hand, if the closing bracket were opened, interval I_1 would be expressed as [u_1,v_1[, and the validity of g would depend on the following expression: g is invalid if g $\leq$ u_1 and g is not equal to and is not more specific than v_1. Figures 3 and 4 show graphs u_2 and v_2 respectively, where I_2 = [u_2,v_2[is the following constraint: *the agent relation links only physical entities to action verbs*. This restricts the parameters of the agent relation; such information is normally represented in the canonical basis of a cg system[3].

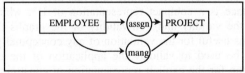

Fig.1. The first graph u_1 of interval I_1.

---

[1] Considering the size of H*, it is less resource-consuming to represent non-validity intervals than validity intervals. This reduces the amount of resources needed to represent H, a compact representation of H* (see section 2.2).

[2] The absurd graph is defined as the graph built with only one concept, the generic concept where the concept type is the absurd type. Since the absurd graph can not be joined to any other graph (by definition of the join operation), it is said to be maximally specific in terms of all the graphs in H*. Therefore, it stands as the bottom element of H*.

[3] As with this example, graphs u and v are bounded through the use of *binding variables*. With I_2, the concept bounded to [T:*x] in u_2 is bounded to [ACTION_VERB: *x] in v_2.

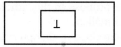

Fig.2. The absurd graph, the second graph v_1 of interval I_1.

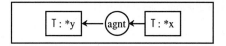

Fig.3. The first graph u_2 of interval I_2.

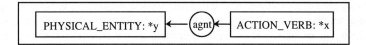

Fig.4. The second graph v_2 of interval I_2.

With such an interval, when all concepts in graph u are universal concepts attached to a single relation (as in Figure 3), and when graph v is the same as u except for the types of the concepts which are more specific than the universal type (as in Figure 4), then the interval represents a *selectional constraint* usually represented in the canonical basis B of the system. So, all the graphs in B can be represented using such intervals.

Furthermore, non-validity intervals allow the expression of selectional constraints not available with conceptual graphs. For example, suppose that the agent relation, which connects physical entities to action verbs (Figure 4), restricts the physical entities to be employees when the action verb is WORK. There is no way of representing this additional selectional constraint with the cg formalism. However, we could add interval I_3, whose u_3 and v_3 graphs are shown in Figures 5 and 6 below.

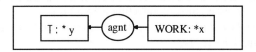

Fig.5. The first graph u_3 of interval I_3.

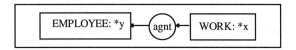

Fig.6. The second graph v_3 of interval I_3.

With non-validity intervals I_2 and I_3, any graph using the agent relation must comply with the selectional constraint of interval I_2. Any graph using the agent relation with the action verb WORK must also comply with the selectional constraint of interval I_3. Selectional constraints expressed by the canonical basis of a cg system constrain the use of relations only; topological constraints represented as non-validity intervals can constrain any syntactically valid combination of concepts and relations. Thus, the representation scheme that we propose in this article is obviously richer than what is currently available in the cg formalism.

Though we did not introduce it, the interval with fully opened brackets $]u,v[$ is also available. Together, the non-validity intervals allow the association of any part (subspace) of the H* hierarchy with graphs that should not be asserted. Consequently, non-validity intervals define validity and non-validity subspaces in H*. The next section introduces H, a compact version of H*.

2.2 Validity Spaces

Since by default we suppose that any graph g is plausible unless proven differently, we can summarize H* by representing only the non-validity intervals (and some validity intervals that exist between them, for connectivity purposes). Let us define H, a validity space, as follows. By definition, we know that $u_i \geq v_i \; \forall i \in [1,n]$, where n is the total number of constraints and I_i is a non-validity interval defined either as $[u_i,v_i]$, $]u_i,v_i]$, $[u_i,v_i[$, or $]u_i,v_i[$. H is a graph defined as $<V,A>$, where V is a set of labeled nodes and A is a set of labeled directed arcs between some of these nodes (taken pairwise). Let $[T]$ and $[\bot]$ be the universal and absurd graphs. Each element of V represents a single conceptual graph such that $V = \{[T], [\bot]\} \cup \{u_i\} \cup \{v_i\} \; \forall i \in [1,n]$. Furthermore, each node that represents an invalid graph is labeled with the - sign, while the other nodes (representing plausible graphs) are labeled with the + sign. The universal graph is labeled with the + sign unless specified differently in some non-validity interval, while the absurd graph is always labeled with the - sign. For all other nodes, if it is associated with at least one closed bracket, it is labeled with a - sign, otherwise it is labeled with a + sign.

Each arc of A connects two nodes (two graphs) together in the following manner: an arc $e_{i,j}$ connects graph n_i to graph n_j if $n_i > n_j \; \forall i,j \in [1,n]$ and $i \neq j$[4]. An arc $e_{i,j}$ represents all graphs in $G_{i,j}$, with $G_{i,j} = \{g \mid g$ is a conceptual graph and $n_i > g > n_j\}$. Arc $e_{i,j}$ is labeled with the + sign if the graphs in $G_{i,j}$ are all plausible, otherwise, it is labeled with the - sign[5]. The algorithm that sets the labels on the arcs is given in Figure 7. Figure 8 shows the validity space H_1 built from our previous example.

For every arc $e_{i,j}$ in A (connecting n_i to n_j) **do:**
    if \$k Œ $[1,n] \mid n_i = u_k$ and $u_k \neq v_k$ (with interval $I_k = <u_k,v_k>$)
      then label($e_{i,j}$) <- -
      else   label($e_{i,j}$) <- +

Fig.7. The algorithm that sets the labels on the arcs of H.

As additional examples, Figure 9 shows a validity space H_2 for a domain where absolutely no constraints are imposed on the derivation of graphs; and Figure 10 shows a validity space H_3 for a domain where only the selectional constraint on the agent relation (constraint I_2) must be enforced.

---

[4]   In this paper, we suppose that all graphs in V are logically different, and that the < operator compares graphs on a logical level, taking into account contractions and expansions.

[5]   A non-validity interval is represented by three arcs connected as a triangle, as shown for I_2 by the bold lines in Figure 8.

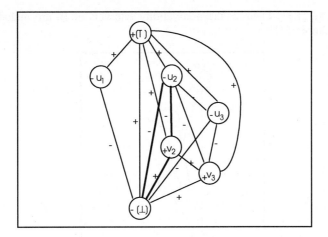

Fig.8. The validity space H_1 produced from our small example.

As shown by the example of Figure 10, this type of representation is compact. In effect, H_3 stratifies H* in three validity subspaces. Subspace S_1, identified by arc $e_{3,4}$, represents all the graphs in H* which are sufficiently specialized to comply with the signature of the agent relation. Subspace S_2, identified by the arcs $e_{2,3}$ and $e_{2,4}$, represents the graphs which do not use the agent relation properly: these graphs shall never be asserted. The third subspace S_3, identified by the arcs $e_{1,2}$, $e_{1,3}$ and $e_{1,4}$, represents all other graphs, which are plausible since no other constraints are imposed upon the application domain. It is obvious that H_3 is much simpler to represent than H*, the generalization hierarchy of all possible graphs.

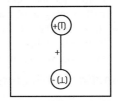

Fig.9. A validity space H_2 where no constraints are modeled.

In order to validate a cg operation resulting in the acquisition or inference of a new graph g, the node in H where g should appear is located. To achieve this, algorithms which search for immediate predecessors of a graph in such a classification structure must be used. [8,9,10] propose particular encodings for classification structures such as H, along with efficient search procedures. However, since the arcs and nodes of H are labeled and since variable bindings must be represented (between concepts of u_i and v_i), these procedures must be adapted. When searching for g through H, the arcs and nodes where g could be located are identified. To be plausible, no - sign must be associated with any of these arcs and nodes.

Finally, one can notice that the validity spaces of Figures 8, 9 and 10 are part of a partial order of inclusion. As a matter of fact, H3 ⊂ H1, H2 ⊂ H1, and H2 ⊂ H3.

We say that $H_j \supseteq H_i$ when all the non-validity intervals of H_i are included in (are covered by) those of H_j.

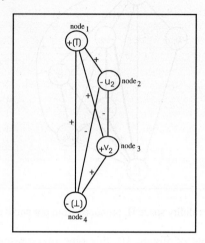

Fig.10. A validity space H_3 where only constraint I_2 is represented.

In summary, the advantages of having such a representation for semantic constraints are: 1) to benefit from having a *unified* representation for *all* types of constraints expressible as non-validity intervals, 2) to have a compact representation which *minimizes* the resources needed to store, use and update semantic constraints, 3) to realize full *integration* of the different constraints expressible as non-validity intervals, resulting in a simplification and validation of the original set of constraints (by the elimination of redundancy and inconsistencies), 4) to obtain a flexible structure where updates are done in this integrated representation instead of being done over many different intervals (if stored individually), and 5) to define a partial order of inclusion over sets of constraints, which is useful for relating domains (sometimes called micro-theories or contexts) to one another.

3 The Representation of Domain Constraints

When modeling an application domain, different constraints on *sets* of individuals need to be represented: constraints on the individuals to which a concept can be instantiated to (called *domain restrictions*), and constraints on relationships between these sets: intersection, union, etc. (called *set constraints*), and subsumption, called *subsumption constraints*. These constraints are popular in database literature. The following subsections introduce each one of them in light of what can be done under the cg formalism in order to represent and enforce them. Though simple, the solutions to the problems presented in the following (as for the previous section) had to provide *easy* and *fast* mechanisms that would insure that a cg knowledge base would be consistent at all times.

3.1 Domain Restrictions

Under domain restrictions, we find two types of constraints: pure *restrictions*, which determine the set of values that a concept can be instantiated to, and *referential integrity*, which impose that certain concepts be individualized at all times.

Restrictions

This semantic constraint aims at restricting the set of individuals represented by a concept, to some *restricted* domain. The first mechanism used in cg theory to define a domain is the *type definition*. Type labeling provides a way of defining a precise domain, since the referent of a concept must always conform to its type label. For instance, if we define the age of an employee as being an integer number between 18 and 70, the following type definition could be used (see Figure 11).

Any individual of type EMPLOYEE-AGE will be described using a conceptual graph that will have the cg of Figure 11 as a generalization. Consequently, this cg is a true consequence of any graph describing an individual of type EMPLOYEE-AGE. We suppose that this verification is done at acquisition time.

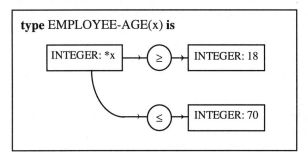

Fig.11. The definition of type EMPLOYEE-AGE.

However, using the canonical formation operators, it may be possible to derive the graph of Figure 11 without creating a new individual of type EMPLOYEE-AGE. In that case, the resulting graph could be false. For example, the graph of Figure 12 is a false graph; it may have been produced by successive join or restrict operations. The graph is false because it clearly appears that the semantics of the ≥ relation is violated. Nevertheless, the system has no way of knowing that it is a false graph, and that it should not be included in the knowledge base.

In order to automatically verify whether or not it is a false graph, we propose to use actors which implement procedures to compute the ≤ and ≥ relations. The type definition of Figure 11 and the graph of Figure 12 would then be expressed as shown in Figures 13 and 14. Although the type definitions of Figures 11 and 13 are equivalent, since by definition no individual will violate its type definition, the graph of Figure 14 can be detected as being false, contrarily to the graph of Figure 12.

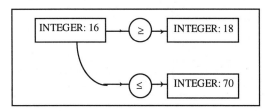

Fig.12. An unacceptable (false) graph produced through restrict and join operations.

In effect, this paper defines actors as procedural attachments which are automatically triggered when their input concepts are all instantiated to individual values. Each actor has a definition which identifies the type of their input and output concepts, called *prototype*. Figures 15 gives the prototype of the ≥ actor. At all times, the prototype of an actor should have a projection into any graph where the actor is used. In other words, we should be able to join the prototype of an actor to any graph where it is used. To join the graph of Figure 15 to the graph of Figure 14, we have to restrict the input concepts of the graph of Figure 15 to [INTEGER: 16] and [INTEGER:18] respectively. This automatically produces (triggers) the instantiation of the output concept which becomes: [BOOLEAN: false]. Since the output concepts of actors are functionally dependent on their input concepts (by definition), the join operation defined in [12] for such a case, would detect that the restricted actors of Figures 14 and 15 are not joinable since the output concepts which need to be joined to one another are not joinable: one is [BOOLEAN: true] (in Figure 14) while the other is [BOOLEAN: false] (from the join operation). Consequently, we know that the graph of Figure 14 violates the semantics of the domain and is a false graph. Any operation which would produce the graph of Figure 14 would be blocked because the domain constraint expressed by the use of actors would be violated if it were to proceed.

Since it may be tedious to specifically determine subdomains using actors, a shorter notation such as [INTEGER: *x ∈ [18,70]] would be useful. Naturally, this shorter notation, which introduces *range expressions*, stands only as a contraction for the more explicit version shown in Figure 13, giving the equivalent type definition of Figure 16. The elaboration of range expressions in the cg theory falls outside the scope of this paper and will be discussed somewhere else.

Referential Integrity

When a generic concept is not instantiated, it represents some information whose value, though typed, is not determined yet. In semantic modeling, this value is said to be null. There are cases where null values are not acceptable. In that case, the corresponding concept must be instantiated (individualized) at all times. For example, let us say that each employee is assigned a unique social insurance number (SIN). So, given an employee, we can find his/her unique SIN; and each SIN determines a single employee. Using functions, Figure 17 shows the corresponding type definition.

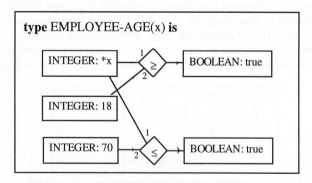

Fig.13. The definition of type EMPLOYEE-AGE using actors.

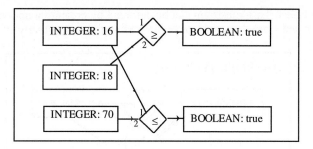

Fig.14. The graph of Figure 12 rewritten using actors.

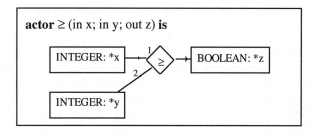

Fig.15. The definition of the prototype of the ≥ actor.

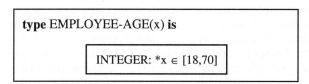

Fig.16. The definition of type EMPLOYEE-AGE using a range expression.

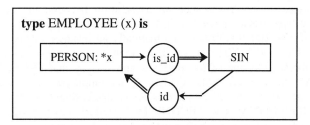

Fig.17. A unique SIN for each employee.

However, with the definition graph of Figure 17, we could acquire employees without determining their SIN. In order to avoid null values for the SIN concept, we propose to use the type definition of Figure 18, where the #? symbol means that this concept must be individualized when acquiring a new employee. For any concept t, we then have the following definition: [t: ∀] > [t:*x] > [t: #?] > [t: #i], where > means *is more general than*, and where i is any element of I (the individuals of the canon) that conforms to type t. This constraint can be enforced by creating a non-validity interval [u,v[where u and v are the graphs of Figures 17 and 18 respectively.

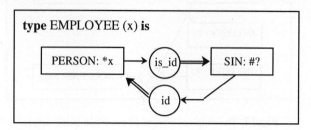

Fig.18. Excluding null values for the SIN concept.

3.2 Set Constraints[6]

In section 3.1, we have presented a way to precisely define the individuals to which a concept can be instantiated to. Now, we need to model the relations that hold between such sets of individuals. For instance, we may want to represent the fact that two different sets of individuals are disjoint, that some set of individuals strictly includes the union of other sets, etc. Most set related relations can be represented using actors. For instance, the $\subseteq$ actor could be defined just as was the $\leq$ actor, letting a procedure enforce that constraint expressed as a subgraph in the appropriate type definitions.

Two types of set constraints apply to specialization [6], namely disjointness and coverage (completeness) constraints. *Disjointness constraints* assert that some types in a generalization hierarchy are disjoint. For instance, we may want to represent the fact that employees are either technicians or professionals, but not both. *Coverage constraints* are set related; they define the inclusion relation between sets of individuals related to a parent-child pair in a type hierarchy. Coverage is either total or partial. *Total coverage* states that every individual associated to a type which is described in terms of subtypes, is mandatory associated to one of these subtypes. *Partial coverage* do not force such an association between the individual and a subtype of the type it is associated with. For example, we may want to know whether an instance of type EMPLOYEE is automatically an instance of one of its subtypes: PROFESSIONAL and TECHNICIAN. If so, then no employee can exist without this explicit membership to one of its subtypes, and we say that PROFESSIONAL and TECHNICIAN *fully* cover EMPLOYEE. If an employee can exist without the information needed to classify him/her as either professional or technician, then we say that PROFESSIONAL and TECHNICIAN *partially* cover EMPLOYEE. Partial coverage is useful for systems which must operate with incomplete knowledge.

Disjointness

In cg systems, the set of individuals represented by a type t is called its *denotation set*, and is written δt. Under the single classification hypothesis (see footnote 7), the

---

[6] For the rest of this paper, we suppose that any individual i in I, the set of individuals of the canon, is associated with only one concept type t. Of course, i conforms to t and to all generalizations of t, but to no specializations of t. This hypothesis is useful in real implementations of cg systems; it simplifies the maintenance of consistency and coherence of the cg knowledge base; it is called the *single classification* hypothesis.

intersection of two denotation sets δt_1 (of type t_1) and δt_2 (of type t_2), obeys the following rule if the two types are said to be disjoint: $\delta t_1 \cap \delta t_2 = \varnothing$. This is theoretically represented in a cg system, when there is no t_3 (other than the absurd type) which is a common subtype of t_1 and t_2. This is how *disjointness* between types is represented. With our previous example, we would have to make sure that the two types PROFESSIONAL and TECHNICIAN, both subtypes of EMPLOYEE, do not have a common subtype other than the absurd concept type. If so, no employee is said to be both a technician and a professional (under the single classification hypothesis[7]).

However, if concept generalization is allowed, as with maximal type expansion, we absolutely need an explicit constraint in order to ensure that this restriction on sets of individuals is enforced. For example, let us suppose that we wish to join two graphs g_1 and g_2 on concepts $[t_1]$ from g_1 and $[t_2]$ from g_2. We know that these two graphs are not joinable on these concepts because no common subtype t_3 exists such that both concepts could be restricted to t_3. However, if maximal type expansion is available as a cg operator [17], then both concepts could be expanded until their type becomes t_0, where t_0 is the common supertype of t_1 and t_2 in the type hierarchy. Graphs g_1 and g_2 could then be joined on their expanded version (over t_0), giving graph g_3, even though they should not have been joined according to the type hierarchy. This problem results from the fact that the disjointness of types is expressed only on type labels (by the type hierarchy), and not on their definition as well. Maximal type expansion allows a definition of some type to be included in a graph where there is a concept of that type. As a result, the type label of this concept is changed to its supertype. Since all types are elements of a type lattice having the universal concept type as top element, there always exists a common supertype for any pair of types. Consequently, this expansion dilutes the semantics associated with a type label by introducing its definition into the graph. The disjointness constraint should then be represented on the type definitions per se, and not just on the type labels.

So, we propose to use a non-validity interval in order to block a join operation which should not take place according to the type hierarchy. In fact, we will add the non-validity interval: [u,v[to H, where u = g, a graph obtained from joining g_1 and g_2 both expanded up to the point where the type of the expanded concepts $[t_1]$ and $[t_2]$ is the universal concept type, and where v is the absurd graph. That way, the result of the join operation applied on maximally expanded graphs will fall within a non-validity interval, and will be blocked. H should include such a constraint for any pair of concept types whose common subtype is the absurd concept type. These constraints can be produced automatically by scanning the type hierarchy. This would resolve what we call the *join-everything-to-everything* problem, also identified in the Quaker-Republican example of [8].

Partial Coverage

When $\delta t_1 \cap \delta t_2 = \delta t_1$ then t_1 is a subtype of t_2. Consequently, the denotation set of a type is at least the union of the denotation sets of all its subtypes. In a cg system, subtyping is used to represent *coverage*. In the cg notation, partial coverage is always assumed. This is useful for systems where incomplete descriptions need to be handled.

---

[7] Historically, in the cg theory, this hypothesis lead to the definition of a type *lattice* instead of a type hierarchy, in order to represent the domain vocabulary.

With our example, we could acquire new employees without classifying them as either professional or technician.

Full Coverage

Full coverage of type t by its subtypes means that no individuals of type t can be acquired directly. With our example, we need to forbid the expression of any graph where the definition graph u, of type EMPLOYEE, would be found without the definition graph of PROFESSIONAL or TECHNICIAN. Let us suppose that v is the definition graph of PROFESSIONAL and w is the definition graph of TECHNICIAN. Let us suppose also that v' (and w') is the result of joining u to v (or to w) over their genus concepts. Full coverage of type EMPLOYEE by its two subtypes PROFESSIONAL and TECHNICIAN would be enforced using the following two constraints: [u, v'[and [u, w'[.

3.3 Subsumption Constraints

In a cg system, subsumption among types appears as a type hierarchy. For example, if type EMPLOYEE subsumes both types TECHNICIAN and PROFESSIONAL as mentioned earlier, then it is automatically implied that δTECHNICIAN ⊆ δEMPLOYEE and that δPROFESSIONAL ⊆ δEMPLOYEE. This means that all technicians and all professionals are employees. It represents a full subsumption relation between EMPLOYEE and each of these types.

If, however, we have an ACTIVE-EMPLOYEE type which represents the employees actually assigned to projects, we then have a *full* subsumption relation between EMPLOYEE and ACTIVE-EMPLOYEE (all active employees being employees), but we have a *partial* subsumption relation between ACTIVE-EMPLOYEE and TECHNICIAN, and between ACTIVE-EMPLOYEE and PROFESSIONAL (some technicians and some professionals are active-employees, but not all of them). We can not express these partial subsumption relations as such in a type hierarchy unless we introduce additional types. We then need to introduce subtypes for technicians and professionals where this distinction is represented. We would introduce an ACTIVE-TECHNICIAN (and ACTIVE-PROFESSIONAL) subtype which would be linked to both ACTIVE-EMPLOYEE and TECHNICIAN (or PROFESSIONAL) types.

4 Semantic Constraints in the Conceptual Graph Literature

In [16], dataflow graphs are used as a complement to conceptual graphs to express functional dependencies. Not only actors represent functional dependencies but they can enforce them by computing the actual output values associated with their input concepts. Though useful for representing certain dependencies constraints, actors are not sufficient to express functional dependencies between graphs. In this paper, we propose to use topological constraints to represent such dependencies[8].

In general, few semantic models allow the representation of a large class of semantic constraints. NIAM [13] is one of the semantically rich conceptual models since it allows the expression of a relatively important set of constraints such as disjointness, subsumption and key constraints. In order to represent a larger set of

---

[8]   Such dataflow graphs are also used as *overlays* in [16] to impose certain constraints on the instantiation (and existence) of concepts and relations. Such overlays are useful mainly to represent transition constraints, which are not covered by the work presented in this paper.

integrity constraints and derivation rules, the NIAM model has been extended in [2,3] using the conceptual graph theory. [4] argues that conceptual graphs offer modeling features that make them useful as a canonical data model, and hence let them serve to schema description, integration, and transformation from a source model (e.g., entity-relationship model) to a target one (e.g., NIAM). We totally agree with that statement. This paper, by introducing a wider range of constraints into the conceptual graph formalism, aims at providing additional tools to reinforce that claim.

Finally, the approach found in [5,12] is the most relevant work about the representation of semantic constraints using conceptual graphs. The authors represent semantic constraints on type definitions by embedding additional graphs to a type definition, which must be satisfied at any time by the individuals of this type. This links the validity of the knowledge base to the creation of new individuals. However, the derivation of sufficient conditions of some type definition by other means than the insertion of new individuals, may result in some violation of some constraints related to the individuals of that type. We believe that semantic constraints should be expressed in such a way as to prevent the representation or *derivation* of any faulty graph.

Furthermore, as this article shows, there are many different types of semantic constraints which need to be represented. The constraints represented [5] only specify constraints on concepts. There may be dependencies that may need to be represented between sets of relations as well. In contrast, this paper proposes a unique paradigm for representing all sorts of constraints.

5 Conclusion and Future Developments

In this article, we showed how the conceptual graph theory can be extended to include the representation of a wide variety of semantic constraints, in fact, most constraints found in database literature. Based on the following three goals: a) constraints should filter out impossibilities, b) constraints should be used to define and relate micro-theories to one another, and c) constraints should therefore be inherited from less to more constrained micro-theories, we came up with the representational framework presented in this paper, which offers the following advantages: 1) a *unified* representation for *all* types of constraints expressible as non-validity intervals, 2) a compact representation which *minimizes* the resources needed to store, use and update semantic constraints, 3) full *integration* of the different constraints expressible as non-validity intervals, resulting in a simplification and validation of the original set of constraints (by the elimination of redundancy and inconsistencies), 4) the continuous enforcement of both domain and topological constraints, and 7) a partial order of inclusion defined over sets of constraints, which is useful for relating micro-theories.

Our main efforts will now be devoted to the design of efficient algorithms for the creation and update of validity spaces, as well as for the search of immediate predecessors. As said before, a lot of relevant approaches such as those found in [8,9,10] will be examined. The efficient comparison of different validity spaces in order to detect partial or full inclusion will also be on our agenda.

Considering the advantages of our approach and the fulfillment of our goals as cited above, we hope that the proposed framework will prove to be useful on the field, as application development using conceptual graphs desperately needs a formal representation of semantic constraints.

6 References

1. ANSI, Committee X3H4.6. *IRDS Conceptual Schema.* X3H4.6/92-091, 1992.
2. Campbell, L. & Creasy, P. A Conceptual Approach to Information Systems Design, *Proc. of the 7th Annual Workshop on Conceptual Structures: Theory and Implementation*, Las Cruces, NM, July 1992, 46-55, 1992
3. Creasy, P. ENIAM: A more Complete Conceptual Schema Language. In: *Proceedings 15th International Conference on Very Large Data Bases*, Amsterdam, August, 1989.
4. Creasy, P. Conceptual Graphs as Canonical Data Model. In: *Supplementary Proceedings of the 2nd Int. Conf. on Conceptual Structures (ICCS-94)*, Maryland, August 94, 70-77, 1994.
5. Creasy, P. & Moulin, B. Approaches to Data Conceptual Modeling. In *Proc. of the 6th Annual Workshop on Conceptual Structures*, E. Way (Ed.), Binghamton, New York: State University of New York at Binghamton, pp. 387-399, 1991.
6. Elmasri, R. & Navathe, S.B. *Fundamentals of Database Systems.* 2nd edition. Benjamin Cummings, 1994.
7. Esch, J.W. Contexts as White Box Concepts. In: *Supplementary Proceedings of the 1st International Conference on Conceptual Structures (ICCS-93).* G.W. Mineau, B. Moulin & J.F. Sowa (Eds.). Dept. of Computer Science, Université Laval, Quebec City, Canada. 17-29, 1993.
8. Levinson, R. Pattern Associativity and the Retrieval of Semantic Networks. *Computers & Mathematics with Applications*, **23**(6-9), 573-600, 1992.
9. Levinson, R. A. UDS: A Universal Data Structure. In *Proceedings of the Second International Conference on Conceptual Structures,*, College Park, Maryland USA: pp. 230-250, 1994.
10. Levinson, R. A. & Ellis, G. Multi-level hierarchical retrieval. *Knowledge Based Systems*, **5**(3), 233-244, 1992.
11. Mineau, G.W. View, Mappings and Functions: Essential Definitions to the Conceptual Graph Theory. In: W. M. Tepfenhart, J. P. Dick, & J. F. Sowa (Eds.), *Conceptual Structures: Current Practices,* (pp. 160-174). Springer-Verlag, 1994.
12. Moulin, B. & Creasy, P. Extending the Conceptual Graph Approach for Data Conceptual Modeling, *Data & Knowledge Engineering*, **8**(3), 223-248, 1992.
13. Nijssen, G.M. & Halpin, T.A., Conceptual Schema and Relational Database Design: A Fact-Based Approach, Prentice-Hall, Englewood Cliffs, NJ, 1989.
14. Pfeiffer, H.D. & Hartley, R.T. Temporal, spatial, and constraint handling in the Conceptual Programming environment, CP. *Journal of Experimental & Theoretical Artificial Intelligence*; **4**(2). 167-183, 1992.
15. Sowa, J. F. *Conceptual Structures: Information Processing in Mind and Machine.* Addison-Wesley, 1984.
16. Sowa, J. F. Knowledge Representation in Databases, Expert Systems and Natural Language, *Proceedings IFIP Working Conf. On the Role of Artificial Intelligence in Databases and Information Systems*, Guangzhou, China, July, 1988.

Representation of Defaults and Exceptions in Conceptual Graphs Formalism

Catherine Faron and Jean-Gabriel Ganascia

LAFORIA-LIP6, University of Paris 6, BP 169
4, place Jussieu - 75252 Paris cedex 05 - France
e-mail: {faron,ganascia}@laforia.ibp.fr

Abstract. A key issue in knowledge representation is the representation of hierarchies of concepts. Conceptual Graphs is particularly well suited to this problem, specially when structured descriptions are required for concepts. Descriptions of concepts are represented by conceptual graphs which are used to define types representing the concepts. Hierarchies of concepts are then represented by Aristotelian hierarchies defined on these types. However, Conceptual Graphs formalism becomes inadequate for the representation of concept hierarchies when exceptions occur in property inheritance between concepts. Other logics formalisms also deal with concept hierarchies while offering ways to account for exceptions but they are less intuitive and readable than conceptual graphs. Moreover, conceptual graphs are often used in knowledge representation. Therefore, we propose to extend Conceptual Graphs formalism in order to allow the representation of default taxonomic knowledge. We first show that the classical "default inheritance" approach appears to be inadequate for Conceptual Graphs formalism. We then adopt a "definitional" approach similar to the one pioneered by Coupey and Fouqueré in $\mathcal{ALN}$ terminological language where defaults and exceptions are seen as parts of the descriptions of concepts.

1 Introduction

A key issue in knowledge representation is the representation of hierarchies of concepts. Hierarchical structures have proven to be useful for organizing and managing knowledge bases because of their searching efficiency [9]. Moreover hierarchical organization is quite natural for several domains such as organism biology, where objects are clustered into classes and sub-classes hierarchically organized into taxonomies. Conceptual Graphs formalism is particularly well suited to represent concept hierarchies, specially when structured descriptions are required for concepts. It is the case of taxonomies in organism biology [13][1], in Chinese characters databases [4] and more generally in every domain whose concepts require structured descriptions that cannot be reduced to simple attribute-value vectors [15]. These descriptions are represented by conceptual graphs which are used to define types representing concepts. Concept hierarchies are then represented by generalization hierarchies of types, according to the notion of Aristotelian hierarchy defined by Sowa [19]. However, Conceptual Graphs formalism becomes inadequate for the representation of concept hierarchies when exceptions occur and, as a matter of fact, exceptions are many in most natural taxonomies. For instance, in the orchidology field, orchids belonging to the *Cephalanthera* kind have pink or white flowers, except those belonging to the *Cephalanthera Pallens* species whose flowers are white and orange-yellow. Then, *by default*, orchids belonging to the *Cephalanthera* kind are said to have pink and white flowers, provided that exceptions may occur. Representing such default inheritance relations between the concepts of a hierarchy requires default knowledge, that is a way to represent default properties admitting exceptions. The problem addressed in this paper is to extend concept definitions in Conceptual Graphs formalism with defaults and exceptions.

Other logics formalisms also deal with concept hierarchies while offering ways to account for exceptions. The classical approaches adopted consist in considering concept hierarchies as inheritance hierarchies whose concepts inherit the properties of their subsumers. Exceptions are then viewed as "exceptions of inheritance": a concept inherits the properties of its subsumers if there is no exception to block the inheritance mechanism. This is called *default inheritance*. However, as we shall see in the following, this approach becomes inadequate when dealing with defaults and exceptions using Aristotelian hierarchies in Conceptual Graphs formalism. Briefly speaking, inheritance mechanism in an Aristotelian hierarchy is based on the generalization relations between graphs defining types and on the type expansion mechanism. Thus type definition prevails over property inheritance: rather than default-inheritance of properties, it is inheritance of default-properties which needs to be defined so as to represent default knowledge using conceptual graphs. In other words, in Conceptual Graphs formalism, exceptions have to be treated in a so-called "definitional" point of view, that is embedded in the definitions of types. Such a definitional approach, where defaults and exceptions can be part of a concept definition, was initially proposed by Coupey and Fouqueré in $\mathcal{ALN}$ terminological language [7]. We show in this paper that a similar outlook can govern the representation of concept hierarchies with exceptions in Conceptual Graphs formalism.

In section 2, we briefly review the classical approaches in logics formalisms, highlighting the default inheritance point of view they share. In section 3, we show its inadequacy to manage default knowledge in Conceptual Graphs formalism. In section 4, we describe the definitional approach of default knowledge, originally adopted by Coupey and Fouqueré in $\mathcal{ALN}$ terminological language. In section 5, we extend this definitional approach to Conceptual Graphs formalism and we propose an extension of type definition encompassing defaults and exceptions.

2 Defaults in classical representations of concept hierarchies

In most representation formalisms, knowledge representation is classically dealt with in an inferential point of view, paying attention to the deduction abilities of knowledge bases. One way to describe the set of elementary deductions between concepts is to structure the knowledge base by hierarchically organizing concepts provided with a property inheritance mechanism. That is why such hierarchies are also called *inheritance hierarchies* [10]. In such a view, exceptions that occur in a concept hierarchy are seen as exceptions to property inheritance and are thus ruled by the inheritance mechanism associated to this concept hierarchy. It is a *default inheritance* mechanism, in the sense that exceptions can block property inheritance between concepts.

2.1 Default inheritance in logic and semantic network representations

Default knowledge handling in a default inheritance point of view comes from default logic and semantic network representations. Reiter has pointed out the inadequacy of first order logic for *default reasoning* modelization [18]. As a solution, he proposes a *default logic*, an enlargement of first order logic with new inference rules called *defaults*. These will allow the inference of facts by default i.e. in the absence of contradictory facts. A strict inheritance hierarchy is represented by a *theory*, that is a set of logical implications between propositions representing domain concepts. A default inheritance hierarchy is then represented by a *default theory* that is a set of logical implications and defaults.

With semantic network, a similar approach suits quite well. Concept hierarchies are represented by semantic networks whose nodes themselves represent concepts and whose arcs are IS_A links representing subsumption between concepts [11]. Concept nodes and IS_A links correspond respectively to propositions and implications in first order logic. To represent default inheritance, Falhman et al. have introduced CANCEL links: a CANCEL link blocks the inheritance of a property between a concept node and one of its subsumers in the hierarchy[12]. It corresponds to a default in Reiter's logic. Correspondences between semantic networks and default theory representations are further formalized in [10] and [14].

2.2 Default inheritance in terminological logics

Terminological logics are a family of representation languages descended from KL-ONE [5]. Brachman's goal in KL-ONE was to design a knowledge representation system combining the best of both semantic networks and frames. Therefore terminological languages focus on both the distinction between definitional and assertional knowledge and the definition of concepts wrt their subsumption relations. Concepts are represented by *terms*; term definitions and their ordering relations form the terminological knowledge included in the TBox of the language while assertions relative to domain objects compose the factual knowledge in the ABox. In a terminological language, concept hierarchies are thus represented by the TBox.

In order to handle default knowledge, several attempts have been made to extend terminological logic systems by adding a form of limited default reasoning [2] [17] [16]. However, Baader and Hollunder [2] as well as Quantz and Royer

[17] are concerned with property inheritance between domain objects and concepts, but not between concepts. The key point is the ABox mechanism used to recognize instances of concepts. Thus exceptions are viewed only at the object level, that is defaults are introduced in incidental rules $C_1 \rightarrow C_2$ stipulating that any instance of C_1 is also an instance of C_2. Therefore instances of concept C_1 may not have all properties normally inherited by instances of C_1 when exceptionally they are not instances of concept C_2. In [2], incidental rules are replaced by defaults of Reiter's logic while in [17] the authors define their own incidental rules based on a preferential model theory.

Padgham and Zhang [16] are concerned with property inheritance in concept hierarchies. They propose an unifying approach of nonmonotonic inheritance in both semantic networks and terminological logics. Each defined concept is viewed as a bipartite description: its "core" describes a set of necessary and sufficient (strict) properties while its "default" describes all its (strict and default) properties which are the "typical" properties of its instances. Each concept is then represented by two terms in the TBox (or two nodes of the semantic network), a T_{core} term and a $T_{default}$ term that are classified, as usual, in the TBox. Here the key point is the representation of subsumption (strict and by default) between concepts by the interpretation of descriptive subsumption relations between the $T_{core}s$ and $T_{default}s$ defining these concepts.

One should note that all these works deal with default knowledge from an inheritance point of view. In the following section, we attempt to adopt this classical approach in order to represent default knowledge in Conceptual Graphs formalism.

3 On default inheritance in Conceptual Graphs formalism

Conceptual Graphs formalism [19], like terminological logic, is descended from semantic networks. In a way, Sowa and Brachman shared a common goal: to remediate the weaknesses of early semantic networks whose links were indifferently used to express definitions, assertions or properties and whose nodes made no clear distinction between individuals and classes [20]. Correspondances between both formalisms are presented in [3]. Conceptual Graphs formalism allows the representation of structured objects and the definition of concepts types and their ordering by generalization. We first detail the representation of concept hierarchies using type hierarchies and then make clear that default inheritance approach is inadequate to handle default knowledge in this formalism.

3.1 Representation of concept hierarchies

Throughout this paper, we will use as an example the taxonomy of contemporary animals (see figure 1). This concept hierarchy, arranged from [16], provides a good example of our representation problem despite the fact that concept definitions do not require the full expression power of Conceptual Graphs formalism. In this example, we represent concepts of contemporary animals by *concept types*

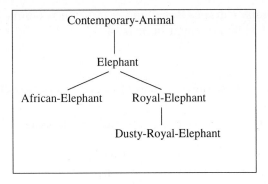

Fig. 1. The taxonomy of contemporary animals

we define using conceptual graphs. Let us note that the term *concept* is here used within two contexts: on one hand, a concept hierarchy designates a taxonomy while a concept designates a taxum; on the other hand, a concept is a pair <*type,referent*> in Conceptual Graphs formalism.

Regarding concept type definition, where a type is defined by its genus and its differentia [19], the genus of a type representing a concept is the type which represents its father in the taxonomy, e.g. the genus of type E that represents taxum Elephant is type CA, itself representing taxum Contemporary-Animal. Types differentiae are conceptual graphs expressing differences between taxa and their fathers in the taxonomy, e.g. the differentia of type E indicates that elephants are contemporary animals that bear a trunk and are grey colored; figure 2 presents the definition of this type. More generally, in a type definition, the differentia represents the differences between the concepts represented by the type being defined and its genus.

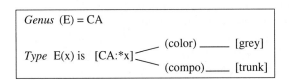

Fig. 2. Definition of type E representing taxum Elephant

Normally, the structure of a taxonomy is represented by the Aristotelian hierarchy that can be defined on types representing the taxa. It is based on the generalization relations between the defined types, themselves based on the generalization relations between their defining graphs. A graph is a generalization of another if the latter is derivable from the former by using canonical formation rules [19]. To compare graphs describing taxa, let us examine their maximally expanded forms [19] [6] presented in figure 3.

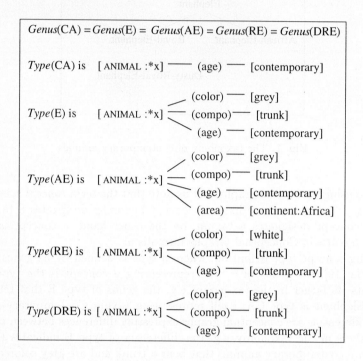

Fig. 3. Expanded graphs describing taxa of contemporary animals

Type E which represents taxum Elephant is a specialization of type CA which is itself the representation of taxum Contemporary-Animal: the defining graph of E is derivable from the defining graph of CA by join (see figure 4a). In the same fashion, we could show that type AE representing taxum African-Elephant is a specialization of type E, the defining graph of AE being derivable from the defining graph of E by join. Now, unlike type AE, type RE which represents taxum Royal-Elephant is not a specialization of type E, neither is type DRE which represents taxum Dusty-Royal-Elephant a specialization of type RE. Indeed, there is no canonical formation rule allowing to replace type GREY by type WHITE in the graph defining E: concept restriction only allows replacing a type by one of its sub-types and WHITE is not a sub-type of GREY (see figure 4b).

Clearly, the Aristotelian hierarchy of the concept types defined by the graphs described in figure 3 makes it impossible to represent properly the taxonomy of contemporary animals. The generalization relations between its types and the taxonomic structure do not coincide, the most specific generalizations of types RE and DRE being respectively types CA and E. As a conclusion, Conceptual

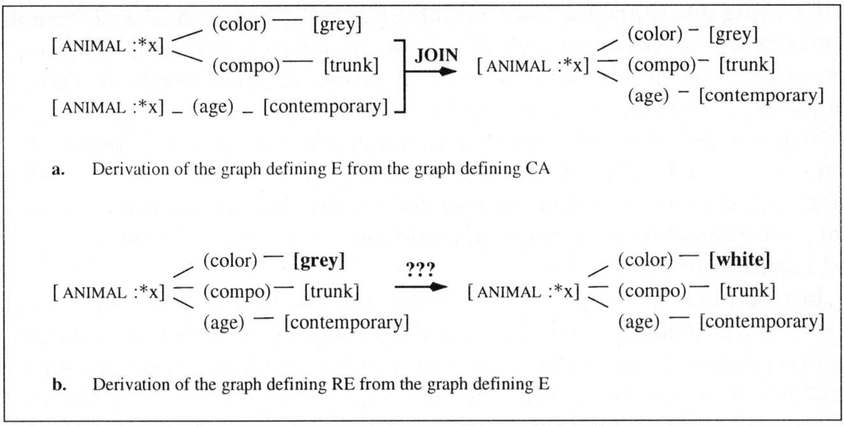

Fig. 4. Generalization relations between graphs of figure 3

Graphs formalism is no longer convenient for the representation of concept hierarchies when exceptions occur. Therefore, we shall attempt to extend it to default knowledge representation and, to do this, we will first try the classical inheritance approach.

3.2 Inadequacy of the default inheritance approach

As mentioned above, Conceptual Graphs formalism and terminological logics are close in the sense that distinctions are clear between concept and object levels and, at the concept level, between concept definition and property inheritance. In both formalisms, property inheritance is based on subsumption between concept definitions. Given these analogies, we could think of handling exceptions based on an inheritance approach.

Now let us consider generalization that represents subsumption in Conceptual Graphs formalism. Given the definitions of types RE and E, the relation of generalization RE ≤ E would signify that the graph defining RE is derivable from the one defining E by replacing in the latter type GREY by type WHITE, using a new canonical formation rule. But such a rule would show several drawbacks. First, contemporary animals bearing a trunk but not grey-colored, yellow butterflies for example, are not all elephants of unusual color, like Royal elephants! It happens that this new canonical formation rule would allow replacing GREY by YELLOW as well as by WHITE and so deriving type B representing taxum Butterfly as well as type RE from type E, thus establishing the relation of generalization B ≤ E! Second, given that this rule will allow replacing GREY by WHITE, it would symmetrically allow replacing WHITE by GREY, thus establishing the relation of generalization E ≤ RE by deriving E from RE. In other words, the antisymmetry of generalization would be lost, implying generalization would no longer be an order relation.

Knowing inheritance is based on subsumption, when we deal with exceptions to inheritance as in the example above, we are led to redefine subsumption that can no more be simply based on generalization. In other words, it leads us to Padgham and Zhang's approach described above, where default subsumption is defined in addition to strict subsumption and where a concept is defined by two terms, a core and a default. But would the results of this method be worthwhile when adapted to Conceptual Graphs formalism? We do not think so, arguing that Conceptual Graphs formalism would lose its best points, that is readability and intuitive understanding.

Instead of excepting property inheritance between concepts, that is blocking it, it is in type definitions i.e. in graphs describing concepts that we will represent the exception to the properties that concepts do not inherit. This is a definitional approach of exception handling. Section 5 presents more precisely the solution we propose. A similar approach has previously been adopted in terminological logic and is presented first.

4 A definitional approach of defaults in concept hierarchies

The so-called definitional approach when representing a concept hierarchy with defaults is descended from Coupey and Fouqueré's work [7]. They propose an extension of the ALN terminological language [7] they call $AL_{\delta\epsilon}$ where definitions of terms in the TBox include representations of default properties, viewed as full properties of concepts.

In ALN, the set of terms includes primitive and defined terms. Defined terms are built of primitive terms and primitive roles using the conjunction and negation connectives: $\neg$ and $\cap$. In $AL_{\delta\epsilon}$, term definition is extended by the introduction of two unary connectives: δ and ϵ. Their application to a term T expresses respectively "T by default" and "exception to T".

Terms are partially ordered by the descriptive subsumption relation:

$$T_1 \subset T_2 \text{ iff } T_1 \cap T_2 = T_1$$

Subsumption relations between T and δT and between T^ϵ and δT are then defined by the conjonctions of these terms:

$$T \cap \delta T = T \qquad T^\epsilon \cap \delta T = T^\epsilon$$

Intuitively, a chain of defaults means nothing else than a single default and an excepted property implicitly is a default property. Formally, δ and ϵ connectives verify the following properties:

$$\delta(\delta T) = \delta T \qquad (\delta T)^\epsilon = T^\epsilon$$

A default on a conjunction of properties is the conjunction of default properties i.e. δ connective is distributive wrt $\cap$ connective. It allows the application of defaults to defined terms as follows:

$$\text{if } T = T_1 \cap T_2 \text{ then } \delta T = \delta(T_1 \cap T_2) = \delta T_1 \cap \delta T_2$$

Inheritance mechanism stays unchanged, based on the descriptive subsumption itself unchanged. It simply is defined on concept definitions that may now include defaults and exceptions. In other words, the definitional approach requires a representation formalism offering a mechanism for concept definition such as can be found not only in terminological logic but also in Conceptual Graphs formalism. We thus attempt to adapt this work to handle default taxonomic knowledge in Conceptual Graphs formalism.

5 Extending Conceptual Graphs formalism

5.1 Extending canonical graph formation with default knowledge

Founding our proposal on $\mathcal{AL}_{\delta\epsilon}$, we provide conceptual graphs with two connectives whe named like these of $\mathcal{AL}_{\delta\epsilon}$; they share the same meanings. First, to allow distinguishing between exception-admitting default properties and strict ones admitting none, we introduce the unary connective δ on the hierarchy of types; the application of δ to a type t, δt, represents a default property. In our example (see figure 5), type GREY in the graph defining type E becomes δGREY while type WHITE in the graph defining type RE becomes δWHITE. Second, to prevent a butterfly from being recognized as an (unusually-colored) elephant, it is not enough to say that Royal elephants are white just as butterflies are yellow; we shall precise that Royal elephants are exceptionally not grey. To this purpose we introduce the unary connective ϵ on the type hierarchy; the application of ϵ to a type t, ϵt, represents an exception to a property. In our example, in the graph defining type RE, concept node [ANIMAL: *x] now have a COLOR relation not only with node [δWHITE] but also with node [$\delta\epsilon$GREY]; in graph DRE, node [ANIMAL: *x] have a COLOR relation with nodes [ϵWHITE] and [$\epsilon\epsilon$GREY].

Connectives δ and ϵ verify the following properties. Let t, t_1 and t_2 be three concept types then:

(1) $t \leq \delta t$	(4) $\epsilon(\delta t) = \epsilon t$
(2) $\epsilon t \leq \delta t$	(5) $\delta(\delta t) = \delta t$
(3) if $t_1 \leq t_2$ then $\delta t_1 \leq \delta t_2$	

Intuitively, "grey by default" designates grey objects or objects exceptionally not grey; formally, every type t verifies properties (1) and (2) which characterize the order relations between t and δt and between ϵt and δt, so that every concept node [δt:r] can be restricted by replacing its type δt by t or ϵt, sub-types of δt in the type hierarchy.

Property (3) indicates that two types representing defaults are comparable if the two types representing the corresponding strict properties themselves are.

Properties (4) and (5) correspond to the properties of simplification defined in $\mathcal{AL}_{\delta\epsilon}$. Property (4) is based on property (2) and on concept restriction: applying connective ϵ to type δt of a concept node [δt:r] is the same as restricting concept type δt by its sub-type ϵt.

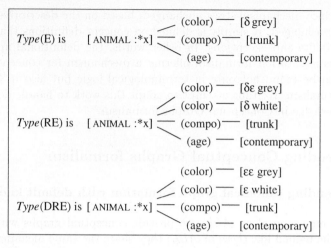

Fig. 5. New descriptions of the taxa of contemporary animals

5.2 Generalization between graphs embedding default knowledge

Let us first examine the generalization relations between the types representing concepts of the contemporary animal taxonomy when these types are defined by the graphs of figure 5. The generalization relations between types CA and E and between types E and AE are maintained, the graph defining E being derivable by join from the one defining CA and the graph defining AE being also derivable by join from the one defining E. Generalization relations between types E and RE and between types RE and DRE are established, the graphs defining RE and DRE being now derivable respectively from the graph defining E by the sequence of a concept restriction and a join (see figure 6a) and from the graph defining RE by the sequence of two concept restrictions (see figure 6b). The expected generalization relations between types CA and E, between types E and AE, etc. are instantly established wrt the generalization relations between their defining graphs. Thus the Aristotelian hierarchy defined on these types actually coincides with the taxonomy to be represented.

Let us now generalize the method we applied to represent the contemporary animal taxonomy. We give a constructive definition of how to embed default knowledge in any conceptual graph.

• A default represents a by-default property. It is defined by applying connective δ to the concept type representing the corresponding strict property. This is an operation of generalization: let G_1 be a graph with a concept node $[\delta t{:}r]$; it is a generalization of graph G_2 identical to itself except at node $[t{:}r]$: G_1 can be obtained from G_2 by replacing type t of node $[t{:}r]$ by its super-type δt.

• An exception first represents an excepted property. It is defined by applying connective ϵ to the concept type representing the corresponding by-default inherited property. This is a concept restriction operation: let G_1 be a graph with

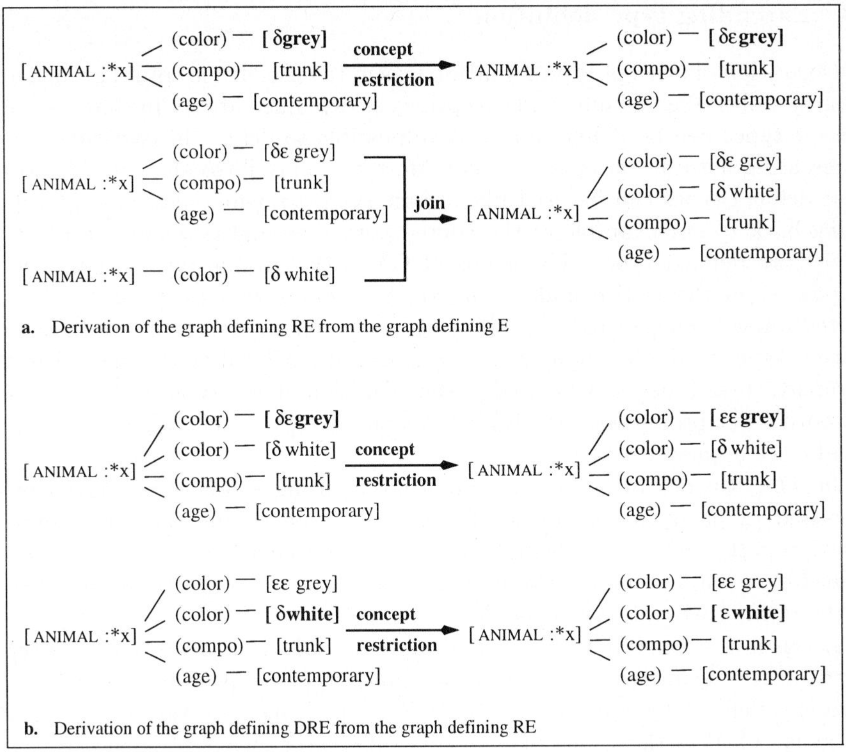

a. Derivation of the graph defining RE from the graph defining E

b. Derivation of the graph defining DRE from the graph defining RE

Fig. 6. Generalization relations between graphs of figure 5

a concept node [ϵt:r]; it is a specialization of graph G_2 identical to itself except at node [δt:r]: G_1 is derivable from G_2 by replacing type δt of node [δt:r] by its sub-type ϵt. Thus an exception applies only to default properties, meaning the strict ones can not be excepted.

• The exception of a property can possibly come with the indication of the unusual property; it is also a specialization operation: a join with a graph representing the unusual property.

One should note that the representation of subsumption remains unchanged; it is based on the generalization relations between types, themselves based on the usual specialization operations defined for canonical graph formation. In such a definitional approach of default taxonomic knowledge handling, it is in the type definitions representing concepts that defaults and exceptions are embedded, while inheritance between concepts stays based on the unchanged subsumption which is represented by the unchanged generalization.

5.3 Extending type definition

The extension of concept type definition lays in the use of conceptual graphs now possibly embedding defaults and exceptions as explained above. In that way, new concept types can be define, formerly impossible to define, in particular those representing natural concepts like taxa. Moreover, once introduced in the lattice, these defined types can be used like atomic types to define new concept types. Going back to our example of the contemporary animal taxonomy, types CA, E, etc. are defined types. The genus of CA is the atomic type ANIMAL; once introduced in the lattice under ANIMAL, CA now serves as genus for E (see figure 2), itself genus of RE, etc. allowing the incremental insertion of these new concept types in the hierarchy of types. Regarding defined type insertion in the hierarchy, these types may be used in the definition of new types by representing not only strict properties but also default and excepted ones when connectives δ and ϵ are applied to them.

In $AL_{\delta\epsilon}$, the definition of any term δt representing a default property can be expressed by using the definition of the term t representing the corresponding strict property: any term t being defined by a conjunction of terms, δt is itself defined by a conjunction of defaults which results from the distribution of δ over all the terms participating in the definition of t.

Let us now express in Conceptual Graph formalism the definition of δt types based on the definition of the corresponding t types. Given any type t of the hierarchy defined by $type\ t(x)\ is\ G$, δt is itself defined by $type\ \delta t(x)\ is\ \Delta G$: the genus of δt is the genus of t and its differentia ΔG is obtained from the differentia G of t by applying connective δ to the type of every concept node of G, the genus of t excepted. For instance, in the contemporary animal taxonomy, the genus of type δE would be type CA and its differentia would be obtained from the differentia of type E by applying connective δ to its types TRUNK and GRAY (see figure 7). For any concept type t, δt is thus introduced in the type hierarchy under the genus of t which is its own genus, and above t, G being derivable from ΔG by the sequence of the restrictions of all its types except its genus.

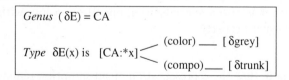

Fig. 7. Definition of type δE

Let us point that in the definition of type δt its genus is not a default: it can not represent a default property allowing exceptions, e.g. an (even by-default) elephant will always be a contemporary animal. Marking out the particular node whose type is the genus of the type being defined, while possible in Conceptual Graph formalism, can not be made in $\mathcal{AL}_{\delta\epsilon}$, the term definition mechanism of the latter being a simple application of the conjunction connective to every term participating in the definition.

6   Conclusion

Our work comes within the framework of information memorization and organization in the fields of organism biology. We have developed a tool for the construction of knowledge based hypermedias integrating both conceptual graphs and textual and graphic descriptions of taxa. Conceptual graphs provides the hypermedia with a formal structure so that consultation and analysis are then based on formal manipulation of the underlying conceptual graphs [13]. Specifically an experiment in the field of orchidology confronted us to the "imperfection" of natural taxonomies and led us to the representation of default taxonomic knowledge in Conceptual Graphs formalism. We implemented methods for the acquisition of taxonomic knowledge now represented in the extension of Conceptual Graphs formalism we proposed. We currently are establishing other algorithms for reasoning with extended conceptual graphs, beginning with object recognition based on the method proposed in [7], where exceptional and typical instances are distinguished. We will carry on with concept classification based on the type insertion protocol described in [6].

Taking into account the similarities between Conceptual Graphs and terminological logics, the extension of Conceptual Graphs formalism we proposed is based on the definitional approach first adopted by Coupey and Fouqueré in $\mathcal{AL}_{\delta\epsilon}$ language. It is the concept definition mechanism that makes this extension possible. The differences among Coupey and Fouqueré's work and ours are due to the differences between term definition in $\mathcal{AL}_{\delta\epsilon}$ and type definition in Conceptual Graphs formalism. Exceptions handling highlights some advantages of type definition in Conceptual Graphs formalism: based on the notion of abstraction, type definition allows to control the introduced default types, particularly the defined ones whose genus can not be a default. Moreover, type definition is based on specialization operations accounting for the exception process: an exception is the restriction of a concept whose type represents the default property and the precision of the unusual property is a join. Our work is a first step in comparing Conceptual Graphs formalism to the terminological languages family. Starting from there, we further aim to undertake a deeper comparison of these languages in a more formal framework.

7   Acknowledgement

The first author thanks Pascal Coupey for the discussions we had and his helpfull suggestions. She also thanks Peter Fauconnier and Mourad Sefrioui for their help when translating this paper in English.

References

[1] E. Aïmeur, J.G. Ganascia. Elicitation of Taxonomies based on the use of Conceptual Graph Operators. In *proc. of the first International Conference on Conceptual Structures. Lecture Notes in AI*, 699, Springer-Verlag, pp. 361-380, 1993.

[2] F. Baader, B. Hollunder. Embedding defaults into Terminological Knowledge Representation Formalism. In *Principles of Knowledge Representation and Reasoning: proc. of the third International Conference*, pp. 306-317, Cambridge, MA, 1992.

[3] B. Biebot, A comparison between conceptual graphs and KL-ONE. In *proc. of the first International Conference on Conceptual Structures. Lecture Notes in AI*, 699, Springer-Verlag, pp. 75-89, 1993.

[4] J. Bournaud, J.G. Ganascia. Conceptual clustering of complex objects: a generalization space based approach. In *proc. of the 3rd International Conference on Conceptual Structures. Lecture Notes in AI*, 954, Springer-Verlag, pp. 173-187, 1995.

[5] R. Brachman and J.G. Schmolze. An overview of the KL-ONE knowledge representation system. In *Cognitive Science*, 9(2), pp. 171-216, 1985.

[6] M. Chein, M. Leclere. A Cooperative Program for the Construction of the Concept Type Lattice. In *proc. of the 2nd International Conference on Conceptual Structures*, proc. supplement, 1994.

[7] P. Coupey, F. Fouqueré. Extending Conceptual Definition with default knowledge. In *Computational Intelligence*, 13(2), 1997.

[8] F. Domini, M. Lenzerini, D. Nardi, W. Nutt. The complexity of concept language. In *Principles of Knowledge Representation and Reasoning : 2d International Conference*, pp. 151-162, Cambridge, MA, 1991.

[9] G. Ellis. Efficient retrieval from hierarchies of concepts using lattice operations. In *proc. of the 1st International Conference on Conceptual Structures. Lecture Notes in AI*, 699, Springer-Verlag, pp. 274-293, 1993.

[10] D.W. Etherington, R. Reiter. On Inheritance Hierarchies With Exceptions. In *proc. of the National Conference on Artificial Intelligence*, Washington, DC, 1983.

[11] C. Fahlman. *NETL: a system for representing and using real world knowledge.* MIT press, Cambridge, MA, 1979.

[12] C. Fahlman, D.S. Touretzsky, W. van Rogger. Cancellation in a parallel Semantic Network. In *proc. of the International join Conference on Artificial Intelligence*, pp.257-263, Vancouver, BC,1981.

[13] C. Faron, Q. Kieu. SATELIT: un outil d'explicitation de connaissances hypermédias. In *Actes des Dixièmes Journées d'Aquisition des connaissances*, Grenoble, 1995.

[14] C. Froidevaux. Taxonomic Default Theory. In *proc. of the European Conference on Artificial Intelligence*, 1986.

[15] G.W. Mineau. Structuration des bases de connaissances par généralisation. Ph.D. d'informatique, Université de Montréal, 1990.

[16] L. Padgham, T. Zhang. A Terminological Logic with Defaults: A Definition and an application. In *proc. of the 13th International Join Conference on Artificial Intelligence*, pp. 663-668, Chambéry, France, 1993.

[17] J. Quantz, V. Royer. A Preference Semantics for Defaults in Terminological Logics. In *Principles of Knowledge Representation and Reasonning: proc. of the 3rd International Conference*, pp. 294-305, Cambridge, 1992.

[18] R. Reiter. A logic for default reasoning. In *Artificial Intelligence*, 13, 1980.

[19] J.F. Sowa. *Conceptual structures: Information Processing in Mind and Machine.* Addison-Wesley Publishing Co, 1984.

[20] W. Woods. What's in a link? Fondation for semantic networks. In York, A.P.N. ed, *Representation and understanding: studies in cognitive science*, pp. 35-82, D.G. Bobrow and A.M. Collins, 1975.

Introduction of Viewpoints in Conceptual Graph Formalism

Myriam Ribière, Rose Dieng
INRIA (project ACACIA)
2004, route des Lucioles BP.93
06902 Sophia-Antipolis Cedex
E-mail: {Myriam.Ribiere,Rose.Dieng}@sophia.inria.fr

Abstract:
To represent knowledge in a context of several experts, it is interesting to model the multiple perspectives that different experts may have on the objects handled in their reasoning. The concept of a particular perspective called viewpoint, is already used in the field of object oriented representation or in the field of databases conception. Taking inspiration of research made in object oriented representation, we propose in this paper an extension of the conceptual graph formalism to integrate viewpoints in the support and in the building of conceptual graphs. We also study mechanisms for managing viewpoints.

Keywords: *Knowledge Representation*, Conceptual Graphs, Viewpoints.

1. Introduction

In knowledge acquisition and modelling, it is often to encounter multi-expertise problem. Indeed, each expert, involved in knowledge acquisition, can have a particular perspective on a handled object, depending on his/her position in the enterprise organization, domain (working context) and the knowledge learned from experience (know-how, competence). Therefore it is difficult to elaborate a unique model taking into account the different terminologies used by the different experts. The notion of viewpoints on the handled object, helps to model knowledge by factorizing information. Viewpoints are also efficient for knowledge extraction; there are viewpoint-based mechanisms allowing accessibility to a subset of information of a common model.

Viewpoints for knowledge acquisition are particularly used in the field of object oriented representation. The interest is based on manipulation of a complex system in terms of accessibility and dynamic knowledge representation. After describing previous work in object oriented representation, we propose an extension of the conceptual graph formalism to support the viewpoint management: we first describe the different steps to integrate viewpoints in support and in conceptual graph, then we explain how to manage viewpoints.

In conclusion, we describe our perspectives and we discuss on the interest of our work in comparison with previous work related to views or viewpoints in conceptual graph formalism.

2. Viewpoint Notion

Research on viewpoint notion was carried out in the seventies. Several interpretations of this notion are possible. The first one is a spatial interpretation [9]: viewpoints correspond to the different perceptions of an object with respect to the observer's position. The second interpretation is a knowledge domain one: viewpoints correspond to the different ways to translate knowledge with respect to the social position, know-how and competence of an expert. In this interpretation, a viewpoint includes a context, and the perception of a person or a group of persons.

This second interpretation underlies several systems developed since 1977. The first one, KRL [1] is dedicated to knowledge representation. LOOPS [12] inherits the approach of KRL. SHOOD [10], ROME [2], TROPES [7] are the second generation of those systems. The definition of the viewpoint interpretation is more restricted in those systems: they consider the different ways by which an expert can see a knowledge base, but they don't explain explicitly what is a viewpoint. In most applications with such systems, viewpoints are restricted to a knowledge domain or a competence domain.

We describe here two example models for the introduction of viewpoint in knowledge representation and for the management of multiple viewpoints. We focus on these models because they are sure as basis of our reflection about introducing viewpoints in conceptual graphs.

2.1 TROPES

TROPES [7] notion of multiple perspectives allows to structure a knowledge base according to different objectives. TROPES is based on the definition of a set of basic concepts, which can be described in a class arborescence [fig. 1], according to different viewpoints.

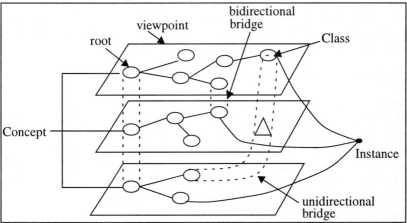

Fig. 1. The TROPES model

Bridges in the structure allow to build an inclusion or equivalence relation between two classes pertaining to different viewpoints. Each class belongs to only one viewpoint. The concept of multi-instantiation enables to instantiate an object with several classes.

This structuration is dedicated to the classification process allowing the manipulation of complex systems and, in particular, an easier access to information.

2.2 ROME

ROME [2] and FROME [3] (the extension of ROME for frames) are dedicated to modelling incomplete and evolving knowledge, in particular through the multiple representation according to multiple viewpoints. FROME uses a simple link for instantiation, but multiple links for representation. So an object [fig. 2] is identified by a main class, which keeps the identity of the object, and can have multiple representation links, which specialize the object through viewpoints.

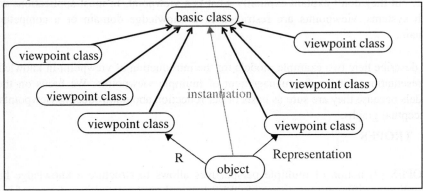

Fig. 2. The ROME model

This technique allows to add information by adding a new viewpoint to an object (by a new link of representation). FROME also uses an exclusion link to prevent an object from having incompatible representation links. In TROPES this exclusion link does not exist, since an object can be instantiated only by one class per viewpoint.

2.3 Objects and Conceptual Graphs

A natural correspondence can be made between formalisms for object oriented representation on the one hand and conceptual graph formalism on the other hand. Indeed, object oriented formalisms often rely on the notions of class and object (i.e. instance). A class is described by slots and methods, that may be inherited through a hierarchy of classes. An object is an instance of a class and can be manipulated in a knowledge base. So a class corresponds to a concept type (in CG formalism), and an object to a concept, having a concept type and a referent. The class hierarchy corresponds to the concept type lattice in the support.

Furthermore previous work already exists on bringing together oriented object language and conceptual graph formalism [5]. Therefore this correspondence convinced us that the integration of viewpoints in conceptual graph formalism can be inspired by research realized in object oriented representation.

3. Viewpoint Integration in Conceptual Graph

In object oriented formalisms, the integration process of viewpoints can be summed up by two phases:
- How to represent viewpoints inside the class hierarchy?
- How to identify an object with different viewpoints?

These phases are situated at two different levels: the first one concerns the class level, which corresponds to the support level in CG formalism. The second one deals with the instance level, which corresponds to the graph level in CG formalism. The figure 3 summarizes our approach. It shows the different entities that take place in the viewpoint integration.

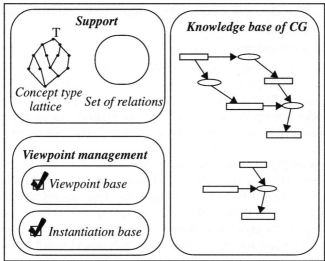

Fig. 3. Viewpoint integration in conceptual graph formalism

We present the different steps of viewpoint integration at different levels in next sections.

3.1 Viewpoint and Support

In this section we define the different elements that we introduced in the support and we describe a viewpoint knowledge base.

3.1.1 Concept Type Lattice

Viewpoints are already underlying in the concept type lattice but they are implicit. The information contained implicitly in the subtype link can be explicitly expressed by a viewpoint. When, in the classic support, a concept type has several fathers, it generally corresponds to the use of several implicit viewpoints for building the lattice: such viewpoints are implicit in the ordering relation upon which the concept type lattice relies.

Definition:
Let C, C' two concept types. If C' is a subtype of C, then a viewpoint P can be made explicit, such that C' is a subtype of C according to the viewpoint P. Once such a viewpoint made explicit, C is called a «basic concept type» and C' an «viewpoint oriented concept type» (noted v-oriented concept).

Note:
- C' itself can also be a basic concept type, if it is decomposed through at least one explicit viewpoint. In this case we must consider the two characteristics (v-oriented and basic) of the concept type for the viewpoint management.
- A v-oriented concept type can be subtype of several basic concept types.

A viewpoint is then the explicit expression of a particular subtype relation existing between two concept types.

3.1.2 Conceptual Relations

In the set of conceptual relations, we introduce the notion of «viewpoint relation», which expresses a more precise subtype link. This viewpoint relation has two concept types in its signature. Esch in [6] uses a superior level concept type TYPE, to manipulate second order concept types. So our viewpoint relation is a second order binary relation between two concept types. The signature of this relation type is (TYPE, TYPE).

In a previous work [11], we defined second order relations with an associated behaviour. A «viewpoint relation» is one of such relations, and its associated behaviour enables to manage consistency between the subtype link in the concept type lattice and the expression of a viewpoint between the same concept types. This behaviour is particularly realized during the process of creation or modification of a viewpoint (i.e. addition or removal of a concept type in the concept type lattice) which has an impact on the subtype link.

Let the following graph G:

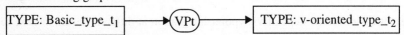

G is read: «the v-oriented type t_2 is a subtype of the basic type t_1 according to the viewpoint VPt»

Definition:
Let C_1, C_2 two concept types. If $C_1 < C_2$, then there may exist a «viewpoint relation» VPt such that C_1 is a subtype of C_2 according to VPt. VPt is a second order conceptual relation of signature (TYPE,TYPE). Then C_1 is a basic concept type and C_2 a v-oriented concept type.

Remark: According to Wermelinger's theory [12], there are several finite concept type lattices (one for each order) and a finite set of finite relational concept type lattices (one for each order).There are several relation type hierarchies: relation types are classified according to their arity and order. Each hierarchy contains all relations with the same signature (i.e. order, arity and arguments' orders).

So, as the relation type SUBTYPE, our «viewpoint relation type» VPt belongs to the hierarchy of dyadic relation types and of signature (TYPE, TYPE), in the lattice of second-order relation types. TYPE belongs to the second-order concept type lattice. In addition, VPt < SUBTYPE.

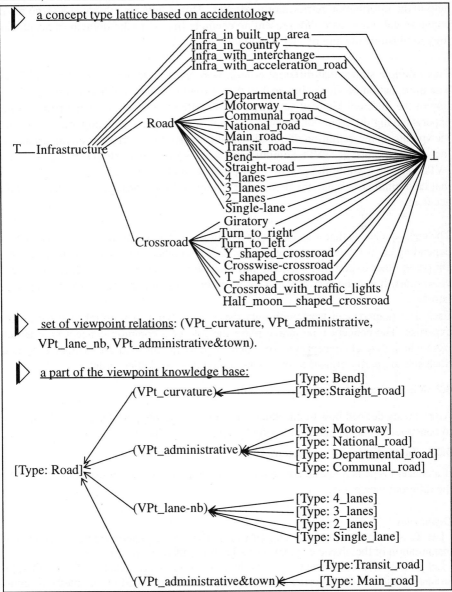

Fig. 4. Example of viewpoint expression in the domain of accidentology

3.1.3 Viewpoint Knowledge Base

Modifications and definitions brought in the two previous sections allow to build a knowledge base, where all subtype links, according to a viewpoint, are described. We call this base a «viewpoint base». We choose to represent viewpoints outside the support, in order to avoid to enforce a systematic use of viewpoints in conceptual graphs. Indeed a viewpoint must be used only if a subtype link contains information which can help in the information retrieval or modelling. This is how we introduce viewpoints in a conceptual graph base. We present an example [fig. 4] in the domain of accidentology to illustrate the aim of such a module [fig. 3].

This example shows that different criteria were used to define the subtypes of Road. The introduction of viewpoints makes these criteria explicit and allows the classification of the different kinds of subtypes. Those different viewpoints come from several experts having different competence domains. Each expert gives importance to some viewpoints on the definition of a road. The infrastructure expert gives an importance to the viewpoints: «VPt_administrative», «VPt_administrative&town» and «VPt_lane_nb», while the psychologist points to «VPt_curvature» or «VPt_lane_nb» that help him/her to make the driver's error explicit to be considered as a factor of the accident.

The consequence is that if an expert only wants to see a part of the concept type lattice depending on his/her competence domain and/or an other competence domain dependent on an other expert, the adequate information to be shown, can be filtered through viewpoints. It may be very efficient for a complex system, having a wide concept type lattice.
Each user (expert) can work with only the concept types that are relevant for his/her expertise. The concept type lattice plays the role of a common concept type lattice, from which several concept type sets, dedicated to the intended user, can be extracted. Such sets are partly organized in a hierarchy but not in a lattice structure.

3.2 Viewpoint and Conceptual Graphs

After having defined how to introduce viewpoints in the support, let us now define how to construct a concept and a conceptual graph with viewpoints.

3.2.1 Instantiation of a Concept Type

If a concept type is a «basic concept type», then it has several representations through the different viewpoints.

Definitions:
- Let Tc a basic type concept, and let r an individual or generic referent. [Tc:r] is an instantiation of the «basic concept type» Tc. It is called «basic concept».
- Let To a v-oriented concept type, and let r an individual or generic referent. [To:r] is an instantiation of the «v-oriented concept type» To. It is called «v-oriented concept».

Each representation of a given basic concept describes the same object as this basic concept, but with different levels of precision due to the specialization characterized by the considered viewpoint. So if a basic concept type is instantiated, then it can be described by the instantiation of any of its subtypes according to the different viewpoints characterizing this object.

Here, we introduce a new relation, called «Repr», which links a concept to one of its representations.

Let the graph R:

R can be read: The concept «v-oriented_concept» is a representation of the concept «basic_concept».

Repr is a first order relation. It can be expressed only if the two concept types are linked through a viewpoint relation in the viewpoint knowledge base. Two concepts linked by a «representation relation» describe the same object (cf. fig.5).

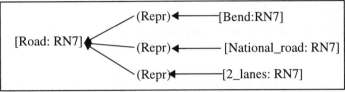

Fig. 5. Example of instantiation graph

3.2.2 Instantiation Knowledge Base

Using this relation of representation, we can now build a knowledge base containing each instantiation of each basic concept type . This base (called «instantiation knowledge base» [fig. 3]) complements the viewpoint knowledge base. In section 5, we detail the interest of both bases.

Two methodologies for building this instantiation base can be thought out:

- We can oblige the user to declare all the instantiations of a concept type in the instantiation knowledge base, before using them in a conceptual graph. In this case, the v-oriented concept types connected by Repr relations with the basic type instantiated, must have the same referent.

- Otherwise, when a concept type C is instantiated, we must verify if it is a subtype of a basic concept type. In that case, we can build an instantiation graph of the basic concept type with the referent proposed for the instantiation of the v-oriented concept type (with verification of the conformity relation). If C is a basic concept type, the adequate instantiation graph must be built.

We can notice that two representations of the same concept can be incompatible (e.g. [National_road:RN7] and [Communal_road:RN7]). Therefore we will introduce in the next section the exclusion relation between two v-oriented concept types.

3.2.3 Additional Relation for Instantiation Management

In this section we detail different second order relations, which can help to manage the representation relation or to make deductions. Those relations depend on the representation relation. They can be expressed in the viewpoint knowledge base [fig. 3].

Several definitions presented in this section will use the following links [fig. 6] (based on the IF...THEN graphs defined by Sowa in [14]).

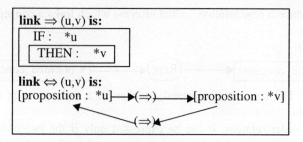

Fig. 6. Definition of implication and equivalence links

3.2.3.1 Exclusion Relation

An exclusion is only an intra-viewpoint relation.

Definition of excl (C,C_1,C_2):
Let C a basic concept type, C_1 and C_2 two v-oriented concept types both subtypes of C. If «Repr» connects the v-oriented concept $c_1=[C_1:ref]$ to the basic concept $c=[C:ref]$ then «Repr» cannot connect the v-oriented concept type $c_2=[C_2:ref]$ to c.

Graph representation:
relation excl (C,C_1,C_2)

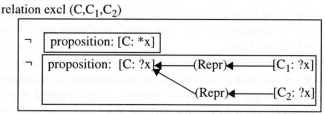

This relation allows to avoid incompatible instantiations. We precise the basic concept type C, because an exclusion relation cannot exist between two v-oriented concept types subtyping two different basic concept types.

Relation behaviour:
Monotonic reasoning: before adding in the instantiation base a new «Repr» relation between a new v-oriented concept Co and a basic concept, it is checked whether none of the types of the v-oriented concepts already representing this basic concept in the instantiation base has any «excl» relation with the type of Co. If one such exclusion relation is found, the new «Repr» relation with Co is rejected.

3.2.3.2 Equivalence Relation

Equivalence relation is an inter-viewpoints relation.

Definition of equiv (C,C_1,C_2):
Let C a basic concept type and C_1 and C_2 two v-oriented concept types both subtypes

of C. If «Repr» connects the v-oriented concept $c_1=[C_1:ref]$ to the basic concept $c=[C:ref]$, then «Repr» necessarily connects the v-oriented concept type $c_2=[C_2:ref]$ to c and vice versa.

Graph representation:

relation equiv (C,C_1,C_2):

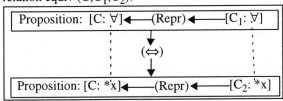

Relation behaviour:
If in a viewpoint graph, there exists an «equiv» relation between, on the one hand C_1 the type of $c_1=[C_1:ref]$, a v-oriented concept connected with Repr to a basic concept $c=[C:ref]$, and on the other hand C_2, then a v-oriented concept type $c_2=[C_2:ref]$ is created if needed and a «Repr» relation connecting c_2 and $c=[C:ref]$ is created.

3.2.3.3 Inclusion Relation

Definition of incl(C,C_1,C_2):
Let C a basic concept type and C_1 and C_2 two v-oriented concept types both subtypes of C. If «Repr» connects the v-oriented concept type $c_2=[C_2:ref]$ to the basic concept $c=[C:ref]$, then «Repr» necessarily connects $c_1=[C_1:ref]$ to c (but not vice versa).

Graph representation:

relation incl (C,C_1,C_2):

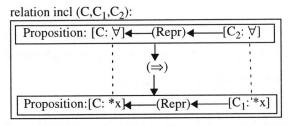

Relation behaviour:
If in the instantiation base, there exists $c_2=[C_2:ref]$ a v-oriented concept connected through Repr to its basic concept $c=[C:ref]$, and if in the viewpoint base, there exists an «incl» relation between C_1 a v-oriented concept type, and the v-oriented concept type C_2, then the v-oriented concept $c_1=[C_1:ref]$ is created if needed in the instantiation base and a «Repr» relation connects c_1 to c.

4. Viewpoint Management in Conceptual Graph Formalism

The two bases described in the previous section allow to have an independent viewpoint management through the conceptual graph formalism. A user can choose to use or not the viewpoint management. So we call «viewpoint management» the set created by the two bases. [fig. 7] shows the different interactions between the different entities. There are two sorts of interactions:
  • interaction for consistency management
  • interaction for construction of conceptual graph.

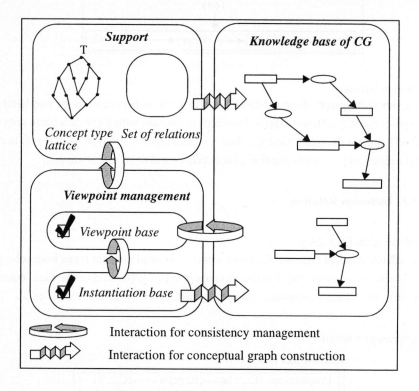

Fig. 7. Viewpoint management in conceptual graph formalism

4.1 Interaction for Consistency Management

There are three kinds of interactions for consistency management:
  • The interaction between the support and the viewpoint base is to manage consistency between the concept type lattice (which expresses all the subtype links between concept types) and the viewpoint base (which expresses all the subtype links according to viewpoints). We must manage at the same time the creation, suppression, and modification of concept types in the two parts.

This is an example of property that must be satisfied:

If $[TYPE:T_1] \leftarrow (VPt) \leftarrow [TYPE:T_2]$ in the viewpoint base then $T_2 < T_1$ in the concept type lattice (even more, T_2 is a direct subtype of T_1 in the concept type lattice).

- The interaction between the two bases of the viewpoint management, with the different relations (exclusion, inclusion and equivalence). We must check also during the creation of a «Repr» relation, if the involved v-oriented concept type is subtype of the basic concept type, according to a viewpoint. The conformity relation must also be satisfied.

- The interaction between the viewpoint management module and the knowledge base of CG allows the verification of the relation signatures. Indeed if a conceptual relation has a v-oriented concept type as maximal concept type in its signature, the other arguments must be in the common viewpoints. It allows also consistency management between information in viewpoint management module and conceptual graph base [fig. 8].

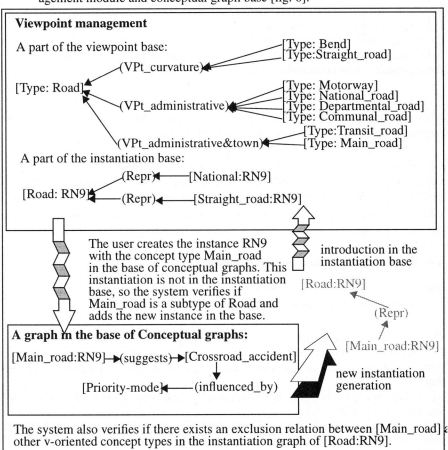

Fig. 8. Example of interaction

4.2 Interaction for Conceptual Graph Construction

The support allows to build a conceptual graph without viewpoint. Using both the instantiation base and the support permits conceptual graph construction with viewpoints. All the instantiation through viewpoints made in a graph, must be stored in the instantiation base before their use (we choose the first methodology described in the section 3.2.2).

4.3 Rules on Conceptual Graph and Viewpoints

The integration of viewpoints has an influence on the definition rules on CG.
There are four canonical formation rules: Copy, Restrict, Join and Simplify [13], that can normally be used for the formation of classic CG. But we can define rules derived from them and specific for the viewpoint management: in particular the Restrict and Join rules can be redefined according to a viewpoint. So we created a new rule, based on Sowa's Restrict rule:

Definition: Restrict w.r.t. a given viewpoint
Let v a viewpoint. The restriction w.r.t. the viewpoint v is obtained by replacement of the concept label with a more specific label: replacement of a generic referent with an individual referent and/or replacement of the concept type with a subtype according to the viewpoint v (the conformity relation must be verified).

Sowa's Join rule is based on the identity of concept. As we have defined an equivalence relation on concept types, this relation helps us to define a new join relation with respect to the equivalence relation in the viewpoint base:

Definition: Join w.r.t the equivalence relation inter-viewpoints
Let $c_1=[C_1:ref]$ a v-oriented concept of a graph G_1 and $c_2=[C_2:ref]$ a v-oriented concept of a graph G_2. If there exists an «equiv» relation between C_1 and C_2, then we can join G_1 and G_2 by identification of c_1 and c_2. A relevant concept among c_1 and c_2 must be chosen for the join.

4.4 Projection Operation and Viewpoints

The projection of a graph G into a graph G' is not really modified, but we can also focus the projection according to a viewpoint v:

Definition: projection operation w.r.t. a viewpoint
Let G and G' two conceptual graphs. The viewpoint projection according to a viewpoint v, of G into G' is an application $\pi:G \to G'$, where πG is a subgraph of G', such that:
- For each concept c in G, πc is a concept in G' and type(πc) is a subtype of type(c) according to the viewpoint v.
- For each conceptual relation r in G, type(πr)=type(r). If the ith arc of r is linked to a concept c in G, the ith arc of πr must be linked to πc in G'.

After a classic projection operation, we can also verify whether the result of this classic projection corresponds in fact to a projection w.r.t. a viewpoint: if there exists a viewpoint v such that for each concept c of G, type(πc) is a subtype of type(c) according to this viewpoint v, then πG is a subgraph of G' according to the viewpoint v.

This precision allows to know if two different experts say the same thing in different viewpoints. They can describe the same thing with their own concepts that depend on their knowledge, competence domain, and perception of the situation or the object to describe.

This projection operation with the viewpoint notion permits comparison of expertise through the conceptual graph formalism. So it is very interesting to know how to represent those different representations and perceptions.

5. Conclusion

In this paper, we described the interest of viewpoints in knowledge representation and we proposed an introduction of viewpoints in conceptual graph formalism.

5.1 Comparison with TROPES and ROME

The creation of the viewpoint base relies on several elements in TROPES model like the class arborescence and bidirectional, unidirectional bridges. Indeed our viewpoint base contains conceptual graphs that describe partially the class arborescence of TROPES thanks to the introduction of the viewpoint relation. Our equivalence and inclusion relations play the same role as bidirectional and unidirectional bridges. Therefore our instantiation base uses the simple instantiation of ROME and the description of multi-representations via a representation relation like in ROME.

5.2 Related Work and Perspectives

The view notion was already used in [4], but instead of introducing viewpoints in conceptual graphs, the aim was rather to use conceptual graph formalism as a meta-language to analyze and compare different specifications of requirements with respect to each other. The multiple views were in fact multiple schemas of representation.

In [8], Martin associates concepts or conceptual graphs to elements of structured documents. He uses viewpoints to permit different representations of the same document element in a knowledge base in order to compare them. Our aim is to define viewpoints to help knowledge representation for multi-expert knowledge acquisition and also to have an accessible and evolutive knowledge base through viewpoints. Our approach is also aimed at making comparison of expertises.

We can notice the interest of exploiting second order relations, with an associated behaviour, as suggested in a previous work [11]. Other relations can be exploited. Our perspective is to extend the viewpoint management to the «composed-of» relation, and to test all the viewpoint management on an application.

References

1. D.G. Bobrow, T. Winograd, *An overview of KRL, a Knowledge Representation Language*, Cognitive Science, vol.1 n°.1, p. 3-45, 1977.

2. B. Carré, L. Dekker and J-M. Geib. *Multiple and Evolutive Representation in the ROME Language. Towards an integrated Corporate Information System.* In Proc. TOOLS' 90, Paris, 26-29 June 1990.

3. L. Dekker and B. Carré. *Multiple and Dynamic Representation of frames with Points of View in FROME.* In Proc. RPO'92. La Grande Motte, France, 22-23 June, 1992.

4. H. Delugach, *Specifying Multiple-Viewed Software Requirements With Conceptual Graphs*, Jour. Systems and Software, vol. 19, p. 207-224, 1992.

5. G.Ellis, *Object-Oriented Conceptuals Graphs*, In Ellis et al eds, Conceptual Structures: Applications, Implementation and Theory. Proc. of ICCS'95. Springer Verlag. Santa Cruz, CA, USA, Aug. 1995, p. 144-172.

6. J. Esch, *Contexts, Canons and Coreferent Types.* In Terpfenhart &al eds, Conceptual Structures: Current Practices: Proc. of the ICCS'94. College Park, Maryland, USA, August 1994, Springer Verlag, LNAI n. 835, p. 185-195.

7. O. Marino, F. Rechenmann, P. Uvietta, *Multiple perspectives and classification mechanism in Object-oriented Representation*, Proc. 9th ECAI, Stockholm, Sweden, p. 425-430, Pitman Publishing, London, August 1990.

8. P. Martin, *Exploitation de graphes conceptuels et de documents structurés et hypertextes pour l'acquisition de connaissances et la recherche d'information*, PhD Thesis, University of Nice-Sophia-Antipolis, October 1996.

9. M. Minsky, *A Framework for Representing Knowledge*, in The psychology of Computer Vision, McGrawHill, New York, P.H. Winston (ed), Chap. 6, p. 156-189, 1975.

10. G.T. Nguyen, D. Rieu, J. Escamilla, *An Object Model for Engineering Design*, in Proc. of ECOOP'92, Utrecht, The Netherlands, June/July 1992, Springer-Verlag, p. 232-251.

11. M. Ribière, R. Dieng, M. Blay-Fornarino, A-M. Pinna-Dery, *Link-based Reasoning on Conceptual Graphs*, in Suppl. Proceedings of ICCS'96, Sydney, Australia, August 1996, p 146-160.

12. M. Stefik, D.G. Bobrow, *Object-Oriented Programming: Theme and Variations.* The A.I. Magazine, vol. 6, n° 4, p 40-62, 1985.

13. J. F. Sowa. Conceptual Structures, *Information Processing in Mind and Machine.* Reading, Addison-Wesley, 1984.

14. J. F. Sowa. *Relating Diagrams to Logic.* In Mineau & al eds, Conceptual Graphs for Knowledge Representation: Proc. of ICCS'93. Springer Verlag, LNAI n. 699. Quebec City, Canada, August 1993, p. 1-35.

15. M. Wermelinger. *Conceptual Structures and First-Order Logic.* In Ellis et al eds, Conceptual Structures: Applications, Implementation and Theory. Proc. of ICCS'95. Springer Verlag. Santa Cruz, CA, USA, Aug. 1995, p. 323-337.

Task-Dependent Aspects of Knowledge Acquisition: A Case Study in a Technical Domain

Galia Angelova[1] and Kalina Bontcheva[2]

[1] Linguistic Modelling Laboratory, Bulgarian Academy of Sciences
Acad. G.Bonchev Str. 25A, 1113 Sofia, BULGARIA
galja@lml.acad.bg
[2] Department of Computer Science, University of Sheffield
Regent Court, 211 Portobello Str., Sheffield S1 4DP, UK
K.Bontcheva@dcs.shef.ac.uk

Abstract. This paper summarises principles of manual acquisition of conceptual graphs which evolved within the framework of a natural language processing system and are now enriched and elaborated to facilitate the construction of a larger knowledge base. Our conventions provide the mapping between language structures (at syntactic and semantic levels) and conceptual structures. The task-dependent aspects of our approach are clearly presented. We also discuss the problems of finding suitable domain descriptions to be used as acquisition sources. Finally, we evaluate the manual and automatic knowledge acquisition from the perspective of our current experience.

1 Introduction

Knowledge Acquisition (KA) is approached in two principally different ways: (*i*) *automatic acquisition from texts* – this task relies on lexicons, (partial) morphological, syntactic and semantic analysis. The acquired knowledge usually follows closely the text semantics; (*ii*) *manual acquisition from various sources* – then the knowledge engineer is free to decide which facts are worth to be encoded depending on the specific acquisition goal. The acquired knowledge represents the semantics of the sources or might be a generalised model.

This paper discusses our experience in the manual KA which resulted in a generalised model of the admixture separation domain. The acquired Knowledge Base (KB) of conceptual graphs [14] is successfully used for generation of Natural Language (NL) explanations within the DB-MAT project.

DB-MAT[3] [2, 21] aims at the application of knowledge based methods in translation aid tools for professional translators. The central idea is that the user (human translator) receives linguistic and/or domain knowledge support. The domain knowledge explanations are generated in NL from an underlying KB

---

[3] A joint German-Bulgarian project in knowledge-based Machine-Aided Human Translation, funded by the Volkswagen Foundation (Germany), 1992—1995. Second phase DBR-MAT with Romanian partners in 1996–1998. http://www.informatik.uni-hamburg.de/NATS/projects/db-mat.html

of Conceptual Graphs (CG) [15]. All system resources are hidden behind a user-friendly, menu-driven interface where the translator highlights unknown terms, selects a request from a menu and receives NL explanations. Requests for clarification or new explanations are formulated within the context of the previous explanation. We summarise the DB-MAT paradigm from a few perspectives:

Single KB: Given a problem area, there are many texts to be translated but their domain is *one*. That is why we have a single KB which contains only domain knowledge. The user obtains domain information *via* the concrete document to be translated. In such a framework the basic "cross-points" between the KB and the technical texts are the domain terms. Highlighting a term from the text, the user gets an explanation about its definition, function(s), etc. while the underlying complex algorithms remain completely hidden. In this way, DB-MAT offers a computational model of terminology and its approach can be compared to methods used in knowledge based term banks [4]. At the same time, such a system is a language-based tool for browsing a KB [1].

Knowledge Based System: The KB is clearly separated from the linguistic resources and processed in a relevant manner. The CG operations *projection* and *join* are applied as well as inheritance of characteristics.

Knowledge Representation Formalism: CG were chosen due to (*i*) the contexts that allow knowledge items with different granularity to be considered as one item; (*ii*) well-defined operations; (*iii*) existing experience in the field.

NL Processing: DB-MAT is a NLP system, so the lexicons are central components in the system architecture. For instance, complex (i.e., multi-word or compound) terms are mixed with ordinary words in the concrete texts. The recognition of such terms in the highlighted phrases is provided by the lexicons.

Multilinguality: DB-MAT is a multilingual environment[4]. The single language-independent KB is connected to each language-specific lexicon by linking pointers. So, the internal names of the KB items never appear in the system interface and are used only for identification and reminders during the KA.

NL Generation: DB-MAT is one of the few systems which apply inference rules to the domain knowledge and, thus, provides explanations containing knowledge which is not directly encoded into the KB. At present we are investigating advanced language generation and user modelling techniques which will ensure more fluent explanations sensitive to the context and the user expertise.

Technical Domain: The KA task is performed in the domain of admixture separation, especially oil-separation devices.

Section 2 below presents the general requirements which had to be at least partially satisfied in the KA process. Section 3 summarises our conventions for the CG construction: direction of arcs, concept referents, conceptual relations. In Section 4 we show how domain texts are encoded into conceptual structures. The type hierarchy (TH) is considered in more detail. Section 5 exemplifies the verbalisation of a type definition, while Section 6 discusses KA approaches. In the end, we give some directions for future work and conclusions.

---

[4] At present, two lexicons are incorporated (German and Bulgarian) but the modular structure allows the easy integration of new ones (e.g. Russian).

2 Knowledge Acquisition Desiderata

The KA goal is (*i*) to represent the domain in a task-dependent manner by constructing the domain ontology of important objects and relationships; (*ii*) to encode the knowledge items into the most suitable structures of the representation language and (*iii*) to ensure in this way the correct application of the formal operations and inference procedures [10]. Below we discuss our choices of *what to talk about* and *how to tell it* in our KB so that it meets the expectations of our users.

There are many different levels of detailness when we look at the domain to be described by a KA process [5]. In addition, the ISA relation can be defined from different perspectives of natural and role types (e.g. sex, professions, marrital status for persons) which imply different lattices in the types poset. Due to our system objectives, our KA is influenced by the NLP paradigm and by the technical domain. The following requirements are to be satisfied.

Uniform granularity of the knowledge items:

Granularity of concept types: Usually the encoding of English sentences into CG applies a very simple one-to-one mapping between types and words. For each of the eleven English parts of speech it is as follows: *nouns*, *adjectives* and *adverbs* are encoded as concept types; *verbs* are either encoded as concept types or as conceptual relations. The verb *to be (is a)* forms the type hierarchy; *articles*, some *pronouns* and *numerals* influence the concept referents; *prepositions* are reflected by the conceptual relations; *conjunctions* (e.g., *and, or*) are expressed by the logical graph connections as conjunctions, disjunctions, implications, etc; *interjections* and *particles* (e.g., *oh, ah*) convey pragmatic information and never occur in technical domains.

On the other hand, KA from multilingual texts depends on the fact that in the different natural languages, lexical items "name" knowledge fragments with different granularity. Sowa [16] shows a vehicle type hierarchy where some concept types exist in English, others in Chinese and still others have equivalents in both languages. Thus, the KA task seems nearly undecidable, if we aim at a *language-independent* KB that can be used as a common semantic representation between several lexicons with *general lexica* in a multilingual environment.

However, KA faces less problems in technical domains where the main objects have corresponding material denotats and the domain terms determine the appropriate granularity. Due to the influence between languages, more abstract types like *dispersion* or *emulsion* exist as terms in many languages. Our experience shows that (*i*) the basic granularity of the concept types is defined by the terms as they are collected in more or less existing terminological dictionaries in several languages; (*ii*) terminological gaps are relatively rare especially those concerning basic domain notions; (*iii*) we meet gaps and inclusions of the terms' meanings – e.g., `Abwasser` (*waste water*) and `Wellplatte` (*corrugated plate*) where a German term names a whole situation while a corresponding English word does not exist. But usually we do not observe gaps in the names of the basic processes that are performed over the atomic domain- meaningful objects; (*iv*) the language in technical domains (so called LSP, Language for Special Purposes) is rather restricted as a syntactic structure as well as from semantic and pragmatic point of view.

Our general rule is that the meaning of one term is represented as one concept or one situation. Once the granularity of concept types is established for domain terms, it influences the granularity of the other types.

Granularity of type definitions and conceptual graphs: In general, we encode (*i*) as a type definition the knowledge fragments which concern all possible concept instances and (*ii*) as one conceptual graph the facts which are satisfied by the same concept instance(s). However, due to our task-oriented framework some universally valid facts are encoded as graphs. This is primarily an implementation decision and does not affect the generality and expressivity of the approach (see below the discussion about [OIL-SEPARATOR: every] -> (CHAR) -> [WEIGHT].).

Uniform coverage of the domain: It appears that a good textbook provides a very clear intuitive feeling which lexical usage of a term is a *type definition*, which is a *concept instance* in a certain graph or just an *individual* with the same type label. Where the text looks like a definition in the style of terminological dictionaries, a type and its definition is constructed. In other cases the next concept instance, individual or a context is acquired. In this way the acquired concept types are predetermined by the source texts and their terminology, style, and semantics.

Separation between domain and linguistic knowledge: The lexical names of the KB items are given by links to the lexicons in our star-like architecture *one KB ↔ many lexicons*. Therefore, some conflicts between domain knowledge and lexicon information might arise. For instance, concept referents encode semantic information which should not inhibit the correct grammatical usage of articles and singular/plural numbers for each language. To achieve maximal language independency, we use as few referents as possible. So we have no general rule of how to acquire, e.g., *definite* referents, because the lexicon and the grammar provide the correct lexical structure for each language.

Task-oriented aspects: We also need some conventions about: (*i*) the verbalisation of concept referents; (*ii*) the influence of the conceptual relations and embedded contexts to the language generator; (*iii*) the application of CG inheritance in order to provide more semantic details in the resulting explanation. Therefore, we follow certain principles of encoding the domain knowledge into CG, in order to ensure consistent lexicalisation and explanation. A simple example of such a principle is that we fully explain CG contexts, i.e., their content is either given as a whole or is not included. So, complex processes are encoded as CG contexts to ensure their proper explanation.

In fact, during the system development we agreed on some standard how CG should look like in order to underly explanations with their lexical and syntactic variety. Consequently, our KB is (somewhat) task-oriented. Below we discuss the CG and KA standards in more details.

3   KB Conventions

We represent the processes as concept types and their participants are connected by (specialisations of) conceptual relations such as *effector* **EFCT**, *patient* **PTNT**, *theme* **THME**, *matter* **MATR**, *instrument* **INST**, etc. Below we discuss the directions of the arrows, referents and conceptual relations as encoded in our KB.

3.1   Conceptual Graphs Conventions

The direction of the arrows between concepts and conceptual relations is very important, since it influences the language generator (see [14, Chapter 5]). In this section we list well-known examples in order to illustrate our conventions for direction of arrows. The conceptual graphs encode either the surface syntax (e.g., c1) or the deep semantics of the phrase (e.g., the thematic role in a3).

Event encoded as a concept type:

   a1. [CHASE] -> (AGNT) -> [CAT]. (action to subject)

   a2. [CHASE] -> (THME) -> [MOUSE]. [EAT] -> (PTNT) -> [PIE]. (transitive verb to direct object)

   a3. [GO] -> (INST) -> [BUS]. (intransitive verb to object)

   a4. [EAT] -> (MANR) -> [QUICKLY]. (action to adverb)

   a5. [GO] -> (ALONG) -> [PATH] -> (TO) -> [PLACE]. (action of intransitive verb to indirect object which is another noun phrase, [19, p. 34])

State encoded as a conceptual relation:

   b1. [LEG: SET(*)] -> (PART-OF) -> [PERSON].

   b2. [MAN: Todd] -> (MAR_TO) -> [WOMAN: Merry].

   b3. [ENTITY] -> (LOC) -> [PLACE].

Prepositional phrases:

   c1. [CAT] -> (ON) -> [MAT].

   c2. [TABLE] -> (ON) -> [VASE] -> (ATTR) -> [COLOR: white]. ([19, p. 269], cf. c1). As evident, the direction of arrows in c2 is opposite to c1; we accept c1.

Concepts to concrete individual value:

   d1. [TIME: 19.22] <- (FROM) <- [TIME-PERIOD] -> (TO) -> [TIME: 19.39].

Attribute/Characteristics:

   e1. [BALL] -> (ATTR) -> [RED].

   e2. [BALL] -> (CHAR) -> [COLOR:red].

3.2   Types of Referents Used

CG referents are means for flexible identification of concept types individuals. We use the following types of referents : *generic, plural, every, measure, disjunctive sets*. As evident, from all set types we use only disjunctive sets which encode a disjunction of individuals, while the conceptual relation **OR** encodes disjunctions between concepts and/or contexts.

3.3 Choice of Conceptual Relations

The conceptual relation hierarchy is relatively limited for two main reasons: (*i*) the technical domain imposes its specifity on the KB – e.g., the possible meanings of most common words are rather restricted and the meanings of domain-specific terms are relatively inambigous and stable across the languages. Moreover, there are few animated agents (e.g., no social roles); (*ii*) the explanation task emphasises on domain knowledge so our efforts are not focused on the upper and middle models.

We differentiate *characteristics* CHAR and *attributes* ATTR. The CHAR relation is a distinguishing feature or quality, it is valid for all concept instances and occurs primarily in type definitions. ATTR presents a feature of a concrete instance that is meaningful in the context/graph only.

Some of our conceptual relations are very similar to those in [6] with more specialisations concerning for example the *location* (LOC_IN, LOC_ON, LOC_AT), *material* MADE_OF, MADE_FROM. Patient PTNT is a relation connected to participants in events which undergo changes in their universal features as they are listed by CHAR relations. We also have relation types describing mechanical comparisons – BIGGER_THAN, LIGHTER_THAN, etc.

4 Encoding Domain Knowledge in CG Structures

We encode domain knowledge in various KB structures. Hence, many solutions are influenced by the semantics of the available texts and by the encoding framework as well.

4.1 The Text Sources: A Brief Description

Unfortunately it is virtually impossible to find a single text describing a problem area. In fact, the knowledge engineer needs domain texts starting from the level of "secondary school knowledge", in order to facilitate his cooperation with domain experts. Moreover, due to the multilingual environment, our KB has to be acquired from texts in different languages which overlap and complement each other. In this way, even the type definitions of such basic concepts as *separator* contain knowledge extracted from several texts in several languages.

In the current version of our KB, the basic knowledge was provided by [8], which is a textbook oriented to students in metallurgy, technical chemistry etc. This text describes the physical principles used in admixture separation as well as massive devices used in metallurgy. The textbook contains detailed definitions of the terms and makes possible the understanding of other knowledge sources.

The KB was completed by facts acquired from [20], [9] and [11]. These are encyclopedias oriented to professionals and give more information about contemporary industrial trends (e.g., the application of smaller separators with corrugated fiberglass plates). Thus, our KB incorporates knowledge about separators used in metallurgy and smaller devices applied in technical chemistry. All these devices exploit the same principle of precipitation of particles.

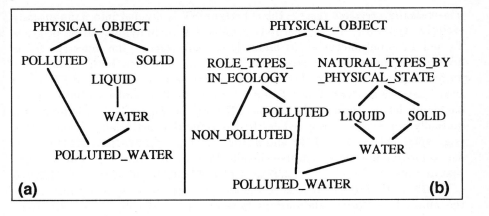

Fig. 1. Type Hierarchies

4.2 Mapping Texts to KB Units

In this section we discuss how the source texts were used for the acquisition of the different CG structures.

Type definitions of concept types: Each type definition contains *positive knowledge* expressing how the concept type can be distinguished from its genus (the supertype). For instance, *oil separator is a device for oil separation based on the principle that oil is lighter than water and consequently it can be collected from the water surface* (see section 5). The type definition also lists characteristics valid for every instance of this type.

When deciding what is to be encoded in a type definition, we analyse whether the semantics of the acquisition texts defines basic type features. Therefore, we can say that our KB is as detailed as the text sources.

Each type definition is verbalised as a whole and according to the explanation style of the selected natural language (e.g., plurals, massnouns).

Type hierarchy: The domain TH is acquired in parallel with the type definitions. The vertical links between types and supertypes evolve from text patterns like "A is B with features C" as follows: the A-type is considered as a *candidate child* of B-type, e.g., OIL-SEPARATOR < DEVICE. Then A is placed in the TH unless there is another type already present as a child of B, e.g., OIL-PROCESSING-DEVICE < DEVICE. In the latter case, one must determine more precisely the immediate supertype of A and further domain information is needed.

The horizontal direction, i.e. relationships between sister types, requires detailed analysis of domain semantics as well. In general, if "A is B with features C" and "D is B with features E" then C and E must distinguish explicitly A and D as immediate subtypes of B, i.e., C and E define the necessary and sufficient conditions for an individual to belong to A or D respectively. There are two kinds of relationships between sister types and their immediate supertypes: (*i*) *non-mutually exclusive types* – the supertype ST is divided into non-mutually exclusive types T1, T2, ..., i.e. it is possible for an individual to be of types T1 and T2 simultaneously; (*ii*) *mutually exclusive types* – the supertype ST is divided into mutually exclusive types T1, T2, ... However, both relationships do not require exhaustive classification, i.e. for each individual I of ST, there is at least one subtype T of ST and I belongs to T. In order to fully explain our point, let us compare the sample TH given on Figure 1 – (a) is built in a mutually non-exclusive way, while (b) has somewhat artificially created types (e.g., ROLE_TYPES_IN_ECOLOGY and NATURAL_TYPES_BY_PHYSICAL_STATE) in order to provide mutually exclusive classification. In this way, in (b), we are sure that an individual is either POLLUTED or NON_POLLUTED and at the same LIQUID or SOLID. Both TH are lattices, so the tree- or lattice-nature is not implied by the mutually-exclusive relationships type-supertype. However, the most common approach for building TH is to mix the mutually exclusive and mutually non-exclusive types, because the exclusive approach at every TH level is feasible for very simple domains only. Usually these issues are not discussed in details in most AI papers, just in contrast to the database design, where all exclusive relationships should be explicitly declared in order to allow special processing [13]. Some bigger AI hierarchies (e.g., Komet/Penman Upper Model [3]) often make mutually exclusive distinctions and contain respective (somehow dummy) type labels like String and Non-string. These distinctions show a systematic attempt to create detailed and accurate classification hierarchies whenever possible [3].

Now, after these general discussions, let us exemplify the principles we followed. We construct *mutually exclusive* types by introducing artificial labels if this is a domain-meaningful distinction (e.g., the NATURAL_TYPES_BY_PHYSICAL-_STATE label on Figure 2). Therefore, the NON_POLLUTED type is not needed, as we are in the domain of pollution. Then type ROLE_TYPES_IN_ECOLOGY from Figure 1 (b) might be ommited as well. However, the type labels are not sufficient to indicate types intersection; that's why we explicitly declare that the distinctions under NATURAL_TYPES_BY_PHYSICAL_STATE are characterised as EXCLUSIVE (see the respective graph on Figure 2). If we want to add a new type - say GAS - it is at first considered as a candidate- subtype of PHYSICAL_OBJECT. After analysis of the other already present subtypes of PHYSICAL_OBJECT, GAS will be moved down as a type mutually exclusive to LIQUID and SOLID. The only sufficient condition for wrong EXCLUSIVE-classification is to encounter a type or individual belonging to more than one type inside an EXCLUSIVE-supertype.

Another important question is the kind of IS-A relation yielding the type subsumption. In [10] some examples are given, how the relations *Subset* and *MemberOf* can be mixed up because of the misleading similarity of their "isa" verbalisation in English. Actually, *Subset* is valid between types (CAT is a MAMMAL) while *MemberOf* is defined between an individual and its type. In our KA experience, however, we faced more complicated language definitions of supertypes, such as "A is B in context C" or "A is B from viewpoint C". For instance, "Physical objects in ecology are polluted or non-polluted". The non-exclusive sister types

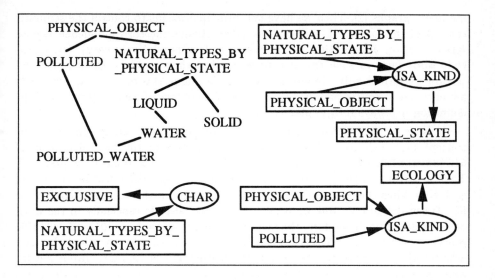

Fig. 2. ISA-KIND Example

are due to such distinct IS-A relationships which, in fact, imply multiple views (from different perspectives) to the same supertype and thus reflect important semantic ambiguities in the domain. For instance, on Figure 2 we might need to add another natural EXCLUSIVE-subtype of PHYSICAL_OBJECT, say ANIMATENESS with subtypes LIVING and NON_LIVING. Then PHYSICAL_OBJECT would be considered simultaneously from ecological, physical and biological perspective. At present we still do not apply the ISA_KIND conceptual relation as exemplified on Figure 2 although it is very tempting to verbalise domain knowledge from different domain-dependent perspectives with corresponding selective mechanism for multiple inheritance.

Conceptual Graphs: In general, we represent knowledge about a phenomenon, state, process or function in one graph. The design of the concrete devices with horizontal or parallel plates, with chambers, tubes for water input/output etc. is given in separate graphs too. Other sample graphs are: "Separators are the most widely used devices for precipitation", "Industry produces oil-water mixtures", "Oil in water is either lighter than water and swims up or it is augmented with mineral particles and falls down because it is heavier than water" (concrete measures are given as well).

In fact, all knowledge that does not fit into our encoding convention for type definitions, is encoded as graphs. The type definition concerns concept features to be verbalised as valid for *every* instance of this type, while the context-sensitive knowledge is presented in graphs and different explanation styles are applied.

Task-dependent conventions concerning the granularity of graphs and type definitions: In DB-MAT the user formulates requests for domain explanations and relevant answers are extracted by projection using corresponding query graphs [2]. Then we have two principal means to control the semantic content of the future NL explanation: (*i*) to change the correspondence *request types* ↔ *query graphs* (this sort of customisation is relatively easy); (*ii*) to encode some facts in such CG structures so that they could be extracted under certain conditions only. Under our current paradigm, this means that such facts are encoded in conceptual graphs, because the content of each context and type definition is always fully explained.

It is clear that certain facts can be positioned in a KB graph or in the concept type definition. For instance, we can have in the type definition of OIL-SEPARATOR that it is [OIL-PROCESSING DEVICE] -> (CHAR) -> [WEIGHT]. Then according to the DB-MAT paradigm this characteristic will be verbalised each time when the user requires domain explanation with the menu item "What is?" (term definition). But we want to avoid trivial facts especially in the most basic explanation. Therefore, the fact is encoded in a graph as: [OIL-SEPARATOR: every] -> (CHAR) -> [WEIGHT]. Then it will be verbalised only when the user asks for all characteristics. Moreover, the feature WEIGHT is more important for massive devices, e.g., in metallurgy. In the description of such devices there is a concrete value of their weight: [OIL-SEPARATOR: #23] -> (CHAR) -> [WEIGHT: 2 t]. So in the process of asking questions, the user will be given more information about WEIGHT when such individuals are concerned.

In this way, the separation of knowledge between graphs and type definitions is used as a task-oriented framework for determining the semantic content of the NL explanation within the contexts of the questions being asked. In other words, we want to apply the same CG projection operation, but we also want to be able to draw the user's attention to one or another aspect depending on what is highlighted in the source text and what request is formulated. We plan further elaboration of our algorithms for semantic extraction in order to tune them better to the user expectations.

Concepts: Each term is represented either as a concept or as a situation. Cases like the German compound Wellplatte (corrugated plate) require additional design of situations which provide the correct lexicon link only. In this way the graph [PLATE] -> (ATTR) -> [CORRUGATED]. will be verbalised as "Wellplatte" only when it is encoded as [SITUATION: [PLATE] -> (ATTR) -> [CORRUGATED]] .

Contexts: At present we do not need negation, so there are no negative contexts. This is due to the technical domain where most factual information is positive. As for the other kinds of contexts, situations are used to encode complex terms with different granularity across the languages and, also, to represent semantic situations in processes, functions, etc.

Conceptual Relations: Events are encoded as concept types. We discuss briefly our empirical techniques for event identification and choice of participants. We identify at first fragments of texts, describing e.g. a process. Afterwards, using some semantic dictionary of verbs, we choose a corresponding verb "naming" the process (it could happen that the selected verb is not mentioned in the text at all). For instance, behind all phrases like "water flows to chambers", "oil swims up to surface" stay some processes that can be roughly named

MOVE. Then the participants and the corresponding conceptual relations are defined by the linguistic dictionary, where the thematic roles of MOVE are listed. Certainly, when replacing FLOW and SWIM by MOVE (or TRANSFER) we lose the lexical expressiveness and turn the NL to an artificial one. At this step we were somehow guided by Roger Schank's ideas that the process semantics can be expressed in a "coordinate system" of (fixed number of) semantic primitives [12]; actually we were influenced by the *spirit* of his NL examples and their *rough* translation into semantic primitives although our semantic primitives (names of processes) are neither restricted nor predetermined. These considerations help us to keep the number of concept types relatively small. The ideal generator, however, with an ideal lexicon, should build exactly explanations that *water flows* (using "movement from A to B") and *oil swim up* (using "movement from A to B in environment C with direction *up*).

Individuals:

To discuss the individuals of material objects we consider again the somewhat artifical TH on Figure 1(b). As far as an individual can change its role from POLLUTED to NON_POLLUTED, it seems better to "attach" individuals to the parent-type, otherwise we could restrict the role changes which individuals might undergo. An example is given in [18, pp. 139-140] where the PERSON Tom grows up from BABY to ADULT. These considerations show us additional reasons why our ISA-KIND relation from Fig. 2 is a good idea: we could distinguish between role subtypes and natural subtypes of a given type. In the simple technical world of DB-MAT, however, we do not encounter such complications. Our individuals of material objects are the concrete devices for oil-separation. Their types have BOTTOM as an immediate subtype in the TH.

Concerning the individuals of more abstract types, we identify them according to the granularity and domain importance of their types. For instance, in [SOLUBILITY_OIL_IN_WATER: {none, low}] we encode the solubility as one of the two possible string values, i.e. NONE and LOW. The latter are individuals since they have no domain subtypes or special domain meaning and, therefore, need not to be concept types. In contrast, the physical states of oil in water (e.g. colloid, dispersed) are also enumerable but are domain terms as well, so they cannot be encoded as a *disjunctive set* referent. In this way we see, unfortunatelly, that the abstract individuals of one domain can be types of abstract individuals in another domain. Therefore, the choice of abstract individuals (which are more or less only labels) is just a matter of convention.

4.3 The Acquisition Cycle

Language vs. knowledge: granularity of concepts/situations. The final granularity is established at the moment when the concept labels are linked to the lexicon entries. This means that when we integrate a new language in our system – say Romanian – we may need new situations which would serve only to provide links to the Romanian lexicon. But as far as the KA relied on multiligual text sources, we hope that the basic domain types are already presented in the KB and no additional types will be necessary.

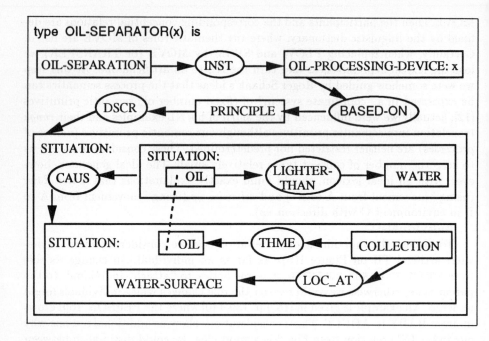

Fig. 3. A Sample Type Definition

Definite referents: Since Russian has no articles, the knowledge acquired from [8] contained no concepts with definite referents. Later we encountered different grammar rules for articles in German and Bulgarian. Therefore, we do not encode definite referents in the KB since they are language-dependent and appear in the explanation depending on syntactic rules and lexicon information.

Task-specific changes: With the evolution of the generator, we revised and elaborated previous KB solutions concerning (*i*) plural referents, set referents and *every* quantifier (also a referent); (*ii*) the task-dependent conventions for granularity of graphs and type definitions (see Section 4.2).

5 Example of NL Explanations

Figure 3 gives the type definition of oil separator as an oil processing device for oil separation based on the principle that oil is lighter than water and consequently it can be collected from the water surface. The description relation DSCR separates the encapsulated information from the information in the outer context, as introduced in [17, p.11]. The conceptual relation LIGHTER_THAN has a corresponding type definition. But it is important in our technical domain and there is also a link to the lexicon and the grammatical information, so LIGHTER_THAN is verbalised as a single lexeme instead of its type definition. If OIL_SEPARATION is related to some lexicon, we receive explanation that oil separator is used for oil separation. Otherwise, if the lexicon links are missing, the generator will be forced to use more general terms as in *oil separator is an oil-processing device for separation* in a subsumption hierarchy where OIL_SEPARATION_DEVICE < SEPARATION_DEVICE < DEVICE.

6 Evaluation

Our domain ontology contains at present about 500 types. The KB comprises about 60 graphs and 20 type definitions. In our evaluation we adopt the notions of *size*, *depth* and *breadth* of semantic coverage [7] although this evaluation scheme concerns mostly NLP systems which construct meaning representations. The *size* of our knowledge resource - measured by number of labels in the ontology - is relatively small if compared to other ontologies with thousands of labels. The *depth* - amount of information per type - is a task-oriented parameter in our case. As far as DB-MAT generates explanations for human users, we do not encode trivial facts like e.g. "separators have colour" and we do not support type definitions of the trivial types in the upper model. To measure the *breadth* - types of information - requires evaluation of a KB of CG plus corresponding lexical and grammatical resources. Our KB covers events and participants, thematic roles, temporal and spatial relations, reference and coreference etc. So in a sense, our KB has more breadth than depth. DB-MAT has a good ontological coverage of its rather narrow world. At present we aim at building a non-trivial size KB.

The primary goal of manual KA is to build a generalised domain model rather than representation of some text semantics. The main problem remains to decide *what to talk about and how to tell it* given the numerous domain sources (multilingual texts, pictures, etc.). These decisions in *manual KA* are rather intuitive and task-dependent, so the enlargement of such a KB requires iterative steps over the acquisition sources. Moreover, returning back to the acquisition technique which was sketched in 4.2. for conceptual relations, we see that two knowledge engineers can acquire different graphs from the same text fragment. Even when they choose the same graph structure, the name of the concept type denoting the process can be a different verb (one of the many possible synonyms). So there will be no way to justify such solutions for further KB comparison. Moreover, there are no formal methods to control and indicate acquisition mistakes. Let us summarise as well the main difficulties of *automatic KA*:

(*i*) after lexical analysis and some disambiguation, automatic KA needs semantic analysis for the identification of events, participants, and corresponding thematic roles. For instance, to recognize AGNT, INST and DEST in the graph "John goes to Boston by bus" requires a very detailed semantic information in the lexicon, attached to every meaning of the verb *go*. But this example is syntactically very simple because the direction of arrows between concepts and conceptual relations is clear if once *go* is defined in the lexicon. In human-oriented domain texts, however, there are sophisticated usages of passive voice, moreover of reflexive verbs, where e.g. in Bulgarian it could be problematic even for a human reader to determine the participants. Therefore, only the simplest message-like texts can be used for KA without syntactical analysis.

(*ii*) automatic KA should resolve the different types of referring expressions. Consider the sentence "oil is lighter than water and consequently it is collected at the water surface". Only an elaborated NLP system can resolve the anaphora attaching *it* to *oil* but not to *water*, thus building the correct coreference link. Also, consider the so called non-personal verbs (which in some languages, e.g. Bulgarian, denote a good scientific style) where the agent/effector is practically missing in the sentence and is implied in the context.

(*iii*) it is not easy to acquire types and their subsumption. Very often the terminological dictionaries define "A is B" where B is not the immediate super-type; the human authors tend to cite the supertype that is most well-known. For instance, we will find "separator is a device" rather than "separator is oil-processing device". Due to these human-oriented explanations, the iterations for constructing the type hierarchy and type definitions will require at least some human aided interventions.

The automatic KA seems more promissing if we expect the existence of corresponding domain texts, specially prepared by domain experts as KA input.

7 Conclusion and Future Work

At present, we are actively evaluating our approach and the KB by comparisons to other existing ontologies. But we also try to study more complicated semantic features of the different conceptual relations and their influence on the inheritance mechanism. For instance, we plan to include a CHAR-distinction countable/uncountable in our KB, because we find such feature is not linguistic but semantic one. Then we can elaborate algorithms for inheritance and verbalisation of uncountable types.

To summarise, we argue that the KA task for the goals of multilingual application can be successful mainly in technical domains after a careful study of the domain and after development of certain principles for knowledge encoding.

Acknowledgements: We are particularly grateful to Zhivko Angelov for our discussions and his comments on the initial drafts of this paper.

References

1. G. Angelova and K. Bontcheva. DB-MAT: A NL Based Interface to Domain Knowledge. In A. Ramsay, editor, *7th International Conference on Artificial Intelligence: Methodology, Systems, Applications (AIMSA '96)*, number 35 in Frontiers in AI and Applications, pages 218 – 227, Sozopol, Bulgaria, Sept. 1996. IOS Press.
2. G. Angelova and K. Bontcheva. DB-MAT: Knowledge Acquisition, Processing and NL Generation. In P. Eklund, G. Ellis, and G. Mann, editors, *Conceptual Structures: Knowledge Representation as Interlingua*, number 1115 in Lecture Notes in AI. Springer-Verlag, Berlin, 1996.
3. J. A. Bateman, R. Henschel, and F. Rinaldi. The generalized upper model 2.0. Technical report, GMD/Institut für Integrierte Publikations- und Informationssysteme, Darmstadt, Germany, 1995.

4. C. Galinski and K.-D. Schmitz, editors. *Terminology and Knowledge Engineering, TKE'96*. Indeks Verlag, 1996.

5. F. Lehman. Big posets of participatings and thematic roles. In P. Eklund, G. Ellis, and G. Mann, editors, *Conceptual Structures: Knowledge Representation as Interlingua*, number 1115 in Lecture Notes in Artificial Intelligence. Springer-Verlag, 1996.

6. S. Myaeng, C. Khoo, and M. Li. Linguistic Processing of Text for a Large-Scale Conceptual Information Retrieval System. In W. Tepfenhart, J. Dick, and J. Sowa, editors, *Proceedings of 2nd Int. Conf. on Conceptual Structures (ICCS'94)*, number 835 in LNAI, College Park, 1994. Springer-Verlag.

7. S. Nirenburg, K. Mahesh, and S. Beale. Measuring semantic coverage. In *Proc. 16th International Conference on Computational Linguistics COLING'96*, Copenhagen, Denmark, August 1996.

8. V. Pushkarev, A. Juzhaninov, and S. Men. *Cleaning of Waste Waters Containing Oil*. Moscow, Metallurgy, 1980. (In Russian).

9. Roempp. *Roempp Lexicon UMWELT*. Thieme, Stuttgart, 1993.

10. S. J. Russell and P. Norvig. *Artificial Intelligence: a Modern Approach*. Prentice Hall, 1995.

11. H. Rutfer and K.-H. Rosenwinkel. *Taschtenbuch der Industrie-abwasserreinigung*. Oldenbourg Verlag, 1991.

12. R. Schank, N. Goldman, C. Rieger, and C. Riesbeck. *Conceptual Information Processing*. North Holland, 1975.

13. J. Smith and D. Smith. Database abstractions: Aggregation and generalization. *ACM Trans. of Database Sytems*, 2(2):105 – 113, 1977.

14. J. Sowa. *Conceptual Structures: Information Processing in Mind and Machine*. Addison Wesley, 1984.

15. J. Sowa. Conceptual Graphs Summary. In T. Nagle, J. Nagle, L. Gerholz, and P. Eklund, editors, *Conceptual Structures: Current Research and Practise*. Ellis Horwood, 1992.

16. J. Sowa. Lexical Structure and Conceptual Structures. In J. Pustejovsky, editor, *Semantics in the Lexicon*. Kluwer Academic Publishers, 1993.

17. J. Sowa. Processes and Participants. In P. Eklund, G. Ellis, and G. Mann, editors, *Conceptual Structures: Knowledge Representation as Interlingua*, number 1115 in Lecture Notes in Artificial Intelligence. Springer-Verlag, 1996.

18. J. Sowa. *Knowledge Representation: Logical, Philosophical, and Computational Foundations*. PWS Publishing Co., Boston, MA, 1997. Forthcoming.

19. W. Tepfenhart, J. Dick, and J. Sowa, editors. *Conceptual Structures: Current Practises*, number 835 in Lecture Notes in Artificial Intelligence, Berlin, 1994. Springer-Verlag.

20. Ullmann. *Ullmanns Encyklopaedie der technischten Chemie. 4th Edition*. Weinheim, 1991. Band 6.

21. W. v.Hahn and G. Angelova. Providing Factual Information in MAT. In *Proc. Int. Conf. Machine Translation: Ten Years On*. Cranfield, 1994.

Uncovering the Conceptual Models in Ripple Down Rules

Debbie Richards and Paul Compton

Department of Artificial Intelligence
School of Computer Science and Engineering
University of New South Wales
Sydney, Australia
Email: {debbier, compton}@cse.unsw.edu.au

Abstract: The need for analysis and modeling of knowledge has been espoused by many researchers as a prerequisite to building knowledge based systems (KBS). This approach has done little to alleviate the knowledge acquisition (KA) bottleneck or the maintenance problems associated with large KBS. For actual KA and maintenance we prefer to use a technique, known as ripple down rules (RDR) that is simple, yet reliable, and later see what models can be produced from the knowledge for the purpose of reuse. Tools based on Formal Concept Analysis have been added to RDR to uncover and explore the underlying conceptual structures.

1 Models and their Role in Knowledge Acquisition

Since Newell's [21] paper on "The Knowledge Level" there has been increasing awareness and acceptance of the need to model knowledge at a level above its symbolic representation. This notion was further explored by Clancey [3] who used task and problem solving methods analysis to divide problems into "heuristic classification" and "heuristic construction". Following Van de Velde [33] approaches which have been built on Newell's knowledge level model include:- Generic Task Framework [2], KADS and CommonKADS [30], Role-Limiting Methods [19], Components of Expertise and the Componential Methodology [32], Protege and Protege II [24], KIF and Ontolingua [22]. All of these approaches impose a particular structure on the knowledge to enable the current problem to be mapped into the appropriate class of problem situation. The structure chosen will depend on the features of the task and the domain of expertise. While each of the approaches mentioned above are different, Van de Velde [33, p.1218] considers three concepts to be generally included as part of a knowledge level model. These are the domain model, the task model and the problem solving method.

While the knowledge level approach is superior to the previous transfer of expertise approach to KA, matching the problem solving method or methods to the problem and adapting them to suit the domain is no mean feat [40]. The use of methodologies such as KADS requires extensive involvement of a knowledge engineer and the modeling process is complex, often doing little to alleviate the KA bottleneck or the problems associated with maintaining large KBS [20].

In the above approaches the purpose of modeling is to allow the capture of expertise in a structured and systematic way. Clancey defines a model as "merely an abstraction, a description and generator of *behaviour patterns over time*, not a mechanism equivalent to human capacity." [4, p. 89]. Since models are at best imperfect representations that vary between users and the same user over time [11] we prefer to use simple, yet reliable techniques for KA. However, we are interested in modeling as an end in itself due to their "explanatory value as psychological descriptions" [4, p.89] and their usefulness in instruction [30].

This paper considers the combination of two approaches, Formal Concept Analysis (FCA) [36] and Ripple Down Rules (RDR) [5], that reduce KA to tasks that can easily be performed by an expert with minimal *a priori* modeling or involvement of a knowledge engineer. In FCA the expert defines a context, which is a Object-Attribute crosstable such as the one shown in Figure 1. From the context, formal concepts are derived, which are ordered to provide a complete lattice of sub and super concepts. These concepts can be represented as a line diagram, also known as a *Hasse* diagram, and can be used to derive implications for use in a knowledge base [37]. This paper investigates starting from the opposite direction by using existing rules in an RDR KBS, acquired manually from an expert, to define contexts and then generate the concepts using FCA to make explicit the models implicit in the rules.

KA using RDR requires the expert to assign a conclusion to a case and then to select the feature/s in the case that were used to make that decision. Each time a new case is seen the expert checks the conclusion assigned by the system and if the expert does not agree a new conclusion is assigned and the features which differentiate the current misclassified case from the case associated with the rule that inappropriately fired are selected by the user. FCA requires some consideration of the whole domain as does Repertory Grids [11] and does not consider incremental maintenance. RDR on the other hand develops the whole system on a case by case basis and automatically structures the KBS in such a way to ensure changes are incremental. The classification of cases and identification of features is very simple and probably less demanding for experts than the development of crosstables.

	Has wings	flys	suckles young	warm-blooded	cold-blooded	breeds in water	breeds on land	has scales
Bird	X	X			X		X	
Reptile					X		X	X
Amphibian					X	X		
Mammal			X	X			X	
Fish					X	X		X

Fig. 1. Context of "Vertebrates of the Animal Kingdom"

FCA and RDR both see that KA should be primarily a task performed by the expert. RDR is founded on the realisation that experts do not offer explanations of why they made a decision, rather they offer a justification and that justification will depend on the situation [5]. Given the inconsistency that exists in users' perceptions, the effort involved in developing the models required by many of the KL-Model approaches mentioned may be pointless. Using FCA and RDR reduces modeling to the tasks of classifying objects (cases) and the identification of the salient features. Interestingly, they approach classification from alternative perspectives. FCA is concerned with finding the similarity between objects, the conjunction of sets of attributes. RDR looks at differences between cases (objects) and is conceptually close to research using repertory grids [11] which is based on Personal Construct Psychology (PCP) [17] and the use of a discernability matrix in Rough Sets [23].

Another view held by FCA and RDR is the importance of context on affecting human behaviour and its impact on the concepts (or rules) formulated and their interpretation [5], [38]. The formalisation of context by FCA, the RDR approach to context and the merging of these approaches are described later in the paper.

While KA and maintenance are easily performed by the expert without an KE [7] it was found that RDR did not easily lend itself to certain modes of use where a model of the knowledge was one of the purposes of building the KBS, such as explanation, *what-if* analysis and critiquing [26]. Other work has looked at deriving causal models from heuristic systems for diagrams and concluded that extra knowledge was needed to provide causal explanation [18]. The explanation provided by the RDR rule pathways has been shown [6] to be superior to a conventional rule trace because the exception structure shows how the knowledge has evolved, why a case has both succeeded and failed and what alternative pathways are possible. However, explanation often required the presentation of a model of the concepts that a rule or collection of rules represents. For the purpose of critiquing it was necessary to determine how close one rule pathway was to another to see if the user's conclusion was within acceptable limits of the system's conclusion or how the two differ.

The issue of the reuse of knowledge and the different ways in which a user may want to use and view the knowledge, prompted the investigation of a technique that would allow the relationships between rules and conclusions to be found. FCA has been shown to be a robust method of finding, ordering and displaying formal concepts [37]. Since the original submission of this paper we have been looking at the extensional definition of the subsumption relationship as adopted by Wille. As Zalta [39] points out intensional subsumption implies extensional subsumption but the converse is possibly but not necessarily true. As a result, in further work we have conducted on comparing concepts (in the form of pathways, new rules, conclusions, etc) we have only been using the intensional definition of concepts to determine the closeness of the concepts to other and to identify sub and superconcepts in the hierarchy. We continue to use Wille's method for finding concepts and drawing line diagrams even though we ignore the extensional definition of concepts for many of our purposes because they were too restrictive. This latter work is still in progress and is not described in this paper. This paper concentrates on what we have done using FCA. We first briefly describe FCA and RDR individually and then describe how FCA tools have been added to RDR. A small case study is included to demonstrate the possible benefits of combining these techniques for comparing the conceptual models of multiple experts.

2 Formal Concept Analysis

Formal Concept Analysis, first developed by Wille [34], is "based on the philosophical understanding of a concept as a unit of thought consisting of two parts: the extension and intension (comprehension); the extension covers all objects (entities) belonging to the concept while the intension comprises all attributes (or properties) valid for all those objects" (37, p.493). A set of objects and their attributes, known as the extension and intension respectively, constitute a formal context which may be used to derive a set of ordered concepts. The following description of FCA follows Wille [34].

A formal context (κ) has a set of objects G (for *Gegenstande* in German) and set of attributes M (for *Merkmale* in German) which are linked by a binary relation I which indicates that the object g (from the set G) has the attribute m (from the set M) and is defined as: $\kappa = (G,M,I)$. Thus in Figure 1 we have the context κ of animals with G = {bird, reptile, amphibian, mammal and fish} and M = {has wings, flys, suckles young, warm-blooded, cold-blooded, breeds in water, breeds on land, has scales}. The crosses show where the relation I exists, thus I = {(bird,has wings), (bird,flys), (bird,cold-blooded), (bird,breeds on land), (reptile, cold-blooded),...,(fish, has scales)}.

A formal concept is a pair (X,Y) where X is the *extent*, the set of objects, and Y is the *intent*, the set of attributes, for the concept. The derivation operators:

$$X \subseteq G : X \longmapsto X' := \{m \in M \mid gIm \text{ for all } g \in X \}$$
$$Y \subseteq M: Y \longmapsto Y' := \{g \in G \mid gIm \text{ for all } m \in Y\}$$

are used to construct all formal concepts of a formal context, by finding the pairs (X'',X') and (Y',Y''). We can obtain all extents X' by determining all row-intents $\{g\}'$ with $g \in G$ and then finding all their intersections. Alternatively Y' can be obtained by determining all column-intents $\{m\}'$ with $m \in M$ and then finding all their intersection. This is specified as:

$$X' = \bigcap_{g \in X} \{g\}' \qquad\qquad Y' = \bigcap_{m \in Y} \{m\}'$$

Less formally, we take the set of objects, G, to form the initial extent X which also represents our largest concept. We then process each attribute sequentially in the set M, finding the intersections of the extent for that attribute with all previous extents. Once the extents have been found for all attributes, the intents X' for each extent X may be found by taking the intersection of the intents for each object within the set. Thereby we determine all formal concepts of the context κ by finding the pairs (X,X').

Having found the concepts it is necessary to find the subconcept-superconcept relation between concepts so that they may be ordered and represented as a labelled line diagram. We can use the subsumption relation $\leq$ on the set of all concepts formed such that $(X_1,Y_1) \leq (X_2,Y_2)$ iff $X_1 \subseteq X_2$. For a family (X_i,Y_i) of formal concepts of κ the greatest subconcept, the join, and the smallest superconcept, the meet, are respectively given by:

$$\bigvee_{i \in I} (X_i,B_i) := \left(\left(\bigcup_{i \in I} A_i\right)'', \bigcap_{i \in I} B_i\right) \qquad \bigwedge_{i \in I} (X_i,B_i) := \left(\bigcap_{i \in I} A_i, \left(\bigcup_{i \in I} B_i\right)''\right)$$

From Lattice Theory, we are able to form a complete lattice, called a concept lattice and denoted $\mathfrak{B}(\kappa)$, with the ordered concept set. The concept lattice provides "hierarchical conceptual clustering of the objects (via the extents) .... and a representation of all implications between the attributes (via its intents)" [37, p.497]. The line diagram in Figure 6 is our implementation, known as MCRDR/FCA, which involved enhancing a current MCRDR for Windows system using Visual Basic. The concepts are shown as small boxes and the sub/superconcept relations as lines. Each concept has various intents and extents associated with it. In MCRDR\FCA it is possible to display the concept number, attribute/s or object/s belonging to each node or all three dimensions can be displayed concurrently, as in Figure 6. It is also possible to click on an individual node to see the concept number and all of its extents and intents. The labeling provided has been reduced for clarity. All intents of a concept δ are reached by ascending paths from δ and all extents are reached by descending paths from the concept δ.

3 Ripple Down Rules

Ripple down rules were proposed in answer to the problem of maintaining a large clinical pathology KBS [5]. To avoid the problem of side-effects that occur when maintaining a typical production rule KBS, rules are never changed or deleted in an RDR KBS. If a case is misclassified a new rule is added as a refinement to the rule that gave the wrong conclusion. This new rule is reached only if the same sequence of rules is followed. In single classification RDR, we define an RDR as a triple <rule,X,N>, where X are the exception rules and N are the if-not rules [28], see Figure 2(a). When a rule is satisfied the exception rules are evaluated and none of the lower rules are tested. This study has used Multiple Classification RDR (MCRDR) [16] which is defined as the triple <rule,C,S>, where C are the children rules and S are the siblings. All siblings at the first level are evaluated and if true the list of children are evaluated until all children from true parents have been exhausted. The last true rule on each pathway forms the conclusion for the case.

Figure 2(b) shows an example of an MCRDR. MCRDR was chosen for this study since the ability to provide multiple conclusion for a given case is more appropriate for many domains and, more importantly, because the problem of how to handle the false "if-not" branches [27] does not exist.

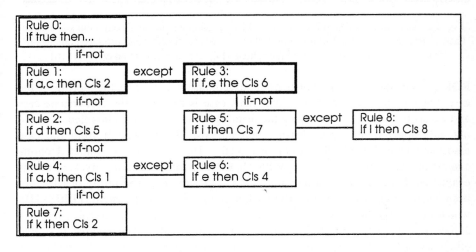

Fig. 2(a). A single classification RDR KBS. The highlighted boxes represent rules that are satisfied for the case {a,c,d,e,f,h,k}

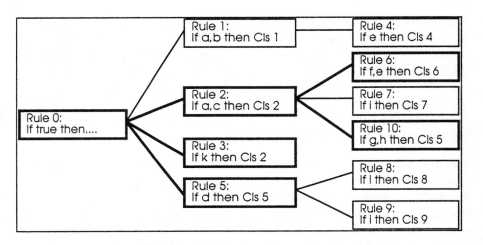

Fig. 2(b). An MCRDR KBS. The highlighted boxes represent rules that are satisfied for the case {a,c,d,e,f,h,k}

Path 1 [(Rule 0,...), (Rule 2, Class 2), (**Rule 6, Class 6**)]
Path 2 [(Rule 0,...), (Rule 2, Class 2), (**Rule 10, Class 5**)]
Path 3 [(Rule 0,...), (**Rule 3, Class 2**)]
Path 4 [(Rule 0,...), (**Rule 5, Class 5**)]

Fig. 2(c). Each of the rules in Fig. 3(a) and 3(b) represent a pathway through the knowledge base. Four pathways from 3(b) are shown above. The rules that fire and produce conclusions from the case {a,c,d,e,f,h,k}are highlighted. (Fig. 2(b) and (c) are taken from [16] .

RDR handles the issue of context by its exception structure and the storing of the case that prompted a rule to be added. This case is known as the cornerstone case and assists the expert in identifying the features in the current misclassified case that not only apply to the new classification but also differentiate it from the case associated with the rule that fired incorrectly. Thus the rule is validated online against the case when it is added. With MCRDR multiple cases are involved in this evaluation, but it has been shown that this is efficient [16]. The utility of RDR has been demonstrated by Pathology Expert Interpretive Reporting System (PEIRS) [10] which went into routine use in a large Sydney hospital with about 200 rules and over a four year period (1990-1994) grew to over 2000 rules. The compactness and efficiency of both RDR and MCRDR have been demonstrated in simulation studies [16].

4   Combining FCA and RDR

As discussed in Section One, certain modes of use required the ability to understand the underlying relationships and models inherent in the RDR rules. The use of Formal Concept Analysis appeared to be a means of discovering these models. To test the benefits and suitability of FCA for such a purpose, MCRDR for Windows was enhanced with FCA tools. The following discussion refers to this implementation, known as MCRDR/FCA. The screen in Figures 3 and 4 use a 60-rule Blood Gases KBS, known as 105, that had been developed from the cornerstone cases associated with the 2000+ PEIRS rules[1].

The first step was to use the rules to generate a context. The RDR KBS was converted to a flat structure by traversing the list structure for each rule picking up the conditions from the parent rule until the top node with the default rule was reached. From this flattened KBS the user chooses either the whole KB or a more narrow focus of attention from which to derive a formal context. When the whole KB is chosen the rules and rule clauses form the extents and intents, respectively. Such a global view is only feasible for small, if not very small, KBS. As with any graphical representation, as the number of rules being modeled grew, the line diagram became too cluttered to be comprehensible.

---

[1] It would have been interesting to use the 2000+ PEIRS rules, but this was not done at this time because they are not in the MCRDR format used by this study.

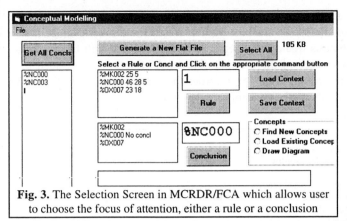

Fig. 3. The Selection Screen in MCRDR/FCA which allows user to choose the focus of attention, either a rule or a conclusion

This was the case with the Blood Gases KBS. Therefore, to limit the concepts to a manageable size that could be understood by looking at a matrix or a line diagram the user was asked to narrow their focus of attention to a particular rule or conclusion. The decomposition of a concept lattice into smaller parts is a strategy that has previously been found useful [35]. Our approach is similar to that proposed by Ganter [13] where the context is shortened to find subcontexts and subrelations.

There are currently 13 different ways a context may be derived. The two main methods are choosing a conclusion or a rule, see Figure 3. If a conclusion was chosen, all rules using that conclusion were selected and added as objects to the set G, forming the extents of the context. As each extent was added the clauses of the rules were added to the set M of attributes to form the intents of the context, first checking to see if any attributes had already been added by previous rules. Where the relation I held, that is object g had attribute m, a cross was marked in the appropriate row and column. If the user chose a particular rule then that rule was added as the first object with the rule clauses as the initial intension. Every clause in each rule in the flattened RDR rule base was searched for a match on the initial set of attributes. If a match was found, that rule was added to the extension and all new attributes (clauses) found in the matching rule were also added to the intension.

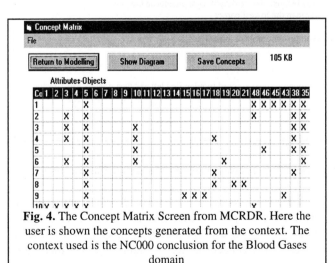

Fig. 4. The Concept Matrix Screen from MCRDR. Here the user is shown the concepts generated from the context. The context used is the NC000 conclusion for the Blood Gases domain

Treating the rule clause, which is an attribute-value pair, as a boolean or condition attribute, is similar to the technique known as *conceptual scaling* [14] which has been used to interpret a many-valued context into a (binary) formal context. A many-valued context, such as that represented in an RDR KBS, is a quadruple (G,M,W,I) where I is a ternary relation between the set of objects G, the set of attributes M and the set of attribute values W (merkmalsWerte in german). Essentially, each attribute is treated as a separate formal context with the values as attributes associated with each of the original objects. A scale is chosen, such as a nominal scale (=) or an ordinal scale (≥), to order these attributes. From the many contexts, one for each attribute, the concepts are derived.

```
For each concept
      Find parent concepts from successor list
      Find children concepts from predecessor list
End For
Locate top concept in fixed position (top and centre)
Get x-factor - find how many branchings in the predecessor list in the second column
Get y-factor - find how many branchings in the successor list in the second column
For all children of the top concept
      position one y-factor lower than the parent  and one x-factor to the right of the previous sibling  remove any
          object labelling from the parent that is found in the child
End For
For all children of the top concept
      Add-children(child)
End For
Add the last concept in fixed position
Reduce attribute labelling

Add-children(parent)  (note the child passed to this procedure is now the parent)
Find the children (parent)
For all children
      If    child is the last concept or has already been located
      Then  process next child
      Else  find out how many parents it has
              If    only one parent
            Then If    child is an only child (its parent has one child only)
                   Then locate the child directly below the parent
                   Else  x-coord = (Parent x-coord - ((NoChildren of Parent * x-factor)/2)
                                    + ((number of child being processed-1)*x-factor)
                   End if
            Else  locate the child midway between left and right most parents on the x-axis
                  and one y-factor lower than the lowest parent
            End If
            If the parent contains some of the children's objects
            Then  remove
            End If
            If    coordinates used
            Then  until a free position is found change x-coord moving first left and then right by half x-factor
            End If
            Flag as located
            Add_children (child)
      End If
```

Fig. 5. The algorithm used to find the concept coordinates for the Hasse diagram

The crosstable generated in the above process was then used to construct all formal concepts of the formal context, using the process described in Section 2. Appropriate ordering of concepts is difficult as a given concept may be a subconcept of different superconcepts. This can be seen in the concept lattice in Figure 4 where there are a number of sets of concepts. Concept ordering can improve the matrix aesthetically. To allow drawing of the *Hasse* diagram it was necessary to compute the predecessors and successors of each concept. Predecessors were found by finding the largest subconcept of the intents for each concept. Successors were found by finding the smallest superconcept of the intents. The successor list was used to identify concepts higher in the diagram, the parents, and the predecessor list identified concepts lower in the diagram, the children. As Wille [37] points out, there is not one fixed way of drawing line diagrams and often a number of different layouts should be used because concepts can be viewed and examined in different ways depending on their purpose and meaning. An algorithm for the graph layout is described in Figure 5. If the line diagram drawn by the system is not to the liking of the user, they may move a node anywhere they like providing the node is not moved higher than any of its parents or lower than any of its children.

5   A Case Study that uses of RDR for KA and FCA for Modeling

A small case study has been performed to demonstrate how FCA can be used for modeling knowledge acquired using RDR. The domain chosen concerns the adaptation and management of the *Lotus Uliginosis cv* Grasslands Maku for pastures in the Australian state of New South Wales. The knowledge was recorded by advisors who were representing local groups of farmers and agribusiness people involved in "Co-learning" about Lotus. Four advisors independently added rules to single classification RDR systems. The domain will be referred to as Lotus and the four KBS will be known as Lotus 1, 2, 3 and 4. Knowledge in the domain is evolving and RDR is being used to accelerate and consolidate knowledge development in the field into a corpus of knowledge about the crop [15]. Such a domain seems eminently suitable for this study since it is small enough to be comprehensible and there is a need to discover the concepts and models that exist but are still emerging.

In order to model each of the KBS it was necessary to reenter the rules into MCRDR. The KBS from the farmers had been developed using XRDR for Windows, a single classification RDR implementation. As stated previously, MCRDR had been chosen as the starting point for adding FCA as there are no false branches which may be difficult to interpret and since many domains, such as Lotus, contain cases that belong to multiple independent classes. It would have been preferable for the individual KBS to have been captured in MCRDR format, since it is always difficult for a non-expert to know what conclusions to assign. While it was easy to assign the classification found in the XRDR version, it was not easy to know which other conclusions may also have been applicable or whether a conclusion should have been stopped. Another problem for a non-expert is the selection of features which justify the conclusion being added or deleted. Given these limitations the knowledge reentered may not be a completely accurate picture of the knowledge entered by the advisor but adequate for the purposes of this case study. Once in MCRDR format, a formal context of each complete KBS was generated and the concepts derived using the processed described in the previous section.

There are two main tools to aid in the comparison of conceptual models. The first is the concept matrix. For space we do not show the four line diagrams associated with each KBS, but we state that we were able to see that each KBS shared a number of concepts and that different concepts could also be identified. By looking at the matrix the farmers are able to see not only what attributes (intents) and conclusions (extents) others consider important but also the relationships between them and how it affects other conclusions.

The second tool for modeling is the line diagram. The diagrams showing the concept labeling only are generally well laid out with minimal line crossing, using the algorithm described in Figure 5. However, in order to understand the models it is necessary to provide complete labeling showing the concept, extents and intents associated with each node. To save space, only the Lotus1 and Lotus4 KBS are shown in Figure 6. The nodes in both diagrams have been moved to fix some overlap of labels and improve readability.

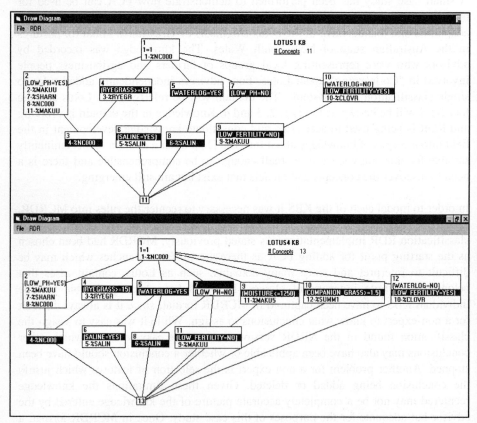

Fig. 6. The line diagrams for Lotus 1 and 4 KBS in MCRDR/FCA

The line diagram provides a more hierarchical understanding of the sub and super relationships in the domain. At a glance we can see that eight concepts are shared by the two KB and that the main difference between the two KBS are the concepts If (MOISTURE ≤ 1250) then the conclusion is %MAKU5 - *Maku OK at rainfalls above 1250mm* and If (COMPANION_GRASS ≥ 1.5) then the conclusion is %SUMM1- *Grass competition is excessive for lotus-keep grass down in summer* which are contained in concepts 9 and 10, respectively, in Lotus4. The diagrams also show that even though the knowledge has been structured slightly differently both farmers consider that when (LOW_PH = YES) and (RYEGRASS ≥ 15) the conclusion should be %NC000 No Conclusion.

To facilitate comparison it was important to ensure the attributes shared by all contexts were in the same order before the concepts were determined. In the case study all rules were used for comparison since the KB size was small enough but it was also possible to choose a particular attribute (rule-clause), rule or conclusion to focus on. It would be useful to view the whole KB to identify variations and then reduce the context by selecting the rules or conclusions that differed.

6 Discussion

The case study has shown that the incorporation of FCA into RDR allows models to be found and compared without the need for prior understanding or explication of that model. This is particularly useful in domains where knowledge is emerging or in the common situation where it is difficult for experts to describe how they arrive at a conclusion. We see that KA using RDR offers a more realistic and reachable goal than approaches that depend on the user to predefine a model.

While these preliminary results appear promising there is still much more work to be done. As mentioned in Section 4, the formulation of concept lattices from many-valued contexts requires their interpretation into a formal context. While this conversion has been relatively straightforward and sufficed for our case study, currently a rule that should be part of the context for a selected focus of attention may be missed if the clause does not match on a conclusion or attribute already selected. The use of different conceptual scales may provide a solution and needs further investigation. Some work has been done using a distance-weighted nearest neighbour algorithm to assign a score to clauses to find if clauses are related at all and to what extent. It may be possible to incorporate these techniques in determining which rules should be added to a context.

The usefulness of FCA to support the reuse of knowledge in a wide range of modes, such as explanation or tutoring, is under investigation. We have begun new work that uses the concepts developed to assist the user to understand how a selected concept, which could be a new rule or conclusion, fits in relation to other concepts in the KBS. This is useful for KA, explanation and critiquing. As mentioned in Section

1, for that work we are only using the intensional definition of the concepts derived using Wille's techniques because of the limitations of the extensional definition [39] and because it is too restrictive in the finding of sub and superconcepts.

The work presented in this paper has simply added to RDR the ideas in FCA that have already been available in other software, such as ConImp and Toscana. The main difference is found in the use of rules to generate contexts. RDR lends itself well to conversion to a formal context because each rule represents a rule pathway. Some work [26] has already been done on the use of rough sets to find relationships in the knowledge base and a comparison will be made between the dependencies and concepts generated using FCA and those found in the cores and reducts computed using rough sets. Other investigations include: a comparison of concept lattices to concept maps [12]; the use of *attribute exploration* for acquisition of formal contexts [36] and review of work done on combining the use of repertory grids and FCA [31].

Acknowledgments

RDR research is supported by Australian Research Council grants. The authors would like to thank Zvi Hochman for providing the LOTUS KBS and related information, to Phil Preston for the MCRDR and the Hierarchy software and to Rudolf Wille for the inspiration of this work.

References

[1] Burmeister, P. (1996) Formal Concept Analysis with ConImp: Introduction to the Basic Features A translation of ConImp -Ein Programm zur Formalen Begriffanalyse In G.Stumme, R.Wille (eds.) *Begriffliche Wissenerarbeitung:Methoden und Anwendungen*, Springer Verlag.

[2] Chandrasekaran, B. (1986) Generic Tasks in Knowledge-Based Reasoning: High Level Building Blocks for Expert System Design IEEE Expert pp: 23-30. Fall 1986.

[3] Clancey, W.J., (1985) Heuristic Classification *Artificial Intelligence*, 1985. **27**: p.289-350.

[4] Clancey, W.J., (1993) Situated Action: A Neurological Interpretation Response to Vera and Simon *Cognitive Science*, **17**: pp.87-116.

[5] Compton, P. and Jansen, R., (1990) A Philosophical Basis for Knowledge Acquisition. *Knowledge Acquisition* 2:2

[6] Compton, P., Edwards, G., Kang, B., Lazarus, L., Malor, R., Menzies, T., Preston, P., Srinivasan, A. and Sammut, C. (1991) Ripple Down Rules: Possibilities and Limitations 6th Banff AAAI Knowledge Acquisition for Knowledge Based Systems Workshop, Banff (1991) 6.1 - 6.18.

[7] Compton, P., Edwards, G., Kang, B., Lazarus, L., Malor, R., Preston, P., and Srinivasan, A. (1992*).* Ripple down rules: Turning knowledge acquisition into knowledge maintainance *Artificial knowledge in Medicine 4* pp:463-475.

[8] Compton, P., Preston, P. and Kang, B. (1995) The Use of Simulated Experts in Evaluating Knowledge Acquisition, *Proceedings 9th Banff Knowledge Acquisition for Knowledge Based Systems Workshop* Banff. Feb 26 - March 3 1995.

[9] Eades, P. (1996) Graph Drawing Methods *Conceptual Structures: Knowledge Representation as Interlingua* (Eds. P.Eklund, G. Ellis and G. Mann) pp:40-49, Springer.

[10] Edwards, G., Compton, P., Malor, R, Srinivasan, A. and Lazarus, L. (1993) PEIRS: a Pathologist Maintained Expert System for the Interpretation of Chemical Pathology Reports *Pathology 25*: 27-34.

[11] Gaines, B. R. and Shaw, M.L.G. (1989) Comparing the Conceptual Systems of Experts *The 11th International Joint Conference on Artificial Intelligence* :633-638.

[12] Gaines, B. R. and Shaw, M.L.G. (1995) Collaboration through Concept Maps CSCL'95 Proceedings September 1995.

[13] Ganter, B. (1988) Composition and Decomposition of Data *In Classification and Related Methods of Data Analysis* (Ed. H. Bock) pp:561-566, North-Holland, Amsterdam.

[14] Ganter, B. and Wille, R. (1989) Conceptual Scaling In *Applications of Combinatorics and Graph Theory to the Biological Sciences* (Ed. F. Roberts) pp:139-167, Springer, New York.

[15] Hochman, Z., Compton, P., Blumenthal, M. and Preston, P. (1996) Ripple-down rules: a potential tool for documenting agricultural knowledge as it emerges. p313-316. In *Proc. 8th Aust. Agronomy Conf.*, Toowoomba 1996.

[16] Kang, B., Compton, P. and Preston, P (1995) Multiple Classification Ripple Down Rules: Evaluation and Possibilities *Proceedings 9th Banff Knowledge Acquisition for Knowledge Based Systems Workshop* Banff. Feb 26 - March 3 1995, Vol 1: 17.1-17.20.

[17] Kelly, G.A, (1955) *The Psychology of Personal Constructs* New York, Norton.

[18] Lee, M. and Compton, P. (1995) From Heuristic Knowledge to Causal Explanations *Proc. of Eighth Aust. Joint Conf. on Artificial Intelligence AI'95*, Ed X. Yao, 13-17 November 1995, Canberra, World Scientific, pp:83-90.

[19] McDermott, J. (1988) Preliminary Steps Toward a Taxonomy of Problem-Solving Methods *Automating Knowledge Acquisition for Expert Systems* Marcus, S (ed.) Kluwer Academic Publishers, pp: 225-256.

[20] Menzies, T.J. and Compton, P. (1995) The (Extensive) Implications of Evaluation on theDevelopment of Knowledge-Based Systems in *Proceedings 9th Banff Knowledge Acquisition for Knowledge Based Systems Workshop* Banff. Feb 26 - March 3 1995,.

[21] Newell, A. (1982) The Knowledge Level *Artificial Intelligence* 18:87-127.

[22] Patil, R. S., Fikes, R. E., Patel-Schneider, P. F., McKay, D., Finin, T., Gruber, T. R. and Neches, R., (1992) The DARPA Knowledge Sharing Effort: Progress Report In C. Rich, B. Nebel and Swartout, W., *Principles of Knowledge Representation and Reasoning: Proceedings of the Third International Conference* Cambridge, MA, Morgan Kaufman.

[23] Pawlak, Zdzislaw (1982) Rough *Sets International Journal of Information and Computer Sciences, 11,* pp:341-356.

[24] Puerta, A. R, Egar, J.W., Tu, S.W. and Musen, M.A. (1992) A Mulitple Method Knowledge Acquisition Shell for Automatic Generation of Knowledge Acquisition Tools *Knowledge Acquisition 4*(2).

[25] Richards, D ., Gambetta, W. and Compton, P (1996) Using Rough Set Theory to Verify Production Rules and Support Reuse *Proceedings of the Verification, Validation and Refinement of KBS Workshop, PRICAI'96* 26-30 August 1996, Cairns, Australia, Griffith University.

[26] Richards, D and Compton, P (1996) Building Knowledge Based Systems that Match the Decision Situation Using Ripple Down Rules, *Intelligent Decision Support '96* 9th Sept, 1996, Monash University.

[27] Richards, D., Chellen, V. and Compton, P (1996) The Reuse of Ripple Down Rule Knowledge Bases: Using Machine Learning to Remove Repetition Proceedings of Pacific Knowledge Acquisition Workshop PKAW'96, October 23-25 1996, Coogee, Australia.

[28] Scheffer, T. (1996) Algebraic Foundation and Improved Methods of Induction of Ripple Down Rules *Proceedings of Pacific Knowledge Acquisition Workshop PKAW'96*, October 23-25 1996, Coogee, Australia.

[29] Schon, D.A. (1987) *Educating the Reflective Practioner* San Francisco: Jossey-Bass.

[30] Schreiber, G., Weilinga, B. and Breuker (eds) (1993) KADS: A Principles Approach to Knowledge-Based System Development *Knowledge-Based Systems* London, England, Academic Press.

[31] Spangenberg, N and Wolff, K.E. (1988) Conceptual Grid Evaluation In H.H. Bock ed. *Classification and Related Methods of Data Analysis* Elsevier Science Publishers B.V. North Holland.

[32] Steels, L. (1993) The Componential Framework and Its Role in Reusability In David, J.M., Krivine, J.-P. and Simmons, R., editors *Second Generation Expert Systems* pp: 273-298. Springer, Berlin.

[33] Van de Velde, W. (1993) Issues in Knowledge Level Modeling In David, J.M., Krivine, J.-P. and Simmons, R., editors *Second Generation Expert Systems* pp: 211-231. Springer, Berlin.

[34] Wille, R. (1982) Restructuring Lattice Theory: An Approach Based on Hierarchies of Concepts In *Ordered Sets* (Ed. Rival) pp:445-470, Reidel, Dordrecht, Boston.

[35] Wille, R. (1989a) Lattices in Data Analysis: How to Draw them with a Computer In *Algorithms and Order* (Ed. I. Rival) pp:33-58, Kluwer, Dordrecht, Boston.

[36] Wille, R. (1989b) Knowledge Acquisition by Methods of Formal Concept Analysis *In Data Analysis, Learning Symbolic and Numeric Knowledge* (Ed. E. Diday) pp:365-380, Nova Science Pub., New York.

[37] Wille, R. (1992) Concept Lattices and Conceptual Knowledge Systems *Computers Math. Applic. (23)*6-9:493-515.

[38] Wille, R. (1996) Conceptual Structures of Multicontexts *Conceptual Structures: Knowledge Representation as Interlingua* (Eds. P.Eklund, G. Ellis and G. Mann) pp:23-39, Springer.

[39] Zalta, E.N. (1988) *Intensional Logic and the Metaphysics of Intentionality*, Cambridge, Massachusetts, MIT Press.

[40] Zdrahal, Z and Motta, E. (1995) An In-Depth Analysis of Propose and Revise Problem Solving Methods *9th Knowledge Acquisition for Knowledge Based Systems Workshop* Banff, Canada, SRDG Publications, Departments of Computer Science, University of Calgary, Calgary, Canada pp:38.1-38.20.

Knowledge Modeling Using Annotated Flow Chart

Robert Kremer, Dickson Lukose and Brian Gaines

Knowledge Science Institute
Department of Computer Science
University of Calgary
2500 University Drive N.W.
Calgary, Alberta, Canada T2N 1N4
Email: {kremer, lukose, gaines}@cpsc.ucalgary.ca

Abstract

This paper describes a user modeling notation called the Annotated Flow Chart (AFC) that is highly intuitive and very easy for the domain experts to use. This notation is a form of "extended" flow chart. This notation is defined using Constraint Graphs. Modelling constructs represented in AFC is then mapped to MODEL-ECS (this is an executable conceptual modeling language based on Conceptual Graphs and Actors formalisms), to enable rapid prototyping of executable knowledge models. This paper describes the mappings between constructs in AFC and MODEL-ECS. An example in knowledge modeling is used to illustrate the application of AFC in rapid prototyping of knowledge models.

1. Introduction

MODEL-ECS is an executable knowledge modeling language based on Conceptual Graphs and an Actors formalism [8]. It is a powerful and flexible graphical tool in the hands of a competent knowledge engineer. But any tool with kind of power and flexibility pays a price: The graphical notation is fairly complex, difficult to understand by non-knowledge engineers, and somewhat cumbersome to handle. Since knowledge modeling languages could be usefully employed by (non-knowledge engineer) domain experts, simplicity and ease of use are two primary requirements for a knowledge modeling language. It therefore seems worthwhile to investigate the possibility of combining the power of a language like MODEL-ECS with the simplicity and ease of use of a simpler, but not-quite-so-expressive, language.

KADS [10] is a popular knowledge modeling methodology from which springs the VITAL-OCML [2] visual design language (see Figure 2 for an example). This is an interesting possibility as the aforementioned simpler language because, if one strips away some of the input/output annotation, it is not much more than a traditional, almost-universally-understood, flow chart. What's left is a flowchart with decomposition; that is, a simple flowchart where processes may be broken down into simpler processes. If one could combine the simplicity of language like this with the processing power of MODEL-ECS, one would achieve a very useful knowledge modeling language indeed.

However, even among non-knowledge engineer domain experts, there are varying degrees of knowledge modeling expertise. Therefore, the ideal situation would be to allow a great deal of modeling language flexibility (for the domain expert), rather than just a single, "hard coded" language. This can be done using a graphical language like

Constraint Graphs [3, 4] that provides a simple method for defining a wide variety of graphical languages. Constraint graphs even allows the "layered" language definitions where the domain expert may start by using a simple flowchart language, but when the model is developed to the point where decomposition of processes is desired, the defined language may easily be extended to include decomposition constructs. The defined language may later be again extended to include input/output and performer constructs as the need arises. When the model reaches a sufficient level of maturity, it can be translated into MODEL-ECS where it may be further elaborated, tested, and even executed by a knowledge engineer.

Section 2 of this paper describes Constraint Graphs (Subsection 2.1) and MODEL-ECS (Subsection 2.2). Section 3 describes the fundamentals of knowledge modeling by using the Annotated Flow Chart (AFC). Section 4 describes the translation of AFC models into its equivalent counterpart in MODEL-ECS, and Section 5 describes the translation of the "Propose and Revise" problem solving method from AFC to MODEL-ECS. Finally, Section 6 concludes this paper with a summary and an outline of future work.

2 Background

In this section, the authors will attempt to describe the two types of graph formalisms that are used for building this knowledge engineering environment. These two graph formalisms are: *Constraint Graphs (CoGs)* and *MODEL-ECS*. CoGs is used to define a graphical notation that is intuative and easy for a domain expert to use, and MODEL-ECS is a graphical based executable conceptual modelling language that is based in *conceptual graphs* and *actors*. The following two subsections will describe each of these formalisms, respectively.

2.1. Constraint Graphs

Constraint Graphs is a meta-graphical formalism in which other graphical formalisms, such as Conceptual Graphs, can be defined. There are only four fundamental component types in Constraint Graphs:

- **Node**: a standard graph labeled node with arbitrary surround shape, color, etc.
- **Context**: a type of node which can contain a subset the components of a graph
- **Arc**: an arc, or connector, between other components of the graph
- **Isa Arc**: an arc subtype which defines any type lattices superimposed on the graph

These are arranged in a type lattice at right:

Since Constraint Graphs is designed to be a medium to represent other graphical formalisms, it must be very general. In particular, arcs are generalized to accommodate the peculiarities of some languages. Unusual rules for arcs are:

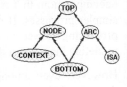

- Arcs terminals may be left "dangling"; that is, not connected to any component. The "dangling" arcs may be later connected to a component, or may be left as a placeholder indicating unknown information:

- Arcs may connect between arbitrary components. That is, arcs are *not* restricted to connect nodes to nodes: arcs may connect other arcs to each other or arcs to nodes. In the following example, the unlabeled arc connecting the two labeled arcs indicates that the *name* arc is a a subtype of the *attribute* arc:

- Arcs need *not* be binary. Arcs must be at least unary, but may be of arbitrary arity. For example, the following arc is an unlabeled trinary arc:

To some languages, this generalization of the concept of an arc is unacceptable. All this flexibility may be removed in a Constraint Graphs language definition using ad-hoc constraints (to be described shortly).

Isa is a subtype of *arc* and is used to superimpose one or more multiple inheritance type hierarchies on the graph. Therefore, declaring one component as a subtype of another component is a simple matter of drawing an isa arc between the two components. Of course, the type lattice so formed implies attribute inheritance from supertypes to subtypes (including the visual attributes displayed in the graph) and the normal typing rules. In particular, arcs inherit their termination restrictions (the component types they may legally terminate on) from their supertypes. The rule is:

Each of an arc's terminals may terminate only on components that are subtypes of the components at the corresponding terminal in all the arc's supertypes.

This rule allows arcs to be specialized (restricted) as one moves down the type hierarchy, and also allows the arity to *increase* as one moves down the type hierarchy.

The overall effect then, is that Constraint Graphs models higher-order typed predicate calculus: nodes correspond to constants and variables, arcs correspond to predicates (or relations). Since arcs can terminate on other arcs, this corresponds to higher order predicates (predicates which take predicates as arguments).

In addition to the above, Constraint Graphs allows the user to annotate components with arbitrary programs to further constraint that component. This feature is known as *ad-hoc constraints* and is necessary to accommodate certain quirks of some graphical formalisms. But ad-hoc constraints are also used for many internal purposes such as to restrict *isa* arcs (and their subtypes) to be binary, non-dangling, and acyclic. There exists a easy-to-use library of constraints from which end users may select pre-existing constraints to apply to arbitrary components.

The Constraint Graphs user interface uses the graph itself (including the superimposed type lattices) to always guarantee the legality of the graph. Additionally, the information in the graph is used to:

- automatically infer arc types: When a user draws an arc from one component to another, the system infers all possible typings. If the typing is unique, the arc is automatically typed; if the typing is ambiguous, the user is prompted to select a type; if there is no possible typing, an error message is displayed.

- provide type information to various popup menus used in the system: Each component has a popup menu for users to access their operators. The list of operations is restricted according to the type information in the graph.

In general, Constraint Graphs is used by a *language designer* to create a language definition. A trivial definition of Conceptual Graphs is:

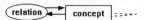

Which says that there exists two node types called *concept* and *relation*; there also exists three arc types called *coreferent* (dotted), *concept-to-relation* and *relation-to-concept*. The three arcs are restricted to terminate on the nodes shown in the definition. The language designer is able to restrict an *end user* to only draw objects of the types in the definition. Therefore, the end user can only create syntactically correct Conceptual Graphs. (But there's nothing to stop the user from creating nonsensical conceptual graphs). This is where conceptual graph processors like the Deakin Toolset [5] or CGKEE [9] comes in play. The graphical user interface is then grounded to the methods in the conceptual graph processors, and all objects created, deleted, and updated are the grounded to the conceptual graph knowledge base that is maintained by these processors.

2.2. MODEL-ECS

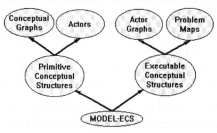

MODEL-ECS is composed of two forms of abstractions, and a set of complex modeling constructs. The two forms of abstractions are: the *Primitive Conceptual Structures*; and the *Executable Conceptual Structures*. There are two types of Primitive Conceptual Structures. They are: *Conceptual Graphs*; and *Actors*. There are also two types of Executable Conceptual Structures. They are: *Actor Graphs*; and *Problem Maps*. The following subsections will elaborate each of these abstractions and describe the complex modeling constructs.

2.2.1. Primitive Conceptual Structures

Conceptual Graphs. Conceptual graphs are finite, connected, bipartite graphs. They are bipartite because there are two different kinds of nodes (i.e., concepts and conceptual relations), and every arc links a node of one kind to a node of the other kind [11, p. 72]. In diagrams (i.e., display form), a concept is drawn as a box, a conceptual relation as a circle, and an arc as an arrow that links a box to a circle. A *generalization hierarchy* is defined over all the conceptual graphs.

Constructs	Descriptions
(FBS)	conceptual relation to indicate *sequence* (Finish Before Start)
(SWS)	conceptual relation to indicate *concurrence* (Start When Start)
[PM]	concept representing a *Problem Map*
[KS]	concept representing a *Knowledge Source*
[TRUE_TEST]	actor graph to indicate *positive* test
[FALSE_TEST]	actor graph to indicate *negative* test
[LT]	actor graph to indicate *less than* test
[GT]	actor graph to indicate *greater than* test
[LTEQ]	actor graph to indicate *less than or equal* test
[GTEQ]	actor graph to indicate *greater than or equal* test
[ASSERT]	actor graph to *assert* a predicate into the working memory
[RETRACT]	actor graph to *retract* a predicate into the working memory
[INCREMENT]	actor graph to *increment* the value of an integer variable
[DECREMENT]	actor graph to *decrement* the value of an integer variable

Table 1: Conceptual Constructs

Actors. There are two types of actors (objects) [6]. They are: the *class (type) actor*; and the *instance actor*. Class actors are defined as abstractions while instance actors are defined as concrete objects. An *actor hierarchy* is defined over all the class actors. An actor responds to an incoming message by executing a method corresponding to the message.

2.2.2. Executable Conceptual Structures

Actor Graphs. Actor Graphs are the most primitive Executable Conceptual Structures. Actor Graphs use conceptual graphs to represent the declarative knowledge associated with an inference step, and an actor to represent the procedural knowledge [6]. An *actor graph hierarchy* is defined over all the actor graphs.

Problem Maps. A Problem Map is an Executable Conceptual Structure which can be used to represent task knowledge in a KADS model [7]. A Problem ap is formed by specifying a partial temporal ordering of the Actor Graphs, which represent the primitive inference steps [1]. A *problem map hierarchy* is defined over all the problem maps.

Syntactic Representation	Semantics
[PM: α] -> (FBS) -> [PM: β]	do α followed by β
[PM: { [PM: α], [PM: b] }]	do either a or β, non-deterministically
[KS: [TRUE_TEST: ϕ]]	proceed if ϕ is true
[KS: [FALSE_TEST: ϕ]]	proceed if ϕ is false
[KS: [LT] -> (OP1) -> [NUMBER: μ] ->(OP2)->[NUMBER:v]]	proceed if μ is less than v
[KS: [GT] -> (OP1) -> [NUMBER: μ] - >(OP2)->[NUMBER:v]]	proceed if μ is greater than v
[KS: [LTEQ] -> (OP1) -> [NUMBER: μ] ->(OP2)->[NUMBER:v]]	proceed if μ is less than or equal to v
[KS: [GTEQ] -> (OP1) -> [NUMBER: μ] ->(OP2)->[NUMBER:v]]	proceed if μ is greater than or equal to v
[KS: [ASSERT: ϕ]]	proceed after successfully asserting predicate ϕ into the working memory
[KS: [RETRACT: ϕ]]	proceed after successfully retracting predicate ϕ into the working memory
[KS: [INCREMENT: ξ]]	proceed after successfully incrementing the variable predicate ξ
[KS: [DECREMENT: ξ]]	proceed after successfully decrementing the variable predicate ξ

Table 2: Syntax and Semantics of Modeling Constructs

2.2.3. Complex Modeling Constructs

The main ingredients for building complex modeling constructs are: actor graph, problem map, conceptual relations (to enable expression of temporal relationships), and special purpose actor graphs (for expressing condition and iteration). The conceptual modeling constructs are listed in Table 1. MODEL-ECS distinguishes between two types of predicates. The first is known as an *ordinary predicate*, which is any conceptual graph, and the second is *variable predicate* that is also represented using conceptual graphs, but this conceptual graph is always of the following structure:

$$[VARIABLE: variable] \rightarrow (EQ) \rightarrow [NUMBER: number]$$

where *variable* is any alphanumeric symbol, and *number* is any natural numbers. Assuming that ϕ and φ denote ordinary predicates (i.e., conceptual graphs), ξ denotes a variable predicate, μ and v denotes *variables*, and α and β denote problem maps, the above conceptual constructs can be used to construct simple knowledge sources and problem maps shown in Table 2. Using the simple modeling constructs outlined in Table 2, we are able to build complex constructs as shown in Table 3. Again, assume

Modeling Constructs	Executable Conceptual Structure Representation
$\alpha;\beta$	[PM: α] -> (FBS) -> [PM: β]
if ϕ then α else β	[PM: { [KS: [TRUE_TEST:ϕ]] ->(FBS)-> [PM:α], 　　　　[KS: [FALSE_TEST:ϕ]] ->(FBS)-> [PM:β] }]
while ϕ do α	[PM:{[KS:[TRUE_TEST:ϕ] 　　　] ->(FBS)-> [PM:α]->(FBS)-> [PM:*], 　　　[KS:[FALSE_TEST:ϕ]] }]
repeat α until ϕ	[PM:α] ->(FBS)-> [PM:{[KS:[FALSE_TEST: ϕ] 　　　　　　　] ->(FBS)-> [PM: α], 　　　　　　[KS: [TRUE_TEST: ϕ]] }]
case $\phi:\alpha$, $\varphi:\beta$	[PM:{[KS:[TRUE_TEST:ϕ]] ->(FBS)-> [PM:α], 　　　[KS: [TRUE_TEST:φ]] ->(FBS)-> [PM:β] }]
for ξ = ψ to ζ do α	[PM:[KS:[ASSERT:[VARIABLE: ξ]->(EQ)->[NUMBER: ψ]] 　　　] ->(FBS)-> [PM:{[KS:[GT] ->(OP1)->[NUMBER: ξ] 　　　　　　　　　　　　　->(OP2)-> [NUMBER: ζ]], 　　　　　　[PM:α] ->(FBS)-> [KS:[INCREMENT: 　　　　　　　　　　　　[VARIABLE: ξ]->(EQ)->[NUMBER] 　　　] 　　　　　　　　　　　　　] ->(FBS)-> [PM: 　　*] } 　　　　　　] ->(FBS)-> [KS: [RETRACT: 　　　　　　　　　　　　[VARIABLE: ξ] ->(EQ) -> [NUMBER] 　　]]]

Table 3: Complex Modeling Constructs in MODEL-ECS

that ϕ and φ are ordinary predicates, ξ is a variable predicate, and α and β are problem maps.

3. Modeling using Annotated Flow Chart

Annotated Flow Chart (AFC) is a graphical language that can be used to model problem solving methods (PSMs). AFC is used to model process, but also annotates the processes with input and output objects. In addition, AFC allows processes to be annotated with "performers" (i.e., entities charged with the responsibility of actually carrying out the process). The core of AFC (just the process part, without the input, output, and performer annotations) is very similar to a simple flowcharting language, so it is fairly intuitive to a wide audience. (It should be noted that AFC has an important *part-of* relationship that doesn't normally occur in flowcharting languages.)

Figure 1 shows a Constraint Graphs definition of the AFC visual language. The *Source*, *Sink*, and *Proto Process* nodes are hidden from the end user, but the rest of the nodes and arcs are visible as object types. The heavy red directed arcs in Figure 1 describe the type lattice of the objects. The rest of the arcs are all part of the AFC visual language.

Table 4 presents the arc termination

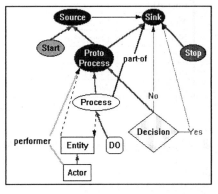

Figure 1: Constraint Graphs Definition of AFC

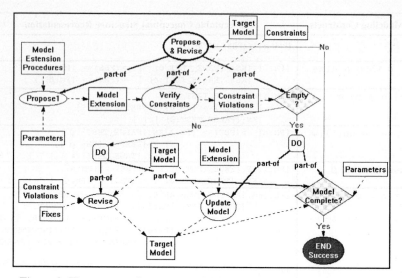

Figure 2: The annotated process model for a "Propose and Revise" PSM

information from Figure 1 in tabular form, explicitly listing all the legal and user-visible types. The Constraint Graphs definition is more concise because it takes advantage of the type hierarchy to describe termination on a collection of subtypes.

The entire AFC language might be somewhat confusing and intimidating to some users. As has already been mentioned, a subset the AFC graphical language elements can be used as a very simple flowcharting tool which may be far more appropriate in some cases. Fortunately, Constraint Graphs makes subsetting of a language extremely easy. The definition may be partitioned by placing the language elements in different *levels*. Detailed discussion of subsetting is beyond the scope of this paper. Figure 2 depicts the ACF representation of a common problem solving method known as "Propose and Revise".

ARC	FROM	TO
next	Process, DO, Decision, Start	Process, DO, Decision, Stop
part-of	Process, DO, Decision	Process, DO, Decision, Stop
Yes	Decision	Process, DO, Decision, Stop
No	Decision	Process, DO, Decision, Stop
input	Entity	Process, DO, Decision
output	Process, DO	Entity
performer	Actor	Process, DO, Decision

Table 4: Legal Arc Terminations According to Figure 1

4. Mapping From AFC to MODEL-ECS

In this section, the authors will describe the mapping between AFC constructs and its equivalent MODEL-ECS constructs. There are five AFC construct types that have a direct mapping to its equivalent counterpart in MODEL-ECS. They are the *conditional*

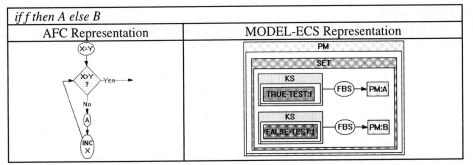

Figure 5: Mapping of the *conditional construct* from AFC to MODEL-ECS

construct, the *while construct*, the *repeat construct*, the *case construct*, and the *for construct*. The following subsections will discuss the mappings between these five control structures in AFC to its equivalent representation in MODEL-ECS.

4.1. Conditional Construct

The *conditional construct* representation in AFC can conventionally be written as *if f then A else B*, where *f* is a *condition* and both *A* and *B* are *actions* (possibly composite) (see Figure 5). The conditional construct maps to MODEL-ECS in the following manner: condition *f* maps to knowledge source [KS: [TRUE_TEST: *f*]], as well as [KS: [FALSE_TEST: *f*]], and both action *A* and action *B* map to two problem maps, [PM: *A*] and [PM: *B*], respectively. There is a conceptual relation (FBS) that links [KS: [TRUE_TEST: *f*]] to [PM: *A*] (which indicates that is *f* is *true* then perform *A*. There is also another conceptual relation (FBS) that links [KS: [FALSE_TEST: *f*]] to [PM: *B*] which indicates that if condition *f* is *false* then proceed to execute problem map [PM: *B*]. This two constructs are then represented in a set within a problem map [PM:{....}]. This yields the following linear form representation:

```
[PM: {[KS:TRUE_TEST: f]]->(FBS)->[PM: A],
     [KS: [FALSE_TEST: f]]->(FBS)->[PM: B] } ]
```

4.2. While Construct

The *while construct* representation in AFC is commonly written as *while f do A* , where *f* is a *condition* and *A* is an *action* (possibly composite). The while construct

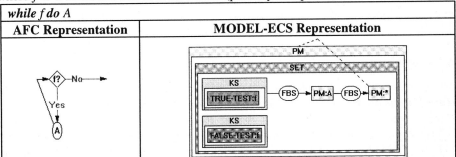

Figure 6: Mapping of the *while construct* between AFC and MODEL-ECS

repeat A until f	
AFC Representation	**MODEL-ECS Representation**
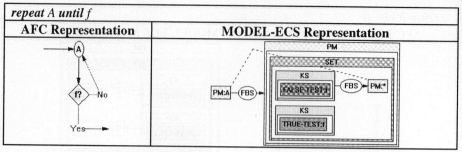	

Figure 7: Mapping of the *repeat construct* between AFC and MODEL-ECS

maps to MODEL-ECS, as depicted in Figure 6, in the following manner: condition *f* maps to [KS: TRUE_TEST: *f*]] and [KS: [FALSE_TEST: *f*]], and action *A* maps to problem map [PM: *A*]. There is a conceptual relation (FBS) that links [KS: TRUE_TEST: *f*]] to [PM: *A*], and from this problem map there is another conceptual relation (FBS) that links to another pseudo problem map [PM: *], where there is a co-referent link between this problem map and the problem map that represents the *while construct*. This implies that, whenever control arrives at the pseudo problem map [PM: *], it is equivalent to arriving at the beginning of the *while construct*. In this way, the control iteratively performs action *A* until condition *f* is no longer *true*. Thus, the linear for representation is:

```
[PM: { [KS: [TRUE_TEST: f]] ->(FBS)-> [PM: A] ->(FBS)-> [PM: *]],
       [KS: [FALSE_TEST: f]] } ]
```

4.3. Repeat Construct

The *repeat construct* representation in AFC is conventionally written as ***repeat A until f***, where *f* is a *condition*, and *A* is an *action* (possibly composite). The repeat construct maps to MODEL-ECS (see Figure 7) representation in the following manner: as before, condition *f* maps to [KS: TRUE_TEST: *f*]] and [KS: [FALSE_TEST: *f*]], and action *A* maps to problem map [PM: *A*], but, in this construct, there is a conceptual relation (FBS) that links [KS: [FALSE_TEST: *f*]] to a psuedo problem map [PM: *] (i.e., which represents the *repeat construct*). In this way, action *A* is executed first, and as long as condition *f* is *false*, action *A* will be iteratively repeated. On every iteration, the condition *f* is check. When condition *f* is *true*, control exits from the *repeat construct*. In linear form, the representation is as follows:

```
[PM: A] ->(FBS)-> [PM: { [KS: [FALSE_TEST: f]] ->(FBS)-> [PM: *]],
                         [KS: [TRUE_TEST: f]] } ]
```

4.4. Case Construct

The *case construct* representation in AFC, which is conventionally written as ***switch (case f:A; case g:B; ....)***, is basically an extension of the *if f then A else B* construct. Here, *f* and *g* are *conditions*, and *A* and *B* are *actions*, maps to MODEL-ECS representation in the following manner: all conditions (i.e., like *f*, and *g*) map to [KS: [TRUE_TEST: *f*]], etc., and a conceptual relation (FBS) links each of these knowledge source to a problem map (which should be executed if the condition is *true*).

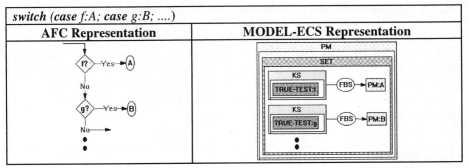

Figure 8: Mapping of the *case construct* between AFC and MODEL-ECS

4.5. For Construct

The *for construct* representation in AFC, which is usually written as *for x = y to z do A*, maps to MODEL-ECS (see Figure 9) in the following manner: firstly, the *variable* predicate is initialize using the following knowledge source:

```
[KS: [ASSERT: [VARIABLE: x] ->(EQ)-> [NUMBER: y] ] ]
```

There is also another knowledge source to *increment / decrement* the value of this variable predicate. The increment knowledge source is used when $y < z$, and the decrement knowledge source is used when $y > z$. The control exits from the "body" of

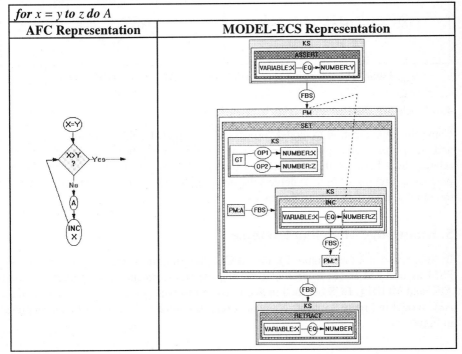

Figure 9: Mapping of the *for construct* between AFC and MODEL-ECS

the **for construct** whenever the following knowledge source returns *true*:

```
[KS: [GT] -> (OP1) -> [NUMBER: x] -> (OP2) -> [NUMBER: z] ]
```

Otherwise, the problem map [PM: *A*] is executed, followed by the increment / decrement knowledge source, and finally, control is passed to a pseudo problem map [PM: *] which is coreferenced to the "body" of the **for construct**. On exit from the "body" of the **for construct**, control will be passed to the following knowledge source to retract the variable predicate from the memory:

```
[KS: [RETRACT: [VARIABLE: x] ->(EQ)-> [NUMBER] ] ]
```

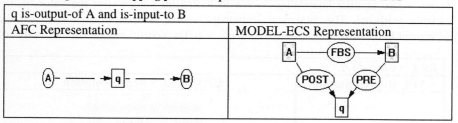

q is-input-to A	
AFC Representation	MODEL-ECS Representation

Figure 10a: Mapping process *input* between AFC and MODEL-ECS

q is-output-to A	
ACF Representation	MODEL-ECS Representation

Figure 10b: Mapping process *output* between AFC and MODEL-ECS

q is-output-of A and is-input-to B	
AFC Representation	MODEL-ECS Representation

Figure 10c: Mapping process *link* between AFC and MODEL-ECS

4.6. Modeling the Process

Using the AFC notation to model "processes", one must be able to illustrate the *inputs* and the *outputs* of a process, as well as be able to illustrate the (possible) inter-relationships between these inputs and outputs that may form the *link* between two processes. Figures 10 (a to c) depicts the mapping between AFC and MODEL-ECS for the input, output, and link, of processes.

5. Knowledge Modeling Example

In Subsection 3.2 (see Figure 2), the AFC representation of the *Propose and Revise* PSM as outlined in [2] is presented. Using the transformations (mappings) between AFC and MODEL-ECS outlined in Section 4 of this paper, one is able to transform the AFC model in Figure 2 into its equivalent representation in MODEL-ECS, as depicted in Figure 11.

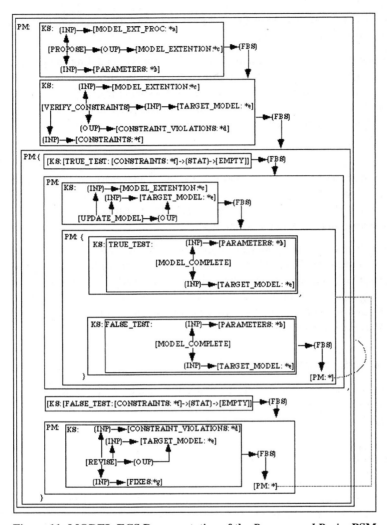

Figure 11: MODEL-ECS Representation of the *Propose and Revise* PSM

The two *conditions* (*Constraint Violation Empty?*, and *Model Complete?*) map to two **conditional constructs**. There are four processes (*Propose, Verify Constraints, Revise,* and *Update Model*) in the AFC representation that map to their equivalent MODEL-ECS knowledge sources.

6. Conclusion and Future Work

This paper briefly outlines the shortfall of the modeling languages developed for the KADS methodology: the language is too complex for the domain experts to use to model the domain processes. Due to the difficulties in using the modeling languages, a highly trained knowledge engineer has to transfer the domain process models from the domain experts into the knowledge based systems. This defeats the purpose and objective of methodologies like KADS.

This paper then proceeds to identify a highly intuitive modeling language called the Annotated Flow Chart (AFC) that is able to carry out the modeling functions necessary when one is using the KADS methodology for modeling a domain process. This notation is then mapped to MODEL-ECS (formal executable conceptual structure based language). This enables the domain experts to do rapid prototyping of their domain process models.

The main contributions of this paper are: the use of Constraint Graphs to define AFC; the mapping between AFC notations to its counterpart in MODEL-ECS; and demonstration of how models represented in AFC can be transformed to MODEL-ECS representation. Two examples serve to illustrate the principles: Visual Programming; and Knowledge Modeling).

The next stage of the development in this research is towards automating the mapping process between AFC and MODEL-ECS. Currently, this process is done manually. Automation will enable the domain experts to instantly view the execution of the model they constructed in AFC. Another extension that is required is the incorporation of more sophisticated temporal relationships between the process in the AFC as well as the MODEL-ECS languages.

Acknowledgements

The authors take this opportunity to thank the members of the Knowledge Science Institute for all the constructive discussions which has contributed very much towards this paper.

References

1. Cross, T. and Lukose, D., The representation of non-linear hierarchical executable plans, in *Proceedings of the 1st Australian Workshop on Conceptual Structures*, Armidale, N.S.W., Australia, 1994

2. Jonker, W., and Motta, E., Functional design activity using $K_{BS}SF$ and VITAL-OCML. *VITAL Project Report DD221 (part III)*, PTT Research, Groningen, March 1993.

3. Kremer, R., Toward a Multi-User, Programmable Web Concept Mapping "Shell" to Handle Multiple Formalisms, in *Proceeding of the 10th Banff Knowledge Acquisition for Knowledge-Based Systems Workshop*. Banff, Canada. October, 1996.

4. Kremer, R., A Graphical Meta-Language for Knowledge Representation. (In Progress.) PhD Dissertation. University of Calgary, Canada, 1997.

5. Lukose, D., Conceptual Graph Tutorial, *Technical Report*, Department of Computing and Mathematics, School of Sciences, Deakin University, Geelong, Australia, 3217, 1991.

6. Lukose, D. Executable Conceptual Structures, in G.W. Mineau, B. Moulin and J.F. Sowa (Eds.), *Conceptual Graphs for Knowledge Representation*, Lecture Notes in Artificial Intelligence (699), Springer- Verlag, Berlin, Germany, 1993.

7. Lukose, D. Using Executable Conceptual Structures for Modelling Expertise, in *Proceedings of the 9th Banff Knowledge Acquisition For Knowledge-Based Systems Workshop (KAW'95)*, Banff Conference Centre, Banff, Alberta, Canada, 1995.

8. Lukose, D., Complex Modelling Constructs in MODEL-ECS, (in this volume), 1997.

9. Munday, C., Cross, J., Daengdej, J., and Lukose, D., CGKEE: Conceptual Graph Knowledge Engineering Environment User and System Manual, *Research Report No. 96-118*, Department of Mathematics, Statistics, and Computing Science, University of New England, Armidale, N.S.W., Australia, 2351, 1996.

10. Schreiber, G., Wielinga, B., and Breuker, J. *KADS: A Principled Approach to Knowledge-Based System Development*, Academic Press, London, UK, 1993.

11. Sowa, J.F., *Conceptual Structures: Information Processing in Mind and Machine*, Addison Wesley, Reading, Mass., USA, 1984.

Complex Modelling Constructs in MODEL-ECS[#]

Dickson Lukose

Distributed Artificial Intelligence Centre (DAIC[*])
Department of Mathematics, Statistics, and Computing Science
The University of New England, Armidale, 2351, N.S.W., AUSTRALIA

Email: lukose@peirce.une.edu.au
Tel.: +61 (0)67 73 2302   Fax.: +61 (0)67 73 3312

Abstract: MODEL-ECS is a *graphically* based *executable* conceptual modelling language, that is suitable for operationalising the Problem Solving Methods (PSMs) modelled using the Knowledge Analysis and Design Support (KADS) methodology. The main contribution of this paper is in the development of all the necessary constructs for modelling *executable* Problem Solving Methods (PSMs). These constructs include: the *conditional construct*; the *while loop*; the *repeat loop*; the *case construct*, and the *for construct*. *Knowledge passing* between Knowledge Sources in a PSM is achieved using *coreference links*, and *line of identity*. The advantage of using MODEL-ECS comes from the *expressibility* provided by Conceptual Graphs, and the *executability* (thus rapid prototyping) that is provided by the Executable Conceptual Structures. With these capabilities, MODEL-ECS can be used not only for knowledge modelling, but also for visual programming.

1. Introduction

In recent years, several Conceptual Graph [19] based modelling languages have been developed (e.g., CG-DESIRE [17] CG-KADS [16]; and MODEL-ECS [10]). All of thee modelling languages are based on some form of "extended" conceptual graphs formalism. For example, MODEL-ECS uses Executable Conceptual Structures [8] [9], which combines conceptual graphs formalism, actors formalism, and the STRIPS architecture [5] to realise execution [7]. Martin [14], Mineau [15], and Dieng [3] also developed knowledge acquisition, engineering and modelling methods based on conceptual graphs. Conceptual graphs are becoming the language of choice for implementing graphical modelling languages for some of the following reasons: Conceptual graph is a representation scheme that can effectively represent both the "abstract" as well as the more "specific" representation of the Problem Solving Methods (PSMs) modelled by the knowledge engineer; and conceptual graphs together with executable conceptual structures has been demonstrated in Lukose et al. [11] as being able to represent, not only the abstract and specific view of the PSMs, but also that the final conceptual model is executable.

Lukose et al [12] demonstrated the capability of the conceptual graphs and the executable conceptual graphs as a uniform representation language for representing *Task Models*, *Models of Cooperation*, *Expertise Model,* and finally the *Conceptual Model* of the Knowledge Analysis and Design (KADS) methodology [18]. But, one must not be mislead into believing that the conceptual graphs and the executable conceptual structures are all good, and all encompassing, representational scheme that solves all modelling problems. There are several crucial research problems to be solved

---

[#] This paper was written while the author was on sabbatical at the Knowledge Science Institute, University of Calgary, Canada.
[*] To find out more about DAIC, refer to our URL: http://turing.une.edu.au/~daic/

to enable the liberal exploitation of the MODEL-ECS for modelling and prototyping. Firstly, its ability to handle complex modelling constructs: *conditional constructs case constructs*, and loops (i.e., *repeat loops, while loops*, and *for loops*). Secondly, its ability to handle *refinements* and *knowledge passing*. Apart from this, there is certainly the issue of having an efficient implementation of a conceptual graphs processor to realise rapid prototyping. This paper will describe the following: the *primitive conceptual structures* that make up MODEL-ECS; how *refinements* and *knowledge passing* is handled in MODEL-ECS, and finally, the *complex modelling constructs* available in MODEL-ECS.

In this paper, Section 2 will identify the primitive conceptual structures used in MODEL-ECS, partial ordering of actor graphs using the temporal conceptual relations to form Problem Maps, describe refinement with the use of nested Problem Maps, and finally, describe the use of *coreference links* and *lines of identity* to facilitate *knowledge passing* between Knowledge Sources within a context and between different contexts. Section 3 describes the necessary syntax and semantics of the primitive modelling constructs. Section 4 provides a detailed description of each of the complex modelling constructs, together with examples of PSMs using these constructs. Finally, Section 5 concludes this paper by describing the future direction of this research.

2. Primitive Conceptual Structures

There are several forms of conceptual structures that are used in MODEL-ECS. They are: Conceptual Graphs [19], Actors, Actor Graphs, Knowledge Source and Problem Map [7] [10]. In brief, Actors are similar to class definitions in the conventional object-oriented modelling languages. An actor definition will define the interface (i.e., *messages* that instances of this actor can handle, and the *methods* to handle these messages). To be able to define an Actor Graph, one has to firstly define the *Actor Type*, and the *Concept Type*. Then, by combining these two forms of abstractions, together with the *precondition*, the *postcondition* and the *delete* lists of conceptual graphs, as shown in Figure 1, we are able to define an Actor Graph of a particular type.

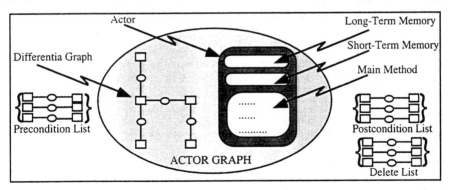

Fig. 1. Graphical Representation of an Actor Graph

Each Actor Graph are represented as a referent to a "knowledge source" concept type (i.e., [KS]). For example, if A is an Actor Graphs, then it is represented as a knowledge source as in [KS: A]. A Problem Map is formed by specifying a partial temporal ordering of the knowledge sources (which represent the primitive inference steps) [2]. Two conceptual relations are used to specify the partial temporal ordering of the knowledge sources. They are called FINISH_BEFORE_START (FBS) and START_WHEN_START (SWS). The (FBS) relation is used to represent a temporal constraint between two knowledge source. The (SWS) relation is used to link two knowledge source when there is no temporal constraint between them [2]. The simplest Problem Map is a single knowledge source. For example, consider a Problem Map that is made up of only one knowledge source (i.e., OBTAIN). This Problem Map is represented as shown below:

```
[PM: [KS: [obtain] -
                    -> (OBJ) --> [ENTITY:*m]
                    -> (SRCE) --> [PLACE:*n]
                    -> (AGNT) --> [ENTITY:*o]
                    -> (RCPT) --> [ENTITY:*o] ] ].
```

Usually, a Problem Map is composed of more than one knowledge source, connected by the temporal conceptual relations FBS or SWS, as described earlier. One of the main advantages of using Problem Maps to represent PSMs is its ability for information hiding. That is, the ability to represent a Problem Map within another Problem Map. To enable nesting, the conceptual relations FBS and SWS are extended to point from or into not only Knowledge Sources (i.e., [KS]), but, also to Problem Maps (i.e., [PM]). Consider the Problem Map shown below, which is made up of six Actor Graphs (i.e., ag_1,....., ag_6).

```
[PM: [PM: [KS:ag_1] -
                      -> (FBS) -> [PM:  [PM: [KS:ag_2] -> (FBS) -> [KS:ag_3]
                                   ] -> (SWS) ->
                                   [PM:  [KS:ag_4] -> (SWS) -> [KS:ag_5] ] ]

     ] -> (SWS) -> [KS:ag_6] ]
```

For simplicity purpose, let us assume that each of the Actor Graphs (i.e., ag_1, ..., ag_6) take a duration of p to execute, where p is the time between t_n and t_{n+1}. Table 1 outlines the execution of these Actor Graphs, where at time t_n both Actor Graphs ag_1 and

t_n	t_{n+1}	t_{n+2}
ag_1	ag_2	ag_3
ag_6	ag_4	
	ag_5	

Table 1. Execution Sequence

ag_6 are executed, at time t_{n+1} the Actor Graphs ag_2, ag_4 and ag_5 are executed, and finally at time t_{n+2} the Actor Graph ag_3 get executed. The level of nesting is not necessarily limited to one. It is possible to have n-levels of nesting, as shown below:

```
[PM: [PM: [KS:*w] -
                   -> (FBS) -> [PM: [KS: *x] -> (SWS) -> [KS: *y] ]
     ] -> (FBS) -> [KS: *z] ]
```

We could re-represent the above Problem Map as shown below:

```
[PM: *a]                                          - nested at Depth 0
```

where a is the following Problem Map:

```
[PM: *b] -> (FBS) -> [KS: *z]                      - nested at Depth 1
```

where z is a Knowledge Source, while b is another Problem Map as shown below:

```
[KS: *x] -> (SWS) -> [KS: *y]                      - nested at Depth 2
```

There is one major issue to be addressed with nested Problem Maps. That is, the flow of information from a sub-component (i.e., concepts) of a knowledge source nested within a Problem Map to concepts in another Problem Map or Knowledge Sources in different context or at different levels of nesting. Knowledge passing of this form is achieved by the use of context, *coreference links* and *line of identity* [19]. In the sample Problem Map shown in Figure 2, the concept *PLACE* which is the object of *SELECT* (in context depth 2) is identical to the *PLACE* which is the object of *SURVEY* (in context depth 3). These two concepts (i.e., *PLACE*) are linked to each other by the concept *PLACE* (in context depth 1), by line of identity. Similarly, the concept *HOUSE* which is the object of *BUILD* is linked to the concept *HOUSE* which is the object of *DESIGN* by a different line of identity.

With *nesting* and *knowledge passing*, MODEL-ECS is ready to tackle more complex modelling constructs that are mainly to do with conditions and loops. Section 3 and 4 describes the necessary vocabulary to model complex constructs which enable MODEL-ECS to be liberally exploited for building executable conceptual models.

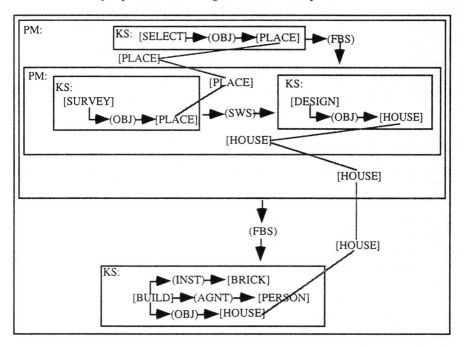

Fig. 2. Knowledge Passing between context by using line of identity

3. Primitive Modelling Constructs

The main ingredients for building complex modelling constructs are the following primitive constructs: *Actor Graph*, *Problem Map*, *Conceptual Relations* to model partial ordering, and *Special Purpose Actor Graphs* for expressing condition and iteration. MODEL-ECS distinguishes between two types of predicates. One is known as *ordinary predicate* that is represented using any conceptual graphs, and the other is known as *variable predicate* that is also represented by conceptual graphs, but these predicates will always have the following format:

[VARIABLE: <variable>] -> (EQ) -> [NUMBER: <number>]

where, <variable> is any alphanumeric symbol; and <number> is any natural number. Conceptual Relations, Concepts, and special purpose Actor Graphs necessary for building complex Problem Maps, are listed in Table 2. Assuming that ϕ and φ denote ordinary predicates (i.e., conceptual graphs), ξ denote a variable predicate, μ and v are numbers, and α and β denote Problem Maps, the above conceptual constructs can be used to construct simple Problem Maps as in Table 3.

Conceptual Constructs	Descriptions
(FBS)	Conceptual Relation to indicate *sequence*.
(SWS)	Conceptual Relation to indicate *concurrence*.
[PM]	Concept to represent *Problem Map*.
[KS]	Concept to represent *Knowledge Source*.
[TRUE_TEST]	Actor Graph to indicate *positive* test.
[FALSE_TEST]	Actor Graph to indicate *negative* test.
[LT]	Actor Graph to check *less than* condition.
[GT]	Actor Graph to check *greater than* condition.
[LTEQ]	Actor Graph to check *less than* or *equal to* condition.
[GTEQ]	Actor Graph to *check greater* than or *equal to* condition.
[ASSERT]	Actor Graph to *assert* a predicate into the working memory.
[RETRACT]	Actor Graph to *retract* a predicate from the working memory.
[INCREMENT]	Actor Graph to *increment* a variable predicate.
[DECREMENT]	Actor Graph to *decrement* a variable predicate.

Table 2. Conceptual Constructs

Syntactic Representation	Semantics
[PM: α] -> (FBS) -> [PM: β]	Do α followed by β.
[PM: { [PM: α], [PM: β] }]	Do either α or β, non deterministically.
[TRUE_TEST: ϕ]	Proceed if ϕ is true.
[FALSE_TEST: ϕ]	Proceed if ϕ is false.
[LT] ->(OP1)->[NUMBER:μ] ->(OP2)->[NUMBER:v]	Procees if μ is less than v.
[GT] ->(OP1)->[NUMBER:μ] ->(OP2)->[NUMBER:v]	Procees if μ is greater than v.
[LTEQ] ->(OP1)->[NUMBER:μ] ->(OP2)->[NUMBER:v]	Procees if μ is less than or equal to v.
[ETEQ] ->(OP1)->[NUMBER:μ] ->(OP2)->[NUMBER:v]	Procees if μ is greaterthan or equal to v.
[ASSERT: ϕ]	Proceed after successfully asserting the predicate ϕ into the working memory (applies to both ordinary and variable predicate).
[RETRACT: ϕ]	Proceeds after successfully retracting the predicate ϕ from the working memory (applies to both the ordinary and variable predicate).
[INCREMENT: ξ]	Proceeds after successfully increments the numeric value associated with the variable predicate ξ.
[DECREMENT: ξ]	Proceeds after successfully decrements the numeric value associated with the variable predicate ξ.

Table 3. Syntax and Semantics of Modelling Constructs

There are six forms of complex modelling constructs in MODEL-ECS. They are: *Sequential Construct, Conditional Construct, While Construct, Repeat Construct, Case Construct,* and *For Construct.* In the next section, the author will attempt to describe in details each of these modelling constructs with an example to illustrate its use in knowledge modelling.

4. Complex Modelling Constructs

4.1. Sequential Construct

Modelling Construct: $\alpha; \beta; \ldots\ldots$
MODEL-ECS: [PM: α] -> (FBS) -> [PM: β] -> (FBS) -> ....

This is the simplest form of nested Problem Map. Here, both α and β are nested Problem Maps, where the execution of α must be completed before the execution of β can commence. α and β themselves can be made up of Knowledge Sources, or a combination of Knowledge Sources and Problem Maps.

4.2. Conditional Construct

Modelling Construct: *if ϕ the α else β*
MODEL-ECS: [PM: {[KS: [TRUE_TEST: ϕ]] -> (FBS) -> [PM: α],
                   [KS: [FALSE_TEST: ϕ]] -> (FBS) -> [PM: β]
              }]

Here, ϕ is a conceptual graph representing the condition for executing Problem Maps α or β. This construct will simply execute Problem Map α if ϕ is true, if not, it will execute the Problem Map β. For example, consider a Problem Map that first searches for a place (from a set of available places), then if it has found a place, it will carry out the land surveying process and the designing of a house. If not, control is returned back to the search process. Thus, here we are confronted with a situation which, based on a condition, control will be passed to different sections of the Problem Map. Modelling a Problem Map of this type is not difficult. Using the "*if ϕ the α else β*" construct, where α and β are the two sub-Problem Maps, and ϕ represents the condition, we are able to model a complex Problem Map as shown in Figure 3. In this Problem Map ϕ is represented by the following conceptual graph:

         [FOUND] -> (OBJ) -> [PLACE] g_1

The Problem Map can be represented in linear form in the following manner:

         [KS: g_3] -> (FBS) -> [PM: g_4] g_2

were g_3 represents the Knowledge Source *SEARCH* as shown below:

         [SEARCH] -> (OBJ) -> [PLACE: *x] g_3

and g_4 represents a the conditional Problem Map, as shown below:

         [PM: { g_5, g_6}] g_4

where g_5 is a nested Problem Map as shown below:

         [KS: TRUE_TEST: g_1]] -> (FBS) -> [PM: g_7] g_5

and g_6 is also a nested Problem Map as shown below:

         [KS: [FALSE_TEST: g_1]] -> (FBS) -> [KS:*] g_6

Problem Map g_7 is shown below:

         [KS: g_8] -> (SWS) -> [KS: g_9] g_7

234

where g_8 is the Knowledge Source shown below:

[SURVEY] -> (OBJ) -> [PLACE: *x] g_8

and finally, g_9 is a following Knowledge Source:

[DESIGN] -> (OBJ) -> [HOUSE] g_9

The Knowledge Source *TRUE_TEST* will check the existence of this conceptual graph in the state space. If the Knowledge Source *SEARCH* has successfully found a place, an image of g_1 will be inserted into the state space. Thus, when the control is passed to *TRUE_TEST*, it will be able to successfully identify the existence of an image of g_1, and thus control is passed to the Problem Map that performs the surveying and designing process. On the other hand, if *TRUE_TEST* is unable to find an image of g_1 in the state space, then control is passed to *FALSE_TEST* (which, in this case, will not be able to find an image of g_1 in the state space, thus returning a value "true"). Thus, control is passed to the Knowledge Source which is linked to *SEARCH* via a line of identity. Therefore, control is once again passed to the Knowledge Source called *SEARCH*. The execution of this Problem Map will continue until it finds a place, and successfully carries out the survey and design process.

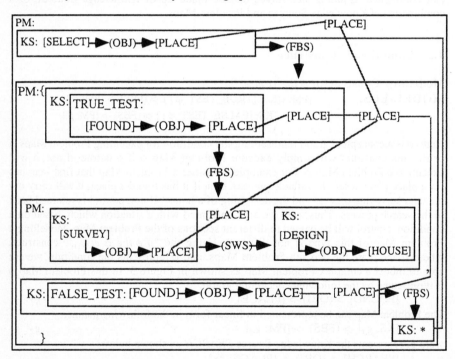

Fig. 3. MODEL-ECS representation of Conditional Construct

4.3. While Construct

Modelling Construct: *while φ do α*

MODEL-ECS: [PM: { [KS: [TRUE_TEST:φ]]->(FBS)->[PM:α]->(FBS)->[PM: *],
[KS: [FALSE_TEST: φ]]
}
]

The "*while φ do α*" construct is a nested Problem Map. This construct simply repeats executing the Problem Map α as long as the conceptual graph represented by φ is true. This construct is made up of two knowledge sources (i.e., *TRUE_TEST*, and *FALSE_TEST*), and a complex Problem Map (i.e., [PM:α]). The functions of each of the knowledge sources are as follows:

- the knowledge source *TRUE_TEST* checks the "state space" to determine whether φ exists in the "state space", and if so, execution simply proceeds to the next knowledge source in the sequence (i.e., in the above construct, execution proceeds to the Problem Map [PM: α]); and
- the knowledge source *FALSE_TEST* checks the "state space" to ensure that φ does not exist in the "state space", and if so, execution simply proceeds to the next knowledge source (i.e., in the above case, the execution terminates at this point).

To illustrate modelling the "*while φ do α*" construct, the author will utilise a variation of the Propose-and-Revise (P&R) PSM which was described in Marcus et al. [13]. The P&R PSM constructs a solution to a problem by iteratively extending and revising partial solutions. The assumption being that the solution can be described in terms of assignments of *values* to *parameters*, and that the *PROPOSE* phase can be carried out by applying a domain dependent knowledge source which allows the derivations of values of unbound parameters from parameters which have already been assigned [20]. The P&R PSM used in this paper is adopted from Fensel [4], and is called the *Select-Propose-Check-Revise Method (SPCR-Method)*. It is depicted in Figure 4. Descriptions of the four sub-tasks are as follows:

- the *SELECT* sub-task chooses the parameter which should be processed next;
- the *PROPOSE* sub-task proposes a value for the selected parameter;
- the *CHECK* sub-task checks the currently given partial assignment (i.e., the old one which is enriched by the new parameter-value pair) as to whether it fulfils the given constraints; and
- the *REVISE* sub-task corrects the partial assignment if constraint violations were detected by the *CHECK* sub-task.

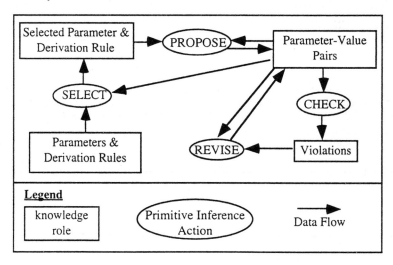

Fig. 4. Select-Propose-Check-Revise Method (adopted from Fensel [4])

The MODEL-ECS equivalent of the *SPCR-Method* is shown by the Problem Map depicted in Figure 5. It consists of three primitive inference structures (i.e., *SELECT*, *PROPOSE*, and *CHECK*), a complex inference structure (i.e., *REVISE*), and a complex modelling construct (i.e., Problem Map representing the "*while ϕ do α*" modelling construct). The interesting part of the Problem Map shown in Figure 5 is the complex modelling construct that represents the "*while ϕ do α*" construct. It is reproduced below:

[PM: {   [KS: [TRUE_TEST: *d]] -> (FBS) -> [PM: *pm_revise] -> (FBS)->[PM: *],
         [KS: [FALSE_TEST: *d]] -> (FBS) -> [KS: [SELECT: *e]]
      }]

Here, the author did not reproduce the *line of identity* (i.e., *coreference links*) that links the knowledge source *SELECT* in the context of *depth 2*, to the *outermost context*. The function of the "*while ϕ do α*" construct is as follows: The *SYSTEM* firstly checks the "state space" to determine whether violations (i.e., represented by the co-reference link *d) exist. If so, execution proceeds to the next knowledge source. In this case, it is a complex Problem Map (i.e., *pm_revise*).

When the execution of the nested Problem Map *pm_revise* is completed, control is once again passed to the knowledge source *TRUE_TEST*. This repeat-loop terminates only when *TRUE_TEST* returns a false answer. When this happens, it indicates that appropriate revisions have been done to eliminate all the violations identified by the Knowledge Source *CHECK*. The construct passes control onto the knowledge source *FALSE_TEST*, to make sure that there are no violations (thus, returning a value "true"), then, control finally flows back to the knowledge source *SELECT*, which will select the next set of parameter and derivation rules; and the process continues. The *SPCR-Method* terminates when there are no more parameter and derivation rules to select.

4.4. Repeat Construct

Modelling Construct:*repeat α until ϕ*
MODEL-ECS:[PM:$α$]->(FBS)->[PM:{[KS: [FALSE_TEST:$ϕ$]]->(FBS)->[PM:$α$],
                                  [KS: [TRUE_TEST: $ϕ$]]
                               }]

The "*repeat α until ϕ*" is another form of complex modelling construct that is based on nested Problem Maps. In this construct, the Problem Map $α$ is executed first, then, a *FALSE_TEST* is performed on the predicate $ϕ$. If the *FALSE_TEST* returns "true" (i.e., which indicates that an image of the conceptual graph $ϕ$ is not found in the state space), the control is passed to the Problem Map $α$. This loop continues until *FALSE_TEST* returns "false", in which case control is then passed to *TRUE_TEST* (which will return "true", since we are testing the same $ϕ$). For example, in Figure 6, the condition $ϕ$ is represented by the following conceptual graph:

[FOUND] -> (OBJ) -> [PLACE] g_1

This Problem Map can be represented in linear form as shown below:

[PM: g_3] -> (FBS) -> [PM: {g_4 , g_5}] g_2

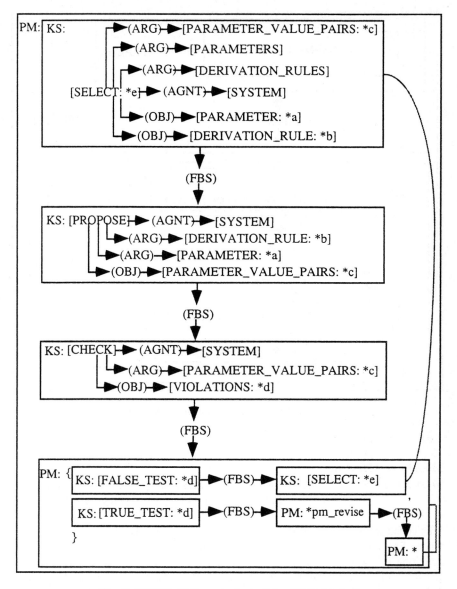

Fig. 5. MODEL-ECS equivalent of the SPCR-Method

where g_3 is a Problem Map (in this example, it is a Problem Map with only one Knowledge Source called SELECT), which is represented as shown below:

[KS: [SELECT] -> (OBJ) -> [PLACE]] g_3

and finally, g_4 and g_5 are as shown below:

[KS: [FALSE_TEST: g_1]]-> (FBS) -> [PM: g_3] g_4
[KS: [TRUE_TEST: g_1]] g_5

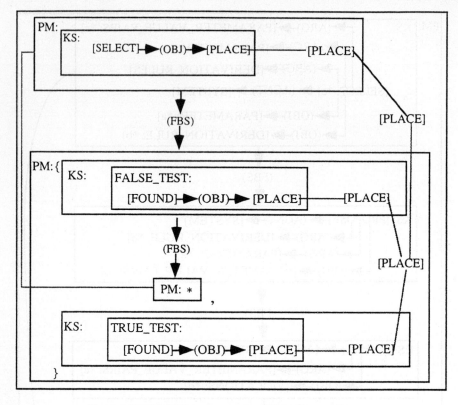

Fig. 6. MODEL-ECS representation of the Repeat Construct

4.5. Case Construct

Modelling Construct: *case* $\phi : \alpha, \varphi : \beta$

MODEL-ECS: [PM:{[KS: [TRUE_TEST: ϕ]] -> (FBS) -> [PM: α],

                [KS: [TRUE_TEST: φ]] -> (FBS) -> [PM: β]

                }]

This construct is also based on the nested Problem Map. Here, we have a situation where Problem Map α is executed if condition ϕ is "true". If it is "false", the condition φ is tested. If it returns "true" then Problem Map β will be executed. If the test for both of these conditions return "false", then this whole Problem Map returns control to the next executable structure. An example of the MODEL-ECS representation of this form of complex modelling construct is shown in Figure 7. In this example, the two conditions are represented by conceptual graph g_1 and g_2, respectively:

        [FOUND] -> (OBJ) -> [PLACE] g_1

        [COMPLETE] -> (OBJ) -> [HOUSE_PLAN] g_2

In linear form, the Problem Map can be represented as follows:

        [PM: {g_4, g_5}] g_3

where g_4 is a Problem Map shown below:

[KS: [TRUE_TEST: g_1]] -> (FBS) -> [PM: g_6] g_4
and g_5 is another Problem Map as shown below:
[KS: [TRUE_TEST: g_2]] -> (FBS) -> [PM: g_7] g_5
The Problem Map g_6 is shown below:
[KS: g_8] -> (SWS) -> [KS: g_9] g_6
The Problem Map g_7 consist of only one Knowledge Source, shown below:
[KS: g_{10}] g_7
Finally, the Knowledge Sources, g_8, g_9, and g_{10}, are all shown below:
[SURVEY] -> (OBJ) -> [PLACE] g_8
[DESIGN] -> (OBJ) -> [HOUSE] g_9
[BUILD] -> (OBJ) -> [HOUSE] g_{10}
The execution of this problem map is as follows. If the agent has found a suitable place (i.e., condition graph represented by graph g_1), then carry out the land survey (represented by Knowledge Source g_8) and house design (i.e., represented by Knowledge Source g_9) process; if not, then check if the house plan is completed (i.e., condition represented by graph g_2). If so, then proceed to build the house (i.e., represented by Knowledge Source g_{10}). If not, then control will be passed to the next executable structure (in this example, there is no other executable structure, thus the execution of the Problem Map will terminate).

4.6. For Construct

Modelling Construct: *for x = y to z do α*
MODEL-ECS:
[PM: [KS: [ASSERT: [VARIABLE: x]->(EQ)->[NUMBER:y]]
        ] ->(FBS) -> [PM: { [KS: [GT] ->(OP1) ->[NUMBER: x]
                            ->(OP2)-> [NUMBER: z]],
            [PM:$α$]->(FBS)->[KS: [INCREMENT:[VARIABLE:x]
                                            ->(EQ)->[NUMBER]]
                    ] -> (FBS) -> [PM: *] }
        ] -> (FBS) -> [KS: [RETRACT: [VARIABLE: x] ->(EQ) -> [NUMBER]]]]]
The *for construct* is implemented in slightly different manner compared to the previous constructs. Here, there is a variable predicate ξ that is initialised to an integer value y, and this predicate is *incremented* or *decremented* until it gets to an integer value z (incremented if $y < z$, and decremented if $y > z$). Also, in every loop, there is a conditional check that takes place before the execution of $α$ to ensure that value of x has not reached z. The variable predicate ξ is usually defined as follows:
[VARIABLE: x] -> (EQ) -> [NUMBER: y]
and asserted into the working memory using the following knowledge source:
[KS: [ASSERT: [VARIABLE: x] -> (EQ) -> [NUMBER: y]]]
If $y < z$, and we are incrementing x, then the conditional check is usually defined as follows:
[KS: [GT]->(OP1)-> [NUMBER: x]
        ->(OP2)-> [NUMBER: z]]
The last action that is performed by this *for construct* is to retract the variable predicate from the working memory, and this is done using the following knowledge source:
[KS: [RETRACT: [VARIABLE: x] ->(EQ)-> [NUMBER]]]

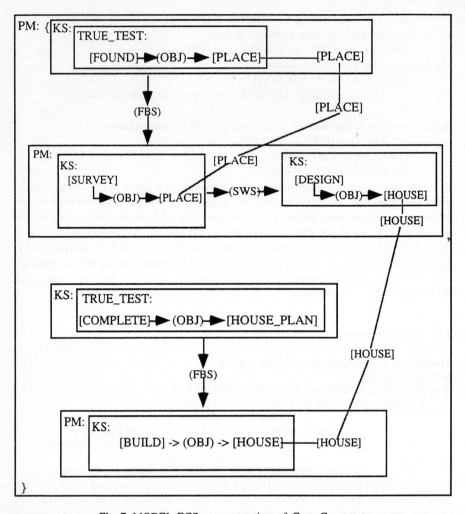

Fig. 7. MODEL-ECS representation of Case Construct

An example of a problem map using the for construct is depicted in Figure 8. This problem maps represents the following:

> **for** $x = 1$ **to** 10 **do**
> {   *load* plane with goods;
>   *fly* to California
> }

The interpretation of this problem map is left to the reader.

5. Conclusion

In this paper, the author described the Executable Conceptual Structures necessary to build *primitive* and *complex modelling constructs* for the *graphical* based executable conceptual modelling language called MODEL-ECS. The major contribution of this

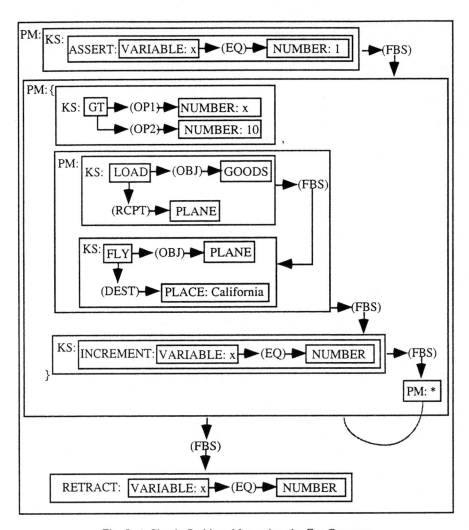

Fig. 8. A Simple Problem Map using the For Construct

paper lies in the development of some necessary extensions to MODEL-ECS, to enable a formal modelling process for building complex PSMs. The extensions described in this paper are the *conditional modelling constructs*, and *loops constructs* (i.e., *repeat construct*, *while construct*, *case construct*, and *for construct*). In this paper, the author describes the *syntax* and *semantics* of primitive and complex modelling constructs of MODEL-ECS. This paper also provided examples of using these constructs to model complex PSMs. To facilitate complex modelling, *nested* Problem Maps are necessary. With nesting, the need from *knowledge passing* between Knowledge Sources within a context, as well as between contexts is essential to enable information transfer to take place while the Problem Map is being executed. Knowledge passing in complex Problem Maps is achieved using *coreference links* and *line of identity*.

By using MODEL-ECS, knowledge engineers can build executable versions of the KADS Conceptual Model, without carrying out the Design Modelling Process. This capability is essential for rapid-prototyping. Since the Conceptual Models are executable, the knowledge engineer will be able to verify the correctness of the model by executing it. Further, the knowledge engineer could easily modify the Conceptual Model with the use of Canonical Formation Rule [19].

The future direction of this research involves carrying out evaluation processes to identify the limitations of the modelling constructs described in this paper. The next area of development is to associate each knowledge source with a representative "icon" (an example of this may be the "User Modelling Notation" [6] or the "Desire Notation" [1]. When this is done, the knowledge modeller will not have to deal to conceptual graphs any longer. Thus, conceptual modelling tasks will involve the use of meaningful "icons" (i.e., which can be executed). This form of extension is not only useful for modelling PSMs, but it is also useful as a visual programming tool, and most importantly for enterprise modelling. This is just a brief indication of the future direction of this research.

References

[1]   Brazier, F.M.T., Langen, P.H.G. van, Treur, J., Wijngaards, N.J.E. and Willems, M., DESIRE: Designing an elevator configuration. In: Schreiber, A.Th., and Birmingham, W.P. (Eds.), Special Issue on Sisyphus-VT. *International Journal of Human-Computer Studies*, 1996, Volume 44, pp. 469-520.

[2]   Cross, T. and Lukose, D., The representation of non-linear hierarchical executable plans, in *Proceedings of the 1st Australian Workshop on Conceptual Structures*, Armidale, N.S.W., Australia, 1994.

[3]   Dieng, R., Specifying a Cooperative System through Agent-Based Knowledge Acquisition, in *Proceedings of the 9th Banff Knowledge Acquisition For Knowledge-Based Systems Workshop*, Banff Conference Centre, Banff. Alberta, Canada, 1995.

[4]   Fensel, D., Assumptions and Limitations of a Problem Solving Method: A Case Study, in *Proceedings of the 9th Banff Knowledge Acquisition For Knowledge-Based Systems Workshop*, Banff Conference Centre, Banff, Alberta. Canada, 1995.

[5]   Fikes, R. E., and Nilsson, N. J., STRIPS: A new approach to the application of theorem proving to problem solving, *Artificial Intelligence* 2(3-4), 1971.

[6]   Kremer, R., Lukose, D., and Gaines, B., Knowledge Modeling using Annotated Flow Chart, (in this volume)., 1997.

[7]   Lukose, D., Executable Conceptual Structures, in G.W. Mineau, B. Moulin and J.F. Sowa (Eds.), *Conceptual Graphs for Knowledge Representation. Lecture Notes in Artificial Intelligence* (699), Springer-Verlag, Berlin, Germany. 1993.

[8]   Lukose, D., Planning Knowledge Acquisition Techniques and Mechanisms, in *Proceedings of the ICCS'94 Workshop on Knowledge Acquisition using Conceptual Graphs*, M.L. Mugnier, M. Willims, B. Gaines and D. Lukose (Eds.), Maryland, USA., 1994.

[9]   Lukose, D., Problem Map: A Hybrid Knowledge Representation Scheme, In *Proceedings of the 7th Australian Joint Conference on Artificial Intelligence*, C. Zhang, J. Debenham, and D. Lukose (Eds.), Armidale, N.S.W., Australia. 1994.

[10] Lukose, D., MODEL-ECS: Executable Conceptual Modeling Language, in *Proceedings of the Knowledge Acquisition for Knowledge Based Systems Workshop (K AW'97)*, Banff, Canada, 9 - 14 November, 1996.

[11] Lukose, D., Cross, T., Munday, C., and Sobora, F., Operational KADS Conceptual Model using Conceptual Graphs and Executable Conceptual Structures, in *Proceedings of the International Conference on Conceptual Structures (ICCS'95)*, USA., 1995.

[12] Lukose, D., Mineau, G., Mugnier, M-L., Möller, J-U., Martin, P., Kremer, R., and Zarri, G.P., Conceptual Structures for Knowledge Engineering and Knowledge Modelling, in *Proceedings of the International Conference on Conceptual Structures*, USA., 1995.

[13] Marcus, S., Stout, J., and McDermott, J., VT: An Expert Elevator Designer that uses Knowledge-Based Backtracking, *AI Magazine* 9(1), 1988.

[14] Martin, P., Knowledge Acquisition using Documents, Conceptual Graphs and a Semantically Structured Dictionary, in *Proceedings of the 9th Banff Knowledge Acquisition For Knowledge-Based Systems Workshop*, Banff Conference Centre, Banff, Alberta, Canada, 1995.

[15] Mineau, G.W., Establishing a Semantic Basis: Toward the Integration of Vocabularies, in *Proceedings of the 9th Banff Knowledge Acquisition For Knowledge-Based Systems Workshop*, Banff Conference Centre, Banff, Alberta, Canada, 1995.

[16] Möller, J-U., Operationalisation of KADS Models by using Conceptual Graph Modules, in *Proceedings of the 9th Banff Knowledge Acquisition For Knowledge-Based Systems Workshop*, Banff Conference Centre, Banff, Alberta, Canada, 1995.

[17] Möller, J-U., and Willems, M., CG-DESIRE: Formal Specification Using Conceptual Graphs, in *Proceedings of the 9th Banff Knowledge Acquisition For Knowledge-Based Systems Workshop*, Banff Conference Centre, Banff, Alberta, Canada, 1995.

[18] Schreiber, G., Wielinga, B., and Breuker, J., *KADS: A Principled Approach to Knowledge-Based System Development*, Academic Press, London, UK., 1993.

[19] Sowa, J.F., *Conceptual Structures: Information Processing in Mind and Machine*, Addison Wesley, Reading, Mass., USA., 1984.

[20] Zdrahal, Z., and Motta, E., An In-Depth Analysis of Propose & Revise Problem Solving Methods, in *Proceedings of the 9th Banff Knowledge Acquisition For Knowledge-Based Systems Workshop*, Banff Conference Centre, Banff, Alberta, Canada, February 26 - March 3, 1995, Paper No: 38.

Modeling Cause and Effect in Legal Text

Judith P. Dick

ActE

120 Eglinton St. E. Ste 215

Toronto, ON. Canada M4P 2E1

dick@canlinks.com

Abstract: Causative relations in the *Palsgraf* v. *Long Island Railway Co.*(1928) 248 N.Y. 339; 162 N.E. 99. case are analysed using John Sowa's conceptual graphs (cgs) and Harold Somers's cases in order to contribute to our knowledge of conceptual relations for use in knowledge representation. The model of causative relations is intended to be adapted for use in other domains using cgs. Our long-term goal is to facilitate automatic extraction of knowledge for intelligent systems retrieving legal information.

1. Introduction

The most formidable barrier to achieving intelligent information retrieval is deriving knowledge from text automatically. Unfortunately, there is neither a clear way through the barrier, nor a quick way over it[1]. Until one is found, we chip away at the problem by clarifying our ontological representations, and text analysis in order to identify generalizable relations and predictable patterns in the hope that they will be reused. Experiments in automatic text extraction add to our NLP (natural language processing) techniques[2], but the problem of constructing knowledge representations (Krs) brings us back to the need to represent meaning to enable machine reasoning.

This is an attempt to chip away at the representation problem by analyzing causal relationships occurring in legal text. Concepts of cause and effect are common to all domains. The patterns derived from legal text can be useful elsewhere. We do not promise a breakthrough in automatic extraction, but isolation of causal patterns whose occurrence may be anticipated in other texts in future. Most of all, by analyzing fundamental relationships we help uncover the meaning in the text in a way that will allow its accurate representation for use in reasoning.

1.1 Outline

The classic negligence case, *Palsgraf* v. *The Long Island Railway Co.*[3] is used to test the capability of Sowa's conceptual graphs (cgs) to represent causation. The exposition of the arguments in *Palsgraf* is incomplete here, although an explanation is given in order make the examples comprehensible. It is not possible in this space to deal with the principles of negligence in adequate depth.

---

[1] For a review of work see Cowrie and Lehnart (1996).

[2] For a review of work on lexicons see Guthrie *et al* (1996).

[3] *Palsgraf* v. *Long Island Railway Co.*(1928) 248 N.Y. 339; 162 N.E. 99.

In attempting the causal representation, we look first at the conceptual relations available in Sowa's writings. Next we proceed to analyse Somers's case system. Finally, we look at excerpts of the representation of the facts of *Palsgraf* to evaluate the causal elements and to arrive at conclusions about causes which will help us to predict their occurrence, to represent them accurately and ultimately to be able to extract them automatically.

2. The Palsgraf Case

The facts are that the Plaintiff (P), Helen Palsgraf,

> "was standing on a platform of the defendant's railroad after buying a ticket to go to Rockaway Beach. A train stopped at the station bound for another place. Two men ran forward to catch it. One of the men reached the platform of the car without mishap, though the train was already moving. The other man, carrying a package, jumped aboard the car, but seemed unsteady as if about to fall. A guard on the car, who had held the door open, reached forward to help him in, and another guard on the platform pushed him from behind. In this act, the package was dislodged, and fell upon the rails. It was a package of small size, about fifteen inches long, and covered by a newspaper. In fact it contained fireworks, but there was nothing in its appearance to give notice of its contents. The fireworks when they fell exploded. The shock of the explosion threw down some scales at the other end of the platform, many feet away. The scales struck the plaintiff, causing injuries for which she sues."[4]

In this appeal, the court dismissed the complaint. Cardozo's majority opinion turned on the decision that D owed no duty of care to P. Liability was limited by establishing the need for a duty rather than by looking at causes. The Railway was not held responsible for the injury done, because the duty was not extant. The majority judges (four of six) begged the question of what causation. The facts provide a pattern of a situation that is judged not causative.

In the dissenting opinion of Andrews, J., the guard's action was the proximate cause of P's injury. The general duty of care involving the security of any person is enough to establish liability. The injury to the P can be adjudged to have happened as a consequence of the negligent accident of the package falling, caused by the guard's action. The injury would not have happened had the package not fallen. There are no collateral or contributory causes.

Palsgraf is a classic cockeyed case[5]. That is to say, the events are highly improbable. It is the pattern of facts which, given the dearth of reasons for judgment, later courts look to for knowledge of what is causation. We will need to be able to show the relationship between the push and the injury, between the explosion and the injury, and so on. Furthermore, we shall have to represent facts clearly enough to support discussions of the foreseeability events. The point of the decision is to place some limitation on liability for freak accidents. The question is how are we to represent conceptual relations in a case where the facts are eminently significant in determining liability.

---

[4] Ibid., 99, as recounted by Cardozo, C.J.
[5] Prosser 1953, 19.

In constructing the Kr, we use Sowa's original definitions and follow the general practice of the community of users, in order to take advantage of knowledge, and to prepare compatible representations. We look forward to sharing Krs in the future and developing ontological resources. They are the key to automatic knowledge extraction. First, lets look at the causal relations in cgs.

2.1. Conceptual Relations

In Sowa's catalog [14] some causative relations are defined. They were a first cut at analysis, but they are the common currency of the cgs community. If they are adequate to our purpose we need look no further. First is the cause relation (CAUS)

> cause. (CAUS) links [STATE: *x] to [STATE: *y] where *x has a cause *y.
> Example: *If you are wet, it is raining*.
> [STATE: [PERSON: You]<-(EXPR)<-[WET]]->(CAUS)->[STATE: [RAIN]].

State is defined in the same catalog

> [STATE] < T. States have duration, as opposed to events, which are in flux.
> [STATE]-
>   (DUR)-> [TIME-PERIOD]
>   (LOC)-> [PLACE].

The weather example and the definition of [STATE] are consistent, but make the definition of (CAUS) unsuitable for our use. Our definition must relate events to each other. The definition of [EVENT] is also directly below the T, so is independent of [STATE].

> [EVENT] < T. Events include acts by animate agents as well as happenings like explosions, where an agent may not be present.

This one suits our need Note the example of explosion which corresponds to one of the events in *Palsgraf*. The concepts [EVENT] and [STATE] are independent of each other in the type hierarchy. We must adapt it. However, we then run the risk of being less able to share our work because of incompatibilities.

We need to represent causal relationships with events on both sides. It is likely that events can cause states, and perhaps that states can cause events. Furthermore, to say that x is "caused by" y tells us too little about the nature of the relationship. In due course, we will attempt a definition of causation. For the present, let's just mark (CAUS) as needing some further explication. What is a cause?

There is no "effect" in Sowa [15] but there is a "result" relation.

> result. (RSLT) links an [ACT] to an [ENTITY] that is generated by the act.
> Example: *Erich built a house*.
> [PERSON: Erich]<-(AGNT)<-[BUILD]->(RSLT)->[HOUSE].

Erich in his agentive role, is a causal factor, but the flavor of the example is more like a factitive, a manufacturing relation than a causality relation. Causal events, like explosions, often do not have agents. This relation expresses one aspect of causation but not precisely what we should like to see in a representation of effect.

Elsewhere Sowa uses the intersentential relation, consequence (CNSQ) without definition or example [15, pp 7], saying that it may defined in terms of more primitive relations. He does not relate it to [RESULT]. It appears likely that it will be closer in meaning to a causally related act and event than (RSLT). The concept [ACT] is defined as a subtype of the concept [EVENT], a comfortable choice for us in terms of relating the two types causally. Anything defined with an [EVENT] as a participant may presumably use an [ACT] also.

Ideally, we should be able to represent any state, act, or event on either side of the causal relationship. We should not be restricted to using the relation as an intersentential device. We want to be able to link cause and effect even if they do not occur in the same sentence, or even in the same paragraph; but we also want to be able to extract cause-effect pairs from sentences. We want to link several causes in a chain. We may find it necessary to distinguish variant types of relationships.

At first glance, "effect" appears to be the easier concept to recognize. We can clearly see in *Palsgraf*, the package falling, the explosion and the injury. Results stand out. We have to work backward, to uncover the causes. One event's cause may be another event's result, as the fallen package caused the explosion but was the result of the push   One entire event may be the cause of another—in Andrews judgment, the push event was the cause of the injury, but does that mean the intervening incidents can be safely ignored or omitted? It depends on the language and the interpretation. How close is proximate? Causes can be implicit, not explosions of course, but certainly psychological events, assumed background occurrences, and linguistic presuppositions. It is clear that causal linkages may be even more obscure.

Looking now at the nature of causes, if one overlooks the fact that a cause is defined above as a state, there are several relevant relations. First is the agent, the subject-like relationship between an animate and an act.

agent. (AGENT) links [ACT] to [ANIMATE], where the ANIMATE concept represents the actor of the action. Example: *Eve bit an apple.*
[PERSON: Eve]<–(AGNT)<–[BITE] –>(OBJ)–>[APPLE].

It is clear that an agent can *cause* if it can act. Eve *caused* the apple bite. We need to know about other sources of causation. Initiator is the only other similar relation I was able to locate.

initiator. (INIT) links an (ACT) to an [ANIMATE] who is responsible for initiating it, but who does not perform it directly.
Example: *Tony boiled the potatoes.*
[PERSON: Tony]<–(INIT)<–[BOIL]–>(OBJ)–>[POTATOES].

It is not clear what distinguishes an initiator from an agent, or why. It appears to be a semantically motivated decision, in that the initiator is not the actual doer. Also, room is left for the insertion of an instrument or path case. Something had to do the boiling, a stove, a pot. If that was the intent, then why is the role not instantiated in the surface structure example? Insofar as causation is concerned, this animate initiator appears not to be allowed to get his hands dirty. If he digs, he must use a shovel, but could his hands not be interpreted as a tool? If so, can he still be an initiator, or is he an agent, or do we really care since they are both animate? Does the distinction tell us there must have been an instrument? There is no apparent reason to maintain a gap if there is no realization of the instrument role in the sentence. Again, if he can perform an act, the initiator also can certainly be a cause.

The goal relation is another enigma. It appears in a complex graph illustrating the use of anaphora and quantifiers. It links one embedded graph to the verb "try" in another graph [15; pp. 26]. "Goal" is a useful semantic designation for attempted results. The passenger tried to get somewhere, that was his "goal". "Goal" is useful in clausal analysis to relate larger graphs to each other,. perhaps also for inchoate causative acts. For example, the "goal" of the trainman was to get the passenger aboard. If the attempt failed, it is inappropriate to discuss cause, since the effect is not actuated. However, in a state of incompleteness, we want to examine the guard's action with regard to causation in order to determine the degree of his carelessness. We could make better use of this relation. Too bad it is not more precisely defined.

Finally, there is the "purpose" relationship, although it does not appear in the catalog but in a schema for the concept [DEMONSTRATE] [14; pp. 262].

purpose. [DEMONSTRATE]->(PURP)->[DEMAND: {*}].

The example says that the purpose of the act of demonstrating is a set of demands. The example and accompanying text seem to indicate that "purpose" means "reason". This difficult semantic realization is reminiscent of the Peircean Thirdness subcategory described by Sowa as "depending on the motives of the agent".[6] Demands are not the cause of the demonstration. A cause has Secondness characteristics. It is relational. Yet it is often difficult to distinguish between a cause and a purpose or a reason for an effect. The difficulty has to do with the wording. It may be helpful to regard the reason as a less potent entity. It is characterized by mental activity, or abstraction, rather than movement or physical change.

It is not clear how the relation is intended to be used in comparison with the conceptual relations (CAUS) and (GOAL). Indeed it is not clear that either (GOAL) or (PURPOSE) was intended to be a part of the cgs language. However, we believe that it is necessary to keep track of example relations and to avoid inconsistencies and ambiguities resulting from the use of similar relationships in different contexts.

The conceptual relations we have seen so far are not adequate for *Palsgraf*, certainly not in a way that would allow us to represent the "causes" discussed in the arguments. Moreover, beyond *Palsgraf* if we have precise and expressive conceptual relations we can agree on, for commonly occurring ideas like cause and effect, we can develop lexicons, which will lend themselves to use in text analysis.

Effects are easily recognized, but causes may be covert. In the catalog of conceptual relations there is provision for animate causal agents. However, inanimate objects can be agents of cause as well, for example, the scales injured Helen Palsgraf. There is no provision for representation of such a role so far, or it is not clear that there is. Again, the original catalog was written as a demonstration, with full expectation that it would be refined and expanded. It has given us a model of mixed level relational analysis which is tremendously useful in text analysis. However, the relations are not adequate and it is time we developed

6 Sowa 1994, 100.

and refined them. As a first step toward a resolution, we shall discuss enriching the list by the inclusion of a set of relations expressing deep case structures.

2.2. Case Analysis

Case gives us the most predictable analysis of patterns occurring in natural language. It is key to the representation of meaning. Moreover, it enables us to write logically significant language representations. Sowa has accepted and incorporated case theory, in the development of cgs but not fully exploited it. He incorporated the basic common cases and then left the way open for future development. There are many different case systems in use. They share a common core and then have their own variations. It is important not to construct a new case for every variation, but to see the fundamental nature of the basic syntactic functions. Once these are consolidated, some variation is allowable. Well developed case systems are adaptable, variations can be converted from one to the other.

Beginning with syntactic cases, there is a long history of case analysis. Among the extant systems there is a base set. Somers [13], discusses the history of case preliminary to presenting his own system, designed for use in computer applications. His cases have not received widespread acceptance. However, Fillmore, who is generally credited with the formulation of deep Case theory has endorsed the concept and been developing syntactic-semantic case relations along the same lines [5] [6].

This paper builds on earlier work and corrects some usages, adapts and refines others. I have found no reason to abandon Somers's system overall and there has been little development toward a fuller case representation within cgs to this point. Negotiation of nominal differences is a trivial matter. The semantic realizations in Somers's cases make them particularly useful for computer processing of text. They enable us to sort out causes as Source cases and effects as Goal cases, clearly related, and clearly distinguished. From Source to Goal is a directional relation Somers sometimes represents with an arrow.

Somers's takes advantage of valency "the potential that a word possesses for combining with other words both syntactically and semantically"[7]. Given the valency of each entry in the lexicon, it is possible to analyze the clause or the sentence, the correct parts of speech, within the specified semantic constraints, recognized. Conversely, selections may be made to fill the available slots in the valency pattern. It is this capability which allows one to regard analyzing the base relations, such as causation, as a step toward knowledge extraction. In order to get there we need more information about the valency patterns of the various classes of verbs. This aspect will be investigated in future.

Complications come when a given term has variant valency patterns, expressing multiple senses, or when slot fillers are optional and ill defined. Performance of good matches is dependent on the accuracy and completeness of the lexicon. It is desirable that the lexicon is be as close to exhaustive as possible and that it contain both syntax and semantics, with some seminal morphological information to allow for partial matches, with stemming and unexpected variants. Phonology is a plus if the system is to have a voice recognition component.

---

[7] Leech 1981, 190.

Feature analysis done with a view to generalized results helps us map semantic realizations to syntactic relations, promoting the cause of predictable results in automatic knowledge extraction. Somers's cases with their built-in semantic component explicitly combine syntactic and semantic components in each cases relation. The organization of the cases and their relationships to one another makes them a powerful analytical tool as has already been demonstration [4].

The specific cases useful for representation of causal agents are found in the Source category of semantic components. The Active designation indicates the syntactic category of the first case, similar to the "agent" relation in Sowa, and to the traditional grammatical "subject" case. Following Sowa's style its definition may be written

active-source. (ACTS) links and [ENTITY] or an [ANIMATE] to an [EVENT] where the entity or animate is the instigator of the event.
Example: *The guard pushed the passenger.*
[GUARD]<–(ACTS)<–[PUSH]–>(ACTG)–>[PASSENGER].

There are two features to be marked in the Active-Source (ACTS) case. The first is ±animate. Indicating that, as we have seen, a causal agent may be animate or an inanimate object or event. For example, a weighing scale or an explosion is inanimate and so its case feature marking is "-".

Example: *The explosion damaged the scale.*
[EXPLOSION]<–(ACTS-)<–[DAMAGE]–>(ACTG)–>[SCALE]

If the Active-Source is +animate, and the slot filler is, therefore, animate then the volitivity feature must be addressed. Was the change brought about intentionally?

The volitive feature, is also to be marked plus or minus "±". The guard in the example above, would be associated with a case relation with two markings, (ACTS+-). Clearly intentionality is significant in law. For example, we want to indicate that the guard negligently rather than willfully caused the package to drop. We can expect to find statements indicating decisions and capability, as well as planned activity when the subject is acting intentionally. As a result of the differences in meaning there will be predictable syntactic variants. For example, imperatives make sense only when directed to an animate and volitive entity. You can sensibly direct a cook to add orange juice, but if you order an orange to add juice, your command is nonsensical. So, the unexpressed subject of an imperative will be some nearby object with a +volitive marking, recognizable without precoding by some abstraction method.

In this analysis, we are assuming for the first time, that if there is a cause, and then an effect, a change must have occurred. It is likely that this is the import of the (CAUS) relation defined by Sowa, and so accounts for the definition in terms of states only. If that is true, then there is clearly no conflict with the cgs relation and Somers's case is compatible with Sowa's representation of fundamental causality.

Some grammarians regard the primary causation case to be a path or instrument case rather than an agentive case [13; pp. 125]. Somers argues, and I have found it to be true, that the most satisfactory analysis of causation results from the choice of the Source case as default, with the addition of the appropriate path case (here Active-Path) only if accuracy requires that the agent and the instrument to be distinguished, if there is a tool.

> active-path. (ACTP) links an [ENTITY] to an [EVENT] where the entity is the instrument or the means by which the event is brought about.

For example, if *the guard caused the explosion by using the fireworks*, then the guard would take the Source case and the fireworks the Path.

> [CAUSE]–
>   (ACTS++)–>[GUARD]
>   (ACTG)–>[EXPLOSION]
>   (ACTP)–>[FIREWORKS].

Otherwise the feature marking in the Active-Source case makes the necessary distinctions.

All effects are clearly Goal cases, the results of the change wrought by the object in the Active-Source case through the action or verb. An intended result is expressed as Active-Goal (ACTG).

> active-goal. (ACTG) links an [ENTITY] or [ANIMATE] to an [EVENT] where the entity or animate designates the end point of the act.

This is the case used to represent effects which are either events or entities. For example, the explosion in the [CAUSE] above, was an intended result.

The example used by Sowa in defining the relation "cause. (CAUS)", regarding a change of state, is represented by the Objective cases which are not the traditional objective cases but deal with objects, things. They are passive in contrast with those designated "Active". The Objective-Source (OBJS) case would represent the original state of an object, and Objective-Goal (OBJG), the result state. Both have implicit feature marking " -concrete". The representation is condensed slightly but is precise. Causation is explicit in the case definition.

> Example: *If you are wet, it is raining*.
> [You]<–(OBJG)<–[WET]–>(OBJS)–>[RAIN].

Somers regards similar "meteorological verbs" as exceptional. For example, he classes "is raining" as syntactically monovalent but semantically avalent [13; pp. 19]. We may wanted to abandon the state analysis above. The Objective case transition from Source to Goal represents a passive realization, like a process, or an event of becoming or growth.

> objective-source. (OBJS) links a [STATE] or a [MATR] to an [EVENT] where the state designates the original state in the change-of-state, an abstraction, and material designates a substance which undergoes some change.
> Example: *Water becomes ice because of the cold*.
> [WATER]<–(OBJS)<–[BECOME]–>(OBJG)–>[ICE]–>(AMBS)–>[COLD].

> objective-goal. (OBJG) links a [STATE] or an [ENTITY] to an [EVENT] where the state or entity is the end of the event. If it is a state, it will be the terminal state in a change of state, an abstraction. If it is an entity, it will be the concrete result of a process, factitive, something made.
> Example: *Lazarus becomes an object of charity*.
> [LAZARUS]<–(ACTS)<–[BECOME]–>(OBJG)–>[OBJECT_OF_CHARITY].

This characteristic is noticeable in the weather examples. This appears to solve the problem we noticed above in Sowa's definition, of (CAUS) lending itself to a very restricted use and impractical when applied to the real world situation of *Palsgraf.*

Also, Somers recommends that the agentless instrument of change, also a type of causal agent, be relegated to the Objective-Path case.

> objective-path. (OBJP) links an [ENTITY] to an [EVENT] where the entity is the passive means which enables the event to take place.
> Example: *The scales injured Plaintiff Palsgraf.*
> [SCALES]->(ACTS-)->[INJURE]->(OBJG)->[P: Palsgraf].
> [SCALES]->(OBJP)->[INJURE]->(OBJG)->[P: Palsgraf].

The first version indicates that the scales are the Active instigator of the injury even though inanimate. The second version says that the scales passed from their original inactive state to the result state of causing injury to P. (OBJG) Objective-Goal describes the result state, the affect of the change, and is implicitly marked "-concrete", indicating an abstract concept, that is, state. The second version is closer to the meaning we need.

The idea of "reason" which is closely allied to "cause" is represented in Somers's system by Ambient-Source. Ambient cases describe abstractions and are somewhat peripheral, less central to the action the function described by the verb. Ambient-Source (AMBS) is similar to the "purpose" relation in Sowa. It occurs in the originating position.

> ambient-source. (AMBS) links and [ENTITY] to an [EVENT] where the entity designates a reason or abstract cause for the event.
>
> ambient-goal. (AMBG) links an [EVENT: *x] to an [EVENT: *y] where the *y is an intended aim or an unintended consequence of *x.

The correlative Goal case, Ambient-Goal (AMBG) can be used to represent aims, intended but unrealized goals. Alternatively, they represent unintended consequences. They relate to abstractions, more in the nature of explanations. It might be argued that the unintended consequences of the guard's original action be grouped in a contextual graph as (AMBG-), consequences, marked unintentional.

```
[PUSH]-
       (ACTS++)->[GUARD]
       (ACTG)->[PASSENGER]
       (AMBG-)->[DROP] [EXPLODE] [FALL] [INJURE].
```

Such a representation would telescope the chain of causation to a single cause with multiple results. The [PUSH] event is the prime cause. None of the other events functions as a cause for the ensuing events. The nature of the results is not in the spirit of the abstract trait of the Ambient case, although it does indicate distance from the action.

The Ambient cases are seductive. One is tempted to put in that category causes which do not fit neatly into one's conception of the agentive and instrument cases but which have a causal function. "Reason" is Somers's explication. It has the same mental connotations as discussion in relation to the conceptual relation "purpose" above, a lack of dynamism, potency.

2.3. Further Analysis

Here we are looking at excerpts from the representation of the *Palsgraf* summary to see if the analysis of causation is adequate to support the judicial arguments in the case. It is a multilevel Kr, a coarse-grained partial parse. With clausal analysis and case relations it is possible to derive the essence of an utterance without interpreting every modal, preposition, and particle. We are seeking accurate information, not grammatical completeness. Frequently occurring lexemes, determiners like "the" and modals like "been", although significant syntactically, convey little information so may be suppressed. Zipf [18] showed that the least commonly occurring words carry the greatest amounts of information. Moreover, multi-level analysis is conjectured to be the most productive method of tackling the problem of extracting meaning. The long range goal is to further the cause of knowledge extraction. The short range goal is to unambiguously support the representation of the reasoning as reported in the decision.

The central problem for the court is to establish liability consistent with fault.

The one unquestioned fact is that injury did occur where it should not have. The occurrence of an accident proves there was some negligence. Now the court must deal with the assignment of liability. The sequence of events may be represented as

```
[PUSH:#1]-
    (ACTS)->[GUARD: B]<-(POSS)<-[D: LONG_ISLAND]
    (ACTG)->[PASSENGER: # 2]->(POSS)->[PACKAGE: #1]-
            <-(CONT)<-[NEWSPP]->(CONT)->[FIREWORKS]
[FALL]-
    (ACTS)->[PACKAGE: #1]
    (LOCG)->[RAILS]
[EXPLODE]-
    (ACTS)->[PACKAGE]
    (ACTP)->[IMPACT]
    (OBJG)->[EXPLOSION]
[THROW_DOWN]-
    (ACTS)->[EXPLOSION]
    (ACTG)->[W_SCALES]
    (LOCL)->[PLATFORM]->(LOCL)->[DISTANCE: #3]
[STRIKE]-
    (ACTS)->[W_SCALES]
    (ACTG)->[P: Palsgraf]
[INJURE]-
    (ACTS)->[STRIKE]
    (ACTG)->[P: Palsgraf].
```

The case relations show the causation within each event. This representation is somewhat simplified.[8] Some elements of description have been omitted here in order to focus attention on causes. The [DISTANCE] entity has fuller description elsewhere, which relates the position of the [PUSH] incident to the position of the [W_SCALES] as several feet distant from each other. If there is a chain of causation, there will be Active cases in successive events with slot fillers from earlier ones. For example, in our example, in the event THROW_DOWN, the impetus is

---

[8] More details exposition of the use of Somers's cases may be found in Dick 1992, and later work.

(ACTS-)–>[EXPLOSION] from the preceding event [EXPLODE]. There is also a link to the next event through the effect on the weigh scales which then become the instigator of the [STRIKE] event. This is the pattern of the representation.. One event simply follows another and the causal links are explicit to preceding and succeeding events. This is a very clear and simple representation to construct automatically, using valency patterns from a lexicon with reliable syntactic information.

The semantics of the passage strengthen the causal implications. [THROW_DOWN] makes a strong connection and if we add the predicate

```
[DISLODGE]–
        (ACTP)–>[PUSH:#1]
        (OBJG)–>[PACKAGE:#1]–>(UNDER)–>[ARM]<–(POSS)<–
        [PASSENGER: #2]
```

there is another connective link with a undeniable semantic expression of causation. Because of the [PUSH] event, the package is dislodged. Note that there is no explicit statement here of negligence.

. We must deal with the facts several times, in different contexts, as the arguments are presented. Cardozo states that is no duty of care between D and P. We see in the representation that there is no contact between them. It is clear from the representation as well that there is no contact between the guard B and Mrs. Palsgraf. The representation supports Cardozo's reasoning. He has avoided discussion of causation in attributing liability. We have interjected no arbitrary causative relations of our own. Even the very condensed representation presented as an example [PUSH] in the previous section, which flattens the chain of events to a simple list of sequential actions without representation of causal links would have been acceptable.

Cardozo's argument is based on his opinion that the conduct of the second guard B did not involve any risk of foreseeable danger to Mrs. Palsgraf, who was a considerable distance down the platform[9] with her two young daughters.

The [PUSH] event, is described in his reasons as having been done without intent to harm, and inadvertently. This is a curious construction but means that the [PUSH] act was done inadvertently *with regard to* the consequent injury. B did not The explication conveys the sense of the argument.

```
[REASONS: #1]–
    [PUSH: #2]–
        (ACTS)–>[GUARD:B]–
                (ATTR)–>[~[INTENT]–>(ATTR)–>[HARM]]
                (ATTR)–>[~[FORESEE]–>(ACTG) ]–>[INJURE: #1]]
            (ATTR) –>[INADVERTENT]–>[INJURE: #1] . . . .
```

Cardozo averred B is not negligent because the accident was not foreseeable. Our lexicon of legal concepts has an entry for [FORESEE]. It is a composite of the

---

[9] Unreported details of the case are derived from the *Record,* set out in Scott and Simpson (1950), 891-940 are quoted in Prosser (1953) 2.

basic definition from *Blacks Law Dictionary* [1] and elements from other cases in our knowledge base dealing with the concept. This is an excerpt. [10]

```
[FORESEE]–
    (DEFN)–>[[[EXPECT] or [ANTICIPATE] or [KNOW] or [THINK]]–
        (TEMPL)–>[TIME: foresee]
        (DATPSYL)–>[EVENT: {*}]—
            (CHRC) –>[INJURE] or [HARM]
                (TEMPL)–>[TIME: event]->[TIME: forsee+],
        (AMBS) –>[[ACT:{*}] or [OMISSION: {*}]]–
            (LOCL) –>[TIME: foresee],,,].
```

Since we have no evidence from the facts that B expected or anticipated, and so on, any injurious event occurring at all, Cardozo's statement is supportable by the representation we have so far. Since the guard B did not have a duty of care to P, the Railway was not liable for her accident. Cardozo has limited liability to exclude the defendant Railway Company from obligation to the injured P.

Andrews, J. dissenting says that because P suffered an injury, and there is a standard duty of care for the safety of any person the Railroad be considered liable. The argument over the legal principle of duty of care does not concern us here as it does not relate to causation. He goes on to focus the issue of proximate cause,which does. It is otherwise described as remoteness. Given the strange sequence of events, is the connection between the D and the injury close enough to warrant liability? He decides that the cause of the PUSH is proximate, the action not remote from the injury, because there is *an uninterrupted natural sequence of events connecting* the [PUSH] with the [INJURY].

Our representation works here because Andrews wants to work closely through each step of the sequence of events, and it follows in order. Since the order of events is sequential, the simple contiguous representation works. we would have to have a more explicit relation or linkage if there were intervening or peripheral events under discussion. In *Palsgraf*, each event contributes to the accident. Also, we are not bothered by contributory events. Where it is necessary to decide which of a number of incidents caused or contributed to the cause of an event, we would need more explicit representation of the causal relations. Such a representation is difficult because like streams flowing into a river, once mixed, the causality is difficult to distinguish. It is not clear that our presently defined relations are up to the job of such a representation. We may need to find more explicitly distinct links.

Within Andrews argument, within the modal indicating his opinion, not fact the (CAUS) relation from Sowa has been "overloaded" to mean that event B could not have occurred had event A not have occurred. The representation is not shown as it is not true to the language of the text. It is a device I have used temporarily until a deeper understanding of the problem makes adequate representation possible. And it otherwise duplicates the example already included.

With regard to the long-term goal of deriving knowledge from text, where a sequence of events occurs we may quite properly look for causation between pairs of them and construct longer chains where appropriate. Common sense tells us that a cause never succeeds an effect, therefore, when there is a sequence events, we may

---

[10] Legal concepts are open-ended. In LOG+, definitions are cumulative, functioning as cores for clusters of cases.(Dick 1992)

expect causation. In the more difficult case where cause is stated out of sequence, temporal indicators help—cause will precede effect.

Furthermore, if we can agree that in the nature of the causal relationship, the effect could not exist had the cause not occurred, we have a handle on the problem of determining what a cause is. First, it is necessary to establish whether or not the effect is factual. If it is, then the cause must exist, whether or not its expression is apparent. The most difficult problem is to recognize and identify implicit causes. A single cause can be more easily unearthed than a combination. Finally, causes are easy to identify where single actions or events are occurring than where interaction is taking place. Sowa suggests that early development of extraction systems may be augmented by human analysts, tutors. They could be used in this area to sort out the complexities of interacting causal events..

Andrews J. discusses proximate cause[11] as meaning essential[12]. The injury to the P can be adjudged to have happened as a consequence, albeit remote, from the negligent accident of the package falling, caused by the guard's action. The test is, would the injury have happened had the [DISLODGE] and [FAL] events not occurred.

Nothing in *Palsgraf* intervenes between [PUSH] and the [INJURY] to interrupt the chain of events. For this reason, the dissent makes its case for proximate cause, even though the initiating action is a long way from the [INJURY] event. The causal chain is continuous. The events are all lined to each other fore and aft. The chain could be built, checked and processed automatically, given a table of accurate valencies for the event words involved. We do not anticipate any representational difficulties in dealing with such a barrier. The text events should some indication of interruption in space or activity when a causal chain is interrupted. A new relation is unlikely to be needed.

However, we may anticipate another problem because what the dissent has done is shift the focus to the P. In Cardozo's decision, the investigation of a duty of care focuses on the D. Proximate cause keeps P's involvement in the forefront. It is possible that an accurate representation must provide two perspectives of the same events. Our linear representations are still difficult to adjust to multi-perspective situations. Nested contexts and modal components are very difficult to use to achieve the result of shifting values and are not entirely satisfactory. Each argument is handled separately within the modal operator indicating it is a judge's opinion of the facts, rather than events as they happened. Ideally we should be able to present the facts, and in direct conjunction with them, a variant perspective from each opinion. So far this has proven too difficult to accomplish because of contradictions. The multi-perspective representation requires a very details representation of the facts, not suitable for presentation here.

J.C. Smith illuminates the reasoning by contrasting *Palsgraf* with *Menlove* case[13]. Menlove, a farmer stacked his hay in such a way that it spontaneously ignited. The fire spread to Vaughan's farm nearby, and cottages on that property were destroyed by it. Although there was no intent to harm Menlove

---

[11] Proximate cause simply stated, requires both a natural, uninterrupted chain of events, and an essential cause. *Black's Law Dictionary* (1968).
[12] *Palsgraf* v. *Long Island Railway Co* (1928) 162 N.E. 99 at 104.
[13] *Vaughan* v. *Menlove* (1837) 3 Bing, N.C. 468; 132 E.R. 490.

was liable. The difference between the two cases is that judicial awareness in *Palsgraf* goes with the causal act, and in *Menlove* with the resultant act[14]. The railway man does not engage liability because he is said not to have been able to have foreseen the possibility of damage occurring. Menlove was liable, even though the damage was unforeseeable, because it was the direct result of his completed act.

The only apparent way of making this distinction is by strengthening the assertion of causality. In both sequences, the chain is uninterrupted. In representing the *Menlove* arguments it may be necessary to make the original act a cause for each ensuing act. The nonsensical result is to write the [SPREAD] event for the fire with a causation relation, designated either (CAUS), or alternatively [ACTS-] for the [STACK] event, such that the stacking of hay was the causal event which resulted in the spread of the fire!

It is also possible to make an additional linking list, attached to a larger bracketed context, such as the case identification, or as a (CHRC) a characteristic of an initiating event which identifies all the elements in the causal chain. Neither of these methods works well because the representation deviates too widely from the text and meaning is distorted in several respects. As well some diligence is required in searching to make sure they are available for consideration in contexts as needed.

The best solution so far achieved is to represent the causality by the case relations described with the addition of refined versions of the original Sowa causation concepts, and use of additional brackets to enclose causal chains within their own contexts. Once again, we can represent knowledge as it is presented to us, but the quality of the Kr depends on the quality of the information.

The other issue with regard to causation is determining its extent. In attempting to limit liability, the judges want to draw a finish line for causes "somewhere short of the freakish and the fantastic"[15]. Our concern is to draw the line at the end of the event in a given representation. If things stop happening, the cases do not show any more causal links. Similarly, where time is represented in intervals, at the end of a given interval, the action, event or chain ceases. Alternatively, if the causal relations are made more emphatic, for example with overloaded (CAUS) relations, or links which do not coincide with the sequential representation of events, then one must be sure to draw a closing line.

It is almost impossible to establish boundaries for this, like the foreseeability requirement. It is accepted wisdom, in legal interpretation, that an act can have limitless effect, "in a very real sense the consequences of an act go forward to eternity, and back to the beginning of the world".[16] The best guidelines for knowing when one has come to the point where causation should be terminated derive from analyzing proximate causes. There should be proximity of time and space and some strongly integrating factor indicating a sequential thread of events. When that the unities are not there, the chain has ended. These too are conditions which can be checked automatically, given a representation in which causal case relations are used to advantage.

---

[14] Coval and Smith (1986) examine in detail the language of negligence.
[15] Prosser, 1953, 27.
[16] Prosser 1953, 24.

3. Conclusion

A representation of the causal chain in *Palsgraf* has been done to elucidate the use of causal relations in text with a view to moving toward automatic processing. The previously defined cgs relations were shown to be lacking and a minimal set of additional case relations has been suggested as a supplement. Relevant parts of the representation have been shown to support the judicial arguments in the case. An attempt has been made to show how we can start to work out automatic representation using valency in lexicons and case relations in the construction of simple, linked causal chains.

References

1 *Black's Law Dictionary* rev. 4th ed. . West Publishing, St. Paul MI, 1968.
2 S.C. Coval and Joseph C. Smith. *Law and Its Presuppositions: Actions, Agents and Rules*. Routledge & Kegan Paul, London, 1986.
3. Jim Cowie, and Wendy Lehnert. "Information Extraction." *Communications of the ACM*, 39(1), January 1996, 80-91.
4. Judith P. Dick. *A Conceptual, Case-Relation Representation of Text for Intelligent Retrieval*. Technical Report CSRI-265. Computer Systems Research Institute, University of Toronto, Toronto, 1992.
5. Charles J. Fillmore. "The Case for Case". Bach, Emmon Werner and Harms, Robert Thomas (editors). *Universals in Linguistic Theory*. Holt, Rinehart and Winston, New York, 1968.
6. Charles J. Fillmore. "Toward a Modern Theory of Case." David A. Reibel, and Sanford A. Schane (editors). *Modern Studies in English: Readings in Transformational Grammar*. Prentice-Hall, Englewood Cliffs, NJ, 1969.
7. Charles J. Fillmore. "The Case for Case Re-opened." In Cole, Peter and Sadock, Jerold M (editors). *Syntax and Semantics 8: Grammatical Relations*. Academic Press, New York, 1977.
8. Louise Guthrie, *et al.* "The Role of Lexicons in Natural Language Processing." *Communications of the ACM*, 39(1), January 1996, 63-72.
9. Lewis N. Klar, (editor). *Studies in Canadian Tort Law*. Butterworths, Toronto, 1977.
10. Geoffrey Leech,. *Semantics: The Study of Meaning*. 2nd ed. Rev. Penguin Harmondsworth Middlesex, England, 1981.
11. William F. Prosser. "Palsgraf Revisited." *Michigan Law Review*, 52(1), Nov. 1953, 1-32.
12. William F. Prosser. *Cases and Materials on Torts*. By William L. Prosser, J.W. Wade, *et al.* 4th ed. Foundation Press, Westburg NY, 1994.
13. Harold L. Somers. *Valency and Case in Computational Linguistics*. Edinburgh Univ. Press, Edinburgh, 1987.
14. John F. Sowa. *Conceptual Structures: Information Processing in Mind and Machine* Addison-Wesley, Reading MA, 1984.
15. John F. Sowa. "Notation of Conceptual Graphs" Draft chapter from forthcoming book, SUNY, Binghamton NY, 1987.

16. John F. Sowa. *Knowledge Representation: Logical, Philosophical, and Computational Foundations.* Preliminary Edition. ICCS'94 University of Maryland, College Park MD, 1994.

17. Cecil A. Wright. *Canadian Tort Law; Cases, Notes and Materials.* By Allen M. Linden and Lewis N. Klar. Butterworths, Toronto, 1990.

18. George KinglselyZipf. *Human Behavior and the Principle of Least Effort.* Addison-Wesley, Reading, MA, 1949.

Information Systems Modeling with CGs Logic

Ryszard Raban

School of Computing Sciences
University of Technology, Sydney
PO Box 123, Broadway NSW 2007, Australia
richard@socs.uts.edu.au

Abstract. This paper shows how to apply the conceptual graphs logic to information systems modeling. The graphs are not used as just another graphical representation of information system requirements, but the full power of this graphical logic system has been employed to fully and precisely capture type definitions, referential integrity constraints and global constraints. Such a logic based definition of an information system takes its semantics from a set of facts represented by fully instantiated simple conceptual graphs. An incremental validity checking procedures for adding new facts to a fact base have been defined and illustrated by an example.

1. Introduction

Recent efforts in applying conceptual graphs to enterprise modeling concentrated on converting traditional notations, like E-R, NIAM or OMT, into conceptual graphs [9], [5], [1], [2], [3]. And while this work has proved conclusively that conceptual graphs subsume most of the exiting modeling techniques, one might argue that the full power of the conceptual graphs logic has not been utilized in it.

To model an information system an analyst has to capture its concepts, its structure as well as its referential integrity and global constraints. Conceptual graphs formalism is capable of handling all those aspects of modeling. Equally, it is capable of capturing semantics of the models through the conceptual graphs version of the model theory as described in [7]. This paper shows how the conceptual graphs logic can be used to comprehensively and precisely represent information systems requirements, and how to use this representation to check the integrity of data stored as simple conceptual graphs.

Using conceptual graphs and the model theory for conceptual modeling of information systems creates some terminological problems. A conceptual model is what in the model theory is called a set of formulas making statements about the world [6]. On the other hand, a model in the model theory is a set of structures that represent a possible world. The term 'model' means two distinctly different things in the two contexts it is used. In order to avoid any confusion, it is assumed for the purpose of this paper that what is called a conceptual model in the information system context, and a set of formulas in the model theory context will be called an information system definition. A model as understood in the model theory will be called an information system fact base. Both information system definition as well as information system fact base will be represented by conceptual graphs as suggested in [7].

The first part of this paper defines set referents in order to establish their precise meaning when used in computing joins, projections and projective extents. Then, using a running example, it shows how to create a system definition, evaluate denotations and validate information system facts using the definition.

2. Referent Notation in Conceptual Graphs

Conceptual graphs are usually drawn in one of two forms, either as diagrams [7], [8] or as linear form expressions [7], [10]. In the both cases, no complete standard of acceptable forms of referents has emerged as yet. A precise and comprehensive method of representing all types of referents is very important in determining semantics of conceptual graph definitions of information systems. For that reason a variation of the method which was initially suggested in [Sowa93a] has been adopted for the purpose of this paper. This notation uses six basic types of referents:

individual referents:
- **names** which can be used with concepts of type WORD and of all its subtypes. For example, [PERSON]->(NAME)->[WORD: John].
- **measures** which can be used with concepts of type MEASURE, of all its subtypes. For example,
[PERSON]->(CHRC)->[HEIGHT]->(MEAS)->[MEASURE: 180 cm]
or in the contracted form [PERSON] ->(CHRC)->[HEIGHT: @ 180 cm].
- **individual markers** which can be used with concepts that are not of type WORD or type MEASURE, and that are not measurable concepts either. For example, [PERSON: #1082]->(NAME)->[WORD: John].

generic referent:
- **generic marker** which can be used with any concept, and is the default if no referent is present in a concept. A generic marker means an unspecified instance of the concept. For example, [PERSON: *]->(NAME)->[WORD: John].
A generic marker can be restricted to a type consistent individual referent, that is either a name or measure or individual marker;

universal referent:
- **universal quantifier** which can be used with any concept. A universal quantifier means that all individual concepts of a specific type have properties described by a graph. For example, [PERSON: ∀]->(NAME)->[WORD].
A universal quantifier can be restricted to a generic marker or to a type consistent individual referent, that is either name or measure or individual marker;

null referent:
- **null referent** which can be used with any concept. It is mainly utilized to define optionality when included into a disjunctive set referent. For example, graph [PERSON: *]->(NAME)->[WORD: * I NULL] can be used to say that there is a person with an optional name.

Projection of a graph with concept [PERSON: NULL] is successful if a graph with concept [PERSON: NULL] removed has a projection, and a graph with concept [PERSON: NULL] replaced by concept [PERSON: *] does not have a projection.

In addition to these singular referents it is convenient to adopt after [7] two types of set referents. These, however, can be viewed as composite referents because they are always expandable to equivalent representations using the basic types of referents only. The set referents are:

- **collective set** which is a set of any concept type consistent referents except for universal quantifiers and NULL referents. A concept with a set referent can be expanded into a set of individual concepts with all relations linked to the original concept repeated for every individual concept introduced. For example, graph [PAIR: *]->(PART)->[PERSON: {#1082, *}] is equivalent to [PAIR: *]-
  (PART)->[PERSON: #1082]
  (PART)->[PERSON: *].

All possible ways a concept with a collective referent set can be restricted come from valid restrictions that can be performed on its expanded form.

- **disjunctive set** which is a list of any concept type consistent referents except for universal quantifiers separated by the I sign. For example, graph
  [PERSON: #3201]->(LIKE)->[PERSON: {#1082 I NULL}]
  is equivalent to a disjunction of two graphs:
  ¬[¬[[PERSON: #3201]->(LIKE)->[PERSON: #1082]]
    ¬[[PERSON: #3201]->(LIKE)->[PERSON: NULL]]].

A concept with a disjunctive set referent can be restricted by selecting any subset of its elements and/or replacing its generic marker with a type consistent individual referent. Type restriction can be done the usual way.

The shorthand referents defined below can also be used for convenience:

- **counted generic set** which means a set with <u>an exact number</u> of referents. For example, [GROUP] >(HAS)->[PERSON: {*}@5] means that a group has exactly 5 people.

When a graph g containing a concept o with a counted generic set referent is projected into another graph h, the projection is successful only if a graph h contains exactly the number of instances of concepts of $type(o)$ as specified in the generic set referent.

Any partially defined counted generic set, can be represented as a collective set of the individual markers and a counted generic set of the remaining generic elements. For example, [PERSON: {#123, #320, *}@5] can be shown as [PERSON: {#123, #320, {*}@3}].

- **variable generic set** which means a variable number of elements within a specified range. For example, [PERSON: {*}@1...n] is equivalent to [PERSON: {* I {*}@2 I{*}@3 I ... I{*}@n};

- **generic set** which means a set with any number of elements. For example, [PERSON: {*}] is equivalent to [PERSON: {* I {*}@2 I{*}@3 I ...}. Note that {*} does not mean a set of one generic element. To specify a set like that, one of the following notations should be used: ∗ or {∗}@1, except that the later one means exactly one referent.

3. Defining Information System in Logic

Let an information system be defined by system definition $D = \langle S, T, R \rangle$ where
- T is a type hierarchy,
- R is a relation hierarchy, and
- $S = \{g_1, g_2, ..., g_n\}$ is a schema which is a consistent and satisfiable set of simple and compound conceptual graphs built out of concepts and relations described by type labels from T and R, referents being either generic markers and/or names and/or measures[1], and coreference links.

Let us illustrate this on a simple case study. An information system definition will be built for a college that provides a three year course to a small number of students. Due to its exclusive character a number of places available in every year of studies is limited. There can be only up to 50 students in the first year, up to 40 in the second, and no more than 35 in the third. In the final year every student must have a supervisor for the final project. A teaching staff member cannot supervise more than five students. For simplicity let's assume that every person registered in the system must have a name and an optional telephone number. A student identification number and a year of study must be stored for every student, and a staff identification number and a date of hire must be registered for every teaching staff member.

The type hierarchy T in this case contains the following type labels: OBJECT > PERSON > STUDENT and PERSON > STAFF. The relation hierarchy R is made of three relations CHRC, NAME and SUPER. Using these types, the system requirements of the case study can be represented by the following set S of conceptual graphs:

($g1$) {¬[[PERSON: ∗x] ¬[[OBJECT: ?x]->(NAME)->[WORD]
                                  (CHRC)->[PHONE: ∗ I NULL]]],

($g2$) ¬[[STUDENT: ∗x] ¬[[PERSON: ?x]->(CHRC)->[STUD-ID]
                                  (CHRC)->[STUD-YEAR]]],

($g3$) ¬[[STAFF: ∗x]   ¬[[PERSON: ?x]->(CHRC)->[STAFF-ID]
                                  (CHRC)->[DATE-HIRED]]],

($g4$) ¬[[STUDENT: ∗x]->(CHRC)->[STUD-YEAR: 3]
                         ¬[[STAFF]->(SUPER)->[STUDENT: ?x]]],

($g5$) ¬[[STAFF: ∗x] ¬[[STAFF: ?x]->(SUPER)->[STUDENT: {∗}@0...5]]],

($g6$) [STUDENT: {∗}@0...50]->(CHRC)->[STUD-YEAR: 1],

($g7$) [STUDENT: {∗}@0...40]->(CHRC)->[STUD-YEAR: 2],

($g8$) [STUDENT: {∗}@0...35]->(CHRC)->[STUD-YEAR: 3]}.

---

[1] Universal quatifier is not used explicitly in defining schemas. It can, however, be represented by an oddly enclosed dominant concept in a line of identity. For example,
[PERSON: ∀]->(NAME)->[WORD]
is equivalent to
¬[[PERSON: ∗x] ¬[[PERSON: ?x]->(NAME)->[WORD]]].

Graphs *g1*, *g2* and *g3* define types **PERSON, STUDENT** and **STAFF**. Graphs *g4* and *g5* impose referential integrity constraints on relation **(SUPER)**. And finally, student yearly quotas are represented by graphs *g6*, *g7* and *g8*. In this way, a complete set of information requirements with type definitions, referential integrity and global constrains has been represented by a set of conceptual graphs.

4. Semantics of System Definition

D represents a logic based definition of an information system. A set of propositions in schema S contains statements about the system, its concepts, relations and constraints. This conceptual graphs logic has to be complemented with semantics. For this purpose, let us define a fact base $W = \langle F, I \rangle$, where

- $F = \{f_1, f_2, ..., f_n\}$ is a set of simple graphs asserting atomic facts. Graphs in F are fully instantiated concepts and relations defined in D, and it is assumed that every object instance appears in the set only once,
- I is a set of individual referents (markers, names and measures) used in F.

Since schema S is satisfiable there has to be a fact base W such that the denotation of S, δS, is **true** in F. The evaluation game based method of evaluating denotations for conceptual graphs has been given in [Sowa84]. A fact base W which satisfies a schema S of D is called a fact base of information system D. A set of all fact bases of a given information system D is denoted by $\mathcal{W}_D$.

Any fact base of D from which no single graph, concept or relation can be removed without making δS **false** in the reduced fact base is called a minimal fact base of D. For every definition D there is a set of minimal fact bases. The minimal fact bases usually represent the set up data of an information system, like for example exchange rates, company name or default values. Fact base

($W0$) $\langle \{[STUD-YEAR: 1], [STUD-YEAR: 2], [STUD-YEAR: 3]\}, \{\} \rangle$

is the only minimal fact base of the case study information system.

During its life time an information system is constantly updated and evolves through a possibly infinite sequence of information system fact bases $W_0, W_1, W_2, ...$, where W_0 is one of the minimal fact bases of D. An information system maintains integrity if and only if for all i, $W_i \in \mathcal{W}_D$. The rest of this paper will demonstrate how the denotation evaluation procedure is used to maintain the integrity of an evolving information system defined by D. The simplest approach in evaluating changes could be to check if a modified fact base still satisfied the definition. This, however, means evaluating the entire fact base every time a change is made. This will become more and more costly and inefficient as the fact base grows larger and larger. For that reason this paper adopts different strategy based on checking only increments by which a fact base is being changed.

In order to establish increments by which an information system fact base can be changed, let us have a look at different kinds of concepts populating a fact base. Following the ontological criteria of the foundation and semantic rigidity introduced in [Guarino92] it is obvious that some concepts can exist independently while others necessarily depend on existence of other concepts. In the conceptual graphs formalism concepts of names and measurable types are semantically rigid but they represent dependent qualities. These concepts are called later properties. They describe the independent concepts which are terms called later objects. This leads to the following high level structure of type hierarchy T: T > OBJECT, T > PROPERTY, PROPERTY > NAME-TYPE, PROPERTY > MEASURABLE-TYPE. And since properties convey the knowledge about objects, properties can only be related to objects. On the other hand, objects can also be related to each other representing the relational knowledge. This leads to two types of relations: these linking objects with properties called later type defining relations, and those linking objects with other objects called later structural relations.

Properties cannot be added to an fact base on their own since they are dependent on objects that are described by them. Objects, however, when added into a fact base must have all required properties as well as all structural relations with other objects as required by referential integrity constraints. It means that it is reasonable to assume that a fact base will be changed by adding individual objects to it.

5. Evaluating denotations

The evaluation game decides whether a set of graphs in $S = \{g_1, g_2, ..., g_n\}$ is **true** or **false** in a set of facts $F = \{f_1, f_2, ..., f_n\}$. The game is played by a proposer and a skeptic who take turns in a sequence of moves consisting of project, select and reduce operations always performed in this order. The result of the game is not decided by an outcome of any particular play, but instead by the fact which side has a winning strategy in the game. If a proposer has a winning strategy then the denotation is **true**, if a skeptic has one then the denotation is **false**.

As it has been said before, the aim of this paper is not to evaluate a whole fact base but only incremental changes to a fact base. Every change is always represented by a simple graph. And therefore, a set of information system definitions is always evaluated against a single graph f. It means that instead of projective extents there are just simple projections in the project moves, and that players cannot anymore control the evaluation game by selecting a favorable projection from a projective extent. This section shows how the game is played in those simplified conditions.

In presenting the game, πg will be used to denote a projection of g in a fact f, and $g\#$ will denote a graph obtained by copying individual markers from instantiated dominating concepts into concepts in a graph g. Also, it is assumed that a compound graph takes always the form of $\neg[p_1, p_2, ..., p_n]$, and its referent $\{p_1, p_2, ..., p_n\}$ must consist of at least a single simple graph or a couple of compound graphs. This does not reflect on generality of the discussion as $\neg[\neg[p]]$ can always be reduced to just p.

For a set $S = \{g_1, ..., g_n\}$ of graphs evaluated for a graph f the game is played as shown in TABLE I and II.

	PLAYER	MOVE	OUTPUT GRAPH	
	START		$\{g_1,..., g_n\}$	
1	Prop	Proj	$\{g_1,...,\pi g_i, ..., g_j\#, ..., g_n\}$	
2	**Skept**	**Sel**	$\{g_h\}$ - g_h is a simple graph	$\{\pi g_i\}$ - g_i is a simple graph
3	Prop	Red	$\{\}$ **Skept wins**	$\{\pi g_i\}$ **Prop wins**

Table I. Evaluation of simple graphs

Since a skeptic has the first selection it initially controls the game. Any graphs in S with the denotations **false** wins the game for a skeptic. To block this possibility all graphs in S must be **true** in f. To ensure that, firstly (as shown in TABLE I), all simple graphs must have projections in f. And secondly, for all compound graphs a proposer must have a winning strategy. Since a proposer is now in control (move 5 in TABLE II) it can win either by selecting a simple graph with no projection in f, p_k in this case, or if all of them have projections in f, select one of the compound graphs, p_s in this case, for which the denotation of its referent is **true**. To prove it the graph has to be evaluated further. But at this point the evaluation of a set of graphs being the referent of the compound graph selected must go through exactly the same steps as for a schema S the game started with.

	PLAYER	MOVE	OUTPUT GRAPH	
	START		$\{g_1, ..., g_n\}$	
1	Prop	Proj	$\{g_1,...,g_i, ..., g_j\#, ..., g_n\}$	
2	**Skept**	**Sel**	$\{g_p\}$ - g_p is a compound graph $\neg[p_1,..., p_m]$	
3	Prop	Red	$\{p_1,..., p_m\}$	
4	Skept	Proj	$\{p_1,..., p_s\#, ..., \pi p_r, ...,p_m\}$	
5a	**Prop**	**Sel**	$\{p_k\}$ - p_k is a simple graph	$\{\pi p_r\}$ - p_r is a simple graph
6a	Skept	Red	$\{\}$ **Prop wins**	$\{\pi p_r\}$ **Skept wins**
5b	**Prop**	**Sel**	$\{p_s\}$ - p_s is a compound graph $\neg[q_1, ..., q_l]$	
6b	Skept	Red	$\{q_1, ..., q_l\}$	
7b	Prop	Proj	$\{q_1, ..., q_l\}$ **which is the starting problem**	

Table II. Evaluation of Compound Graphs

The above evaluation game analysis is useful in establishing the most efficient strategy for evaluating the denotations of definition graphs. Basically, both a proposer and a skeptic have to find and select a false graph to win a game. Since a proposer controls selection in oddly enclosed negative contexts and a skeptic in evenly enclosed ones, for a set of graphs to be **true**:

- a proposer must always be able to find at least one false graph in oddly enclosed negative contexts being evaluated, and
- all graphs in evenly enclosed negative contexts being evaluated must always be **true**, so a skeptic cannot find and select a false graph and win a game.

Therefore, when looking for a false graph it is reasonable to progress from simple graphs, and then progressively to more complex ones. On the other hand, while trying to prove that a number of simple graphs are all **true**, it is desirable to join them first and then compute projective extent once for the join, instead of computing it for each of the graphs separately.

These simple strategies have been used in the procedures for type, referential integrity and global constraints checking described in the next sections.

6. Enforcing Typing

All type defining graphs are in the form of $\neg[\, p \, \neg[q]]$ where q contains only type defining relations between an object (genus) and its properties (differentia). An object in graph p is of a type that is being defined. For example, graph $g1$ defines a person as an entity with a name and a phone as properties.

The addition of a new object must create a new information system fact base which still satisfies the information system definition. To ensure that, all type defining implications must be **true**. This, in turn, requires that in all type defining compound graphs in the form $\neg[\, p \, \neg[q]]$ either p is **false** or the both p and q are **true**. In order to illustrate the type checking, and later the integrity checking, the case study system defined earlier will be used. Let us consider adding a new student

(f1)   [STUDENT: #123]-
                    (NAME)->[WORD: John]
                    (CHRC)->[STUD-YEAR: 3]

to the following fact base

(W1)   {{[STUD-YEAR: 1], [STUD-YEAR: 2], [STUD-YEAR: 3],
            [STAFF: #432]-
                    (NAME)->[WORD: Mary]
                    (CHRC)->[PHONE: 663-6754]
                    (CHRC)->[STAFF-ID: 723]
                    (CHRC)->[DATE-HIRED: 95/10/20],
            [STAFF: #521]-
                    (NAME)->[WORD: George]
                    (CHRC)->[STAFF-ID: 320]
                    (CHRC)->[DATE-HIRED: 92/02/12]},
        {#432, #521}}

of the college system. Fact base W1 satisfies the system definition and the additions of the new fact must lead to a new fact base that also satisfies the definition.

The premise of graph $g3$ has an empty projection in $f1$, so graph $g3$ is satisfied in $f1$. False premises rule out all irrelevant type defining graphs which have nothing to do with the fact actually being checked. The premises of graphs $g1$ and $g2$ have projections in $f1$, and therefore are they are relevant type defining graphs and their conclusions have to be satisfied in $f1$. In order to check the conclusions individual markers from the projections must be transferred to the respective premises of *g1* and *g2*, and then copied to the conclusions along lines of identity. Finally, maximal join of the resulting conclusion graphs is performed leading to:

(*g9*)   [PERSON: #123]-
                   (NAME)->[WORD]
                   (CHRC)->[PHONE: * | NULL]
                   (CHRC)->[STUD-ID]
                   (CHRC)->[STUD-YEAR].

By using the join not only properties of the type being checked but also all properties inherited from its supertypes are included into the checking process. Obviously, if a result of the join is satisfied in a new fact, so are all the graphs that have been used to create the join. In order to check whether a new fact graph conforms with the type definition, a fact graph has to be maximally joined with the combined type defining graph. Such a maximal join of *g9* and *f1* produces the following graph:

(*g10*)   [STUDENT: #123]-
                   (NAME)->[WORD: John]
                   (CHRC)->[STUD-ID: *]
                   (CHRC)->[STUD-YEAR: 3].

A new fact graph is valid if all its concepts are covered by the maximal join, and all concepts of a combined definition graph are also covered by the join. A concept *c* in a graph *u* is said to be covered by a maximal join *w* of graph *u* and graph *v* if there is a concept *c'* in *v* such that *c* and *c'* were joined together into a concept in *w*. Otherwise, a concept is uncovered. Concepts in a fact graph not covered by the maximal join represent some spurious elements in the fact graph, elements that are not present in the type definitions. On the other hand, all concepts of a combined definition graph not covered by the maximal join are all those elements required by the type definitions that are not present in the fact graph.

Graph *g10* shows that combined definition graph *g9* contains one uncovered concept. This is [STUD-ID] which appears with a generic referent indicating that fact *f1* does not contain one required property. And therefore, fact graph *f1* is invalid. The problem can be rectified by adding the missing referent like in the graph below:

(*f2*)   [STUDENT: #123]-
                   (NAME)->[WORD: John]
                   (CHRC)->[STUD-ID:123]
                   (CHRC)->[STUD-YEAR: 3].

It is left to the reader to follow the previous steps and check that the modified fact is valid. The validity checking and correction process for a fact graph *f* can be summarized by the following procedure:

```
1°   select all type defining graphs with premises having non empty projections in a fact graph f;
2°   transfer individual markers from the projections to respective premises;
3°   copy the individual markers to conclusion graphs along lines of identity;
4°   create a graph u by joining all the conclusions of the selected graphs;
5°   create a graph w by performing a maximal join of u and f;
6°   if all concepts in u are covered by w and all concepts in f are covered by w then
                a graph f is type valid and procedure ends;
        else
                remove from a graph f all concepts not covered by w, if any;
                attach to a graph f all concepts in u which are not covered by w, if any;
        end if;
7°   goto step 5°;
```

7. Referential Integrity Checking

The type checking procedure creates object instances, and it always deals with in a single object at a time, and therefore does not need to make references to any objects instances already present in a fact base. The referential integrity checking procedure creates new relations between a new object instance and existing ones. It requires constant interaction with a fact-base in order to retrieve potential participants in the new relations as well as perform referential integrity constraints checking. In effect, graphs containing a mix of new and existing object instances will be used in the process. Let us introduce some notions related to such graphs.

A graph with one or more relations between new object instances and generic objects is called an *incomplete* graph - there are some relations that are still to be created. A graph with no relations between new object instances and generic objects is called either a *specific* graph if all its objects are instantiated, or a *partial* graph if there are still some generic objects in it. Note that a partial graph has to have at least one existing concept instantiated. Let us also define operator σ which for a graph g creates a graph or a set of graphs obtained by removing all new object instances, their properties and relations from a graph g. For a partial graph g, every graph produced by σg contains at least one instantiated object. In order to check if a given specific or partial graph g is consistent with a fact set F, operation *specify*, defined below, is used.

1°   let U be a set of graphs σg;
2°   **for** every graph u in U **do**
        if there is a non empty projective extent of $\Pi(u, F)$ **then**
                maximally join the projection in $\Pi(u, F)$ with the graph g and replace g with
                the result;
        else
                mark the operation as unsuccessful and end procedure;
        end if;
        end for;
3°   mark the operation as successful;

Step 1° extracts from a graph g all its parts which refer to existing elements of a fact set F. All graphs in set U have at least one object instantiated as the input graph is either specific or partial. This in turn means that every projective extent calculated in step 2° is either empty or have a single graph in it. An empty projective extent raises contradiction and makes the operation unsuccessful. When a projective extent is not empty, it contains a single graph which is joined with the input graph instantiating any generic objects that might be there. In effect, successfully performed operation *specify* produces a fully instantiated input graph consistent with a fact base.

In the referential integrity checking process only referential integrity defining graphs with premises satisfied in a fact graph f extended by relevant elements of a fact set F (in short $f+$) have to be considered. But the checking procedure may expand a fact graph with new object instances and relations, therefore it is not possible to tell up front which referential integrity defining graph will eventually become relevant. So, the checking process starts with the first referential integrity defining graph found to have its premise **true,** and after successful checking of its conclusion moves to a next graph which has true premise for a now possibly modified fact graph.

While checking a premise it is not sufficient to just look for a projection in a fact graph f. To make sure that also all relevant parts of a fact set F are considered, it is required to maximally join the fact graph with the premise graph and produce a projection of the premise in the join. If the projection is incomplete the premise cannot be satisfied in $f+$ because it contains a missing relation with a new object instance that will not be found in a fact base. Otherwise, there is a possibility that the premise can be satisfied. Procedure *specify* attempts to match the projection with a fact base and retrieve relevant instances for all generic objects in the projection. Successfully performed *specify* means that the premise is satisfied in $f+$, and that all individual markers from the fully instantiated projection can be transferred to the premise and copied to the conclusion graph along lines of identity. If the premise turns out to be **false,** it means that the referential integrity defining graph under consideration is not relevant at this stage of checking. The process of checking a premise p for a fact graph f and a fact set F is formally defined by the following procedure:

```
1°  if   there is a maximal join of f and p  then
              let u be the maximal join;
              case a projection πp of p in a join u is
                  when incomplete           ⇒   the premise is false;
                  when specific or partial  ⇒
                      if specify(πp) is successful then
                              transfer individual markers from πp, modified by specify, to p;
                              copy all individual markers to q along lines of identity;
                              the premise is true;
                      else
                              the premise is false;
                      end if;
              end case;
    else
              the premise is false;
    end if;
```

In the example, there are two graphs $g4$ and $g5$ that represent referential integrity constraints. At this stage, only the premise of graph $g4$ is **true** since it is satisfied in fact graph $f2$. In this case projection of the premise in a maximal join of the premise and $f2$ produces a graph consisting of the new object instance only, and therefore *specify* performed on the projection has to be successful.

Checking a conclusion graph uses basically the same method as the one employed in the premise checking. The only significant difference in this case is that if a projection of the conclusion in a maximal join of the conclusion and a fact graph turns out to be incomplete, some additional steps are taken to create missing relations with new object instances. Projective extent is used to retrieve all possible options for the relations, and to make sure that all possible dependencies between existing object instances are properly captured and presented to the user for selection. Also, an option to create new object instances is provided. The conclusion checking procedure is defined formally as follows:

```
1°   create a maximal join u of f and q,
2°   if the maximal join exists then
         case a projection πq of q in a join u is
             when specific or partial        ⇒
                 if specify(πp) is successful then
                     the conclusion is true;
                 else
                     the conclusion is false;
                 end if;
             when incomplete        ⇒
                 mark all generic concepts in πp which are related to new object instances;
                 let V be a set of graphs σπp;
                 for every graph g in V do
                     create an empty set A;
                     for every graph h in projective extent Π(h, F) do
                         create a set of all individual markers of marked objects and add it to A;
                     end for;
                     add an option new to the set A, and present it to the user for selection;
                     if the user selects the set of individual markers then
                         add to the fact graph f the instantiated marked objects and the
                         relations between them and respective new objects;
                     else
                         ask the user to create new object instances of types consistent with
                         types of the marked objects in g;
                         add the new object instances and respective relation tothe fact graph f;
                     end if;
                 end for;
                 the conclusion is true;
         end case;
     else
         if specify(q) is successful then
             the conclusion is true;
         else
             the conclusion is false;
         end if;
     end if;
```

Back to the example. The premise of graph *g4* turned out to be **true** so its conclusion must now be checked. After transferring individual markers to the premise graph and copying them into the conclusion graph, a projection of the conclusion in a maximal join of the conclusion and fact graph *f2* is performed giving the following result:

(*g11*)   [STUDENT: #123]<-(SUPER)<-[STAFF: *].

Graph *g11* is incomplete since it contains a relation between a new object instance and a generic concept. It indicates a missing relation that has to be created. The student can be supervised by an existing staff member or a new one. In order to retrieve existing staff members a graph σ*g11* is created. This graph contains just one generic concept [STAFF: *] and its projective extent into the fact set *F1* produces all instances of existing staff members. A set of their individual markers supplemented by the option *new* is produced as a disjunctive set referent *(g11)*   [STUDENT: #123]<-(SUPER)<-[STAFF: {#432 I #521 I *new*}]

and presented to the user for selection.

If the user selects the option *new*, a new instance of object [STAFF] must be created and added to a fact base before the process can proceed further. But, if an existing instance (for example, [STAFF: #432]) is chosen then a new relation is created immediately leading to new fact graph *f3* for which graph *g4* is now **true,** and therefore the related integrity constraint satisfied.

(f3)   [STUDENT: #123]-

            (NAME)->[WORD: John]
            (CHRC)->[STUD-ID:123]
            (CHRC)->[STUD-YEAR: 3]
            (SUPER)<-[STAFF: #432].

After the fact graph has been modified to satisfy *g4*, graph *g5* becomes relevant since its premise has now a non-empty projection in fact graph *f3*. So after propagation of the individual marker, maximal join of the conclusion of graph *g5* and fact graph *f3*, the following projection of graph *g5* in the join is obtained:

(g12)   [STAFF: #432]->(SUPER)->[STUDENT: {#123, {*}@0...4}]

This graph shows one peculiarity of the maximal join operation performed on a fact graph with generic set referents. When a new object instance in a fact graph is joined with an object containing a generic set referent, it is not necessary that all instances indicated by the generic set referent are instantiated in the fact graph since the remaining instances might be present in a fact base. In such a case, the two concepts are joined producing a partially defined generic set as shown in graph *g12*.

Graph *g12* can be successfully processed by procedure *specify* since there is only one student supervised by [STAFF: #432]. So the conclusion of graph *g5* is **true**, and therefore fact graph *f3* conforms to the referential integrity constraint represented by graph *g5*. In the example, there is no more referential integrity defining graphs to check, so fact graph *f3* complies with all referential integrity constraints.

In summary, the type checking procedure for a fact graph *f* can be defined as:

1°create a set *C* of all referential integrity constraints defining graphs;
2°**if** there is a graph *g* in *C* which has its premise **true then**
        perform the conclusion checking procedure on the conclusion of a graph *g*;
        if the conclusion is **false then**
            indicate an offending new relation to the user and backtrack to the point where
            the relation has been created and prompt the user to take corrective actions;
      else
        remove a graph *g* from *C*;
      end if;
    else
        a graph *f* fulfills referential integrity constraints;
        end procedure;
    end if;
3°**goto** 1° **end if**;

---

8. Global Constraints Checking

The global constraints are represented by simple graphs that are checked exactly the same way as premises of referential integrity constraint graphs. Here again a projection of a global constraint graph in its maximal join with a new fact graph, and subsequent application of procedure *specify* is used. For the new graph to be valid, all global constraints that can be maximally joined with it have to be satisfied.

In the example, there are three global constraint graphs - *g6*, *g7* and *g8*. Only one of them, *g8*, can be maximally joined with the fact graph *f3*. The truth value of the other graphs will not be affected by the new graph as they must have been already **true** in fact base *W1* when the existing objects were being added.

Therefore, the only global constraint graph to be checked in this case is *g8*. In order to establish if it is still going to be **true** after the new fact is added to the fact base, maximal join of graph *g8* and fact graph *f3* has to be performed and a projection of graph *g8* into the join created. The projection creates the following result:

(*g13*)   [STUDENT: {#123, {*}@0...34}]->(CHRC)->[STUD-YEAR: 3].

Graph *g13* is partial and procedure *specify* performed on the graph *g13* is successful since there is only one student enrolled in the third year. Thus fact graph *f3* is valid and can be added to fact base *W1* giving new modified fact base *W2*.

(*W2*)   {{[STUD-YEAR: 1], [STUD-YEAR: 2], [STUD-YEAR: 3]
          [STAFF: #432]-
                         (NAME)->[WORD: Mary]
                         (CHRC)->[PHONE: 663-6754]
                         (CHRC)->[STAFF-ID: 723]
                         (CHRC)->[DATE-HIRED: 95/10{/20]
          [STAFF: #521]-
                         (NAME)->[WORD: George]
                         (CHRC)->[STAFF-ID: 320]
                         (CHRC)->[DATE-HIRED: 92/02/12]}
          [STUDENT: #123]-
                         (NAME)->[WORD: John]
                         (CHRC)->[STUD-ID:123]
                         (CHRC)->[STUD-YEAR: 3]
                         (SUPER)<-[STAFF: #432], {#123, #432, #521}}

9. Conclusion

This paper shows how the conceptual graphs logic can be applied to information systems modeling. Conceptual graphs logic has been used to create information system definition, and a set of simple, fully instantiated conceptual graphs have been used to represent facts stored in the system. Incremental methods of checking typing, referential integrity constraints and global constraints have been introduced and illustrated by an example.

At this stage the information system definition has been limited to simple graphs and compound graphs in the form ¬[*p* ¬[*q*]] which while sufficient in defining types might prove too restrictive for representing referential integrity constraints and global constraints. Of course more complex constraints can always be checked by the evaluation game introduced by [Sowa84] and discussed in detail in the paper. This however might be too costly for checking incremental changes to an information system fact base.

The paper covered the checking procedure for adding facts to a fact base. This leaves integrity checking problems related to deleting facts from and updating facts in a fact base. These issues will be addressed in future work.

References

[1] Creasy, P., and Ellis, G. A Conceptual Graphs Approach to Conceptual Schema Integration. In Mineau, G.W., Moulin, B., and Sowa, J.F., editors, *Conceptual Graphs for Knowledge Representation*, pages 126-141. Springer-Verlag, Berlin, 1993.

[2] Boksenbaum, C., Carbonneill, B., Haemmerle, O., and Libourel, T. Conceptual Graphs for Relational Databases. In Mineau, G.W., Moulin, B., and Sowa, J.F., editors, *Conceptual Graphs for Knowledge Representation*, pages 345-360. Springer-Verlag, Berlin, 1993.

[3] Gardyn, E. Multiple Information Models. In Ellis, G., and Eklund, P., editors, *Proceedings of the 1st Australian Conceptual Structures Workshop*, pages 148-164. University of New England, Armidale, Australia, 1994.

[4] Guarino, N. Concepts, Attributes and Arbitrary Relations./Data and Knowledge Engineering 8 (1992), pages 249-261.

[5] Moulin, B. and Creasy, P. Extending the Conceptual Graph Approach for Data Conceptual Modelling. *Data and Knowledge Engineering*, 8(1992), pages 223-248.

[6] Makowsky, J.A. Model Theory and Computer Science: An Appetizer. In Abramsky, S., Gabbay, D.M., and Maibaum, T.S.E., editors, *Handbook of Logic in Computer Science*. Vol 1, Clarendon Press, Oxford, 1992.

[7] Sowa, J.F. *Conceptual Structures - Information Processing in Mind and Machine*. Addison-Wesley, Reading, Massachusetts, 1984.

[8] Sowa, J.F. Relating Diagrams to Logic. In Mineau, G.W., Moulin, B., and Sowa, J.F., editors, *Conceptual Graphs for Knowledge Representation*, pages 345-360. Springer-Verlag, Berlin, 1993.

[9] Sowa, J.F., and Zachman, J.A. Extending and Formalizing the framework for Information Systems Architecture. *IBM Systems Journal* 31, 3.

[10] Wermelinger, M. Conceptual Structures Linear Notation - Proposal for Pierce. In Eklung, P., and Ellis, G., editors, *Proceedings of the Fourth International Workshop on PIERCE: A Conceptual Graphs Workbench*, pages 106-125. University of California Santa Cruz, CA, August, 1995.

Modelling and Simulating Human Behaviours with Conceptual Graphs

Corinne BOS[1)2)], Bernard BOTELLA[3)], Philippe VANHEEGHE[1)2)]

1) ISEN	2) LAIL	3) Dassault Electronique
41 Bd Vauban	URA CNRS D1440	55 Quai Marcel Dassault
59046 Lille (France)	Cité scientifique -BP48	92214 Saint Cloud (France)
cpa, pva@isen.fr	59651 Villeneuve d'Ascq (France)	Bernard.Botella@dassault-elec.fr

Abstract

This paper describes an application of conceptual graphs in knowledge engineering. We are developing an assistance system for the acquisition and the validation of stereotyped behaviour models in human organizations. The system is built on a representation language requiring to be at expert level, to have a clear semantics and to be interpretable. The proposed language is an extension of conceptual graphs dedicated to the representation of behaviours. Tools exploiting this language are provided to assist the construction of behaviour models and their simulation on concrete cases.

1. Introduction

To face critical situations in a complex environment involving a human organization, it seems to be efficient to define and dispatch in advance roles to the different members of the organization. Carefully defined stereotyped behaviours, responding to some events and requiring to be precisely followed, are then associated with each of these roles so that people can react in a fast and correct manner. The different behaviours are built and synchronized precisely to enable the organization as a whole to respond correctly to the critical situation. Such stereotyped behaviours can be found, for example, in civil security domains to describe emergency procedures in case of accident or disaster. A stereotyped behaviour can be seen as a vague plan composed of ordered and synchronized actions, associated with a single role (for example witness, control room operator, succour chief...). Fig.1 gives an example of the procedure to be followed by the witness of disaster in a french chemical factory. The witness must identify the disaster, activate the alarm and warn the operators of the control room about the disaster. The warning of the control room operators will lead them to execute their corresponding procedures. The ringing of the alarm will lead the guardian and some other people to execute the corresponding procedures.

```
identify the disaster
press the closest alarm button
phone to the control room number 373
communicate the following message:
                            the location of the disaster
                            the nature of the disaster
                            the importance of the disaster
```

Fig.1 An example of an emergency procedure

We are building (simulation, control,...) systems aiming at training people involved in such organizations or aiming at studying and evaluating the ability of different organizations or procedures to face critical situations. In the design of these systems, we are more precisely investigating the acquisition of these stereotyped behaviours that will be then coded in the systems. The classical development of such systems (focusing on the stereotyped behaviour aspects) can be described as follows (Fig.2):

- The elicitation of knowledge consists in collecting the knowledge of one or several experts, this is done by multiple discussions between the expert and a knowledge engineer and it needs a lot of time and rigor. During this phase, the engineer uses semi-formal models of expertise to guide the elicitation and to describe the results. However, the results are paper format documents.
- These documents are then used for different application developments, they are part of the requirements and lead to the coding of programs corresponding to the behaviours and taking into account the application constraints.
- Finally, these different pieces of code are integrated with other components to build the complete application.

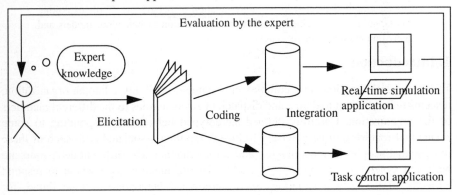

Fig.2 A too late expert evaluation

The main drawback of this approach comes from the difficulty for the expert to describe precisely and completely the behaviours. Hence, the coded behaviours turn out to be too simplistic and the final application is often judged as non realistic by experts though they have validated elicitation results. The second drawback is that elicitation results being described only semi-formally, they contain a lot of ambiguities, inconsistencies or incompleteness. This leads software engineers to make conscious or unconscious choices during the coding phase without any expert validation. All these problems are only discovered when the totally developed application is demonstrated on concrete cases to the experts. At this point, the determination of the responsibilities and the correction of problems have a great cost.

To address these drawbacks, and following the current research in knowledge engineering [7], we propose to define a formal language to model the behavioural expertise. To answer the second drawback, this language will allow to guide the elicitation phase and to obtain formally validated results. To answer the main drawback, we want the language to be executable, so that it is possible to simulate behaviours from an initial scenario. The objective of this proposition is to involve the expert in the modelling process, to show him the realism level of the built models by putting them to work on concrete situations. All this work should be done at expert level, that is to say without any impact from the intended application development constraints. The new approach to develop applications is thus described by Fig.3.

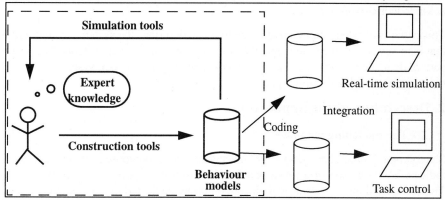

Fig.3 The proposed system

Two types of tools can be provided on the proposed language:
- Construction tools assisting the knowledge engineer to build the behaviour models with edition, consultation, static verification functions and behaviour base management,
- Simulation tools allowing to put behaviours to work, to visualize the situation evolutions, to apply dynamic verifications, and to evaluate the behaviour robustness.

In this approach, the elicitation phase produces semantically clear, validated and exploitable behaviour models that can then be used to develop applications. In accordance with the discussion about the use of Conceptual Structures for knowledge engineering [11], we think that Conceptual Graphs are a good candidate as the base of the proposed language. The reasons are that:
- Experts appreciate graphic notations. The very simple graphic concepts of Conceptual Graphs (boxes, ovals and arrows) and the use of labels chosen in the expert language make modelled knowledge understandable by the expert with no effort, whereas they are afraid by textual formal languages. This is very important since we want to involve experts in the modelling process.
- This graphic notation is adapted to represent not only the control of behaviours (like other formalism: State-Transition graphs, organigrams...), but also all the required knowledge (formulas defining conditions and effects of actions, states describing the current world, concepts and relations defining the domain).
- The semantics, that can be associated with each graph, allows an automatic treatment of behaviours for checking and simulation and makes modelled knowledge unambiguous.

The language we propose can be related to MODEL-ECS [12] as an extension of conceptual graphs to represent and operationalize knowledge models, but it differs in two aspects: first the language we are building is dedicated to acting processes whereas MODEL-ECS aims at modelling reasoning processes. Secondly, the knowledge we are trying to model being not well set up, we want the language to accept indeterminacy, the simulation tools must then take into account this indeterminacy and generate the possible world evolutions.

This paper is divided into two parts. The first one describes the system we are building that is to say the defined language, the construction and simulation tools. The second part is dedicated to the special use of conceptual graph in our system concerning representation and reasoning aspects.

2. Description of the application

2.1 The modelling language

Stereotyped human behaviours are inaccurate plans composed of actions, which indicate how a person or a set of people must react to respond to a critical situation. The first aim of our work was to define an ontology dedicated to the modelling of such behaviours. This ontology is divided into two parts: behavioural part and world description part. The behavioural part provides all types required to model human behaviours in any domain. These types should be realistic, intuitive and close to the expert so that the language can be understood. They should also have a clear semantics so that they can be exploited by the simulation algorithm. The world-description part provides all types required to describe a state of the world. This part deals with the representation of interveners and instruments, of their spatial localization and attributes, and of situation verbs describing actions and states. It is strongly related to the domain (for example, civil security domain). The beginning of the paper addresses the modelling language of the proposed system. We mainly speak about the behavioural types of the predefined ontology. First, we describe how to model a behaviour. Then, we discuss about some representation problems raised by a realistic modelling of actions. Finally, we present how to describe a progress by synchronizing and conditioning some specialization of action models.

Behaviour modelling

A stereotyped behaviour characterizes the role of a type of person. For example, the behaviour of Fig.4 must only be applied by operators. Concerned persons must then apply the behaviour if and only if some applicability conditions are satisfied. For example, an operator should execute the behaviour if he is in the control room and as soon as the siren starts ringing (Fig.4). Such applicability conditions can be expressed by states and/or events:

- A *state* is a characterization of the world, being described by a property that is true during a temporal period. This property is expressed by a graph nested in the state concept[1].
- An *event* is a punctual change of the world. We do not consider a list of

---

1. Concepts with same fillings are coreferent concepts, they refer to the same instance

predefined events because it would require to give names to all possible changes of state in the world (Readability and extensibility would be reduced). An event is indirectly expressed by the conjunction of four base types of events: *tff* event (a state becomes false), *ftt* event (a state becomes true), *ftf* event (a state is punctually true) and *tft* event (a state is punctually false).

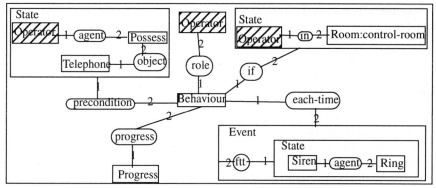

Fig.4 Description of a behaviour

Once these applicability conditions are satisfied, the behaviour must be applied. However, it can only be executed in a correct manner if some prerequisites are respected. For example, to execute his behaviour, the operator must possess a phone (Fig.4). These conditions are called the *preconditions* of the behaviour, they are represented by states. The procedure of a behaviour is finally described by a *progress*, that is to say by a plan of temporally synchronized actions. The progress enumerates the actions which must be executed by the intervener.

Action modelling

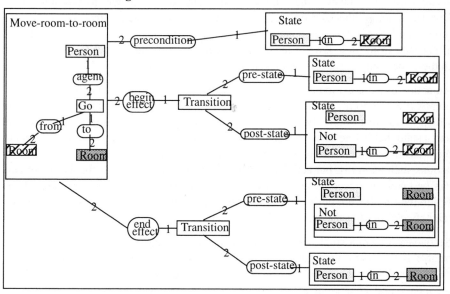

Fig.5 The model of "move-room-to-room" action

Action modelling incites to consider three main aspects: action preconditions (conditions requiring to be true so that the action can be performed), action progress (the way the action is performed) and action effects (action results = the state to which the action leads). Fig.5 gives the model of the move-room-to-room action. Preconditions are expressed by states linked to the action by *precondition* relations. The way the action is performed can simply be represented by an *abstract* of the action. Such an action abstract is described by a graph nested in the action concept, this graph being composed of a situation verb and of the parameters of the action (its agent, its location, needed instruments...). The way the action is performed can also be described by a progress linked to the action by a *progress* relation. The progress of an action is a plan of more detailed actions.

The representation of action effects raises two classical problems. The first one is the frame problem [13], which is the problem of determining what does not change when an action occurs. The second one is the ramification problem [9], which is the problem of enumerating what changes when an action occurs. The frame problem, also called persistence problem, is often resolved with STRIPS-like action models [8] by assuming that only the changes of the world are described. What does not change in the world when an action is performed is not described. In our representation approach, we make a similar assumption: states of the world persist from a time-point to the following time-point unless the contrary can be proved.

In a STRIPS-like action model, all states (represented by first-order predicates) that must be added to the world and all states that must be removed from the world when an action is performed must be explicitly defined, this is the ramification problem. In the case of real actions, effects often depend on the description of the world before the performance of the action. These effects are contextual effects. In STRIPS, to take into account this dependence, there must be as many action models as possible initial worlds. In some other approaches [6], [9] and [16], only main effects of actions are represented. Secondary and contextual effects are computed from this little number of effects and from static laws of the domain (represented by first-order logical formula). In the same manner, we only represent main effects in actions.

An action effect corresponds to a change of the world between two consecutive time-points; it is represented by a *transition*. A transition is described by two states: a state which is true just before the transition, linked by a *pre-state* relation to the transition and a state that is true at the moment of the transition, linked by a *post-state* relation to the transition. Moreover, to be closer to reality, we do not consider instantaneous actions such as in [8], but actions with duration. A consequence of this is that effects can be taken into account at various moments of the action: at the beginning of the action (begin effects), during the action (intermediary effects), at the end of the action (end effects) or after the action (delayed effects).

Secondary or contextual action effects depend on static laws of the domain. The trouble is that static laws can not express directly the properties of the world which evolve and how they evolve. Hence, it is difficult to exploit them to compute secondary and contextual action effects. That is why we represent secondary and contextual effects in a generic manner in dynamic laws of the domain. Dynamic laws express all the properties of the world which are considered to be able to evolve by experts and also express the way these properties evolve in accordance with domain static laws. All other properties are considered to be immutable. Dynamic laws are time-dependent rules, linking two consecutive time-points. A transition which occurs at a time-point (the left-hand transition) entails another transition to occur at the same time-point (the right-hand transition) if some conditions are satisfied. For example, a transition "a person gets into a room" entails another transition "the person becomes near an object" under a condition "the object is in the room" (Fig.6).

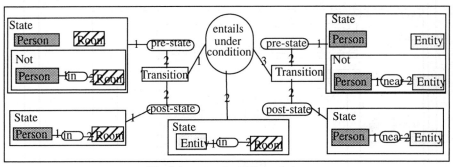

Fig.6 An example of dynamic law

Progress modelling

A progress is a set of *action* model specializations:
- partially ordered by symbolic *temporal relations* derived from Allen's relations [1] and including two granularity levels [2]
- synchronized by events with *as soon as* relations
- conditioned by states or events (conditioned actions are gathered in progresses of *conditional blocks*)

A progress might contain *objectives* (states that must be reached at some time-points or that must be maintained during some temporal periods) It might also contain indeterminacy: inaccurate temporal relations (disjunctions of base temporal relations), *optional blocks* (an action must be made or not), *choice blocks* (an action can be chosen among several actions) and indeterminate parameters of actions; various reasons can lead experts and knowledge engineers to express such indeterminacy (see [2] for more details).

Fig.7 gives an example of a short part of a progress. As soon as the person is near a disaster, he must identify it. After or just after the identification (inaccurate temporal relation), he must go to the control room only if the disaster is grave. Move-room-to-room concept is a specialization of the corresponding action model of Fig.5. Some parameters of the abstract of the action model have been specialized, either they have been linked by coreference to other concepts of the progress or their generic markers have became individual markers.

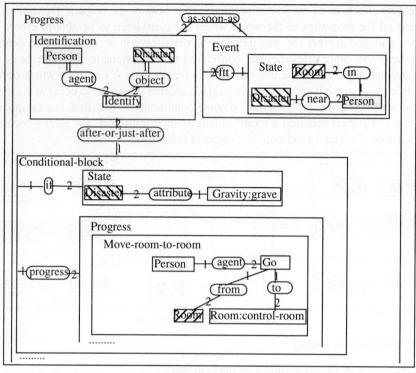

Fig.7 An example of progress

2.2 Modelling assistance: construction

Modelling can be assisted by defining several modelling levels to complete the predefined ontology. The predefined ontology constitutes the first modelling level. This level does not depend on a domain, it is dedicated to the modelling of behaviours in any domain. The second modelling level will be representative of a domain, for example civil security domain. In the civil security domain, the third level will be representative of a particular industrial place, for example a chemical factory. A fourth level could also relate to a particular scenario which happens in an industrial place, for example "a fire propagates in repositories". Once a level is settled, it can no longer be modified. The predefined ontology is settled. The second level is built from the predefined ontology by specializing concept and relation types. Then, the third level is built from the second level, and so on... Some concept and relation types of the predefined ontology should not be specialized. It is essentially the case of behavioural types which have a particular semantics exploited by the simulation algorithm. However, world-description-types should be specialized according to the domain.

In order to facilitate the construction of a new level from a settled level, concept and relation types of the settled level should be explained and documented. A solution to document a type is to associate with it a text of explanation. But conceptual graph model also provides several means of defining types, we call these different means "metaconcepts". Metaconcepts permit to give all invariant properties between the defined type and other hierarchy types, these properties being expressed by graphs. At any level, they contribute to document types and to assist the construction of graphs.

Signatures and acceptions are metaconcepts which permit to constrain the syntax of the language. A signature of a relation type fixes the arity of the relation and shows two maximal concept types this relation can link. Contrary to [3], we consider that a relation type has not necessary a single signature, but can have a set of signatures. For example, the progress relation can link a behaviour to a progress to describe the behaviour procedure or it can link an action to a progress to describe the action procedure. In this case, the logical interpretation of the signature set of the progress relation is:

$$\forall x \forall y (progress(x, y) \rightarrow (action(x) \wedge progress(y)) \vee (behaviour(x) \wedge progress(y))) \ (1)$$

An acception of a concept type shows a maximal concept type to which this concept can be linked by a maximal relation type. A concept type can have many acceptions. For example, a behaviour can be linked to a progress by a progress relation, it can be linked to a state by a precondition relation and it can be linked to an event by an each time relation. Signature and acceptions allow the checking of graph syntax. They can also make easier the construction of graphs. In this way, they are considered as base graphs which can be specialized and joined to build larger graphs.

Static-rules (i.e metaconcepts such as necessary conditions or sufficient conditions) related to world-description concept types correspond to static laws of the domain, they can be translated into dynamic laws of the domain. It is the case of the dynamic law of Fig.6, which is generated from the proximity property between two entities that are in the same room (Fig.8). All the transitions which could make the left-hand part of the static rule true are automatically generated. For example, in the case of the static rule of Fig.8, an entity might get into a room whereas another entity was already in the room. Then, experts and knowledge engineers choose realistic transitions among these generated transitions. For example, they could consider that people can move, but that some objects can not be moved. In this case, the entity which gets into the room can only be a person or a movable object. The dynamic law of Fig.6 can then be generated. Another similar dynamic law with movable objects can also be generated.

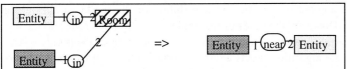

Fig.8 The proximity property of two objects that are in the same room

2.3 Evaluation assistance: simulation

Simulation requires additional knowledge with regard to the knowledge captured for modelling. An initial scenario should start the simulation. An initial scenario corresponds to a "concrete case". From this initial scenario and from modelling (behaviour graph + hierarchies + metaconcepts + dynamic laws of the domain), the simulation algorithm generates all the possible evolution of the world. There are several possible evolution of the world because of indeterminacy in behaviours. A possible evolution of the world is a totally ordered set of instantaneous descriptions of the world. We call these instantaneous descriptions situation-points. Hence, we can say that a possible evolution of the world is a sequence of situation-points. A situation-point is described by a conceptual graph since experts and knowledge engineers should be able to consult sequences of situation-points in order to get simulation results.

The simulation algorithm is based on the semantics given to the behavioural part of the ontology. On one hand, it must generate readable sequences of situation-points. On the other hand, it must detect automatically some cases of inconsistencies by checking execution. Finally, it must allow users to perturb execution so that they can judge model robustness. Users can interact with the simulation algorithm. They can ask for a result sequence, make request on this result sequence, consult the checking report, perturb or stop execution...

The first task of the simulation algorithm is to choose a candidate behaviour. A behaviour is candidate if its applicability conditions are recognized in the current situation-point. The second task of the algorithm is to check if the chosen behaviour is applicable. The behaviour is applicable if its preconditions are true in the current situation-point. Then, the algorithm chooses a candidate action from this chosen behaviour. An action is candidate if it has no more temporal dependencies (it begins the behaviour or the actions that precede it have already been finished), if the events linked to the action by as soon as relations are recognized in the current situation-point and if conditions (events and states) that condition the action are satisfied in the current situation-point. After that, the algorithm checks if the action is executable. An action is executable if its precondition are satisfied in the current situation-point. Finally, if the action is executable, it executes the action. It builds a new situation-point which is the update of the current situation-point with the effects of the action. This update must take into account the dynamic laws of the domain in order to maintain the consistency of the new situation-point with regard to the domain. After this update, the algorithm makes another choice, it chooses another action or another behaviour and so on.

In parallel with all these tasks, the algorithm must manage indeterminacy. If the knowledge contained in behaviour models had been complete, the simulation algorithm applied on a complete description of the world would only generate one sequence. However, the cases of incompleteness are numerous and the algorithm must take into account all these cases to be able to generate all the possible sequences. Each time that the algorithm meets a case of indeterminacy, it can build several sequences (the user can choose which sequence he wants to consult). For example, if a fireman must stop fire before or after he helps injured people, in one generated sequence, he will stop fire before helping injured people and in another generated sequence, he will help injured people before stopping fire. The algorithm also checks execution. For example, it detects if some compulsory actions have never been executed because their preconditions have never been satisfied or because some events have not occurred. It can also detect if some objectives have not been reached or maintained. In this manner, it can detect automatically some cases of inconsistencies.

3. Some additional features of our conceptual graph use

3.1 Representation features

Positive nested graphs

Nesting permits to create several representation levels. The main advantage of nesting a graph g inside a concept c is the ability to establish a link between c and g (i.e it is possible to establish a link between c and not only all the concepts in g but also all the relations in g). It is impossible to create such a link with all concepts and relations at the same level. In our representation language, nesting is necessary in several cases: graphs nested in concepts of "state" type express the properties of the world, invariant during the validity of the state, graphs nested in concepts of "progress" type express plans of actions and graphs nested in concepts of "action" type express the abstract of the action. We have chosen to consider the nested graph definition expressed in [4]. The label of a concept vertex is written (t,m,d). t and m are respectively the type and the marker of the concept like in the base model. d is called internal partial description, it may be equal to **, the generic description or to a set of graphs. Mugnier and Chein have defined operations on nested graphs: projection, simplification, concept restriction, relation restriction, join and disjunctive sum. A first-order logic semantics has also been defined in [5].

In some other cases, nesting is not necessary, but can be advantageous to structure graphs and hence to make them more readable. For example, ftt states, tff states, tft states and ftf states can be nested inside the event they describe (Fig.4). The condition and the progress of a conditional block can also be nested inside the conditional block concept (Fig.7). Whereas in the previous paragraph, nesting semantics was given by the type of the nesting concept, in these two cases, several graphs are nested in the same concept with different nesting semantics. We must use typed nested graphs to express nesting semantics. The internal partial description d is then replaced by a set $\{(e1,G1),(e2,G2),....,(ek,Gk)\}$ with ei the nesting types and Gi the typed nested graphs [4]. In our representation of typed nested graphs, a nesting type is given by the relation that links the nested graph to the nesting concept, for example, the ftt relation (Fig.4) or the if and progress relations (Fig.7).

Negation

We just need to express negation in two cases: to describe false properties in situation-points and to describe false properties in the states appearing in behaviours. We have decided to represent negation with a closed-world assumption in situation-points and with a negation nesting in states.

A situation-point is described by a not necessary connected, ground and positive conceptual graph. Negation is indirectly expressed by a closed-world assumption. Any conceptual graph that can not be demonstrated from the situation-point is assumed to be false. This involves not to use generic concepts in situation-points. Suppose that [Person: *] is a generic concept in a situation-point and that the only individual markers of person type are Smith and Brown, then the closed-world assumption of the situation-point contains [Not:[Person: Smith]], [Not:[Person: Brown]] and [Person: *].

The closed-world assumption of the situation-point is then inconsistent. A consequence of this is that all concept referents must be individual markers. Names must be given by users to concrete concepts referents (i.e concepts which really exist in the world, for example, an entity or a place) and since asking users to give names to abstract concepts (i.e reifications of relations) would have no sense, abstract concept referents are automatically numbered.

A graph nested in a state concept contain negations that are explicitly represented. We have chosen to write negation with a particular nesting: negation nesting. The logical interpretation of a graph inside a negation nesting is the negation of the usual logical interpretation of conceptual graphs given by the operator ϕ of Sowa [15]. A graph G characterizing a state is composed of a simple, but not necessary connected graph (the positive part of the graph, written P) and of several negation nesting (the negative parts of the graph, written N_i). The logical interpretation of the graph G is then:

$$\phi(G) = \phi(P) \wedge \neg\phi(N1) \wedge \ldots \wedge \neg\phi(Nn) \qquad (2)$$

Representing explicitly negation in situation-points instead of assuming the closed-world would increase expressiveness in situation-points. However, reasoning operations would be more complicated. We have tried to reach the better compromise between expressive power and reasoning. In this way, we have judged that expressiveness is more important in behaviour descriptions than in situation-points. Indeed, situation-points are just useful for simulation, that is to say for the evaluation of modelling.

3.2 Reasoning features

Verification

Recognizing events and testing preconditions and conditions require to be able to demonstrate state validity from situation-points. The projection operation is necessary but not sufficient to perform this task. We can wonder how to use projection and what projection failure means, what are the consequences of projection success if some concepts of graphs are coreferent concepts and what several projection results mean.

A graph characterizing a state is composed of a positive part and of negative parts. Projection is used to prove the positive part and a negation as failure rule is exploited to prove each of the negative parts. A graph characterizing a state can be demonstrated from a situation-point if:
- there exists a projection of its positive part in the graph characterizing the situation-point and
- there does not exist any projection of one of its negative parts in the graph characterizing the situation-point.

The projection of the positive part of the state in the situation-point will instantiate some concepts that were generic in the state. In the behaviour graph, these concepts may be linked by coreference to other concepts. These coreference links must be "propagated" in the behaviour. For example, in Fig.4, in order to get the possible interveners of the behaviour, the operator concept is "projected" in the current situation-point. Suppose that Smith is a possible intervener, then the precondition of the behaviour will be specialized by replacing the generic marker of the operator concept by the Smith individual marker. It will be this specialized precondition that will be tested in the current situation-point to check whether the behaviour is applicable.

If at least one concept of the positive part of the state is generic, there might be several possible projections of this positive part in the current situation-point. Several projection results are differently interpreted according to the nature of the state that is tested. The various interpretations are given by the semantics of the behavioural ontology. In the case of behaviour interveners, several projection results lead to different specialized behaviours which must all be applied. In Fig.4, if Smith and Brown are two possible interveners, Smith and Brown must execute the behaviour. However, if the state characterizes a precondition, several projection results lead to several possible sequences of situation-points and hence bring indeterminacy. In Fig.4, if Smith possesses two phones, either he will use the first one, or he will use the second one to execute his behaviour.

Update

Update is divided into two steps: it consists in building a new situation-point from the current situation-point with the effects of an action and then it consists in taking into account dynamic laws to maintain the consistency of the new situation-point with regard to the domain.

In a first step, transitions representing action effects are applied. Transitions can be seen as graph rules [14] relating two consecutive time-points. Generic concepts of the pre-state, coreferent with concepts of the post-state are interpreted as universal quantifiers such as in [14]. However, two cases are possible, a property that was true becomes false (begin effect in Fig.5) or a property that was false becomes true (end effect in Fig.5). In the two cases, the algorithm checks whether the pre-state of the transition is satisfied in the current situation-point. If it is satisfied, the transition is applied in forward chaining for each projection result of the positive part of the pre-state in the current situation-point. Coreference links are "propagated" from the pre-state to the post-state and update must make the resulting specialized post-state valid in the new situation-point. In the first case, the algorithm makes a physical join (i.e it joins concepts with same referents) of the specialized post-state with the current situation-point. New concrete concepts are created. New abstract concepts are created and their referents are automatically numbered. The resulting situation-point is simplified (i.e redundant relations are removed). In the second case, the algorithm removes from the current situation-point all projections of all negative parts of the graph characterizing the specialized post-state.

In a second step, after having applied the effects of the action, the algorithm must take into account the dynamic laws of the domain. Like transitions, dynamic laws can be seen as rules relating two consecutive time-points. Generic concepts of the left-hand transition and of the conditions, coreferent with generic concepts of the right-hand transition, are interpreted as universal quantifiers. A dynamic law applies to a

situation-point SPi if:
- the pre-state of the left-hand transition can be demonstrated from SPi-1
- the post-state of the left-hand transition can be demonstrated from SPi
- application conditions can be demonstrated from SPi-1
- application conditions can be demonstrated from SPi.

The application of the dynamic law consists then in updating in a recursive manner SPi with the right-hand transition.

4. Conclusion

In this paper, we have described an application of conceptual graphs in knowledge engineering aiming at assisting the acquisition and the validation of behaviour models in human organizations. Topics of current and ongoing research include theoretical topics, implementation and experimentation.

We are finishing to settle a formal semantics of the behavioural ontology expressed in a first order reified temporal logic to be able to prove the correctness and the completeness of the simulation algorithm. Another theoretical topic of ongoing research could be the improvement of expressive power in situation-points by removing the closed world assumption. This topic would involve research about reasoning operations. We are also working on the representation of graph differences to express in a more clear and precise manner state changes (i.e events and transitions). On the implementation side, we are developing tools aiming at assisting experts and knowledge engineers to build behaviour models. These tools exploit the graphical notation and the terminological aspects of conceptual graphs. We are also developing tools aiming at assisting experts and knowledge engineers to evaluate behaviour models. These tools implement the simulation algorithm and manage the interaction of experts and knowledge engineers with simulation. Construction and simulation tools are being implemented on the CoGITo platform [10].

Finally, we will begin soon an experimentation of the language and the tools in two application domains: civil security domain and military domain.

5. Acknowledgment

The work of Corinne Bos is partly supported by the "Conseil Régional du Nord-Pas de calais".

6. References

[1] J.F.Allen, "Maintaining Knowledge about Temporal Intervals", Communications of the ACM vol 26 N 11, pp 832-843, 1983

[2] C.Bos, B.Botella, "Modelling Stereotyped Behaviours in Human Organizations", 7th Workshop on Knowledge Engineering: Methods and Languages, Milton Keynes, United Kingdom, 1997

[3] M.Chein, M.L.Mugnier, "Conceptual Graphs: Fundamental Notions", Revue d'Intelligence Artificielle vol 6 N 4, pp 365-406, 1992

[4] M.Chein, M.L.Mugnier, "Positive Nested Conceptual Graphs", Research Report LIRMM, #97004, submitted to ICCS'97

[5]   M.Chein, M.L.Mugnier, G.Simonet, "Nested Conceptual Graphs: Projection and FOL semantics", Research Report LIRMM, #97003

[6]   M.O.Cordier, P.Siegel, "A Temporal Revision Model for Reasoning about World Change", In B.Nebel, C.Rich, W.Swartout, eds, Principles of Knowledge Representation and Reasoning (KR'92), pp 732-739, Morgan Kaufmann, 1992

[7]   D.Fensel, F.van Harmelen, "A Comparison of Languages which Operationalise and Formalise KADS Models of Expertise", The Knowledge Engineering Review vol 9, pp 105-146, 1994

[8]   R.E.Fikes, N.J.Nilsson, "STRIPS: a new approach to the application of theorem proving to problem solving", Artificial Intelligence 2, pp 189-208, 1971

[9]   M.L.Ginsberg, D.E.Smith, "Reasoning about Actions I: A Possible Worlds Approach", Artificial Intelligence 35, pp 165-195, 1988

[10] O.Haemmerlé, "Implementation of Multi Agent Systems using Conceptual Graphs for Knowledge and Message Representation: the CoGITo Platform", 3rd International Conference on Conceptual Structures, Santa Cruz, United States, 1995

[11] D.Lukose, G.Mineau, M.L.Mugnier, J.U.Möller, P.Martin, R.Kremer, G.P.Zarri, "Conceptual Structures for Knowledge Engineering and Knowledge Modelling", 3rd International Conference on Conceptual Structures, Santa Cruz, United States, 1995

[12] D.Lukose, "MODEL-ECS: Executable Conceptual Modelling Language", 10th Workshop on Knowledge Acquisition for Knowledge-based Systems, Banff, Canada, 1996

[13] J.Mc Carthy, P.J.Hayes, "Some Philosophical Problems From the Standpoint of Artificial Intelligence", Machine Intelligence 4 eds B.Meltzer, D.Michie, pp 463-502, 1969

[14] E.Salvat, M.L.Mugnier, "Sound and Complete Forward and Backward Chainings of Graph Rules", 4th International Conference on Conceptual Structures, Melbourne, Australia, 1996

[15] J.Sowa, "Conceptual Structures: Information processing in mind and machine", Addison Wesley, Reading Mass, 1984

[16] M.Winslett, "Reasoning about Actions using a Possible Models Approach", Proceedings of AAAI, pp 89-93, 1988

Conceptual Graphs and
Formal Concept Analysis

Rudolf Wille

Technische Hochschule Darmstadt, Fachbereich Mathematik
Schloßgartenstr. 7, D–64289 Darmstadt, wille@mathematik.th-darmstadt.de

Abstract. It is shown how Conceptual Graphs and Formal Concept Analysis may be combined to obtain a formalization of Elementary Logic which is useful for knowledge representation and processing. For this, a translation of conceptual graphs to formal contexts and concept lattices is described through an example. Using a suitable mathematization of conceptual graphs, basics of a unified mathematical theory for Elementary Logic are proposed.

1 Formalization of Elementary Logic

Conceptual Graphs and *Formal Concept Analysis* have been used both for knowledge representation and processing in a large extent. This has caused the desire to combine the two approaches for deriving benefits from both disciplines and their experiences. There is even a fundamental reason for associating Conceptual Graphs and Formal Concept Analysis which lies in their far-back reaching roots in philosophical logic and in their pragmatic orientation; more specifically, both together can play a substantial role in the *formalization of logic* by which reasoning is based on "communicative rationality" in the sense of Pragmatism and Discourse Philosophy (cf. [Ap76],[Ha81],[Wi96a]). This shall be explained further in the sequel.

Until the beginning of this century, *Elementary Logic* was understood and taught by the traditional paradigm of philosophical logic based on "the three essential main functions of thinking — *concepts, judgments*, and *conclusions*" [Ka88;p. 6]. Elementary Logic was therefore presented in three parts: the doctrine of concepts, the doctrine of judgment, and the doctrine of conclusions (cf. [Gr18]). These three philosophical doctrines have the common aim to make clear how human knowledge is formed: *concepts* as basic units of thought are shaping contents of thinking, *judgments* as combinations of concepts and facts are joining thought and reality, and *conclusions* as entailments between judgments are extending known relationships of thought and reality (cf. [Br76]).

The increasing use of computers for knowledge representation and processing forces to rethink the common formalizations of logic and, in particular, to look for a more adequate *formalization of Elementary Logic* allowing a formal treatment of knowledge and its formation. In the paper *"Restructuring Mathematical Logic: An Approach Based on Peirce's Pragmatism"* [Wi96a], formalizations following the paradigm of predicate logic are criticized because they narrow too much the connections to reality and support one-sidedly the mechanization of human thinking. The abstraction of judgments to formal propositions formed by quantifiers, variables, constants, and predicates weakens the ontological ties which are necessary for rational communication and argumentation. One must always be aware that formalizations cannot grasp realities without an eventually serious loss of content; hence formalizations have to keep the connections to their origins so that the consequences of formal treatments may be rationally analysed and interpreted in human communication.

A formalization of Elementary Logic has first of all to serve with an appropriate *formalization of concepts* which establishes the basic connection between thought and reality. For this, Formal Concept Analysis offers a formalization of the philosophical understanding of a concept as a unit of thought constituted by its extension and its intension (cf. [Wi82],[Wi92],[Wi95]). These two components represent the fundamental complementarity of particularity and generality as it is present in the relationship of reality and thought. The formalization of both components allow to fix enough references for interpreting the formalized concepts in human communication and argumentation. The essential point is that extension and intension of a concept are unified on the base of a specified context. This contextual view is supported by Peirce's pragmatism which claims that we can only analyse and argue within restricted contexts where we always rely on preknowledge and common sense.

Thus, Formal Concept Analysis starts with the *formalization of contexts* by introducing a *formal context* as a relational structure $\mathbb{K} := (G, M, I)$ consisting of a set G of *formal objects*, a set M of *formal attributes*, and a binary relation I between G and M indicating which formal object has which formal attribute. A *formal concept* of such formal context $\mathbb{K}$ is then defined as an ordered pair $\mathfrak{c} := (A, B)$ where the formal extension A (called the *extent* of $\mathfrak{c}$ and denoted by $Ext(\mathfrak{c})$) contains all formal objects of $\mathbb{K}$ having the formal attributes collected in the set B and the formal intension B (called the *intent* of $\mathfrak{c}$ and denoted by $Int(\mathfrak{c})$) contains all formal attributes of $\mathbb{K}$ valid for all formal objects collected in the set A. A formal concept (A, B) is a *subconcept* of a formal concept (C, D) (in symbols: $(A, B) \leq (C, D)$) if the extent A is a subset of the extent C (or, equivalently, if the intent B is a superset of the intent D). The ordered set $\mathfrak{B}(\mathbb{K})$ of all formal concepts of $\mathbb{K}$ together with the subconcept-superconcept-relation $\leq$ is always a complete lattice, called the *concept lattice* of the formal context $\mathbb{K}$. The described basic notions of Formal Concept Analysis have initiated the development of an extensive theory and practice of the formal treatment of concepts and concept systems which, to a great extent, is documented in the monograph [GW96].

After offering a formalization of concepts by referring to Formal Concept Analysis, the next in formalizing Elementary Logic must be the *formalization of judgments*. For this, the Theory of Conceptual Graphs, created by J. Sowa [So84], yields a convincing approach. Conceptual graphs can be considered as formal judgments which are tightly linked to natural language. Therefore they are representations of judgments which can support successfully human communication and argumentation. In particular, they are based on concept nodes combined with references to realities so that they can express already by their nodes relationships between thought and reality. The edges of conceptual graphs correspond to function words and case relations which allows an understandable verbalization of the indicated relationships. The formation of conceptual graphs even allows deduction procedures by which new conceptual graphs can be obtained from given conceptual graphs. Thus, the Theory of Conceptual Graphs also offers a *formalization of conclusions* based on the formalization of judgments by conceptual graphs. This formalization is a further development of Peirce's Theory of Existential Graphs (see [Pe83]) to approach the expressibility of first-order, modal and even higher-order logic.

A formalization of Elementary Logic which uses Formal Concept Analysis for the doctrine of concepts and the Theory of Conceptual Graphs for the doctrines of judgments and conclusions forces to consider the concepts (concept types) in conceptual graphs as formal concepts of formal contexts. This yields the base on which both disciplines should be combined. The basic question is how formal contexts can be appropriately introduced for conceptual graphs so that both disciplines combine most successfully. This question will be discussed in the next section through an example. The last section of this paper is devoted to the question how the involved conceptual structures allow mathematizations for obtaining a unified mathematical theory for Elementary Logic and its applications in knowledge representation and processing.

2 From Conceptual Graphs to Formal Contexts

In [So92], J. Sowa gives an actualized introduction to *Conceptual Graphs* which is prefaced by the following statement explaining what conceptual graphs are and for what they are designed:

> "Conceptual graphs are a system of logic based on the existential graphs of Charles Sanders Peirce and the semantic networks of artificial intelligence. The purpose of the system is to express meaning in a form that is logically precise, humanly readable, and computationally tractable. With their direct mapping to language, conceptual graphs can serve as an intermediate language for translating computer-oriented formalisms to and from natural languages. With their graphic representation, they can serve as a readable, but formal design and specification language."

There is a natural way of *transforming* conceptual graphs into formal contexts. This shall be demonstrated by an example of knowledge representation for which we choose a paragraph out of *"The Smithsonian Guide to Historic America"* that informs about Seattle's central business district [Sm89; p.349]:

> "Seattle's central business district, bounded by Yesler Way, Route 5, Stewart Street, and the waterfront, has among its sleek glass monoliths smaller buildings with gargoyles and other lively early-twentieth-century

decorations that give it an interesting texture. The restored **Arctic Building** (1917), on the corner of Third Avenue and Cherry Street, is a Renaissance Revival palazzo in terra-cotta decorated with a set of walrus heads. Eight Indian heads distinguish an upper-story frieze of the brick-and-terra-cotta **Cobb Building** (1910), on the corner of Fourth Avenue and University Street, which reflects the Beaux-Arts orientation of its New York architects, Howells and Stokes. The **Seattle Tower** (1928-1929), 1218 Third Avenue, has a lobby of dark marble walls and a gilt ceiling."

For a formal treatment of the knowledge coded in this text, it may be translated to conceptual graphs as shown in Figure 1 (the dates of the buildings are omitted). The four sentences of the text are represented by four (connected) conceptual graphs in which the nodes are drawn as boxes and the directed edges as circles together with directed line segments (notice that all edges of the example are dyadic!). Each box represents a concept (understood as concept type in [So84]) together with at least one of its objects (individuals), while an edge represents at least one instance of a semantic relation. Concept names are written in each box with capitals (T denotes the universal concept). If a box does not contain an object name (written behind the colon), then the existence of an object is assumed for this box to which an individual marker like #437 may be assigned ({*} behind the colon indicates that there exist several objects whose exact number might also be given). Dotted line segments, representing so-called *coreference links*, indicate that the objects of the joined boxes are the same. A directed edge with label R pointing from the object g to the object h is read in general: "*g has an R which is h*" or "*h which is an R of g*"; for instance, the first line of the last conceptual graph in Figure 1 can be read: "The Seattle Tower has a location which is 1218 Third Avenue" or "1218 Third Avenue which is a location of the Seattle Tower". The abbreviations for the used semantic relations translate as follows: PTNT=patient, AGNT=agent, STAT=state, ATTR=attribute, RCPT=recipient, LOC=location.

For transforming conceptual graphs into formal contexts one has to specify what are the formal objects and the formal attributes. Obviously, the natural candidates for the formal objects are the *objects* (*individuals*) of the nodes of the conceptual graphs. The derived formal context shown in Figure 2 has therefore as formal objects all those objects named and claimed to exist by the nodes in Figure 1 except those to which an ATTR-edge is directed to ({*} gives always rise to (at least) two individual markers; the individual markers #3.6–#3.13 designate the eight Indian heads). To obtain the concepts of the conceptual graphs as formal concepts (except those to which an ATTR-edge is directed to), it is necessary (if there is no further information) to take all *concepts* in Figure 1 (except T) as formal attributes. The context relation, represented in Figure 2 by the crosses, is derived by linking each object to the concept of its node (including those joined by a coreference link) and, in addition, to the concepts whose nodes are joined with the object node by an ATTR-edge. Now, each concept c represented in Figure 1 corresponds to a formal concept of the formal

294

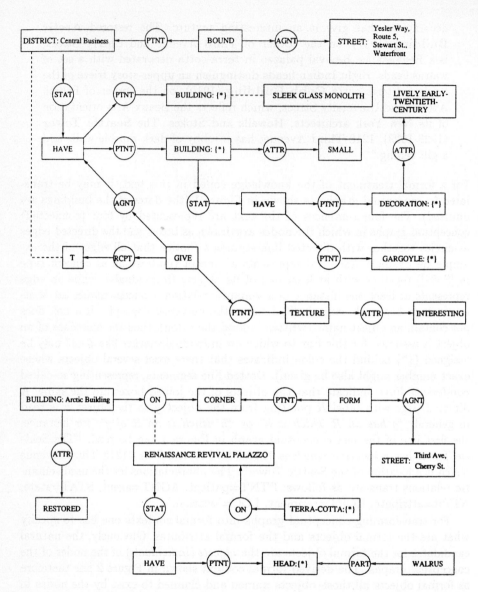

Fig. 1. Conceptual graphs representing a text about Seattle's central business district

context of Figure 2, namely to the formal concept μc whose extent contains all formal objects linked by the context relation to c and whose intent contains all formal attributes linked by the context relation to all formal objects of the extent. By the assignment of c to μc, the concepts of Figure 1 become part of the conceptual hierarchy given by the concept lattice of the formal context of Figure 2.

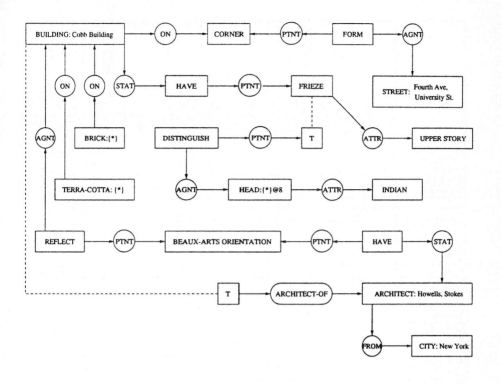

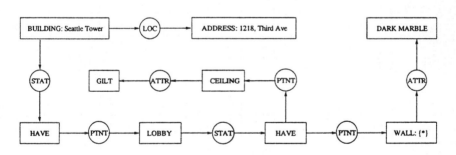

Fig. 1. (cont'd)

Since conceptual graphs combine concepts by semantic relations, we have now to answer the question how the semantic relations can be formalized in such a way that they naturally link formal concepts and formal objects, respectively. Such formalization should, in particular, respect subsumptions between semantic relations (for instance, the relations ON and FROM in Figure 1 subsume under the relation LOC). Thus, it is mostly consequent to formalize *semantic relations* as formal concepts too. For this, as shown in Figure 3, we choose formal objects which are ordered pairs of formal objects from Figure 2. Analogously to the formation of the formal context of Figure 2, the semantic relations (except

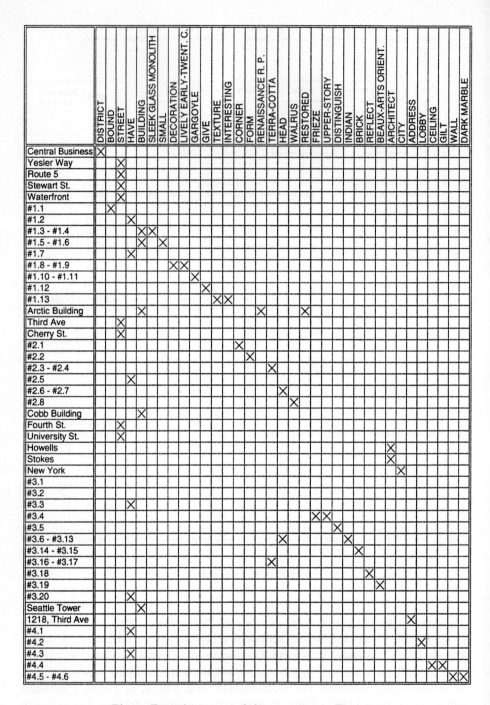

Fig. 2. Formal context of the concepts in Figure 1

	PTNT	AGNT	STAT	RCPT	PART	ON	FROM	LOC	ARCH.-OF
(#1.1, Central Business)	X								
(#1.1, Yesler Way)		X							
(#1.1, Route 5)		X							
(#1.1, Stewart St.)		X							
(#1.1, Waterfront)		X							
(Central Business, #1.2)			X						
(#1.2, #1.3) - (#1.2, #1.4)	X								
(#1.2, #1.5) - (#1.2, #1.6)	X								
(#1.5, #1.7) - (#1.6, #1.7)			X						
(#1.7, #1.8) - (#1.7, #1.9)	X								
(#1.7, #1.10) - (#1.7, 1.11)	X								
(#1.2, #1.12)			X						
(#1.12, Central Business)				X					
(#1.12, #1.13)	X								
(Arctic Building, #2.1)						X		X	
(#2.2, #2.1)	X								
(#2.2, Third Ave)		X							
(#2.2, Cherry St.)		X							
(#2.3, Arctic Building)						X		X	
(#2.4, Arctic Building)						X		X	
(Arctic Building, #2.5)			X						
(#2.5, #2.6) - (#2.5, #2.7)	X								
(#2.8, #2.6) - (#2.8, #2.7)					X				
(Cobb Building, #3.1)						X		X	
(#3.2, #3.1)	X								
(#3.2, Fourth Ave)		X							
(#3.2, University St.)		X							
(Cobb Building, #3.3)			X						
(#3.3, #3.4)	X								
(#3.5, #3.4)	X								
(#3.5, #3.6) - (#3.5, #3.13)		X							
(#3.14, Cobb Building)						X		X	
(#3.15, Cobb Building)						X		X	
(#3.16, Cobb Building)						X		X	
(#3.17, Cobb Building)						X		X	
(#3.18, Cobb Building)		X							
(#3.18, #3.19)	X								
(#3.20, #3.19)	X								
(#3.20, Howells)			X						
(#3.20, Stokes)			X						
(Howells, Ney York)							X	X	
(Stokes, New York)							X	X	
(Cobb Building, Howells)									X
(Cobb Building, Stokes)									X
(Seattle Tower, 1218 Third Ave)								X	
(Seattle Tower, #4.1)			X						
(#4.1, #4.2)	X								
(#4.2, #4.3)			X						
(#4.3, #4.4)	X								
(#4.3, #4.5) - (#4.3, #4.6)	X								

Fig. 3. Formal context of the semantic relations in Figure 1

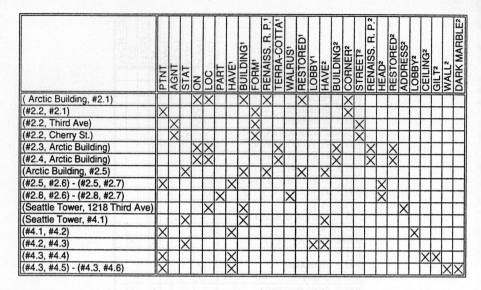

	PTNT	AGNT	STAT	ON	LOC	PART	HAVE[1]	BUILDING[1]	FORM[1]	RENAISS. R. P.[1]	TERRA-COTTA[1]	WALRUS[1]	RESTORED[1]	LOBBY[1]	HAVE[2]	BUILDING[2]	CORNER[2]	STREET[2]	RENAISS. R. P.[2]	HEAD[2]	RESTORED[2]	ADDRESS[2]	LOBBY[2]	CEILING[2]	GILT[2]	WALL[2]	DARK MARBLE[2]
(Arctic Building, #2.1)				X		X			X		X		X			X											
(#2.2, #2.1)	X								X							X											
(#2.2, Third Ave)		X							X									X									
(#2.2, Cherry St.)		X							X									X									
(#2.3, Arctic Building)				X		X				X					X						X	X					
(#2.4, Arctic Building)				X		X				X					X						X	X					
(Arctic Building, #2.5)			X						X	X			X	X													
(#2.5, #2.6) - (#2.5, #2.7)	X						X												X								
(#2.8, #2.6) - (#2.8, #2.7)			X									X			X				X								
(Seattle Tower, 1218 Third Ave)						X			X													X					
(Seattle Tower, #4.1)		X							X				X														
(#4.1, #4.2)	X								X															X			
(#4.2, #4.3)		X												X	X												
(#4.3, #4.4)	X								X																X	X	
(#4.3, #4.5) - (#4.3, #4.6)	X								X																	X	X

Fig. 4. Aggregated context concerning the Arctic Building and the Seattle Tower

ATTR) become the formal attributes of the formal context of Figure 3. The context relation is then derived by linking each ordered pair to those semantic relations whose representing edges in the conceptual graphs point from the first to the second object of the ordered pair; in addition, such pair is also linked to the semantic relations under which already linked relations are subsumed (for instance, (Howells, New York) is linked to FROM and therefore also to LOC). Let us remark that already assumed subsumptions between concepts should be analogously treated in establishing the first formal context as the one in Figure 2. Now, each semantic relation R represented in Figure 1 (except ATTR) corresponds to the formal concept μR of the formal context of Figure 3 (analogously formed as μc). By the assignment of R to μR, the semantic relations of Figure 1 become part of the conceptual hierarchy given by the concept lattice of the formal context of Figure 3.

The formal contexts of Figure 2 and 3 contain all information coded in the conceptual graphs of Figure 1. In particular, we can reconstruct the given conceptual graphs from the two formal contexts. Also further conceptual graphs can be obtained directly from the formal contexts; as a simple example we choose:

[DISTRICT: central business]
$\longrightarrow$(STAT)$\longrightarrow$[HAVE]$\longrightarrow$(PTNT) $\longrightarrow$[SLEEK GLASS MONOLITH:{*}]

As already discussed for assumed subsumptions, further knowledge may be represented by extending the formal attributes, the formal objects, and the context relation, respectively. This indicates that the coding of knowledge in formal contexts as those in Figure 2 and 3 has its advantages.

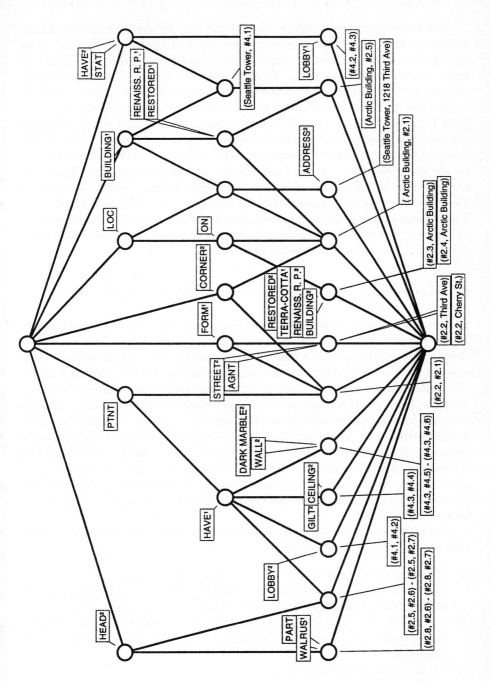

Fig. 5. The concept lattice of the formal context in Figure 4

The formal contexts of Figure 2 and 3 can also be *aggregated* to one formal context so that the interplay between the two formal contexts becomes more transparent. Figure 4 shows the aggregation of the parts of the two formal contexts concerned with the Arctic Building and the Seattle Tower: The formal objects are those ordered pairs from Figure 3 concerned with the two buildings; the formal attributes are of three types: first there are the formal attributes from Figure 3 which apply to the chosen pairs, secondly and thirdly there are the formal attributes from Figure 2 which apply to the formal objects at the first and the second place of the chosen pairs, respectively (the attribute names for the first place got the index 1 and for the second place the index 2). For example, in the formal context of Figure 4, the ordered pair (Seattle Tower, 1218 Third Avenue) has the formal attributes LOC, BUILDING$_1$, and ADDRESS$_2$. The concept lattice of the aggregated context, which is graphically represented in Figure 5, contains all information coded in the conceptual graphs concerning the Arctic Building and the Seattle Tower.

3 Mathematization of Conceptual Structures

In the previous section it is shown how Conceptual Graphs and Formal Concept Analysis can be connected through their conceptual structures. This connection should be elaborated toward a unified mathematical theory of Elementary Logic suitable for knowledge representation and processing. For this, the basic task is the *mathematization* of the occurring conceptual structures. In Formal Concept Analysis, hierarchical concept structures are mathematized as concept lattices of formal contexts as described in the first section. Conceptual graphs have been mathematized by M. Chein and M.-L. Mugnier [CM92] as finite bipartite graphs consisting of concept vertices labelled by concept types together with individual references and relation vertices labelled by relation types. Our mathematization of conceptual graphs will slightly modify the Chein-Mugnier-approach to obtain a better unification of the mathematized conceptual graphs with formal contexts and concept lattices.

The mathematization starts by defining an *abstract concept graph* as a structure $\mathfrak{G} := (V, E, \nu, C, \kappa, \theta)$ for which

1. V and E are finite sets and ν is a mapping of E to $\bigcup_{k=1}^{n} V^k$ $(n \geq 2)$ so that (V, E, ν) can be considered as a finite directed multi-hypergraph with vertices from V and edges from E (we define $|e| = k :\Leftrightarrow \nu(e) = (v_1, \ldots, v_k)$),
2. C is a finite set and κ is a mapping of $V \cup E$ to C such that $\kappa(e_1) = \kappa(e_2)$ always implies $|e_1| = |e_2|$ (the elements of C may be understood as abstract concepts),
3. θ is an equivalence relation on V.

Next, abstract concept graphs shall be related to formal contexts and their concept lattices. For this, we introduce a *power context family* $\vec{\mathbb{K}} := (\mathbb{K}_1, \ldots, \mathbb{K}_n)$ $(n \geq 2)$ with $\mathbb{K}_k := (G_k, M_k, I_k)$ $(k = 1, \ldots, n)$ such that $G_k \subseteq (G_1)^k$. Let $C_{\vec{\mathbb{K}}} := \bigcup_{k=1}^{n} \mathfrak{B}(\mathbb{K}_k)$. Now, we call an abstract concept graph $\mathfrak{G} := (V, E, \nu, C, \kappa, \theta)$ an abstract concept graph over the power context family $\vec{\mathbb{K}}$ if $C = C_{\vec{\mathbb{K}}}$, $\kappa(V) \subseteq \mathfrak{B}(\mathbb{K}_1)$, and $\kappa(e) \in \mathfrak{B}(\mathbb{K}_k)$ for all $e \in E$ with $|e| = k$. A *realization* of such abstract concept graph $\mathfrak{G}$ in $\vec{\mathbb{K}}$ is defined to be a mapping ρ of V to the power set of G_1 for which $\emptyset \neq \rho(v) \subseteq Ext(\kappa(v))$ for $v \in V$ and $\rho(v_1) \times \cdots \times \rho(v_k) \subseteq Ext(\kappa(e))$ for $e \in E$ with $\nu(e) = (v_1, \ldots, v_k)$, and $v_1 \theta v_2$ always imply $\rho(v_1) = \rho(v_2)$. Then, the pair $\underline{\mathfrak{G}} := (\mathfrak{G}, \rho)$ is called a *realized concept graph* of $\vec{\mathbb{K}}$ or, shortly, a *concept graph* of $\vec{\mathbb{K}}$.

Figure 2 and 3 yield an example of a power context family $\vec{\mathbb{K}} := (\mathbb{K}_1, \mathbb{K}_2)$. The small conceptual graph described in the previous section as derivable from this context family corresponds to the concept graph $\underline{\mathfrak{G}} := (\mathfrak{G}, \rho)$ of $\vec{\mathbb{K}}$ for which $\mathfrak{G} := (\{v_1, v_2, v_3\}, \{e_1, e_2\}, \nu, C_{\vec{\mathbb{K}}}, \kappa, \theta)$ with $\nu(e_1) = (v_1, v_2)$, $\nu(e_2) = (v_2, v_3)$, $\kappa(v_1) = \mu(\text{DISTRICT})$, $\kappa(v_2) := \mu(\text{HAVE})$, $\kappa(v_3) = \mu(\text{SLEEK GLASS MONO-LITH})$, $\kappa(e_1) = \mu(\text{STAT})$, $\kappa(e_2) = \mu(\text{PTNT})$, $\theta = \{(v_1, v_1), (v_2, v_2), (v_3, v_3)\}$, $\rho(v_1) = \{\text{central business}\}$, $\rho(v_2) = \{\#1.2\}$, and $\rho(v_3) = \{\#1.3, \#1.4\}$.

An important question is how to obtain all concept graphs of a given power context family $\vec{\mathbb{K}}$. For this, a common approach would be to apply rules for transforming known concept graphs of $\vec{\mathbb{K}}$ to new concept graphs of $\vec{\mathbb{K}}$. As such *transformation rules* one can use the following rules, which are applied to a concept graph $\underline{\mathfrak{G}} := (\mathfrak{G}, \rho)$ of $\vec{\mathbb{K}}$ with $\mathfrak{G} := (V, E, \nu, C_{\vec{\mathbb{K}}}, \kappa, \theta)$:

1. deletion of an isolated vertex or insertion of a new vertex v together with an assigned formal concept $\mathfrak{c}$ of $\mathbb{K}_1$ and an assigned non-empty subset A of $Ext(\mathfrak{c})$ so that $\kappa(v) = \mathfrak{c}$ and $\rho(v) = A$ (in addition, θ is extended by (v, v));

2. deletion of an edge or insertion of a new edge e together with an assigned k-tuple $(v_1, \ldots, v_k) \in V^k$ and an assigned formal concept $\mathfrak{r}$ of $\mathbb{K}_k$ so that $\kappa(e) = \mathfrak{r}$, $\nu(e) = (v_1, \ldots, v_k)$ and $\rho(v_1) \times \cdots \times \rho(v_k) \subseteq Ext(\mathfrak{r})$;

3. substitution of an assignment $v \mapsto \kappa(v)$ for a $v \in V$ by $v \mapsto \mathfrak{c}$ with $\mathfrak{c} \in \mathfrak{B}(\mathbb{K}_1)$ and $\rho(v) \subseteq Ext(\mathfrak{c})$;

4. substitution of an assignment $e \mapsto \kappa(e)$ for an $e \in E$ with $\nu(e) = (v_1, \ldots, v_k)$ by $e \mapsto \mathfrak{r}$ with $\mathfrak{r} \in \mathfrak{B}(\mathbb{K}_k)$ and $\rho(v_1) \times \cdots \times \rho(v_k) \subseteq Ext(\mathfrak{r})$;

5. substitution of θ by another equivalence relation $\tilde{\theta}$ on V for which $v_1 \tilde{\theta} v_2$ always implies $\rho(v_1) = \rho(v_2)$;

6. substitution of an assignment $v \mapsto \rho(v)$ for a $v \in V$ by $v \mapsto A$ with $\emptyset \neq A \subseteq G_1$ so that, for all $e \in E$ with $\nu(e) = (v_1, \ldots, v_k)$, one has $A_1 \times \cdots \times A_k \subseteq Ext(\kappa(e))$ if $\rho(v_i) = A$ for all $v_i = v$ and $\rho(v_i) = A_i$ for all $v_i \neq v$ $(i = 1, \ldots, k)$.

These transformation rules allow to obtain each concept graph $\underline{\mathfrak{G}}$ of $\vec{\mathbb{K}}$ from the empty concept graph: First, by Rule 1, all vertices of V can be inserted together with the mapping ρ and the restriction of κ to V; then, by Rule 2, all edges of E can be inserted together with the mapping ν and the restriction of κ to E and, finally, θ can be added by Rule 5. But, of course, this construction needs the knowledge of $\underline{\mathfrak{G}}$ in advance. Therefore it would be desirable to have a general construction yielding few concept graphs of $\vec{\mathbb{K}}$ from which the interesting concept graphs of $\vec{\mathbb{K}}$ can be easily derived by the transformation rules.

For this, we propose the following construction applied to the power context family $\vec{\mathbb{K}}$: First, for each formal object $(g_1, \ldots, g_k)$ of $\mathbb{K}_k$ $(k \geq 2)$, the smallest formal concept $\gamma(g_1, \ldots, g_k)$ having $(g_1, \ldots, g_k)$ in its extent is determined together with the maximal k-tuples $(A_1, \ldots, A_k)$ of non-empty subsets of G_1 satisfying $(g_1, \ldots, g_k) \in A_1 \times \cdots \times A_k \subseteq Ext(\gamma(g_1, \ldots, g_k))$, and the obtained $(k+1)$-tuples $(\gamma(g_1, \ldots, g_k), A_1, \ldots, A_k)$ are all collected in the set $E_{\vec{\mathbb{K}}}$. Then, we define $V_{\vec{\mathbb{K}}} := \{A \subseteq G_1 \mid A = A_i$ for some $(\gamma(g_1, \ldots, g_k), A_1, \ldots, A_k) \in E_{\vec{\mathbb{K}}}$ and some $i \leq k\}$, $\nu : E_{\vec{\mathbb{K}}} \longrightarrow \bigcup_{k=1}^{n} V_{\mathbb{K}}^k$ by $\nu(\gamma(g_1, \ldots, g_k), A_1, \ldots, A_k) := (A_1, \ldots, A_k)$, $\kappa : V_{\vec{\mathbb{K}}} \cup E_{\vec{\mathbb{K}}} \longrightarrow C_{\vec{\mathbb{K}}}$ by choosing $\kappa(A)$ as the smallest concept $\bigvee_{g \in A} \gamma g$ of $\mathbb{K}_1$ having A in its extent and $\kappa(\gamma(g_1, \ldots, g_k), A_1, \ldots, A_k)$ just as $\gamma(g_1, \ldots, g_k)$. This setting allows to define $\rho(A) := A$ for all $A \in V_{\vec{\mathbb{K}}}$. Finally, θ is taken as the identity relation on $V_{\vec{\mathbb{K}}}$. Obviously, the defined sets and mappings combine to a concept graph $\underline{\mathfrak{G}}(\vec{\mathbb{K}}) := (\mathfrak{G}(\vec{\mathbb{K}}), \rho)$ of $\vec{\mathbb{K}}$ with $\mathfrak{G}(\vec{\mathbb{K}}) := (V_{\vec{\mathbb{K}}}, E_{\vec{\mathbb{K}}}, \nu, C_{\vec{\mathbb{K}}}, \kappa, \theta)$; we call $\underline{\mathfrak{G}}(\vec{\mathbb{K}})$ the *canonical concept graph* of the power context family $\vec{\mathbb{K}}$. The construction of the canonical graph follows the idea to obtain a realized concept graph in which, first, the edges are labelled by the smallest possible formal concepts, secondly, the individual references are as large as possible in respecting the first condition and, thirdly, the vertices are labelled by the smallest possible formal concepts in respecting the second condition. The canonical concept graph of the power context family given by Figure 2 and 3 has as corresponding conceptual graph only a slight variation of that in Figure 1.

The described mathematization unifying Conceptual Graphs and Formal Concept Analysis should only be considered as a proposal for a start of combining both approaches. As important task, the central notion of context for conceptual graphs (cf. [So95]) has to be integrated within a unified mathematical theory. For this, the mathematization of concepts seems to force an extension of formal contexts to triadic contexts or multicontexts (cf. [LW95],[Wi96b]). For a mathematization of conceptual graphs nested by contexts, M. Chein and M.-L. Mugnier offer an approach in [CM95]. Furthermore, the semantics of the mathematical models of concept hierarchies and conceptual graphs should be completed by the syntax of an appropriate logic language for which description logics might indicate the direction (cf. [Ne90]). A first integration of description logics within Formal Concept Analysis is discussed in [Pr96]. For conceptual graphs, logic syntax has been already widely investigated (cf. [So97]). Thus, there are enough preconditions for a successful development of a unified mathematical theory for Elementary Logic based on the Theory of Conceptual Graphs and Formal Concept Analysis. Of course, first of all, the presented approach should be discussed, applied, and constructively criticized so that the proposed theory could be substantially improved.

References

[Ap76]   K.-O. Apel: Das Apriori der Kommunikationsgemeinschaft und die Grundlagen der Ethik. In: Transformation der Philosophie. Band 2: Das Apriori der Kommunikationsgemeinschaft. Suhrkamp Taschenbuch Wissenschaft 165, Frankfurt 1976.

[Br76] W. Brugger (Hrsg.): Philosophisches Wörterbuch. Herder, Freiburg 1976.

[CM92] M. Chein, M.-L. Mugnier: Conceptual Graphs: fundamental notions. Revue d'Intelligence Artificielle 6 (1992, 365–406.

[CM95] M. Chein, M.-L. Mugnier: Représenter des connaissances et raisonner avec des graphes. R.R.LIRMM 003-95. Université Montpellier 1995.

[GW96] B. Ganter, R. Wille: Formale Begriffsanalyse: Mathematische Grundlagen. Springer, Berlin-Heidelberg 1996.

[Gr18] K. J. Grau: Grundriß der Logik. Teubner, Leipzig und Berlin 1918.

[Ha81] J. Habermas: Theorie kommunikativen Handelns. Band 1. Suhrkamp, Frankfurt 1981.

[Ka88] I. Kant: Logic. Dover, New York 1988.

[LW95] F. Lehmann, R. Wille: A triadic approach to formal concept analysis. In: G. Ellis, R. Levinson, W. Rich, J. F. Sowa (eds.): Conceptual Structures: Applications, Implementations and Theory. Springer, Berlin-Heidelberg-New York 1995, 32–43.

[Ne90] B. Nebel: Reasoning and revision in hybrid representation systems. Springer, Berlin-Heidelberg-New York 1990.

[Pe83] Ch. S. Peirce: Phänomen und Logik der Zeichen. Suhrkamp Taschenbuch Wissenschaft 425, Frankfurt 1983.

[Pr96] S. Prediger: Symbolische Datenanalyse und ihre begriffsanalytische Einordnung. Staatsexamensarbeit. FB Mathematik, TH Darmstadt 1996.

[Sm89] The Smithonian Guide to Historic America. Stewart, Tabori & Chang Inc., New York 1989.

[So84] J. F. Sowa: Conceptual structures: information processing in mind and machine. Adison-Wesley, Reading 1984.

[So92] J. F. Sowa: Conceptual Graphs summary. In: T. E. Nagle, J. A. Nagle, L. L. Gerholz, P. W. Eklund (eds.): Conceptual Structures: Current Research and Practice. Ellis Horwood, 1992, 3–51.

[So95] J. F. Sowa: Syntax, semantics, and pragmatics of contexts. In: G. Ellis, R. Levinson, W. Rich, J. F. Sowa (eds.): Conceptual Structures: Applications, Implementations and Theory. Springer, Berlin-Heidelberg-New York 1995, 1–15.

[So97] J. F. Sowa: Knowledge representation: logical, philosophical, and computational foundations. PWS Publishing Co., Boston (to appear)

[Wi82] R. Wille: Restructuring lattice theory: an approach based on hierarchies of concepts. In: I. Rival (ed.): Ordered Sets. Reidel, Dordrecht-Boston 1982, 445–470.

[Wi92] R. Wille: Concept lattices and conceptual knowledge systems. Computers & Mathematics with Applications. 23 (1992), 493–515.

[Wi95] R. Wille: Begriffsdenken: Von der griechischen Philosophie bis zur Künstlichen Intelligenz heute. Diltheykastanie, Ludwig-Georgs-Gymnasium Darmstadt 1995, 77–109.

[Wi96a] R. Wille: Restructuring mathematical logic: an approach based on Peirce's pragmatism. In: A. Ursini, P. Agliano (eds.): Logic and Algebra. Marcel Dekker, New York 1996, 267–281.

[Wi96b] R. Wille: Conceptual structures of multicontexts. In: P. W. Eklund, G. Ellis, G. Mann (eds.): Conceptual Structures: Knowledge Representation as Interlingua. Springer, Berlin-Heidelberg-New York 1996, 23–39.

How Triadic Diagrams Represent Conceptual Structures

Klaus Biedermann

Technische Hochschule Darmstadt, Fachbereich Mathematik
Schloßgartenstr. 7, D–64289 Darmstadt,
biedermann@mathematik.th-darmstadt.de

Abstract. This paper is devoted to explain different kinds of information and knowledge which can be read off triadic diagrams (cf. Fig. 3). Such labelled line diagrams graphically represent the conceptual structure of triadic contexts which can be represented as three dimensional data tables. In greater detail it is elaborated how to read such diagrams and how ordinary (dyadic) conceptual structures can be determined within the triadic diagrams. For the complete order-theoretic understanding of the triadic diagrams, the necessary formal definitions are gradually introduced and illustrated in the discussion of the example about the three synoptic Gospels St. Matthew, St. Mark, and St. Luke.

Introduction: Philosophically, the theory of *Formal Concept Analysis* is based on the *dyadic* understanding of a concept as a *unit of thought* which is constituted by *two* parts: the *extension* consists of all *objects* belonging to the concept and the *intension* comprises all *attributes* which are common to all these objects. Experiences in data analysis and Charles Sanders Peirce's universal categories of *Firstness, Secondness, and Thirdness* have suggested to create the setting of *Triadic Concept Analysis* (cf. [LW95], [Wi95]). There it is possible to say *under* which *conditions* an object has certain attributes. Mathematically, these ideas are formalized by so-called *formal dyadic* and *triadic contexts* which are first to be explained.

For more than fifteen years, Formal Concept Analysis has extensively been applied in data analysis and knowledge processing (cf. [Wi87], [Wi96a]), but applying Triadic Concept Analysis is just at the beginning. The differences between the dyadic and the triadic theory, and particularly the novelties of the triadic approach, can be best seen from its way how data are represented by the characteristic *triadic diagrams*. Therefore, at the center of our investigations stands a triadic diagram showing the Twelve Disciples of Christ, some passages in the New Testament and three evangelists (gospels). This specific knowledge is represented by the triadic diagram which also makes it possible to analyse, investigate and explore the underlying data. In this way, it helps us discuss meaningful contributions of a triadic theory to data analysis. The basic notions of Formal Concept Analysis can be found in [GW96].

Dyadic Contexts and their Concept Lattices: The example of the three gospels is taken from [Wi96b]. There the three *dyadic contexts* represented as data tables in Fig. 1 belong to a *multicontext* which can be thought of as a network of several formal contexts under consideration. The entered crosses in each of the three formal contexts indicate who of the Twelve Disciples of Jesus is mentioned in which of the 36 passages - in fact, these are all passages of the New Testament mentioning at least one of the disciples.

THE PASSAGES

1. Four fishermen called as disciples
2. Peter's mother-in-law healed
3. Preaching in Gallilee
4. The miraculous draught of fishes
5. Matthew the tax collector
6. A girl restored to life and a woman healed
7. Sending out the twelve apostles
8. A sinful woman forgiven
9. Feeding the five thousand
10. Jesus walks on the sea
11. Peter confesses Jesus as Christ
12. Jesus transfigured on the mount
13. Jesus forbids sectarianism
14. A samaritan village rejects the Savior
15. Jesus counsells the rich young ruler
16. Greatness is serving
17. Jesus predicts the destruction of the temple
18. The death of Lazarus
19. The fruitful grain of wheat
20. The way, the truth and the life
21. Jesus promises another helper
22. Seeing and believing
23. The anointing at Bethany
24. Judas agrees to betray Jesus
25. Preparation of the Passover
26. Jesus announces the betrayal of Judas
27. Jesus predicts Peter's denial
28. The prayer in the garden
29. Betrayal and arrest in Gethsemane
30. Jesus faces the Sanhedrin
31. Peter denies Jesus, and weeps bitterly
32. Judas hangs himself
33. He is risen
34. The road to Emmaus
35. Breakfast by the sea
36. Jesus restores Peter

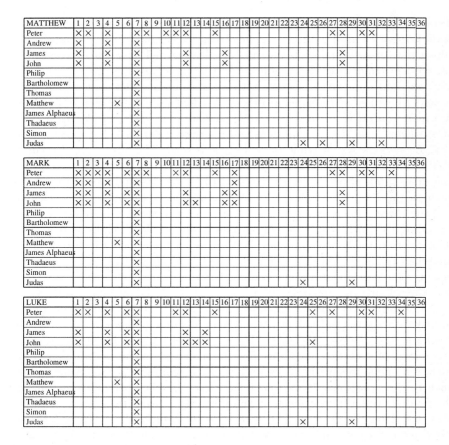

MATTHEW	1	2	3	4	5	6	7	8	9	10	11	12	13	14	15	16	17	18	19	20	21	22	23	24	25	26	27	28	29	30	31	32	33	34	35	36
Peter	×	×		×			×	×		×	×	×			×												×	×		×	×					
Andrew	×			×			×																													
James	×			×			×					×				×												×								
John	×			×			×					×				×												×								
Philip							×																													
Bartholomew							×																													
Thomas							×																													
Matthew					×		×																													
James Alphaeus							×																													
Thadaeus							×																													
Simon							×																													
Judas							×																		×		×			×			×				

MARK	1	2	3	4	5	6	7	8	9	10	11	12	13	14	15	16	17	18	19	20	21	22	23	24	25	26	27	28	29	30	31	32	33	34	35	36
Peter	×	×	×	×		×	×	×			×	×			×		×										×	×		×	×		×			
Andrew	×	×		×			×										×																			
James	×	×		×		×	×					×				×	×											×								
John	×	×		×		×	×					×	×			×	×											×								
Philip							×																													
Bartholomew							×																													
Thomas							×																													
Matthew					×		×																													
James Alphaeus							×																													
Thadaeus							×																													
Simon							×																													
Judas							×																		×					×							

LUKE	1	2	3	4	5	6	7	8	9	10	11	12	13	14	15	16	17	18	19	20	21	22	23	24	25	26	27	28	29	30	31	32	33	34	35	36
Peter	×	×		×		×	×				×	×			×										×		×			×	×		×			
Andrew							×																													
James	×			×		×	×					×		×																						
John	×			×		×	×					×	×	×											×											
Philip							×																													
Bartholomew							×																													
Thomas							×																													
Matthew					×		×																													
James Alphaeus							×																													
Thadaeus							×																													
Simon							×																													
Judas							×																		×					×							

Fig. 1. Formal contexts concerning the three Gospels according to St. Matthew, St. Mark and St. Luke

Without reduction the formal contexts can be transformed into the *labelled line diagrams* in Fig. 2 which, respectively, represent the so-called *concept lattices* of the formal contexts.

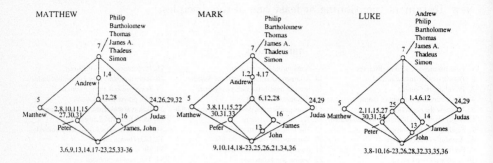

Fig. 2. The corresponding concept lattices

Triadic Contexts: As already pointed out in [Wi96b], the theory of multicontexts is closely linked with the triadic setting of Formal Concept Analysis after which the three formal contexts can also be regarded as a *triadic context* and the *concept trilattice* of this triadic context be understood as the *natural triadic extension* of the involved concept lattices. To elaborate what this means, we repeatedly recall some basic definitions of Triadic Concept Analysis from [Wi95]: A *triadic context* $\mathbb{K} := (K_1, K_2, K_3, Y)$ consists of a set K_1 of *objects*, a set K_2 of *attributes*, a set K_3 of *conditions* and a ternary relation Y between them, i.e. $Y \subseteq K_1 \times K_2 \times K_3$. A triple $(g, m, b) \in Y$ is read: The object $g \in K_1$ *has the* attribute $m \in K_2$ *under the* condition $b \in K_3$.

One may think of a triadic context as a three dimensional cross table. In our example the Twelve Disciples of Jesus serve as objects, the passages as attributes and the three gospels as conditions. The ternary relation Y obviously tells you according to which gospel in which passage who of the disciples is mentioned. It therefore extends and combines the three binary relations, which respectively underlie the formal contexts in Fig. 1. The triadic context can be visualized by placing the dyadic contexts one behind the other.

The three Peircean categories (cf. [Pe35]) can nicely be explained with respect to the example. The Twelve Disciples can be understood as the objects or entities which are given in the first place *"regardless of anything else"* and therefore belong to the category the First. They are involved in different stories which are a kind of *"reaction"* and therefore "assigned" to them or, in other words, there were no stories without disciples while the converse is conceivable. Consequently, the stories are *"Second to some First"*, i.e. they belong to the category the Second. But the evangelists have differently written these stories down as passages. Thus the gospels are different *"representations"*, a *"Third between a First and a Second"* and belong to the category the Third. So, the whole situation is based on a triadic relation and therefore requires Thirdness. For a more general discussion see [LW95].

The Triadic Diagram: Without reducing the original data, the considered triadic context can be transformed into the labelled line diagram in Fig. 3.

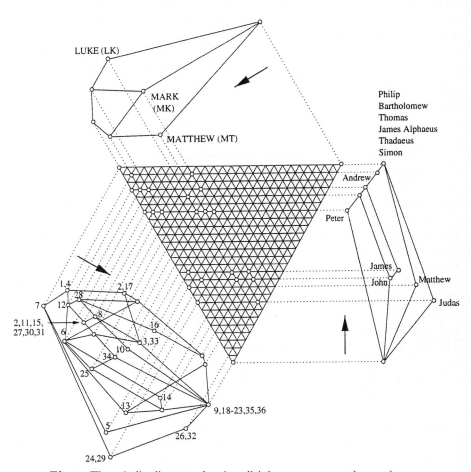

Fig. 3. The triadic diagram showing disiples, passages and gospels

How to read such a *triadic diagram* shall now be explained. The three line diagrams outside the 3-net show the *ordered structures* of disciples, passages and gospels, respectively. Pick out, for example, Peter in the line diagram on the right. If you follow up the horizontal line into the interiour *3-net*, the so-called *geometric structure*, you will find six circles, each of which is connected to the ordered structures of passages and gospels via straight lines. The first three circles you encounter on this route, for instance, establish connections between Peter and each of the three gospels alone while the right one of the other circles links Peter to the Gospels according to St. LUKE *and* St. MARK since the corresponding circle lies above LUKE and MARK (note the arrow!). Likewise the next circle combines MARK and MATTHEW, and the last one on utmost left means that there are simultaneously passages, in which Peter is mentioned in all three gospels. What are these passages? Following the line to the ordered structure of passages one ends up with a circle to which the passages $2, 11, 15, 27, 30,$ and 31 are attached. But as for the gospels, we are not finished here because we also must take into account the passages labelled below like $12, 1, 4,$ and 7. Putting the three parts together yields the triple

$$(\{\text{Peter}\}, \{1, 2, 4, 7, 11, 12, 15, 27, 30, 31\}, \{\text{MT,MK,LK}\}),$$

which is a so-called *(formal) triadic concept* of the triadic context $\mathbb{K}$. Similarly one can find the triadic concept

$$(\{\text{Peter}\}, \{1, 2, 3, 4, 6, 7, 8, 11, 12, 15, 17, 27, 28, 30, 31, 33\}, \{\text{MK}\})$$

which is represented by the right of the six circles. This triple obviously contains all the passages in which Peter is mentioned just in the Gospel according to St. MARK. It therefore has the additional passages $3, 6, 8, 17, 28, 31,$ and 33. Consequently, Peter cannot simultaneously be mentioned by MATTHEW and LUKE in these passages (because otherwise they would have belonged to the first triple). But in which of the two gospels is he not mentioned then?

Note that it is just Peter and none of the other disciples that belong to the two triadic concepts above. If one wishes to compare the passages in which Peter and John are mentioned with respect to the different gospels one must proceed from the circles above Peter and John in the ordered structure on the right to the circles in the 3-net by horizontal lines and from there to the corresponding passages and gospels.

The upper left circle in the 3-net structure has an intersting interpretation. It says that all three evangelists mention in their seventh passage all the disciples which is clear because in this story the Twelve Apostles are sent out. What is the interpretation of the two other circles bounding the 3-net?

Triadic Concepts: The triples represented by the circles in the 3-net structure have a certain property. They are *maximal* in each component or, with respect to the triadic context, they can be understood as the maximal boxes which are completely filled with crosses where permutations of parallel layers in the triadic context are allowed. Thus the boxes need not be connected. Note that the three triples, which bound the 3-net, can have an empty component in which case the corresponding boxes are too flat to contain any crosses. The formal definition of a triadic concept is based upon three *derivation operators* which are responsible for the mentioned maximality: Let $\mathbb{K} := (K_1, K_2, K_3, Y)$ be a triadic context, let $\{i, j, k\} = \{1, 2, 3\}$ and $X_i \subseteq K_i$ and $X_j \subseteq K_j$. Then

$$\langle X_i, X_j \rangle^{(k)} := \{a_k \in K_k \mid x_i, x_j, a_k \text{ are related by } Y \text{ for all } x_i \in X_i \text{ and } x_j \in X_j\}$$

is called the $(k)-derivation$ $operator$, e.g. $\langle \{\text{Peter}\}, \{\text{MK}\} \rangle^{(2)}$ consists of all passages in which Peter is mentioned according to MARK's Gospel (cf. the second of the two triples above). Furthermore, a triple $(A_1, A_2, A_3) \in \mathfrak{P}(K_1) \times \mathfrak{P}(K_2) \times$

$\mathfrak{P}(K_3)$ is said to be a *triadic concept of* $\mathbb{K}$ if $A_j = \langle A_i, A_k \rangle^{(j)}$ for all assignments in $\{i, j, k\} = \{1, 2, 3\}$. A_1 is called the *extent*, A_2 the *intent* and A_3 the *modus* of the triadic concept (A_1, A_2, A_3). The set of all triadic concepts of $\mathbb{K}$ is denoted as $\mathfrak{T}(\mathbb{K})$.

The *triadic diagram* is obviously a graphical representation of all triadic concepts of the triadic context $\mathbb{K}$ and therefore shows the conceptual structure of $\mathbb{K}$. Note that in the triadic case the families of extents, intents, and modi are no longer closure systems, i.e. complete lattices. In our example the intersection of the intents $\{1, 2, 4, 7, 17\}$ and $\{1, 2, 4, 7, 11, 12, 15, 27, 30, 31\}$ is not an intent any more. The extents, intents, and modi ordered by inclusion are just ordered sets. The three derivation operators are *triadic Galois connections between sets*. In general, triadic Galois connections (between three ordered sets) can be understood as a collection of ordinary Galois connections (cf. [Bi97a]). We will implicitly make use of this fact in the later discussion of triadic implications.

According to [LW95], a concept can be understood as *unit of thought* which tends to be *homogeneous* and *closed*. A triadic concept of a triadic context has these two properties because each object in its extent has all the attributes in the intent under all conditions in the modus, which yields the homogeneity. Moreover, extent, intent, and modus cannot be enlarged without violating the ternary relation Y, which yields the closedness of the triadic concept.

In order to become more familiar with the triadic diagram it is recommended to proceed exploring the data in a similar way as already begun. The following questions typically arise on such an exploration and therefore represent different types of questions involving different mathematical notions. Note that each of the following types start out from disciples, passages, gospels and arbitrary combinations of them and continues to ask for new disciples, passages and gospels.

- *Is Peter mentioned in the third passage of MATTHEW's Gospel?*

Simple questions of this type just require yes/no decisions. A short glance at the triadic context yields a negative answer which is a little bit harder to get from the triadic diagram because there a triadic concept must be determined which simultaneously contains Peter, the third passage and MATTHEW. The following questions ask for more information which can be more easily read from the triadic diagram.

- Peter occurs in passage 8 according to MATTHEW's Gospel. *Is it just him or are there also other disciples mentioned? Does Peter also occur in the eighth passage of the other gospels or in other passages in MATTHEW's Gospel?*

All these questions are satisfactorily answered by the following triadic concepts:

$$(\{Peter\}, \{1, 2, 4, 8, 11, 12, 15, 27, 28, 30, 31\}, \{MT, MK\})$$
$$(\{Peter\}, \{1, 2, 4, 8, 10, 11, 12, 15, 27, 28, 30, 31\}, \{MT\})$$

Obviously, it is just Peter who is mentioned in the eighth passage of MATTHEW's and also MARK's Gospel, but not in LUKE's Gospel.

- *In which passages do Peter and John occur according to MATTHEW's Gospel?*

The corresponding derivation operator yields $\langle\{\text{Peter,John}\}, \{\text{MT}\}\rangle^{(2)} = \{1, 4, 7,$
$12, 28\}$. A little more information about Peter and John according to MATTHEW's
Gospel offer the following triadic concepts:

$$(\{\text{Peter, John, James}\}, \{1, 4, 7, 12, 28\}, \{\text{MT, MK}\})$$
$$(\{\text{Peter, John, James}\}, \{1, 4, 7, 12\}, \{\text{MT, MK, LK}\})$$
$$(\{\text{Peter, John, James, Andrew}\}, \{1, 4, 7\}, \{\text{MT, MK}\})$$
$$(\text{All disciples}, \{7\}, \{\text{MT, MK, LK}\})$$

So, in some of the relevant passages also James and Andrew are mentioned
according to MARK's and even LUKE's Gospel.

- *Which are the gospels mentioning Peter and James and what are the corresponding passages?*

First of all, they are mentioned in MARK's Gospel, but also simultaneously
in MARK's and LUKE's, in MARK's and MATTHEW's and also in all three
Gospels. All this and even more is contained in the previous four and the following triadic concepts:

$$(\{\text{Peter, John, James}\}, \{1, 4, 6, 7, 12\}, \{\text{MK, LK}\})$$
$$(\{\text{Peter, John, James}\}, \{1, 2, 4, 6, 7, 12, 17, 28\}, \{\text{MK}\})$$
$$(\{\text{Peter, John, James,Andrew}\}, \{1, 2, 4, 7, 17\}, \{\text{MK}\})$$

It has become clear that, as in the dyadic case, it is possible to investigate the
"near vicinity" of the considered data and in this way also explore their surrounding.

The Triadic Diagram as the Natural Triadic Extension: As far as the
representation of data is concerned, a triadic diagram is a *symmetric* structure,
for the sets of objects, attributes, and conditions are all treated equally since
none of them is preferred to the others. But a triadic diagram is also a very *dense*
structure, for it contains many dyadic conceptual structures, i.e. a great number
of ordinary concept lattices. Similarly, a triadic context $\mathbb{K} := (K_1, K_2, K_3, Y)$
contains the dyadic contexts

$$\mathbb{K}^{ij}_{X_k} := (K_i, K_j, Y^{ij}_{X_k})$$

where $\{i, j, k\} = \{1, 2, 3\}$, $X_k \subseteq K_k$ and $(a_i, a_j) \in Y^{ij}_{X_k}$ if and only if a_i, a_j, and
x_k are related by Y for all $x_k \in X_k$.
In this terminology, the dyadic contexts in Fig. 1 correspond to the contexts

$$\mathbb{K}^{12}_{\{\text{MT}\}}, \ \mathbb{K}^{12}_{\{\text{MK}\}}, \ \text{and} \ \mathbb{K}^{12}_{\{\text{LK}\}}.$$

How can the concept lattice of the first dyadic context be found in the triadic
diagram? First of all, we must consider all those triadic concepts having at least
MATTHEW in their modus. In the 3-net, they can be found on parallel lines

ending at circles above the one labelled with MATTHEW. In a second step all
the triadic concepts with *maximal* extent and intent must be single out because
only for them it is guaranteed that extent and intent form dyadic concepts of
$\mathbb{K}^{12}_{\{MT\}}$. In Fig. 4 the corresponding circles are blackened[1].

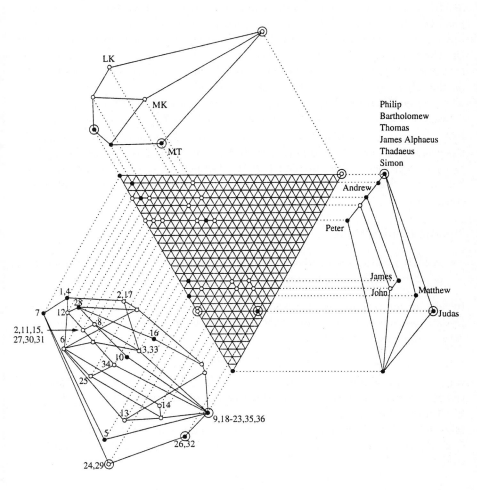

Fig. 4. The concept lattices of $\mathbb{K}^{12}_{\{MT\}}$ and $\mathbb{K}^{23}_{\{Judas\}}$ within the triadic
diagram

In particular, all the triadic concepts with MATTHEW in the modus give immediately rise to dyadic concepts. Increasing their modus while fixing the extent

---

[1] It is still surprising that the dually isomophic closure systems of extents and intents
appear within the ordered structure of all extents and the more complicated structure
of intents.

simultaneously means to decrease the intent (and vice versa) which does not give rise to a dyadic concept. This is obviously a rule to exclude (some of the) irrelevant triadic concepts.

In the triadic diagram, the dyadic concepts of $\mathbb{K}^{12}_{\{MT\}}$ can be immediately compared with respect to the gospels because they are split into three groups according to the modi {MT}, {MT,MK} and {MT,MK,LK}, e.g. MATTHEW and MARK mention Peter, John, James, and Andrew in the first, fourth and seventh passage[2]. In contrast to the seperate concept lattices in Fig. 2 the comparison has already been carried out in the triadic diagram. If one wishes to consider the three gospels simultaneously, the line diagram of $\mathbb{K}^{12}_{\{MT,MK,LK\}}$ must be determined which is fairly easy because the relevant triadic concepts are all circles on the left borderline of the geometric structure.

The triadic diagram does not only allow extensions and combinations of the concept lattices of the above contexts. Apart from $\mathbb{K}^{12}_{\{MT\}}$, the line diagram of $\mathbb{K}^{23}_{\{Judas\}}$ is made visible in Fig. 4 by bigger circles. Obviously, in all three gospels the passages 24 and 29 deal only with Judas, but in the passages 26 and 32 he is just mentioned according to MATTHEW's Gospel. An interesting interpretation has the line diagram of $\mathbb{K}^{23}_{\{Peter\}}$ (Remember the rule to exclude irrelevant triadic concepts.).

Triadic Implications:

- MARK mentions some disciples in his second passage. *Does he also mention other passages with the same disciples?*

In fact, in his seventeenth passage all the disciples who already occured in the second passage turn up again. This statement is an instance of a triadic implication which is expressed by

$$(2 \rightarrow 17)_{MK}$$

where, for simplicity, set brackets are and will sometimes be omitted. Note that the disciples are not further specified but can be determined by $\langle \{2,17\}, \{MK\} \rangle^{(1)}$ = {Andrew, Peter, James, John}.

More general, let $\mathbb{K} := (K_1, K_2, K_3, Y)$ be a triadic context, let $M, N \subseteq K_2$ and $C \subseteq K_3$. Then the *triadic implication (between attributes under the conditions in C)*

$$(M \rightarrow N)_C$$

is said to hold in $\mathbb{K}$ if for all $g \in K_1$ the following condition is satisfied:

$$(\{g\} \times M \times C) \subseteq Y \Rightarrow (\{g\} \times N \times C) \subseteq Y.$$

The set M is called the *premise*, the set N the *conclusion* of $(M \rightarrow N)_C$. If $C = K_3$ the triadic implication is shortly denoted as $M \rightarrow N$.

---

[2] Note that according to MARK's Gospel these disciples occur also in the passage 2 and 17. So, a dyadic concept of $\mathbb{K}^{12}_{\{MT\}}$ is not necessarily a dyadic concept of $\mathbb{K}^{12}_{\{MK\}}$.

New for the triadic case is that under different conditions a premise can have *different* conclusions. In our example we also observe the triadic implication

$$(2 \to 11, 15, 27, 30, 31)_{MT,MK,LU}.$$

Here, the relevant disciple is just Peter. Both triadic implications can be unified writing $2 \to 17 \vee 11, 15, 27, 30, 31$ which has the interpretation that under some conditions passage 17 is implied and under different conditions the passages $11, 15, 27, 30, 31$ are implied, respectively.

Basically, all the triadic implications between attributes under certain conditions $C \subseteq K_3$ can be understood as ordinary implications in the dyadic context $\mathbb{K}_C^{12}$. Since the derivation operators $\langle ., C\rangle^{(1)}$ and $\langle ., C\rangle^{(2)}$ are just the ordinary derivation operators in $\mathbb{K}_C^{12}$, we define $A_2^{(C)} := \langle A_2, C\rangle^{(1)}$ and $A_1^{(C)} := \langle A_1, C\rangle^{(2)}$ for subsets $A_2 \subseteq K_2$ and $A_1 \subseteq K_1$. A triadic implication $(M \to N)_C$ then holds in $\mathbb{K}$ if and only if $N \subseteq M^{(C)(C)}$. So, all implications which can be read from the three concept lattices in Fig. 2 are also valid as triadic implications in $\mathbb{K}$ but there are also those for the combinations of gospels. Moreover, these implications can be unified such that a new syntax seems to be needed for the triadic case. Suitable bases of implications still need to be determined for the triadic case.

The discussed implications might have seemed a littel artificial but they are suitable for an introduction of triadic implications. More meaningful implications can be obtained by changing the role of objects, attributes, and conditions, e.g. $\{MT, LU\} \to MK$ and $(Peter, John \to James)_{MT}$. The triadic implication about the gospels means that whenever a disciple occurs in whatever passage accoring to MATTHEW and LUKE, he also occurs in that passage according to MARK's Gospel. The triadic implication about the disciples can even be extended with respect to the conditions by determining $\{MT\}^{(Peter, John, James)(Peter, John, James)} = \{MT, MK, LU\}$. This yields the triadic implication $(Peter, John \to James)_{MT,MK,LU}$. Not valid in $\mathbb{K}$, for example, is $(Peter, John \to James)_{LU}$.

How triadic implications can be read from the triadic diagram can be found in [WZ97], where a general introduction into Triadic Concept Analysis is given.

Coherence Mappings: MARK mentions John in his passages 1, 2, 4, 6, 7, 12, 13, 16, 17, and 28 which can also be read from the triadic concept $(\{John\}, \{1, 2, 4, 6, 7, 12, 13, 16, 17, 28\}, \{MK\})$.

- *In which passages does John occur according to MATTHEW's Gospel?*

This is answered by simply determining $\{John\}^{(MT)} = \{1, 4, 7, 12, 16, 28\}$. A little more information is offered by the corresponding triadic concept $(\{John, James\}, \{1, 4, 7, 12, 16, 28\}, \{MT, MK\})$ which can be generated from $\{John\}$ and $\{MT\}$ by iterating the derivation operators. This is achieved more generally in the following way:

Let $\mathbb{K} := (K_1, K_2, K_3, Y)$ be a triadic context and let $X_1 \subseteq K_1$ and $X_3 \subseteq K_3$ be arbitrary sets of objects and conditions. The triadic concept

$$\mathfrak{b}_{13}(X_1, X_3) := \left(X_1^{(X_3)(X_3)}, X_1^{(X_3)}, \langle X_1^{(X_3)}, X_1^{(X_3)(X_3)}\rangle^{(3)} \right)$$

is said to be *generated by* X_1 *and* X_3. It is characterized as the triadic concept with largest intent among all triadic concepts with largest extent (and simultaneously smallest modus) having X_1 in the extent and X_3 in the modus.

The above question deals with the comparison of a dyadic concept of $\mathbb{K}^{12}_{\{MK\}}$ with a dyadic concept of $\mathbb{K}^{12}_{\{MT\}}$ sharing common elements in the extents. In [Wi96b], this is described by the *coherence mapping*

$$\lambda_{MK,MT} : \mathfrak{B}(\mathbb{K}^{12}_{\{MK\}}) \to \mathfrak{B}(\mathbb{K}^{12}_{\{MT\}})$$
$$(A_1, A_2) \mapsto (A_1^{(MT)(MT)}, A_1^{(MT)}).$$

This mapping can naturally be raised to the set of all triadic concepts of a triadic context $\mathbb{K}$. For $\{i, j, k\} = \{1, 2, 3\}$ and $X_k \subseteq K_k$, the mapping

$$\lambda_{i,X_k} : \quad \mathfrak{T}(\mathbb{K}) \quad \to \mathfrak{T}(\mathbb{K})$$
$$(A_1, A_2, A_3) \mapsto \mathfrak{b}_{ik}(A_i, X_k)$$

is said to be a *triadic coherence mapping*.

Note that the triadic coherence mapping λ_{i,X_k} is isotone with respect to the ith component. Moreover, the ith and j^{th} component of $\mathfrak{b}_{ik}(A_i, X_k)$ form a dyadic concept in $\mathbb{K}_{X_k}$. Further properties of triadic coherence mappings such as the concatenation of different triadic coherence mappings are still to be investigated.

Concept Trilattices and the Basic Theorem of Triadic Concept Analysis:
The set $\mathfrak{T}(\mathbb{K})$ of all triadic concepts of a triadic context $\mathbb{K}$ is naturally endowed with three quasiorders $\lesssim_1, \lesssim_2$, and $\lesssim_3$ defined by

$$(A_1, A_2, A_3) \lesssim_i (B_1, B_2, B_3) :\Longleftrightarrow A_i \subseteq B_i$$

for $i = 1, 2, 3$ and the corresponding equivalence relations

$$(A_1, A_2, A_3) \sim_i (B_1, B_2, B_3) :\Longleftrightarrow A_i = B_i.$$

The equivalence classes of $\sim_i$ are denoted by $[(A_1, A_2, A_3)]_i$ and can obviously be ordered by $[(A_1, A_2, A_3)]_i \lesssim_i [(B_1, B_2, B_3)]_i :\Leftrightarrow A_i \subseteq B_i$.

Now the tradic diagram can completely be understood. The triadic concepts which are i-equivalent, i.e. coincide in the ith component, are placed on one line in the geometric structure representing $(\mathfrak{T}(\mathbb{K}), \sim_1, \sim_2, \sim_3)$, such as the six triadic concepts having Peter in their extent, and the lines corresponding to one component are all drawn parallel to each other. The hierarchies of extents, intents, and modi are represented in the outer ordered structures $(\mathfrak{T}(\mathbb{K})/\sim_i, \leq_i)$ with $i = 1, 2, 3$.

Up to now, triadic diagrams have always been suitable representations for real data which is due to the observation that the crosses in such triadic contexts are rather rarely spread. The *tetrahedron context* in Fig. 5 is the smallest example known which cannot be represented by a line diagram exclusively using straight lines. So, whenever the *tetrahedron condition* is satisfied, i.e. there are pairwisely distinct triadic concepts $\mathfrak{a}, \mathfrak{b}, \mathfrak{c} \in \mathfrak{T}(\mathbb{K})$ of a triadic context $\mathbb{K}$ with

$\mathfrak{a} \sim_i \mathfrak{b}$, $\mathfrak{b} \sim_j \mathfrak{c}$, $\mathfrak{c} \sim_k \mathfrak{a}$ for some assignment in $\{i,j,k\} = \{1,2,3\}$ and there is also another triadic concept $\mathfrak{d} \in \mathfrak{T}(\mathbb{K})$ with $\mathfrak{d} \sim_i \mathfrak{c}$, $\mathfrak{d} \sim_j \mathfrak{a}$, $\mathfrak{d} \sim_k \mathfrak{b}$ then the line diagram cannot be embedded into the 3-net structure. One also gets in trouble if the so-called *Thomson condition* is violated (cf. [WZ97]). To find suitable representations for such cases requires further investigations.

Fig. 5. The tetrahedron context

The relational structure $(\mathfrak{T}(\mathbb{K}), \lesssim_1, \lesssim_2, \lesssim_3)$ has two further properties. A triadic concept is already determined by two of its componentes, which is known as the *geometric condition*, and, for triadic concepts, increasing two components means to decrease the third component. This condition is called the *antiordinal dependency*. Moreover, for two subsets $\mathfrak{X}_i, \mathfrak{X}_k \subseteq \mathfrak{T}(\mathbb{K})$ where $\{i,j,k\} = \{1,2,3\}$, six operations can be defined making use of the generating operator for triadic concepts:

$$\nabla_{ik}(\mathfrak{X}_i, \mathfrak{X}_k) := \mathfrak{b}_{ik}\left(\bigcup\{A_i \mid (A_1, A_2, A_3) \in \mathfrak{X}_i\}, \bigcup\{A_k \mid (A_1, A_2, A_3) \in \mathfrak{X}_k\}\right)$$

is called the *ik-join* of the pair $(\mathfrak{X}_i, \mathfrak{X}_k)$. These operations correspond to meet and join in the dyadic case.

This situation can also be described on a purely order-theoretic level: A *triorderd set* $(S, \lesssim_1, \lesssim_2, \lesssim_3)$ is a relational structure if the relations are quasiorders on S satisfying $\sim_i \cap \sim_j = id_S$ (geometric condition) and $\lesssim_i \cap \lesssim_j \subseteq \gtrsim_k$ (antiordinal dependency) for all $\{i,j,k\} = \{1,2,3\}$, respectively. To define the *ik-joins* of (X_i, X_k) with $X_i, X_k \subseteq S$ one first of all needs the *ik-bounds*

$$(X_i, X_k)^{(ik)} := \{u \in S \mid u \gtrsim_i x_i \text{ for all } x_i \in X_i, \ u \gtrsim_k x_k \text{ for all } x_k \in X_k\}$$

and then the *ik-limits*

$$(X_i, X_k)^{\overline{(ik)}} := \{v \in S \mid v \gtrsim_j u \text{ for all } u \in (X_i, X_k)^{(ik)}\}.$$

It can be shown that in a triordered set there is with respect to $\lesssim_k$ at most one smallest element among the ik-limits which is called the *ik-join of* (X_i, X_k) (if it exists) and denoted by $\nabla_{ik}(X_i, X_k)$. As in the dyadic case for complete lattices, a *complete trilattice* $(L, \lesssim_1, \lesssim_2, \lesssim_3)$ is a triordered set in which *all* the ik-joins $\nabla_{ik}(X_i, X_k)$ exist for $\{i,j,k\} = \{1,2,3\}$ and $X_i, X_k \subseteq L$.

The *Basic Theorem of Triadic Concept Analysis* (cf. [Wi95]) now ensures that $(\mathfrak{T}(\mathbb{K}), \lesssim_1, \lesssim_2, \lesssim_3)$ is a complete trilattice, the so-called *concept trilattice*, with the ik-joins as defined above. Conversely, any complete trilattice $(L, \lesssim_1, \lesssim_2, \lesssim_3)$ is isomorphic to $(\mathfrak{T}(\mathbb{K}), \lesssim_1, \lesssim_2, \lesssim_3)$ for some suitable triadic context $\mathbb{K}$ for which $\mathbb{K}_L :=$

(L, L, L, Y_L) can be chosen where $Y_L := \{(x_1, x_2, x_3) \in L^3 \mid$ there exists $u \in L$ with $x_i \lesssim_i u$ for $i = 1, 2, 3\}$. The last result gives rise to an embedding theorem for triordered sets and trilattices. In [Bi97b], one can also find an algebraic characterization of trilattices which requires unusual types of equations compared with the lattice equations.

To determine all triadic concepts of a triadic context $\mathbb{K}$, it is helpful to first *clarify* the context by omitting identical (parallel) layers leaving just one as a representative, such as $\mathbb{K}_{\{1\}}^{13}$ and $\mathbb{K}_{\{4\}}^{13}$ or $\mathbb{K}_{\{18\}}^{13}$ and $\mathbb{K}_{\{23\}}^{13}$ in our example. The *reduction* of $\mathbb{K}$ is more difficult than in the dyadic case but, in general, the layers which are completely filled with crosses can be cancelled from the context, e.g. $\mathbb{K}_{\{7\}}^{13}$.

The ik-joins are the perfect analogue to the dyadic operations of meet and join. But they are very special in the sense that they just determine one specific triadic concept. Many of the discussed questions aim at a *comparison of different triadic concepts* once some disciples, passages, and (or) gospels are chosen.

For a triadic context $\mathbb{K} := (K_1, K_2, K_3, Y)$ the *comparison mapping* $\kappa : \mathfrak{P}(K_1) \times \mathfrak{P}(K_2) \times \mathfrak{P}(K_3) \to \mathfrak{P}(\mathfrak{T}(\mathbb{K}))$ is defined by $(A_1, A_2, A_3) \mapsto \{(B_1, B_2, B_3) \in \mathfrak{T}(\mathbb{K}) \mid A_i \subseteq B_i$ for all $i = 1, 2, 3\}$.

Further investigations are needed to clarify whether it is algebraically valuable to raise the comparison mappings to ("power"-) operations between concept trilattices. Such operations are not extended from ordinary (n-ary) operations by replacing the elements in the argument with subsets of the carrier set ([cf. [Br93]). They are naturally power operations.

Summary: For dealing with data, which are linked by a ternary relation, it has turned out that triadic diagrams offer a more holistic view than some dyadic line diagrams which are chosen to represent the same data (cf. Fig. 2). In fact, a triadic diagram does not only represent the conceptual structure of a triadic context, but also the many conceptual structures of dyadic contexts within the triadic context.

The presented triadic diagram just shows the three synoptic Gospels. Because of their great extent of similarity to each other, the diagram is not too complicated for a first discussion. Including the Gospel according to St. John, which is quite different from the others, enlarges the triadic diagram by about one third. For even larger data sets different representations such as *nested triadic diagrams* must be investigated. The mathematical questions raised on our rather pragmatic-oriented appoach to Triadic Concept Analysis are to be pursued further.

References

[Bi97a]   K. Biedermann: Triadic Galois Connections. In: Proceedings of the Conference on General Algebra and Discrete Mathematics (52. Arbeitstagung Allgemeine Algebra), Potsdam 1996.

[Bi97b]   K. Biedermann: An Equational Theory for Trilattices (to appear).

[Br93]   C. Brink: Power Structures. Algebra Universalis, 30 (1993), 177-216.

[GW96]   B. Ganter, R. Wille: Formale Begriffsanalyse: Mathematische Grundlagen. Springer, Heidelberg 1996.

[LW95]   F. Lehmann, R. Wille: A Triadic Approach to Formal Concept Analysis. In: G. Ellis, R. Levinson, W. Rich and J.G. Sowa (ed.). Conceptual structures: applications, implementations and theorey. Lecture Notes in Artificial Intelligence 954. Springer-Verlag, Berlin-Heidelberg-New York, 1995, 32-43.

[Pe35]   Ch. S. Peirce: Collected Papers. Harvard Univ. Press, Camebridge 1931-35.

[Wi87]   R. Wille: Bedeutung von Begriffsverbänden. In: B. Ganter, R. Wille, K. E. Wolff (Hrsg.): Beiträge zur Begriffsanalyse. B.I.-Wissenschaftsverlag, Mannheim 1987, 161-211.

[Wi95]   R. Wille: The Basic Theorem of Triadic Concept Analysis. Order 12 (1995), 149-158.

[Wi96a]   R. Wille: Conceptual landscapes of knowledge: A pragmatic paradigm for knowledge processing. FB4-Preprint, TH Darmstadt 1996.

[Wi96b]   R. Wille: Conceptual Structures of Multicontexts, In: P. W. Eklund, G. Ellis, G. Mann (eds.), Conceptual structures: representation as interlingua. Springer-Verlag, Berlin-Heidelberg-New York, 1996, 23-39

[WZ97]   R. Wille, M. Zickwolff: Grundlagen einer Triadischen Begriffsanalyse (to appear)

Concept Exploration – A Tool for Creating and Exploring Conceptual Hierarchies

Gerd Stumme

Technische Hochschule Darmstadt, Fachbereich Mathematik
Schloßgartenstr. 7, D–64289 Darmstadt, stumme@mathematik.th-darmstadt.de

Abstract. Concept exploration is a knowledge acquisition tool for interactively exploring the hierarchical structure of finitely generated lattices.
Applications comprise the support of knowledge engineers by constructing a type lattice for conceptual graphs, and the exploration of large
formal contexts in formal concept analysis.

1 Introduction

Lattices are a popular mathematical structure for modeling conceptual hierarchies. The existence of greatest common subconcepts and least common superconcepts provides additional algebraic structure to models based on ordered sets,
and makes them more suitable for computation. Exploration tools, as developed
in formal concept analysis,[1] benefit from this algebraic structure (cf. [12]). The
best known and most often used of these tools is *attribute exploration* ([4], [2], [6]),
which uses only infima (greatest common subconcepts) for the computation. It
determines – in interaction with the user – the implicational logic of attributes
of a given formal context, or, more algebraically spoken, the $\bigwedge$-subsemilattice of
a finite lattice generated by some subset.

When, in addition, suprema (least common superconcepts) are considered,
the problem turns out to determine the sublattice of a given lattice (of concepts)
generated by some subset. This is the aim of *concept exploration*. If a priori the
lattice is known to be distributive, then the exploration can benefit from the
much stronger algebraic structure induced by the distributive law. The corresponding exploration tool is called *distributive concept exploration* ([13], [15]).

Already in the first publication on formal concept analysis, [17], the basic
conception of concept exploration is mentioned. Its scheme is demonstrated by
examples in [18] and [19]. U. Klotz and A. Mann further elaborated the method in
[7]. Unfortunately, their work is, with more than hundred pages, very extensive,
so that the results became too complicated for an efficient implementation. This
gave the impulse to develop a more compact tool using existing algorithms which
have been proven successful.

Concept exploration is designed to explore a lattice L that is generated by
a subset $B \subseteq L$. The algorithm generates questions about the lattice (i. e., the

---

[1] An introduction into formal concept analysis can for instance be found in [6], [17],
or [20].

conceptual hierarchy) of the kind "Is c_1 a subconcept of c_2?" that are answered by a (human) user, usually an expert of the field of interest. There are (at least) two interesting applications for such a knowledge acquisition tool.

Firstly in formal concept analysis: Suppose we have a formal context $\mathbb{K}$ which is too large (e. g., infinite) to be completely given. The aim is to determine the structure of the concept lattice, or at least a part of it. Therefore one might consider formal concepts of the concept lattice one is particularly interested in. They are called *basic concepts* (thus the letter B). Concept exploration interactively computes the sublattice L of the concept lattice of $\mathbb{K}$ generated by B. Information on the context $\mathbb{K}$ is acquired from the user.

Secondly for conceptual graphs: The type hierarchy for conceptual graphs is assumed to be a lattice ([10]). This lattice needs to be constructed for modeling a situation by conceptual graphs. Concept exploration can support this creative process: B is now the set of "basic types", i. e., the set of types a knowledge engineer assumes necessary for his purpose. Concept exploration supports him in generating the type lattice L.

2 Concept Exploration

Let L be a lattice generated by a finite subset $B \subseteq L$. The exploration follows the generation process of the lattice. First it computes the set $B^{(1)} := B^{\wedge}$ consisting of all infima of the elements of B. Then it computes $B^{(2)} := B^{\wedge\vee}$, the set of all suprema of $B^{\wedge}$. The next step provides $B^{(3)} := B^{\wedge\vee\wedge}$, and so on. $B^{(k)}$ is considered as a weak sublattice of L in which, for k odd (even), all infima (suprema) are defined, and suprema (infima) are only defined for subsets of $B^{(k-1)} \subseteq B^{(k)}$. If the lattice is finite, then $B^{(k)} = B^{(k+1)} = L$ for some $k \in \mathbb{N}$. Otherwise the exploration would not terminate (which is possible for all B with card$(B) \geq 3$, because the free lattice over the set B is then infinite) and has to be interrupted at some moment; but the exploration can approach L as much as desired. The resulting partial lattice may then be completed by other tools, as described for instance in [17].

For the step from $B^{(k)}$ to $B^{(k+1)}$, where k is even, concept exploration uses $\bigwedge$-*exploration*, a modification of the algorithm of attribute exploration with background implications ([11]). For odd k, the situation is dual to the former, and $\bigvee$-*exploration*, a modification of object exploration is applied. Object exploration can be understood as an attribute exploration of the dual lattice, i. e., with objects and attributes interchanging their roles ([12]). The idea to use attribute and object exploration alternatively for concept exploration is due to P. Burmeister.

The intermediate results of concept exploration are the weak sublattices $B^{(k)}$ as defined above. They are stored as formal contexts (G, M, I), where G and M contain lattice terms over B. For odd (even) k, G contains exactly one term for every element in $B^{(k)}$ ($B^{(k-1)}$) denoting it, and M contains exactly one term for every element in $B^{(k-1)}$ ($B^{(k)}$) denoting it. The relation I reflects the hierarchical order: $(s, t) \in I :\iff s^L \leq t^L$ for $s \in G$ and $t \in M$. The weak partial lattice $B^{(k)}$ is isomorphic to the weak partial lattice $\mathfrak{B}(G, M, I)$ in which infima are defined on $\gamma(G)$ (which is the whole lattice for odd k), and suprema are defined on $\mu(M)$ (which is the whole lattice for even k). Partial concept lattices are described in detail in [16]. In the sequel, we identify the lattice terms with the corresponding elements of L. For L being finite, the final result of the exploration is the context $(L, L, \leq)$.

The first step in concept exploration is a $\bigwedge$-exploration starting with a context having the basic concepts as attributes ($M := B^{(0)} = B$), and having no objects ($G := B^{(-1)} = \emptyset$). The relation I is empty as well. More general, for odd k, the kth exploration step, a $\bigwedge$-exploration, will transform the context $(B^{(k-2)}, B^{(k-1)}, \leq)$ into $(B^{(k)}, B^{(k-1)}, \leq)$. The next exploration step, a $\bigvee$-exploration, then transforms it into the context $(B^{(k)}, B^{(k+1)}, \leq)$, and so on. This yields a first sketch of the algorithm of concept exploration:

(0) Start with $k := 0$ and $(G, M, I) := (B^{(-1)}, B^{(0)}, \leq) = (\emptyset, B, \emptyset)$.
(I) Increase k. Determine $(B^{(k)}, B^{(k-1)}, \leq)$ by $\bigwedge$-exploration.
   If $B^{(k)} = B^{(k-1)}$ then STOP.
(II) Increase k. Determine $(B^{(k-1)}, B^{(k)}, \leq)$ by $\bigvee$-exploration.
   If $B^{(k)} = B^{(k-1)}$ then STOP, else go to Step (I).

Since the situation is perfectly symmetric, we only describe Step (I) in detail. Let k be odd in the sequel.

Step (I) extends the set $G = B^{(k-2)}$ to $B^{(k)}$ by lattice terms, one for each element in $B^{(k)} \setminus B^{(k-2)}$. For elements in $B^{(k-1)} \setminus B^{(k-2)}$, they can just be copied from M to G. For every copied element x, the relation I is extended by

($\dagger$) $(x, y) \in I : \Longleftrightarrow \{x\}' \subseteq \{y\}'$, where $X' := \{g \in G \mid \forall m \in X : (g, m) \in I\}$.

This "copying" is only omitted for $k = 1$, because the first $\bigwedge$-exploration is also used for determining the ordering on B which is not known at the beginning.

For determining the elements in $B^{(k)}$ which are not included in $B^{(k-1)}$, Step (I) is concluded by a $\bigwedge$-exploration.

2.1 Discovering New Concepts by $\bigwedge$-Exploration

The elements in G correspond to infima of elements in $M = B^{(k-1)}$, since the lattice equality $x = \bigwedge(\{x\}')$ holds for every element x in G. For $x \in B^{(k-1)}$, this is obvious, since x is also an element of M. For $x \in B^{(k)} \setminus B^{(k-1)}$, the intent x' is just constructed such that this equality holds.

Since $Y \subseteq x' \Longleftrightarrow x \in Y' \Longleftrightarrow \forall y \in Y : x \leq y \Longleftrightarrow x \leq \bigwedge Y$ holds for every lattice element x in G and every subset Y of M, we can also identify every subset $Y \subseteq M = B^{(k-1)}$ with the infimum $\bigwedge Y$ of its elements. At the beginning of the $\bigwedge$-exploration, there may be subsets $Y \subseteq M$ such that $\bigwedge Y \neq \bigwedge(Y'')$ in L. The aim of $\bigwedge$-exploration is to add new objects to G in order to change the derivation operator $''$ such that $\bigwedge Y = \bigwedge(Y'')$ holds in L for all $Y \subseteq M$. As we will see below, it is not necessary to test all subsets Y, but we can restrict ourselves to *pseudo-intents*, which are defined next.

The following definitions and results are cited from [11], where the original algorithm of attribute exploration of B. Ganter ([4], [6], [5]) was modified to accept additional knowledge in form of background implications.

Let $\mathbb{K} := (G, M, I)$ be a formal context. We assume in the following that M is finite.

Definition. An *implication* $X \to Y$ of $\mathbb{K}$ is a pair of subsets X and Y of M, such that every object having all attributes in X also has all attributes in Y.

Let $\mathcal{L}$ be a set of implications of $\mathbb{K}$ (called *background implications*). The closure operator on the set M of attributes induced by the background implications is denoted by $P \mapsto \overline{P} := P \cup P^{\mathcal{L}} \cup P^{\mathcal{L}\mathcal{L}} \cup \ldots$ with

$$Q^{\mathcal{L}} := Q \cup \bigcup \{Y \subseteq M | X \subseteq Q, \, X \to Y \in \mathcal{L}\}.$$

A subset P of M is called an $\mathcal{L}$-*pseudo-intent* of $\mathbb{K}$ if $P = \overline{P} \neq P''$ and if, for every $\mathcal{L}$-pseudo-intent Q with $Q \subset P$, the inclusion $Q'' \subseteq P$ holds.

Theorem 1. *The set* $\mathcal{B}_{\mathcal{L}} := \{P \to P'' | P \text{ is a } \mathcal{L}\text{-pseudo-intent}\}$ *of implications is an irredundant set of implications such that every implication of* $\mathbb{K}$ *can be deduced from* $\mathcal{L} \cup \mathcal{B}_{\mathcal{L}}$.

The set of all intents and $\mathcal{L}$-*pseudo-intents of a finite context* (G, M, I) *is a closure system on* M*; with the closure operator* $A \mapsto A^{\bullet} := A \cup A^{\circ} \cup A^{\circ\circ} \cup \ldots$, *where* $A^{\circ} := \overline{A \cup \bigcup \{Y \subseteq M | X \to Y \in \mathcal{B}_{\mathcal{L}}, X \subset A\}}$.

In our application, the set $\mathcal{L}$ will store information about the hierarchical structure that is computed in previous exploration steps. Intents will correspond to already existing elements, i.e., elements in $B^{(k)}$, while pseudo-intents correspond to potentially new elements which may be added to $B^{(k)}$ in order to obtain $B^{(k+1)}$, depending whether $\bigwedge P = \bigwedge(P'')$ (which is equivalent to $P \to P''$) holds or not – which will be asked from the user.

$\bigwedge$-Exploration uses Ganter's Next-Closure-algorithm (cf. [4]) to compute all intents and all $\mathcal{L}$-pseudo-intents in a *lectical order*. For the sake of simplicity we assume that $M := \{1, \ldots, n\}$.

Definition. For $X, Y \subseteq M$ and $i \in M$ we define[2]

$$X <_i Y :\iff (i \in Y \setminus X \text{ and } X \cap \{i+1, \ldots n\} = Y \cap \{i+1, \ldots n\}).$$

The *lectical order* on $\mathfrak{P}(M)$ is defined by

$$X < Y :\iff \exists i : X <_i Y \ .$$

For $X \subseteq M$ and $i \in M$ let

$$X \oplus i := ((X \cap \{i+1, \ldots, n\}) \cap \{i\})^{\bullet} \ .$$

---

[2] In contrast to [11], in this definition the ordering on M is reversed, since this matches better the generation process of concept exploration.

Theorem 2. *The lecticly first intent or $\mathcal{L}$-pseudo-intent is $\overline{\emptyset}$. For a given subset $Y \subseteq M$ the lecticly next intent or $\mathcal{L}$-pseudo-intent is the set $Y \oplus i$, where i is the minimal element in $M \setminus Y$ with $Y <_i Y \oplus i$. The lecticly last intent or $\mathcal{L}$-pseudo-intent is M.*

Attribute exploration and $\bigwedge$-exploration are based on the fact that, during the computation, the list of already computed intents and $\mathcal{L}$-pseudo-intents is stable under adding new objects which respect the previously computed implications (and the background implications).

Additionally to the contexts, $\bigwedge$-exploration produces a list $\mathcal{L}$ of implications. In concept exploration, these implications will be used as background implications for later $\bigwedge$-explorations. (For the $\bigvee$-explorations, a similar list $\mathcal{L}'$ will be kept.) At the beginning of the first $\bigwedge$-exploration, $\mathcal{L}$ is empty. The following is the algorithm described in [11] adapted[3] to concept exploration.

Algorithm 1. Set $n := \mathrm{card}(B^{(k-1)})$, $P := \emptyset$.

(a) Ask the user: "Which of the concepts $\ll$*Here all elements in $P'' \setminus P$ are listed*$\gg$ are superconcepts of $\bigwedge P$?" The answer is a set $Q \subseteq P'' \setminus P$.

(b) Add the implication $P \to Q$ to $\mathcal{L}$.[4]

(c) If there is no $g \in G$ with $g' = P \cup Q$ then add the lattice term $\bigwedge P$ to G, and extend I as follows: $(\bigwedge P, m) \in I : \Longleftrightarrow m \in P \cup Q$ for $m \in M$.

(d) Set $Y := P$.

(e) Determine the next intent or $\mathcal{L}$-pseudo-intent P following Y by applying Theorem 2:

- $i := 1$
- While $Y \not<_i Y \oplus i$ increase i.
- $P := Y \oplus i$.

(f) If $P = M$ then STOP.

(g) If $P = P''$ then go to Step (e), else go to step (a).

The algorithm for $\bigvee$-exploration is exactly the same, only G and M have to be interchanged, "superconcepts" has to be replaced by "subconcepts", "infimum" by "supremum", $\bigwedge$ by $\bigvee$, and $\mathcal{L}$ by $\mathcal{L}'$. The implications $P \to P''$ in $\mathcal{L}'$ correspond to equalities $\bigvee P = \bigvee(P'')$.

---

[3] In Step (c), *one* element (the lattice term $\bigwedge P$) is added to G having *exactly* the attributes in $P \cup Q$. In other applications of attribute exploration, there may be more than one object necessary in order to restrict the conclusion of the implication to $P \cup Q$.

[4] Even if Q is empty! This will be needed in the over-next exploration, where P may again be a $\mathcal{L}$-pseudo-intent (instead of an intent) in the extended context. (See Steps (4) and (8) in Algorithm 2.)

2.2 Concept Exploration

Now we are ready to present the algorithm of concept exploration. Step (8) and its dual, Step (4), are described below, as well as Steps (9) and (5).

Algorithm 2. Set $k := 1$, $(G, M, I) := (\emptyset, B, \emptyset)$, $\mathcal{L} := \mathcal{L}' := \emptyset$.

(1) Determine $(B^{(1)}, B^{(0)}, \leq)$ by Algorithm 1.
(2) Increase k.
(3) Copy $B^{(k-1)} \setminus B^{(k-2)}$ from G to M. Extend I as described dually at (†).
(4) Change every implication $X \to Y$ in $\mathcal{L}'$ to $X \to X''$ (where X'' is computed in the modified context). For every $x \in B^{(k-1)} \setminus B^{(k-2)}$ $\left(x \in B^{(1)} \text{ for } k = 2 \right)$ add the implication $\{x\} \to \{x\}''$ to $\mathcal{L}'$.
(5) Determine $(B^{(k-1)}, B^{(k)}, \leq)$ by the dual of Algorithm 1 (for $k > 2$ starting at Step (e) with $Y := B^{(k-3)}$ and answering "All" automatically while $P \subseteq B^{(k-2)}$).
   If M is not changed (i. e., all questions are answered by "All") then STOP.
(6) Increase k.
(7) Copy $B^{(k-1)} \setminus B^{(k-2)}$ from M to G. Extend I as described at (†).
(8) Change every implication $X \to Y$ in $\mathcal{L}$ to $X \to X''$ (where X'' is computed in the modified context). For every $x \in B^{(k-1)} \setminus B^{(k-2)}$ add the implication $\{x\} \to \{x\}''$ to $\mathcal{L}$.
(9) Determine $(B^{(k)}, B^{(k-1)}, \leq)$ by Algorithm 1, starting at Step (e) with $Y := B^{(k-3)}$ and answering "All" automatically while $P \subseteq B^{(k-2)}$.
   If G is not changed (i. e., all questions are answered by "All") then STOP.
(10) Go to step (2).

In this algorithm, Steps (1) and (6)–(9) correspond to Step (I) in the first sketch, Steps (2)–(5) to Step (II). In Step (8) (and dually in Step (4)), the previously computed implications are adapted to the newly generated attributes. More precisely, the conclusions of the implications are extended by the appropriate attributes from $B^{(k-1)} \setminus B^{(k-2)}$. (The premises remain subsets of $B^{(k-2)}$.) The implications $\{x\} \to \{x\}''$ with $x \in B^{(k-1)} \setminus B^{(k-2)}$ are added because $\{x\}''$ is simply the order filter generated by x, which can be determined automatically. Otherwise, these implications had to be confirmed by the user in the next $\bigwedge$-exploration.

The numbering of the newly generated elements (which is relevant for the variable i of $\bigvee / \bigwedge$-exploration) is just done in the way the elements are generated. This provides the optimization in Step (9) (and dually in Step (5) for $k > 2$): All intents and $\mathcal{L}$-pseudo-intents being subsets of $B^{(k-3)}$ remain unchanged from the previous $\bigwedge$-exploration, and need not be determined again. Since all elements of $B^{(k-2)}$ are also represented *as objects* in the current context, all equalities $\bigwedge P = \bigwedge(P'')$ with $P \subseteq B^{(k-2)}$ hold. For completing the set $\mathcal{L}$ of implications, we can thus provide the answer "All" automatically while $P \subseteq B^{(k-2)}$.

For an implementation, it might be useful to sort the list of objects in the context $(B^{(1)}, B^{(0)}, \leq)$ after Step (1) such that the basic concepts come first (and in the same ordering as in $B^{(0)}$). Then the ith object and the ith attribute will always be the same lattice term in the sequel.

There are three possibilities to simplify the exploration dialogue for the user. Firstly, the lattice terms in $P'' \setminus P$ can be listed in a linear extension of the hierarchical order. When the user chooses a term t for the set Q, then all elements in the order filter $\{x \in B^{(k)} \mid x \geq t\}$ will be chosen automatically as well.

Secondly, the lattice terms may be simplified automatically. For instance, the subterm $\bigwedge \emptyset$ may be omitted in any $\bigwedge$-term. This, and its dual, will be applied in the following example.

Thirdly, the user has the option to introduce a *name* for every lattice term added to the context. For instance, with MAN and WOMAN being basic concepts, the user may define HUMAN := MAN $\vee$ WOMAN. This shortens the lattice terms and supports the readability of the questions generated by the process. For the creation of a type lattice for conceptual graphs, this option is essential.

3 Example

Let us demonstrate concept exploration by an example. Imagine a knowledge engineer who wants to model knowledge about ancient Greek musical instruments by conceptual graphs. He uses concept exploration for supporting the creation for an adequate type lattice. For instance, he might want to start with the following four basic types: CHORD INSTRUMENT, KYTHARA, WIND INSTRUMENT, and AULOS. A kythara is a harp which, in Greek mythology, is the symbol of Apollo, while an aulos is an oboe like instrument associated with Dionysus. The newly generated types are named using the naming mechanism described at the end of the previous section. The dialogue of the first $\bigwedge$-exploration (i.e., Step (1) of Algorithm 2) consists of eight questions:

"Which of the concepts CHORD INSTRUMENT, KYTHARA, WIND INSTRUMENT, and AULOS are superconcepts of $\bigwedge \emptyset$?" – "None!" – "Name for $\bigwedge \emptyset$?" – "IN-STRUMENT."

The first $\mathcal{L}$-pseudo-intent is the empty set. The infimum $\bigwedge \emptyset$ of the empty set is always the largest element of the lattice, and can thus be understood as the concept "everything" comprising all objects of the field of interest, which, in this example, are INSTRUMENTs. It is the first object added to G. The relation I remains empty. The (trivial) implication INSTRUMENT $\to \emptyset$ is added to $\mathcal{L}$, because it will be needed in Step (8).

"Which of the concepts KYTHARA, WIND INSTRUMENT, and AULOS are superconcepts of CHORD INSTRUMENT?" – "None!"

The name CHORD INSTRUMENT is added to G, and the (trivial) implication CHORD INSTRUMENT $\to \emptyset$ to $\mathcal{L}$.

"Which of the concepts CHORD INSTRUMENT, WIND INSTRUMENT, and AU-LOS are superconcepts of KYTHARA?" – "CHORD INSTRUMENT!"

The name KYTHARA is added to G. The first non-trivial implication, KY-THARA $\to$ CHORD INSTRUMENT, is added to $\mathcal{L}$. In this way the extension of G and $\mathcal{L}$ continues...

"Which of the concepts CHORD INSTRUMENT, KYTHARA, and AULOS are superconcepts of WIND INSTRUMENT?" – "None!"

"Which of the concepts KYTHARA and AULOS are superconcepts of CHORD INSTRUMENT ∧ WIND INSTRUMENT?" – "None!" – "Name for CHORD INSTRUMENT ∧ WIND INSTRUMENT?" – "AEOLS-HARP."

An aeols-harp is a harp where the vibration of the chords is induced by wind (either natural wind or generated by bellows).

"Is the concept AULOS a superconcept of CHORD INSTRUMENT ∧ KYTHARA ∧ WIND INSTRUMENT?" – "Yes!" – "Name for CHORD INSTRUMENT ∧ KYTHARA ∧ WIND INSTRUMENT?" – "NOTHING."

There is no kythara being a wind instrument. We obtain the "absurd type".

"Is the concept KYTHARA a superconcept of CHORD INSTRUMENT ∧ WIND INSTRUMENT ∧ AULOS?" – "Yes!"

We obtain again the concept NOTHING. *Since there is no lattice term added to G, we are not asked to provide a name.*

The result of the first ∧-exploration is shown in Fig. 1. Observe that the diagram has to be read as ∧-semilattice, since only all infima are defined, but no suprema. In the following ∨-exploration, for instance, the supremum of KYTHARA and AEOLS-HARP will not be identified with CHORD INSTRUMENT, but with a real subconcept of it, namely HARP.

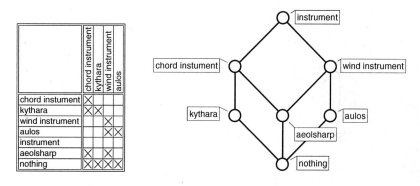

Fig. 1. The result of the first ∧-exploration

In Step (3), the names INSTRUMENT, AEOLS-HARP, and NOTHING are copied from G to M, and I is extended such that it represents the ordering shown in the line diagram in Fig. 1. The list $\mathcal{L}'$ of implications is empty up to now, but in Step (4), for every attribute m, the implication $m \to \{x \mid x < m\}$ is added, for example WIND INSTRUMENT → AULOS, AEOLS-HARP, NOTHING. Then (Step (5)) the next exploration, the first ∨-exploration, starts:

"Which of the concepts CHORD INSTRUMENT, KYTHARA, WIND INSTRUMENT, AULOS, INSTRUMENT, AEOLS-HARP and NOTHING are subconcepts of $\bigvee \emptyset$?" – "NOTHING!"

This justifies the choice of the name NOTHING *in the former* $\bigwedge$-*exploration, since* $\bigvee \emptyset$ *is always the smallest element of a lattice.*

"Which of the concepts CHORD INSTRUMENT, WIND INSTRUMENT, INSTRUMENT, and AEOLS-HARP are subconcepts of KYTHARA $\vee$ AULOS?" – "None!" – "Name for KYTHARA $\vee$ AULOS?" – "ANCIENT GREEK INSTRUMENT."

Here and in the sequel, the concept NOTHING *is omitted in the lattice terms, since it is not relevant for the supremum.*

"Is the concept CHORD INSTRUMENT a subconcept of KYTHARA $\vee$ AEOLS-HARP?" – "No!" – "Name for KYTHARA $\vee$ AEOLS-HARP?" – "HARP."

"Is the concept WIND INSTRUMENT a subconcept of AULOS $\vee$ AEOLS-HARP?" – "No!" – "Name for AULOS $\vee$ AEOLS-HARP?" – "WIND INSTRUMENT WITH VIBRATING PART."

"Which of the concepts CHORD INSTRUMENT, WIND INSTRUMENT, and INSTRUMENT are subconcepts of KYTHARA $\vee$ AULOS $\vee$ AEOLS-HARP?" – "All!"

Here we decide to identify the concept KYTHARA $\vee$ AULOS $\vee$ AEOLS-HARP *with* INSTRUMENT.

The result of this $\bigvee$-exploration is shown in Fig. 2. In the resulting weak partial lattice, all suprema are defined. Infima are only defined for its $\bigwedge$-subsemilattice which is shown in Fig. 1.

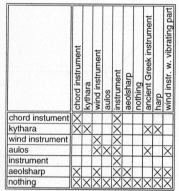

	chord instrument	kythara	wind instrument	aulos	instrument	aeolsharp	nothing	ancient Greek instrument	harp	wind instr. w. vibrating part
chord instument	X				X					
kythara	X	X			X			X	X	
wind instrument			X		X					
aulos			X	X	X			X		X
instrument					X					
aeolsharp	X		X		X	X			X	X
nothing	X	X	X	X	X	X	X	X	X	X

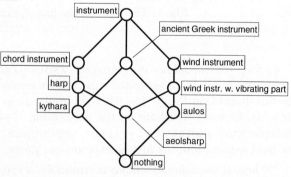

Fig. 2. The result of the first $\bigvee$-exploration

In the next ∧-exploration, the names ANCIENT GREEK CHORD INSTRUMENT, ANCIENT GREEK HARP, ANCIENT GREEK WIND INSTRUMENT, and ANCIENT GREEK WIND INSTRUMENT WITH VIBRATING PART are introduced. The following ∨-exploration provides only one new name, CHORD INSTRUMENT WITHOUT BOW. Finally, the 5th exploration (which is a ∧-exploration again) runs through without generating any question. Hence, the concept exploration is terminated. The resulting lattice is shown in Fig. 3. Of course, all infima and suprema are defined now.

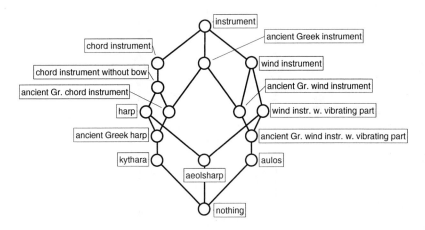

Fig. 3. The result of the concept exploration

The "degree of exactitude", i.e., the decision whether or not to identify two concepts, essentially depends on the purpose the exploration was done for. For instance, a very pedantic user might only accept the two implications KYTHA-RA → CHORD INSTRUMENT and AULOS → WIND INSTRUMENT, and deny all others. Then every question generates a new concept. This process can be continued ad infinitum, converging towards the (infinite) free lattice generated by two two-element chains, as shown in Fig. 4. The ellipses in the diagram indicate which of the elements of the free lattice were identified in the former exploration.

In general, the knowledge engineer can greatly benefit from the line diagram of the lattice freely generated by the ordered set $(B, \leq)$. In fact, the line diagram of FL($\underline{2}+\underline{2}$) in Fig. 4 was used for the exploration of instruments. Unfortunately, there are only very few free lattices over partially ordered sets which can be "drawn" ([8]).

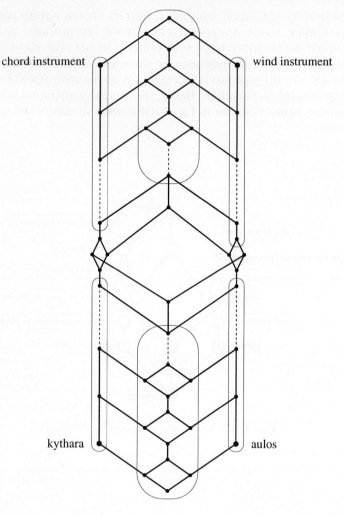

chord instrument

wind instrument

kythara

aulos

Fig. 4. The free lattice FL($\underline{2} + \underline{2}$).

4 Discussion

The main problem for any knowledge acquisition tool like concept exploration is the fact, that, in principal, the exploration may not terminate, because there are infinite lattices generated by only three elements. Hence any such exploration tool must provide the possibility to stop the exploration at any point such that the level of completeness gained so far is known. This was the crucial point for the decision to choose a breath first exploration: After the kth exploration, it is assured that all lattice elements denotable with a lattice term of complexity less or equal than k are generated, and that their hierarchical relationships are determined. A depth first exploration tool (like, e. g., distributive concept exploration) would risk, for arbitrary (e. g., non distributive) lattices, to explore some of the basic concepts so extensively that it will not reach the other ones. Unfortunately, this breath first approach forces the user to confirm all newly generated elements one by one. One might ask to construct another tool that allows to confirm or deny more than one element at a time. E. g., it might be based on A. Day's famous doubling method ([3]) for the computation of free lattices over partial lattices. Since the method is not able to reach all (finite) lattices, but only certain ones (the so-called bounded lattices), the right balance between applying the doubling method and factorizing by suitable congruences has to be kept. First experiments have shown that a too early factorization may prevent further doubling (i. e., generating potentially new elements) resulting in a break down of the exploration, while a too late factorization produces an abundance of potential elements which are not really different.

An interesting question for any exploration tool is, whether the exploration can be continued such that it will not terminate. In our algorithm, this is equivalent to the question, whether, for the intermediate result $B^{(k)}$, the free lattice $\mathrm{FL}(B^{(k)})$ is infinite or not. In [9], V. Slavík presents an effective algorithm for testing its finiteness. This free lattice is generated by concept exploration when all further questions are answered with "None" (i. e., with $Q := \emptyset$). On the other hand, it is always possible to terminate the algorithm in the next exploration step by answering "All" to all further questions.

Sometimes, the user does not want to start an exploration from the scratch, but he already has some background knowledge. In [5], B. Ganter extends attribute exploration to *background knowledge* formulated in predicate formulas. In particular, this comprises background implications as used in our algorithm. Hence his results can also be adapted to concept exploration. Since Ganter's algorithm can handle partial information, this also provides the possibility to use counterexamples. A *counterexample* (or *separating pair*) for $\bigwedge P \neq \bigwedge(P'')$ consists of an attribute belonging to the intent of $\bigwedge(P'')$ and an object not having the attribute but belonging to the extent of $\bigwedge P$. Unlike in distributive concept exploration – where counterexamples are used –, the capability for treating partial knowledge is necessary, because an object may belong to the supremum of two concepts although it does not belong to any of the two concepts. (The dual is true for attributes.) Hence, if complete knowledge is required, the user has to be asked for every newly generated supremum, which objects additionally belong to it. This makes the exploration unnecessarily complex.

Graphical support for the user showing the already computed hierarchy and the next potential elements is very desirable. By drawing only the potential elements of the next $\bigvee/\bigwedge$-exploration one could avoid the problem discussed at the end of the example, that most free lattices are not "drawable" (in the sense of Fig. 4). Unfortunately, there aren't any really satisfying fully automatic drawing algorithms for lattices yet. As with many other lattice oriented tools, concept exploration would certainly benefit from any progress in this field.

As for attribute exploration, concept exploration is designed to be an interactive tool, where a user acts as expert. In order to obtain a completely automatic tool, concept exploration may be combined with a subsumption algorithm (i. e., an (automatic) algorithm able to answer this type of questions) as it was done for the description logic $\mathcal{ALC}$ with attribute exploration ([1]) and distributive concept exploration ([14]), or resort to an online semantic dictionary.

References

1. F. Baader: Computing a minimal representation of the subsumption lattice of all conjunctions of concepts defined in a terminology. In: G. Ellis, R. A. Levinson, A. Fall, V. Dahl (eds.): *Proceedings of the International KRUSE Symposium: Knowledge Retrieval, Use and Storage for Efficiency.* Santa Cruz, CA, USA, August 11–13, 1995, 168–178
2. P. Burmeister: ConImp – A program for formal concept analysis. Technische Hochschule Darmstadt, 1987 (Latest version 1996 for MS DOS)
3. A. Day: Doubling constructions in lattice theory. *Can. J. Math.* **44**(2), 1992, 252–269
4. B. Ganter: Algorithmen zur Begriffsanalyse. In: B. Ganter, R. Wille, K. E. Wolff (eds.): *Beiträge zur Begriffsanalyse.* B. I.-Wissenschaftsverlag, Mannheim, Wien, Zürich 1987, 241–254
5. B. Ganter: Attribute exploration with background knowledge. *Proceedings of the conference on Order and Decision-Making.* Ottawa, Canada, August 5–9, 1996
6. B. Ganter, R. Wille: Formal Concept Analysis: Mathematical Foundations. Springer, Heidelberg 1997 (Translation of: Formale Begriffsanalyse: Mathematische Grundlagen. Springer, Heidelberg 1996)
7. U. Klotz, A. Mann: Begriffexploration. Diplomarbeit, TH Darmstadt 1988
8. I. Rival, R. Wille: Lattices freely generated by partially ordered sets: which can be "drawn"? J. Reine Angew. Math. **310**, 1979, 55–80
9. V. Slavík: Lattices with finite W-covers. (To appear)
10. J. F. Sowa: Conceptual structures: Information processing in mind and machine. Adison-Wesley, Reading 1984
11. G. Stumme: Attribute exploration with background implications and exceptions. In: H.-H. Bock, W. Polasek (eds.): *Data analysis and information systems. Statistical and conceptual approaches.* Studies in classification, data analysis, and knowledge organization **7**, Springer, Heidelberg 1996, 457–469
12. G. Stumme: Exploration tools in formal concept analysis. In: *Ordinal and symbolic data analysis.* Studies in classification, data analysis, and knowledge organization **8**, Springer, Heidelberg 1996, 31–44
13. G. Stumme: Knowledge acquisition by distributive concept exploration. In: G. Ellis, R. A. Levinson, W. Rich, J. F. Sowa (eds.): *Supplementary proceedings of the third international conference on conceptual structures*, Santa Cruz, CA, USA, August 14–18, 1995, 98–111
14. G. Stumme: The concept classification of a terminology extended by conjunction and disjunction. In: N. Foo, R. Goebel (eds.): PRICAI'96: Topics in artificial intelligence. LNAI **1114**, Springer, Heidelberg 1996, 121–131
15. G. Stumme: Distributive concept exploration – a knowledge acquisition tool in formal concept analysis. *Mathematics in artificial intelligence* (submitted)

16. G. Stumme: Partial concept lattices (in preparation)

17. R. Wille: Restructuring lattice theory: an approach based on hierarchies of concepts. In: I. Rival (ed.): *Ordered sets*. Reidel, Dordrecht–Boston 1982, 445–470

18. R. Wille: Bedeutungen von Begriffsverbänden. In: B. Ganter, R. Wille, K. E. Wolff (eds.): *Beiträge zur Begriffsanalyse*. B. I.–Wissenschaftsverlag, Mannheim 1987, 161–211

19. R. Wille: Knowledge acquisition by methods of formal concept analysis. In: E. Diday (ed.): *Data analysis, learning symbolic and numeric knowledge*. Nova Science Publisher, New York, Budapest 1989, 365–380

20. R. Wille: Conceptual structures of multicontexts. In: P. W. Eklund, G. Ellis, G. Mann (eds.): *Conceptual structures: Knowledge representation as interlingua*. LNAI **1115**, Springer, Heidelberg 1996, 23–39

Logical Scaling
in Formal Concept Analysis

Susanne Prediger

Technische Hochschule Darmstadt, Fachbereich Mathematik
Schloßgartenstr. 7, D–64289 Darmstadt, prediger@mathematik.th-darmstadt.de

Abstract. Logical scaling is a new method to transform data matrices which are based on object-attribute-value-relationships into data matrices from which conceptual hierarchies can be explored. The derivation of concept lattices is determined by terminologies expressed in a formal-logical language.

1 Introduction

The aim of formal concept analysis is to explore conceptual patterns in empirical data contexts. Methods have been developed to find conceptual hierachies and to represent them in line diagrams based on concept lattices (cf. [Wi82], [GW96]). These methods can be of great interest for knowledge representation and data mining. Concept lattices can also be relevant as principled ways to structure the type lattices used for conceptual graphs.

In general, there is no immediate, "automatic" way to derive the conceptual structures of data contexts which are based on object-attribute-value relationships. The first approach to transform these data contexts into concept lattices, namely *conceptual scaling*, was presented in [GSW86]. Since then, a great variety of applications in formal concept analysis has been based on this method.

In this paper, conceptual scaling will be rediscussed and compared with a new, alternative method called *logical scaling* in which the derivation of concept lattices is determined by terminologies expressed in a formal-logical language.

2 Contexts derived by conceptual scales

Object-attribute-value-relationships are a frequently used data structure to code real-world problems. In formal concept analysis, they are formalized in *many-valued contexts*. (Note that the word "context" is used in a special way here. It has nothing to do with the contexts of Sowa's conceptual structures (cf. e. g. [So92])).

In [GW89], a *many-valued context* is formally defined as a quadruple (G, M, W, I), where G, M, and W are sets whose elements are called *objects, (many-valued) attributes* and *attribute values* respectively, and I is a ternary relation with $I \subseteq G \times M \times W$ such that $(g, m, v) \in I$ and $(g, m, w) \in I$ always implies $v = w$. An attribute m of a many-valued context (G, M, W, I) may be

$\mathbb{K}$	importance	sports	books	home	events
1	5	su	$\{g,e\}$	am	sp
2	3	sa	$\{e\}$	ki	cu
3	1	wa	$\{s,g\}$	am	cu
4	2	tf	$\{s,e\}$	fs	po
5	4	sk	$\{g,e\}$	hg	sp
6	2	hm	$\{s,g,e\}$	am	cu
7	1	wa	$\{s,g,e\}$	am	cu
8	3	sk	$\{g,e\}$	ki	po
9	4	sk	$\{g,e\}$	hg	sp
10	5	su	$\{e\}$	fs	sp
11	2	jo	$\{s,g\}$	fs	cu
12	1	wa	$\{s,g,e\}$	am	cu
13	4	su	$\{g\}$	hg	sp
14	5	sa	$\{e\}$	hg	sp
15	3	jo	$\{s,e\}$	ki	cu

Fig. 1. Many-valued context *leisure activities*

considered as a partial map of G into W which suggests to write $m(g) = w$ rather than $(g, m, w) \in I$. The context (G, M, W, I) is called n-valued if W has cardinality n. One-valued contexts correspond to formal contexts as defined in [Wi82].

For illustration, we recall one of the first examples of scaled many-valued contexts which was presented in [GSW86]. It deals with a data set extracted from a case study on leisure activities (see [AL83]). 15 persons are listed with attribute values concerning five many-valued attributes:

importance of leisure: very important (5), rather important (4), impor-
(importance)   tant (3), less important (2), unimportant (1);

sports activities: sailing (sa), surfing (su), skiing (sk), walking (wa),
(sports)   hiking in the mountains (hm), jogging (jo), track
and field (tf);

book reading: all combinations of specific text books (s), general
(books)   books (g) and entertaining books (e);

hobbies at home: house and garden (hg), kitchen (ki), art and mu-
(home)   sic (am), family and social life (fs);

visit of events: culture (cu), sports (sp), politics (po).
(events)

The data set is represented by the many-valued context shown in figure 1.

The aim of every *scaling process* in formal concept analysis is to obtain a derived formal context (G, N, J) with the same objects as (G, M, W, I) whose extents can be thought of as the "meaningful" subsets of G. For this derived context, we can obtain a concept lattice as usual (cf. [Wi92] for an introduction). The concept lattice of the derived context is considered to be the conceptual structure of the many-valued context.

"Meaningful" refers to the interpretation of the data which can only be given by an expert of the field the data is from and never by the mathematician alone. This interpretation must always be purpose-oriented and should be founded on theoretical considerations.

The basic idea of *conceptual scaling* is to derive the context by conceptual scales. We assign to each many-valued attribute $m \in M$ a *conceptual scale* $\mathbb{S}_m$ which is a formal (one-valued) context $\mathbb{S}_m := (G_m, M_m, I_m)$ with $m(G) \subseteq G_m$. The choice of these scales is a matter of interpretation. The task is to select $\mathbb{S}_m$ in such a way that it reflects the implicitly given structure of the attribute values as well as the issues of data analysis.

The second step of conceptual scaling is to decide how the different many-valued attributes can be combined to describe concepts. The disjoint union of attribute sets often proves sufficient and we can thus restrict ourselves to looking at this so called *plain conceptual scaling*.

The many-valued context (G, M, W, I), together with the family of scales $(\mathbb{S}_m)_{m \in M}$ is called the *plainly scaled context* and determines the *derived context* which is defined as

$$(G, \bigcup_{m \in M} \{m\} \times M_m, J) \quad \text{where} \quad g\, J\,(m, w) : \iff m(g)\, I_m\, w.$$

In our example, we extract the information about the implicit structures of the attribute values from [AL83] and scale the attribute importance and books

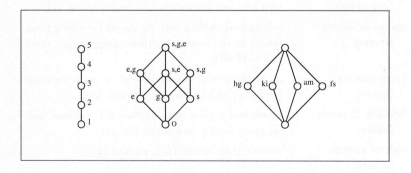

Fig. 2. Concept lattices of the scales $\mathbb{S}_{\text{importance}}$, $\mathbb{S}_{\text{books}}$ and $\mathbb{S}_{\text{home}}$

ordinally and the attribute home nominally.

$$\mathbb{S}_{\text{importance}} := (\{1, 2, 3, 4, 5\}, \{1, 2, 3, 4, 5\}, \leq),$$
$$\mathbb{S}_{\text{books}} := (\mathfrak{P}(\{s, g, e\}), \mathfrak{P}(\{s, g, e\}), \subseteq) \qquad \text{and}$$
$$\mathbb{S}_{\text{home}} := (W, W, =).$$

These scales can be represented by the concept lattices shown in figure 2. Assuming that politics is more similar to culture than to sports, the attribute events is scaled interordinally. The attribute values of sports are understood as structured by a tree-like hierarchy. In this way, we obtain two scales which are represented by the concept lattices in figure 3.

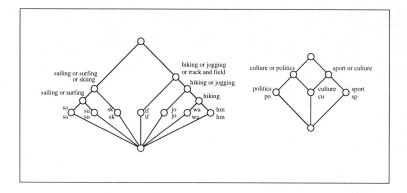

Fig. 3. Concept lattices of the scales $\mathbb{S}_{\text{sports}}$ and $\mathbb{S}_{\text{events}}$

In figure 4, we can see the concept lattice of the corresponding derived context where we can find interesting patterns. For example, none of the persons visiting sporting events reads specific books whereas all culturally interested persons do. More complex results are discussed in [GSW86].

Summarizing, we can say that conceptual scaling yields a global view of the conceptual patterns of data stored in many-valued contexts by applying expert knowledge about the inherent structure of some or all attribute values.

Sometimes, however, it is not the global view that is the desired result but the answer to more specific questions. In this case, we already have a certain conception of relevant combinations of attributes, so it is not necessary to scale all attributes. Instead, we use these relevant combinations of attributes to specify a limited terminology by which we can derive the context. This method is called *logical scaling*.

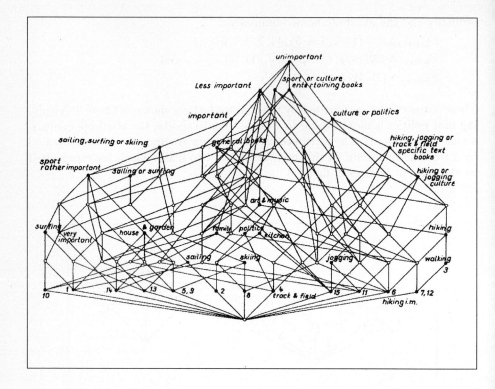

Fig. 4. Concept lattices of the derived context *leisure activities*

3 Contexts derived by terminologies

The basic idea of logical scaling consists of using a formal language to generate unary predicates from the attributes and attribute values of the many-valued context. These predicates form a terminology which determines a derived (one-valued) context.

For example, market researchers are interested in the leisure activities of certain groups of persons which are described by stereotypes of their behavioural patterns. They often examine groups and their stereotypes which are themselves results of data analysis like in [Fo77]. In our example, we define the relevant stereotypes by attributes and attribute values given in the many-valued context *leisure activities*. Then, we analyze how important leisure is for the specified groups. Therefore, the degrees of importance as well as the stereotypes must be described by unary predicates.

The formal language used in this article is based on elements of SQL (structured query language). This is a formal language which is utilized for queries and the administration of relational database management systems as Microsoft Access for example. Using SQL, one can enter the command

leisure unimportant:	(importance ≤ 1)
leisure less important:	(importance ≤ 2)
leisure important:	(importance ≤ 3)
leisure rather important:	(importance ≤ 4)
leisure very important:	(importance ≤ 5)
FAMILY MAN :	(home = fs)
SPORTS FAN :	(events = sp)
INTELLECTUAL:	(events = po $\vee$ cu) $\wedge$ (books $\ni s$)
PRACTITIONER:	(events = sp $\wedge$ home $\neq$ am) $\vee$ (books = {e})

Fig. 5. Terminology *stereotypes*

SELECT Person FROM *activities* WHERE (events = cu AND home = am)

which asks for all persons in the data context *leisure activities* who prefer visiting cultural events and are interested in art and music at home. Here, we write the corresponding predicate in the following way:

$$(\text{events} = \text{cu} \wedge \text{home} = \text{am}).$$

As with many programming languages, SQL is quite intuitive so I refrain from explaining syntax and semantics here (cf. [In91] for details).

For our example, we use SQL to construct the terminology shown in figure 5 where four stereotypes and five degrees of importance of leisure are described. Formally, we define a *terminology* as a tuple $(\mathcal{P}, N, \nu) =: \mathcal{T}$ where $\mathcal{P}$ is a set of unary predicates, N a set of names of attributes and ν a surjective *naming function* $\nu : N \to \mathcal{P}$.

When we *derive* the many-valued context $\mathbb{K} := (G, M, W, I)$ by the terminology $\mathcal{T} := (\mathcal{P}, \widetilde{M}, \nu)$, we obtain the one-valued context $\mathbb{K}^{\mathcal{T}} := (G, N, E)$ where the relation E is given by the semantics of the formal language: for all $g \in G$ and $n \in N$ it holds

$$g \, E \, n : \iff g \text{ satisfies } \nu(n).$$

This derivation can be conducted automatically in every database system using SQL as its query language whereas the choice of the terminology is determined by the research questions.

Figure 6 shows the context derived by our terminology *stereotypes* and the corresponding concept lattice. Looking at the concept lattice, we find interesting information about the dependencies between stereotypes and importance of leisure. For all seven INTELLECTUALS, leisure is not important or less important whereas for the sports fans leisure is much more important. All persons belonging to the stereotype SPORTS FAN consider leisure to be rather or even very important. The FAMILY MEN'S attitudes toward leisure vary.

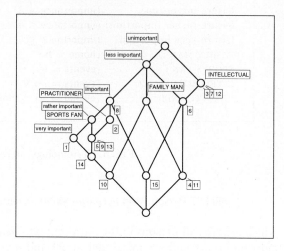

$\mathbb{K}^{\mathcal{T}}$	FAMILY MAN	SPORTS FAN	INTELLECTUAL	PRACTITIONER	leisure unimportant	leisure less important	leisure important	leisure rather important	leisure very important
1				×	×	×	×	×	×
2		×			×	×	×		
3		×					×		
4	×	×				×	×		
5			×	×	×	×	×	×	×
6		×				×	×		
7		×					×		
8					×	×	×		
9		×			×	×	×	×	×
10	×		×	×	×	×	×	×	×
11	×	×				×	×		
12		×					×		
13		×			×	×	×	×	×
14		×	×	×	×	×	×	×	×
15		×			×	×	×		

Fig. 6. Context *leisure activities* derived by the terminology *stereotypes* and corresponding concept lattice

At the same time, this concept lattice does not contain any information about a specific kind of sport. This is because they were considered non-relevant for the expert's, say sociological, theory in use. Thus, the attribute sports was not part of the chosen terminology.

4 Logical scaling based on relational contexts

The language SQL comprises several data types equipped with further structures. For example, for the construction of predicates like (importance ≤ 3) we utilized the fact that SQL offers the data type "integer" equipped with the natural order. Theoretically, it would be possible to define an equivalent predicate using only combinations of $\wedge$ and $\vee$. Nevertheless, we prefer the former since the order is constituent of our natural understanding of the notion "importance" and its values. That is why we want to have relations as elements of our language.

Introducing relations as part of the formalization and the language, we can express explicitly the implicit structures of value sets on which scaling should always be based. Therefore, we represent data in a *relational context* in which all implicitly given relations are formalized explicitly. A more detailed motivation of this notion following the lines of the theory of measurement can be found in [Pre96].

Following [Pri96] we can define a *relational context* formally as a tuple $(G, M, (W, \mathcal{R}), I)$, where (G, M, W, I) is a many-valued context and $\mathcal{R}$ a set of relations on W.

Let us suppose we had specified the information given in [AL83] about the structure of the value sets by relations of our many-valued context *leisure activities*. Then we could have used them to construct predicates and would not have been restricted to the relations provided by SQL.

For example, let us look at the tree-like hierarchy with its five levels which is described by the scale $\mathbb{S}_{sports} := (W_s, M_s, I_s)$. We can describe it by a family $(S_n)_{n=1,...,4}$ of binary relations on the set $W_s \subseteq W$ of sports: two kinds of sports $v, w \in W_s$ are in relation S_n to one another if they belong to the same extension of a concept of level n.

$$(v, w) \in S_n : \iff \exists m \in M_s : v, w \in m^{I_s} \text{ and } |m^{I_s}| = n$$

These relations S_n can be interpreted as similarity of degree n, i. e. as "identical" (S_1), "very similar" (S_2), "rather similar" (S_3), and "similar" (S_4). Providing these relations, we are able to generate the predicate

$$(\text{sports } S_3 \text{ su})$$

which is satisfied by all persons who like a kind of sport which is "rather similar" to surfing, i. e. all persons who are surfing, skiing or sailing. Note that this predicate can be constructed without specifying sailing and skiing explicitly. In SQL, the predicate would be implemented by the following command:

SELECT Person FROM *activities* WHERE sports IN
(SELECT W_{s1} FROM S_3 WHERE $W_{s2} = $ su)

For the construction of predicates, we can also use n-ary relations for $n \geq 2$, in particular operations on values of different attributes. As an example, we imagine an extension of our context *leisure activities* with five many-valued attributes $D_1, D_2, ..., D_5$. They specify, for each person, the disbursements of the person's money for leisure activities in five succeeding weeks. Then, the expert may decide that the importance of leisure should not be measured by the subjective judgement that is given by the many-valued attribute **importance**, but by the average disbursement for leisure activities. Therefore, the expert could describe an attribute called "leisure important" by the predicate below which is satisfied by all persons whose average disbursement for leisure activities is under 15 $ per week:

$$(\frac{1}{5}(D_1 + D_2 + D_3 + D_4 + D_5) \leq 15)$$

This example shows again that the choice of predicates depends on the specific issue of the data analysis. Such a description of "importance of leisure" is certainly determined by the interest of market researchers of finding potential clients with adequate purchasing power and will. Nevertheless, logical scaling always provides the possibility to discuss these decisions because they are listed in the terminology.

5 Discussion

To sum up, formal concept analysis provides different methods to transform many-valued contexts into one-valued contexts from which we can explore conceptual patterns. Both are useful for the treatment of real-world problems.

Conceptual scaling is a well established method which is frequently applied. Supported by the software tools TOSCANA and ANACONDA, the method allows a global view on the conceptual structure of the data context with regard to the inherent structures of the value sets and the research questions.

Logical scaling, on the other hand, is suitable for more specific issues. The expert decides which of the attributes and attribute values of the many-valued context are relevant to his analysis and then declares explicitly how the predicates are to be constructed. This method has several advantages. Firstly, we can construct quite complex predicates by using relations, disjunctions and other elements of our formal language. Secondly, specifying terminologies is more intuitive than defining conceptual scales for people who are not accustomed to work with concept lattices. That is why it might be a great convenience to combine both methods and to use logical scaling for the construction of concrete conceptual scales.

And finally, the explicit declaration of the terminology in view offers the possibility to discuss the decisions at all times because they are always visible. Thus, both scaling methods serve the purpose to support the communication about the data and their conceptual patterns instead of providing "results" automatically.

The choice of the formal language was determined by practical considerations to emphasize the applicability to real-world problems. SQL has a simple structure which is comprehensible without detailed explanations about the syntax and semantics. Additionally, SQL makes it possible to integrate a logical scaling tool into ANACONDA because the program is based on the relational database management system Microsoft Access which has SQL as its query language.

In [Pre96], where the idea of logical scaling was first discussed, we did not use SQL but an attribute logic that is influenced by description logics, a family of knowledge representation languages presented in [SS91]. For this language, questions of algorithmic treating are well investigated.

It might also be interesting to use conceptual graphs and the implicit hierarchy on conceptual graphs to structure terminologies as Sowa hints in [So96]. This may lead to an interesting cooperation between conceptual graphs and formal concept analysis.

References

[AL83]   K. Ambrosi, W. Lauwerth: Ein Klassifikationsverfahren bei qualitativen
            Merkmalen, in: I. Dahlberg, M. R. Schader (ed.): Automatisierung in der
            Klassifikation, Indeks Verlag, Frankfurt 1983, 151 – 160
[Fo77]   H. T. Forst: Anwendung der Cluster-Analyse zur Typisierung des
            Freizeitverhalten von Jugendlichen, in: H. Späth: Fallstudien Cluster-
            Analyse, Oldenbourg Verlag, München Wien 1977, 161 - 178
[GSW86]   B. Ganter, J. Stahl, R. Wille: Conceptual measurement and many-valued
            contexts, in: W. Gaul, M. Schader: Classification as a tool of research,
            Elsevier Science Publishers, North-Holland 1986, 169 - 176
[GW89]   B. Ganter, R. Wille: Conceptual scaling, in: F. Roberts (ed.): Applications
            of combinatorics and graph theory to the biological and social sciences,
            Springer-Verlag, New York 1989, 139 – 167
[GW96]   B. Ganter, R. Wille: Formale Begriffsanalyse. Mathematische Grundlagen,
            Springer-Verlag, Berlin – Heidelberg 1996
[In91]   Informix: Informix Guide to SQL Tutorial, Bohannon Drive 1991
[So92]   J. F. Sowa: Conceptual Graphs Summary, in:T. E. Nagle et. al. (ed.): Con-
            ceptual Structures. Current research and practice, Proceedings, 2nd Inter-
            national Conference on Conceptual Structures, ICCS 1992, 1 – 51
[So96]   J. F. Sowa: Processes and Participants, in: P. W. Eklund / G. Ellis /
            G. Mann (ed.): Conceptual Structures. Knowledge representation as Inter-
            lingua, Proceedings, 4th International Conference on Conceptual Structures,
            ICCS 96, Sydney, Springer, Berlin New York 1996, 1 – 22
[Pre96]   S. Prediger: Symbolische Datenanalyse und ihre begriffsanalytische Einord-
            nung, Staatsexamenarbeit, FB Mathematik, TH Darmstadt 1996
[Pri96]   U. Priß: The formalization of WordNet by methods of relational concept
            analysis, in: C. Fellbaum (ed.): WordNet – An electronic lexical database
            and some of its applications, MIT-Press 1996
[SS91]   M. Schmidt-Schauß, G. Smolka: Attributive concept descriptions with com-
            plements, in: Artificial Intelligence 48 (1991), 1 – 26
[Wi82]   R. Wille: Restructuring lattice theory: an appraoch based on hierarchies of
            concepts, in: I. Rival (ed.): Ordered sets. Reidel, Dordrecht–Boston 1982,
            445 – 470
[Wi92]   R. Wille: Concept Lattices and Conceptual Knowledge Systems, in: Com-
            puters & Math. Applications, vol. 23, no. 5 – 9, 1992

Organization of Knowledge Using Order Factors

Gerard Ellis and Stephen Callaghan

Peirce Holdings International Pty. Ltd. A.C.N. 068-405-872
376 Gore St., Fitzroy, Victoria. 3065, Australia
{ged, stevec}@phi.com.au

Abstract. A central mechanism for the storage and retrieval of conceptual structures is the generalization hierarchy. Improvements on search
in the hierarchy have been made by analyzing the order structure. We
continue this by factorising the order structure into scales. We show how
a membership operation can be improved by scaling of knowledge.
The membership problem is a special case of the insertion problem for
creating hierarchies. We give a $O(k.\log_k n)$ membership algorithm for
a class of orders known as term products that can be generated from
existing term encoding methods.

1 Introduction

Knowledge domains such as natural language semantics, software libraries, mathematics libraries, ontologies and dictionaries stretch the storage and retrieval
capabilities of current knowledge base management systems. The order over
knowledge objects is increasingly being exploited in indexing for these complex
domains. The generalization order over conceptual structures [12] known as the
generalization hierarchy is a key part of conceptual structures theory which is
used for storage and retrieval of conceptual structures. Similarly, subsumption
ordering over logic formulae [2] is used as the basis of description logic systems -
more general formulae subsume more special formulae. Classification algorithms
[10, 9, 11, 2, 5, 6] take a collection of knowledge objects and construct the hierarchy over them. The key to performance of these algorithms is the exploitation
of the order structure to minimize knowledge object comparisons. In [5] it was
illustrated that using lattice embeddings of orders in the form of encodings resulted in improved performance of search for specializations (subsumed objects).
In this paper we show that it is possible to improve further on those methods
for operations such as membership testing   search by using order and lattice
embeddings. Of particular interest are embeddings into orders and lattices which
have properties that allow factorization or scaling that in turn allow divide and
conquer search.

Fig. 1 shows a hierarchy over a collection of program specifications or relation definitions written in conceptual graphs. It is not intended that these
specifications are readable here: they are a reminder that the hierarchical structure has complex objects such as relation specifications on its nodes. Fig. 1
contains binary RelationBetweenSequences, and ternary RelationsBetweenSe-
quencesAndElement where the first and third arguments are sequences and the

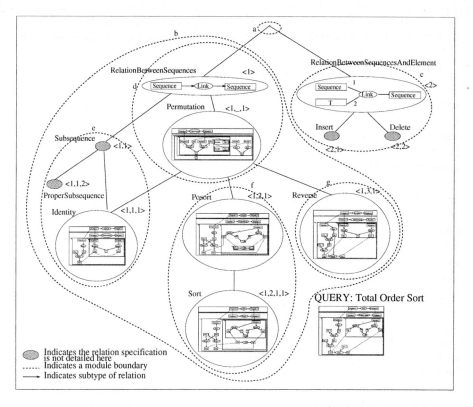

Fig. 1. A hierarchy of conceptual graph program specifications

second argument can be any thing [⊤]. Specializations of RelationsBetween-SquencesAndElement include Insert an element in and Delete an element from a sequence to get a new sequence. Specializations of RelationBetweenSequences include Subsequence and Permutation. Specializations of Subsequence include ProperSubsequence and Identity relations between sequences. Identity is also a specialization of the Permutation relation which also has the Posort (partially ordered sort) and Reverse relations. Posort has a more specialized relation Sort which is totally ordered sorting. These specialization relationships are illustrated by downward edges between the specifications. This example illustrates that the sorts of objects that are intended to be indexed are non-trivial. Further, it is not obvious that traditional techniques such as inverted files, or binary trees are useful for this problem.

A typical problem using these hierarchies is to find the immediate generalizations and immediate specializations of an object. This is known as the classification problem, a term that is also used in description logics for sorting a collection into a hierarchy. For example the only immediate generalization of Posort is Permutation and the only immediate specialization is Sort. More typically it is required for objects that may not already exist in the hierarchy, thus finding the virtual location of the object in the hierarchy. This operation is used in the construction of the hierarchy in an insertion-sort style algorithm [10, 9, 11, 2, 5, 6].

Other problems include finding the specializations of an object not known to be in the hierarchy. A special case of the classification problem operation is membership testing, since an object is a specialization and a generalization of itself. For example, consider the query of Totally Ordered Sort given in Fig. 1, where the problem is to check for membership. The Sort specification currently in the hierarchy is an equivalent specification, so should be returned. There are a number of search techniques that could be used: depth-first, topological and encoded search. In a depth-first searches the immediate specializations of a hierarchy object are compared, if the hierarchy object matches (is a generalization of) the query object. In a topological search a hierarchy object is compared when all of its covering elements have been compared and matched to the query object. Encoded search uses codes similar to those in the labeling of objects in Fig. 1, for example the Sort specification is encoded with $<1,2,1,1>$. These codes reflect the hierarchy structure: the terms in this case are like lists of integers, where underscore '_' indicates unnamed variables. As an example of how the codes are used in search, suppose that during the search it is found that Permutation is a generalization of the query Totally Ordered Sort, then the codes for specializations of Totally Ordered Sort are subsumed by $<1,\_,1>$. Suppose Posort was then found to be a generalization of the query, then the code for the query would be refined to $<1,2,1>$ and specializations (and itself if a member) would have codes subsumed by that code. Further Reverse can now be ignored since its code $<1,3,1>$ does not unify with the current code for the query and hence cannot lead to specializations.

Existing methods are dominated by the need to find the generalizations of the query object. For membership testing or specialization search it would seem that a simple depth-first search would be ideal. In the example in Fig. 1, in the best case, searching from top ⊤, a simple path including RelationBetweenSequences, Permutation, Posort, and Sort would be compared (5 in total). In the worst case, ⊤, RelationBetweenSequencesAndElement (fail), RelationBetweenSequences, Subsequence (fail), Permutation, Reverse (fail), (Identity not compared if failure propagated from Subsequence), Posort, and Sort would be compared (8 in total).

A new search technique based on analysis of the ordering of the objects is illustrated here. The key idea is to embed the order over the database objects into some order or *lattice* with some well-known computable properties. For example, Fig. 2 shows the embedding of the relation hierarchy in Fig. 1 into a *term semilattice* which is the product of smaller semilattices. This embedding could be constructed from existing encoding methods [1, 3, 7]. It is not necessary to have the complete product as shown, but as given in Fig. 1 partial scales can be constructed around the objects shown as dotted lines. Arbitrary embeddings such as this may be found that optimize the density of the embedding, but then inserting a new object into the order is likely to require complete recompilation

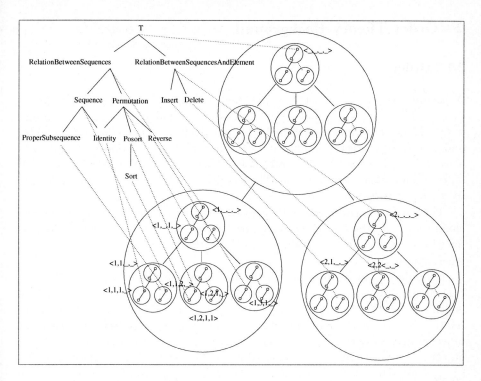

Fig. 2. Embedding the order in Fig. 1 into a term product space

of the embedding. An alternative method is to factorize the order on the basis of the object representations using formal conceptual scales. Ganter and Wille [13] give 14 standardized scales of ordinal type which could be used for this process. Note that scales are not to be confused with modules (subset with only its own top and bottom linked to objects outside subset) described in [1].

Using the new search technique based on this scaling it is possible to more efficiently check for membership and specializations. The new search technique on the above example (shown later) compares only 3 specifications in the worst case and only 6 in the best case. In this example the embedding is fairly sparse, so we expect the search technique to useful for any order that has a dense embedding.

In section 2 the background theory of orders, order products and embeddings is covered, then in section 3 the benefits of a scaled approach to knowledge organization are discussed, with a new algorithm given for the membership problem. This algorithm can be used as the basis of knowledge classification algorithms.

2 Order Theory Background

2.1 Orders

A *partial order* [4] on a set P is a binary relation $\leq$ on P such that, for all $x, y, z \in P$: $x \leq x$; $x \leq y$ and $y \leq x$ implies $x = y$; and $x \leq y$ and $y \leq z$ implies $x \leq z$. These conditions are referred to, respectively, as *reflexivity, antisymmetry,* and *transitivity.* A set P equipped with a partial order relation $\leq$ is said to be a *partially ordered set* or just *ordered set.* The dual of such a relation, denoted $\geq$, is also a partial order relation. The condition $x \leq y$ is read as 'x is subsumed by y'; $x \geq y$ is read as 'x subsumes y'.

The *length* of an ordered set P, denoted $length(P)$, is the length of the longest chain that is a subset of P. (A chain, or *totally ordered set,* C is an ordered set for which $x \leq y$ or $y \leq x$ for all $x, y \in C$, and the *length* of a chain C is $\#C - 1$ where $\#C$ is the size of C).

An object y in an ordered set P *covers* another such object x if $y > x$ (meaning $y \geq x$ and $y \not\leq x$) and there is no $z \in P$ such that $y > z > x$. This is denoted as $y \succ x$. Dually x is covered by y (denoted $x \prec y$). The covering relation is used as the basis for classification. The set of objects that cover an object u in an ordered set P is denoted $\triangledown_P u$. The set of objects covered by an object u in an ordered set P is denoted $\triangle_P u$. The set $\triangledown_P Q$ of a subset Q of P is $\bigcup_{u \in Q} \triangledown_P u$. The set $\triangle_P Q$ of a subset Q of P is $\bigcup_{u \in Q} \triangle_P u$. In all cases we omit the order symbol P from the denotations if it is clear by context which order is intended (e.g. $\triangledown Q$ instead of $\triangledown_P Q$).

A subset Q of P is an *up-set* if $x \in Q$ and $y \in P$ and $y \geq x$ implies $y \in Q$. A subset Q of P is a *down-set* if $x \in Q$ and $y \in P$ and $y \leq x$ implies $y \in Q$. The set $\uparrow_P u$ is u's up-set (or 'up u') in P and consists of all objects in $P \geq u$. The set $\downarrow_P u$ is u's down-set (or 'down u') in P and consists of all objects in $P \leq u$.

The set of maximals of P, denoted $\lceil P \rceil$, is the set of elements in P that have no cover in P. If this set consists of a single element a then a is called the *greatest* element of P. The set of minimals of P, denoted $\lfloor P \rfloor$, is the set of elements in P that have no covered elements in P. If this set consists of a single element a then a is called the *least* element of P.

For any subset S of P an element $x \in P$ is an *upper bound* of S if $s \leq x$ for all $s \in S$. If the set $\{x \in P \mid s \leq x \text{ for all } s \in S \}$ of all upper bounds of S has a least element a then a is called the *join* of S. An element $x \in P$ is a *lower bound* of S if $s \geq x$ for all $s \in S$. If the set $\{x \in P \mid s \geq x \text{ for all } s \in S \}$ of all lower bounds of S has a greatest element a then a is called the *meet* of S.

If an ordered set P has a greatest element this is denoted $\top$ (top). If P has a least element this is denoted $\bot$ (bottom). If P does not have a $\bot$ element it can be *lifted* by adding a $\bot$ element to create an ordered set $P_\bot$ in which $x \leq_{P_\bot} y \iff x \leq_P y \lor x = \bot$. Dually (but not described in [4]), if P does not have a $\top$ element it can be *suspended* by adding a $\top$ element to create an ordered set $P_\top$ in which $x \leq_{P_\top} y \iff x \leq_P y \lor y = \top$. Some of the algorithms described here require that $\top \in P$ and embedding may require that $\bot \in P$.

In Fig. 1, where P is assumed to be this small hierarchy of specifications, $\bigtriangledown$Identity is {Subsequence , Permutation}, and $\triangle$Permutation is {Identity, Posort, Reverse}. $\uparrow$ Identity is {$\top$, RelationBetweenSequences, Subsequence, Permutation, Identity}. $\downarrow$ Permutation is {Permutation, Identity, Posort, Reverse, Sort}.

In description logics [2] the classification problem is determining the subsumption relation $\geq$ over a finite set, given an oracle for determining individual $x \geq y$ relationships. The knowledge classification problem as known in description logics and in conceptual graphs [6] is the problem of finding $\bigtriangledown_P u$ and $\triangle_P u$ for a given query object u so that if u is not already in P, then u may be inserted by adding the appropriate links between u and its cover and covered elements. Typically, $\bigtriangledown_P u$ is computed by finding $\lfloor \uparrow_P u \rfloor = \bigtriangledown_P u$, then searching from that set for $\lceil \downarrow_P u \rceil = \triangle_P u$. For a survey of classification methods see [6].

2.2 Lattices

If for any pair of elements x, y in an ordered set P the join of $\{x, y\}$, denoted $x \vee y$, exists in P then P is a *upper semilattice*. If for any pair of elements x, y in P the meet of $\{x, y\}$, denoted $x \wedge y$, exists in P then P is a *lower semilattice*. Meet and join are defined in section 2.1. If P is both an upper semilattice and a lower semilattice then P is a *lattice*. Let L be a semilattice. An element $u \in L$ is *meet-irreducible* if:

1. $u \neq \top$ (in case L has a top element, which is always the case for finite upper semilattices);
2. $u = v \wedge w$ implies $u = v$ or $u = w$ for all $v, w \in L$.

We denote the set of meet-irreducible elements by $\mathcal{M}(L)$. Let L be a semilattice with a least element $\bot$. Then $u \in L$ is called an *atom* if $\bot \prec u$. The set of atoms of L is denoted by $\mathcal{A}(L)$.

2.3 Order Products

The cartesian product $P = P_1 \times ... \times P_n$ of ordered sets $P_1, ..., P_n$ is itself an order under the relation $\leq_P$ defined by $(x_1, ..., x_n) \leq_P (y_1, ..., y_n)$ if and only if $\forall i (x_i \leq_{P_i} y_i)$. Fig. 3 shows an example of an order product: Fig. 3 (b) is the line diagram of the product of orders shown in Fig. 3(a). Fig. 3(c) is an alternative diagramming convention, known as a *scaled line diagram* [13] for showing the same order product without the need to show all the links explicitly.

In this paper we concentrate on a particular kind of order products called *term semilattices*, of which Fig. 3(c) is an instance. A term lattice is a product of simple term lattices, which are trees of depth 1. We use the notation T_k to represent a simple term lattice consisting of k nodes and k-1 branches. Fig. 3(a) shows the simple term lattice T_3 and T_4. The notation $T_{\langle k_1, k_2, ..., k_n \rangle}$ represents the term semilattice $T_{k_1} \times T_{k_2} \times ... T_{k_n}$. Fig. 3 (c) is $T_{\langle 3, 4 \rangle}$, since it is the product $T_3 \times T_4$. The term semilattice in Fig. 2 is $T_{\langle k_3, k_4, k_3, k_2 \rangle}$. These can be seen as

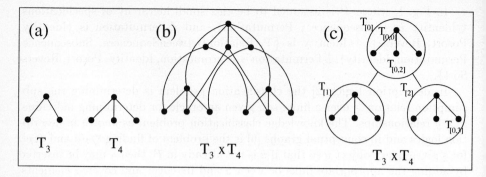

Fig. 3. An Order Product: $T_3 \times T_4 = T_{\langle 3,4 \rangle}$

the signatures of the semilattices, and can be used as type labels for them. For subsemilattices of a term semilattice T we use the notation $T_{[i]}$ for the ith subsemilattice of T, and $T_{[i_1,...,i_n]}$ for the i_nth subsemilattice of subsemilattice $T_{[i_1,...,i_{n-1}]}$.

The new algorithm and proof of correctness applies to term semilattices generally. For complexity analysis we further restrict the focus to term semilattices with a constant branching factor. These are denoted $T_{k,d}$ and defined below. The complexity analysis could be applied to general term lattices using an average branching factor $\frac{k_1+...+k_n}{n}$.

Definition 2.1 *The term semilattice $T_{k,d} = (Nodes(k,d), Edges(k,d))$ is recursively defined as follows:*

1. *$T_{k,0}$ is the graph with a single node and no edges; and*
2. *$T_{k,d}$ is constructed by taking k copies, $G_0, \ldots, G_{k-1}$, of $T_{k,d-1}$ and connecting each node of G_0 to the corresponding node of each of $G_1, \ldots, G_{k-1}$: each node of G_0 covers each of its corresponding nodes in $G_1, \ldots, G_{k-1}$.*

Fig. 5 shows the semilattice $T_{3,3}$. Note that $T_{k,d}$ is the same as $T_{\langle k_1,k_2,...,k_d \rangle}$ where each $k_i = k$. Also $length(T_{k,d}) = d$, $length(T_{\langle k_1,k_2,...,k_n \rangle}) = n$, $length(T_{[i]})$ $= length(T) - 1$ and $length(T_{[i_1,...,i_n]}) = length(T_{[i_1,...,i_{n-1}]}) - 1$. Also in T_k and $T_{\langle k_1,k_2,...,k_n,k \rangle}$, $k-1$ represents the branching factor, denoted by $branch(T)$. Note that $branch(T_k) = \#T_k - 1 = k - 1$. Wherever required we lift a term semilattice T to produce a term lattice $T_\perp$.

Proposition 2.2 *The term semilattice $T_{k,d}$ has k^d nodes, $\frac{d(k-1)}{k}.k^d$ edges and $(k - 1)d$ meet-irreducibles. The term lattice $T_{k,d_\perp}$ has $(k - 1)^d$ atoms and 2^d subsumers for each atom.*

Proof: From the recursive definition of $T_{k,d}$ the recurrence relation for the number of nodes is $\#Nodes(k,d) = k.\#Nodes(k,d-1)$ and so $\#Nodes(k,d) = k^d$. Further, the recurrence relation for the number of edges is $\#Edges(k,d) = k.\#Edges(k,d-1) + (k-1).\#Nodes(k,d-1)$ and so $\#Edges(k,d) = \frac{d(k-1)}{k}.k^d$.

Atoms have no subsumees in $T_{k,d}$, so the atoms of $T_{k,d}$ are the atoms of $G_1, \ldots, G_{k-1}$, and hence the recurrence relation for the number of atoms is $\#\mathcal{A}(T_{k,d}) = (k-1).\#\mathcal{A}(T_{k,d-1})$ and so $\#\mathcal{A}(T_{k,d}) = (k-1)^d$.

For every subsumer of an atom in G_i where $i = 1, \ldots, k-1$, there is a subsumer in G_0, hence the recurrence relation for the number of subsumers of an atom is $\#Subsumers(k, d) = 2.\#Subsumers(k, d-1)$ and so $\#Subsumers(k, d) = 2^d$.

The top nodes of $G_1, \ldots, G_{k-1}$ have one covering element (are meet-irreducibles), the top node of G_0, and the rest of the meet-irreducibles are in G_0. Hence the recurrence relation for the number of meet-irreducibles is $\#\mathcal{M}(T_{k,d}) = (k-1) + \#\mathcal{M}(T_{k,d-1})$ and so $\#\mathcal{M}(T_{k,d}) = (k-1).d$. $\square$

These volumes set bounds on the corresponding volumes for general term semilattices $T_{\langle k_1, \ldots, k_n \rangle}$ provided that we choose $k = max(k_1, \ldots, k_n)$, or we could apply the volumes using the average branching factor $k = \frac{k_1 + \ldots + k_n}{n}$.

2.4 Maps Between Orders

As Fig. 2 illustrates a key concept to the methods discussed here is a mapping from the knowledge hierarchy into some order or lattice, in this case the relation hierarchy on the left is embedded into the term semilattice $T_{\langle 3,4,3,2 \rangle}$. We now define some of the possible mappings (functions in particular) between orders.

Definition 2.3 *Let P and Q be ordered sets. A map $\varphi : P \to Q$ is said to be*

- order-preserving *if $x \leq y$ in P implies $\varphi(x) \leq \varphi(y)$ in Q.*
- an order-embedding *if $x \leq y$ in P if and only if $\varphi(x) \leq \varphi(y)$ in Q.*
- an order-isomorphism *if it is an order-embedding mapping P onto Q.*

An order-embedding is an injective order-preserving function. The embedding of a knowledge hierarchy P into a term semilattice T will not usually be an isomorphism i.e. $T - \varphi(P) \neq \emptyset$. Assume that P includes a bottom element $\perp_P$ (by lifting if necessary). We can extend the function φ to a relation φ' that also maps $\perp_P$ to $\perp_{T_\perp}$ and to all the elements of T that are not an image $\varphi(x)$ for some $x \in P$ i.e. $\varphi' : P \to T_\perp = \varphi \cup (\{\perp_P\} \times (T_\perp - \varphi(P)))$. Then $\varphi'^{-1} : T_\perp \to P$ is a total order-preserving function of $T_\perp$ onto P. Also $\varphi'^{-1} - \{(x, y) : \varphi'^{-1} \mid x \notin \varphi(P)\}$ is the same as φ^{-1} and is injective. So we can search the embedding term lattice without taking any special action for nodes that are not images of the original order P under the original embedding φ: the reverse function φ'^{-1} simply maps these nodes to $\perp_P$. Once we have found an element $x \in T_\perp$ we can immediately find the correspond element $\varphi'^{-1}(x) \in P$. Note that this extended embedding is conceptual: its purpose is rigour in proving the correctness of the new algorithms; we do not propose to store and navigate whole term semilattices as part of the implementation.

3 Classification Techniques based on Embeddings into Factored Semilattices

Classification algorithms [10, 9, 11, 2, 5, 6] in the worst case are dominated by their reliance on computation of $\bigtriangledown_P u$ and are in the worst case related to the size of $(\triangle_P \uparrow_P u) \cup (\uparrow_P u)$. This is because they search from $\top$ or $\lceil P \rceil$ searching for subsuming elements v of the object u being classified, searching the covered set $\triangle_P v$ for further subsuming elements, so most or all elements of $\uparrow_P u$, and further the covered sets of these elements, are typically compared to u. Meet irreducibles are the basis of information in an order from which the elements are joined to form patterns. Our belief is that instead of comparing all subsuming elements of u and their covered elements, $(\triangle_P \uparrow_P u) \cup (\uparrow_P u)$, it should be possible to compare only a small subset, namely $\mathcal{M}((\triangle_P \uparrow_P u) \cup (\uparrow_P u))$ or even a subset of that. This reflects the idea that $\uparrow_P u$ could include all possible generalizations of u and hence all possible ways of constructing u, whereas $\mathcal{M}((\triangle_P \uparrow_P u) \cup (\uparrow_P u))$ contains only those attributes that are used to construct u, or at least a sample of how those components are used in possible non-redundant construction steps to u. Hence by knowing all or most steps to construct u and their position in the order, it is possible to place u in the order. The number of generalizations of a conceptual graph, even for restricted atomic graphs, is in the worst case exponential in the size of the graph, whereas the number of attributes (relations and concepts) are linear in the size of the graph. So it should be possible to index by comparing a linear number of simple attributes rather than an exponential number of complex graphs.

In the following sections the focus is on the simpler membership problem $u \in P$ in order to gain insights into the harder problem of classification. An order can be seen as an address space, and classification can be seen as an addressing problem, where P is a (partial) address space and u is a partial address. The membership problem treats u as a complete address, where unaddressable space is treated as failure to decode the address.

3.1 Member Search: $u \in P$

Ellis [6] gives an algorithm for testing whether $u \in P$ which uses scales on $T_{k,d}$ and which does $O(k. \log_k n)$ comparisons, compared with $O(2^{\log_k n})$ using classification methods [6] and $O(k. \log_k n^2)$ visits using a naive depth-first path search (such as that given in Fig. 4). But this incurs an additional cost of $O(k. \log_k n)$ [6] for propagation of matching information.

The algorithm *member_path* repeated in Fig. 4 tests if $u \in P$ (that is, if there is a path in P from some known subsumer v to u). The element $\top$ (line 5) is a known subsumer of any v, but if other closer subsumers of u are known they can be used. The algorithm selects a subsumer w of u from the covered elements of v, then tests if there is a path from w to u. A path $w_1 \geq w_2 > \ldots \geq w_n$ is constructed where $w_1 = v$ and $w_n = u$ if u is in P.

Consider the search for u ($<2,2,2>$) in the left hierarchy in Fig. 5 using the *member_path* algorithm. The search order is $\top = <\_,\_,\_>$ (assumed subsumer), <1,_,_> (not subsumer), <_,1,_> (not subsumer), <_,_,1> (not subsumer), <_,_,2> (subsumer), <1,_,2> (not compared, known non-subsumer), <_,1,2> (not compared, known non-subsumer), <_,2,2> (subsumer), <1,2,2> (not compared, known non-subsumer), <2,2,2> (subsumer, actually u). Nodes with filled circles like <_,2,2> are subsumers of the query u. Nodes with filled boxes like <1,_,_> are compared and found to be non-subsumers of u. Nodes with non-filled boxes like <1,_,2> are known non-subsumer objects which were checked but not compared if non-subsumer information is propagated. We prove the correctness of the algorithm by first proving the main auxiliary function *member_path* is correct.

```
1   function member_path(u : O, P : ℙO) returns boolean;
2   preconditions ⊤ ∈ P (note that an order Q without ⊤ can
3       can be converted to an order P by adding ⊤)
4   begin
5       v ← ⊤;
7       while ∃w ∈ △ₚv • w ≥ u do
8           v ← w;
9       return u ≥ v;
10  end
```

Fig. 4. A depth-first algorithm for checking a node is a member of a poset

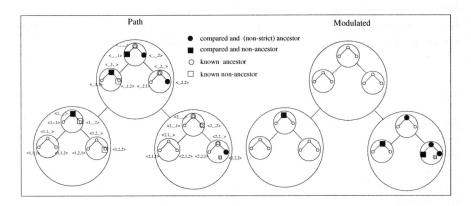

Fig. 5. Comparison of path and scaled search for a node in a term lattice $T_{3,3}$

Proposition 3.1 *Given* $⊤ \in{\uparrow}_P u$*, the member_path algorithm returns the truth value of* $u \in P$*.*

Proof: First we show that $v \in{\uparrow}_P u$ is invariant.

Initially, this is a requirement of $\top$ so we need only show $v' \in \uparrow_P u$ at the end of each iteration. After each iteration, if the while condition is true, $w \in \Delta_P v \wedge w \in \uparrow_P u \wedge v' = w$ and so $v' \in \uparrow_P u$ as required.

If the program terminates, then $(\Delta_P v \cap \uparrow_P u) = \emptyset$ and $v \in \uparrow_P u$. There are two cases:

(i) $u \geq v$: Thus $u = v$ by the antisymmetric property of $\geq$. Further $u \in P$ and the program returns TRUE.

(ii) $u \not\geq v$: Hence $v > u$. Assume $u \in P$. Then $\exists w : P \bullet w \in \Delta_P v \wedge w \in \uparrow_P u$ resulting in a contradiction. Hence $u \notin P$ and FALSE is returned. Hence $member\_path$ is partially correct.

We now show termination. Initially $v \in \uparrow_P u$ and after each iteration $v' \in \Delta_P v$. Since P is finite and acyclic $\uparrow_P u \cap \Delta_P v$ must eventually be empty and hence $member\_path$ terminates. Thus $member\_path$ is totally correct. $\square$

We now analyze the number of objects visited.

Proposition 3.2 *The algorithm member_path visits* $O(k \log_k^2 n)$ *objects in a term lattice* $T_{k,d}$.

Proof: The algorithm is a depth-first search. In the worst case all elements of $\Delta_P v$ are compared before a continuation of the path is selected, and the algorithm may have to search to the bottom of the hierarchy. In the worst case u is subsumed by an atom, so the search is in the worst case $O(average(branch(P)) \times length(P))$ where $O(average(branch(P)))$ is $\frac{\#Edges(P)}{\#Nodes(P)}$. The length (depth) of a term semilattice $T_{k,d}$ is $d = \log_k n$, and at worst there are $(k-1) \cdot \log_k n$ covered elements of an element. Therefore at worst $O(k \log_k^2 n)$ objects are visited. $\square$

3.2 Factored Search: $u \in P$

As indicated in the introduction in section 1 it is possible to more efficiently search an order of complex objects such as a program specification hierarchy, by embedding an order into a semilattice that has a recognizable factorization into a product of smaller simpler orders or lattices. A divide and conquer search on these suborders (or scales) can result in significant reductions in the number of comparisons made.

Fig. 6 gives a factored path algorithm that searches orders, embedded into term semilattices, for a query u, but also indicates the form of a more general algorithm. φ' is the order embedding extended to a relation as described in section 2.4 and the θ is the total function φ'^{-1}. The goal is to find one of the most specific elements of that scale that subsumes u, since this also implies that tops of more general elements of the scale also subsume u. The idea is to get as deep as possible into the hierarchy as fast as possible. The search finds a minimal scale containing an object that subsumes the query. It is called initially for the whole semilattice, and subsequently for a selected scale within the 'current' scale. Line 10 selects first the meet-irreducibles for the 'current' scale: if the top element of none of these subsumes the query then the top element of the current scale is selected. If a scale is found whose top element subsumes the query then the 'while loop' part of the algorithm is reiterated for that scale.

Fig. 5 shows a comparison of a depth-first search and factored search for the element $<2,2,2>$ in a term lattice $T_{3,3}$, where there is an embedding of an order P into this lattice. We focus on the term lattice in the discussion since the extended embedding φ' is invertible (section 2.4). The simplest example of such an embedding is an automorphism $\varphi : P \to P$ where the knowledge hierarchy P is already a term semilattice. The depth-first search visits the nodes $<\_,\_,\_>$ ($\top$), $<1,\_,\_>$, $<\_,1,\_>$, $<\_,\_,1>$, $<\_,\_,2>$, $<\_,2,2>$, $<2,2,2>$.

Now we show how the *factored_path* algorithm performs on the example in Fig. 2 again searching to see if Totally Ordered Sort is a member of the hierarchy. In this search the goal is to find the most specific scale that has as its local top element a generalization of the query, then to search that scale etc. until a path to a most specific scale is found then test for membership or specializations of the query in that scale. There are three scales at the outer level, a, b, and c. Their tops are $\top$, RelationBetweenSequences and RelationBetweenSequencesAndElement. So RelationBetweenSequences can be compared since it is one of the most specific scale tops. Then the scale b is searched. The scales e, f, and g are the most specific scales. The top of scale f is Posort which is a generalization of the query. The scale f is then searched. The most specific scale here contains Sort which is also a generalization of the query. Hence the scale containing Totally Ordered Sort is found. This is the best case of using this method for this search example - only 3 specifications are compared to the query. In the worst case for this search example, RelationBetweenSequencesAndElement (fail), RelationBetweenSequences, Subsequence (fail), Reverse (fail), Posort and Sort are compared (a total of 6). This compares with the more naive depth-first algorithm which compared 5 (best case) and 8 (worst case) specifications respectively on this example - an improvement even for such a sparse embedding. The interesting part of this search is that *top* and Permutation are skipped. Also the branching factor in these scales is reduced and hence the search at each level.

Proposition 3.3 *The* factored_path *algorithm,* $factored\_path(u, T, \theta)$, *where* $\theta = \varphi'^{-1}$, *returns* $\emptyset$ *if there is no object* $s \in T$ *for which* $\varphi'^{-1}(s)$ *subsumes* u, *otherwise it returns a minimal element* $s \in T$ *such that* $\varphi'^{-1}(s)$ *subsumes* u.

Proof: If u is not subsumed by any object in P then $u \not\leq \top_P$ which implies that $u \not\leq \varphi'^{-1}(\top_T)$ so the algorithm returns $\emptyset$ and exits (line 6). Otherwise u is subsumed by some $v \in P$, which implies that $u \leq \top_P = \varphi'^{-1}(\top_T)$ so line 6 initializes S to T and the algorithm executes the while loop.

To prove correctness of the while loop we first show that the predicate $\varphi'^{-1}(\top_S) \geq u$ is an invariant of the loop. Initially $S = T$, so $\varphi'^{-1}(\top_S) = \varphi'^{-1}(\top_T) \geq u$ from line 6 proving that the predicate holds at the start of the loop. Now suppose the predicate holds before an iteration of the loop. After

```
1   function factored_path(u : O, T : TL, θ : TL → O)  returns ℙO;
2   // TL is the set of all term lattices and θ is the total function φ'⁻¹
3   begin
4        if θ(⊤_T) ⋡ u then  return ∅;
5        else
6             S ← T;
7             while length(S) ≠ 0 do
8             begin
9                  Boolean path ← FALSE;
10                     for each i = 1, ..., branch(S) − 1, 0 ∧ while¬path do
11                          if θ(⊤_{S_{[i]}}) ≥ u then
12                               path ← TRUE;
13                               S ← S_{[i]};
14                          endif
15               end
16               return θ(S);
17        endif
18  end
```

Fig. 6. An algorithm for searching for a node using factored path search

the iteration $S' = S_{[i]}$ for some $i \in [1, ..., branch(S) - 1, 0]$, otherwise $\varphi'^{-1}(\top_{S_{[0]}})$ would not subsume u, contradicting the induction hypothesis that $\varphi'^{-1}(\top_S) \geq u$. Also the order of subsumption testing (line 10) ensures that if $\varphi'^{-1}(\top_{S_{[i]}}) \geq u$ for some $i \neq 0$ then S' is set to $S_{[i]}$, not to $S_{[0]}$. Now $\varphi'^{-1}(\top_{S_{[i]}}) \geq u$ (line 11), so $\varphi'^{-1}(\top_{S'}) \geq u$, proving that the predicate holds after the iteration.

If the loop terminates then $length(S) = 0$ implying that S (an instance of $T_{k,0}$) consists of a single object s. $\varphi'^{-1}(s) = \varphi'^{-1}(\top_S) \geq u$ so $\varphi'^1(s) \geq u$. Now suppose that $\varphi'^{-1}(s)$ is not a minimal subsumer of u i.e. there is some $v \in T$ where $v \geq u$ and $\varphi'^{-1}(s) > v$. Let Sm be the minimal sublattice containing both s and v. Then $\top_{Sm[i]} \geq s$ and $\top_{Sm[j]} \geq v$ for some $i, j \in \{0..Branch(Sm) - 1\}$. But $s > v$, so $\top_{Sm[i]} \geq v$, which implies that $\top_{Sm[i]}$ subsumes both s and v, contradicting the hypothesis that Sm is the minimal semilattice subsuming both s and v. So if the loop terminates it returns $\varphi'^{-1}(S)$ where S is a semilattice of length zero containing a single object s where $\varphi'^{-1}(s)$ is a minimal subsumer of u, proving that the loop is partially correct.

At the start of the loop $length(S) = length(T)$ which is finite, and reduces by 1 on each loop iteration. So eventually $length(S)$ must reach zero and the loop terminates, proving that the algorithm is totally correct. □

Fig. 7 gives a membership algorithm that calls the factored path algorithm then tests whether a (single) minimal subsumer of the query u is returned and whether this is subsumed by u. If so then u is a member (or is equivalent to a member) of P.

Proposition 3.4 *The number of objects compared in a factored_path search of a term lattice $T_{k,d}$ of n nodes where $n = k^d$ is $O(k \log_k n)$.*

```
1  function member(u : O, T : TL, θ : TL → O);  returns Boolean;
2  begin
3      S : TL;
4      S ← factored_path(u, T);
5      return (S ≠ ∅ ∧ u ≥ θ(⊤_S);
6  end
```

Fig. 7. An algorithm for searching for a node using factored path search

Proof: On each iteration the top elements of all of the k subscales of scale S (which is initially the same as T) will in the worst case be tested for subsumption $\geq$ of the query u. Each scale $S_{[i]}$ contains a kth of the nodes in scale S. At most one of the subscales $S_{[i]}$ is iteratively searched. Thus the recurrence relation for the number of nodes compared in $factored\_path$ is $\#NodesCompared(n) = \#NodesCompared(n/k)+k$. We prove by induction that $\#NodesCompared(n) = k\log_k n$. For $d = 1$ $n = k$ so $\#NodesCompared(n) = n = k = k\log_k n$. For $d > 1$ we assume $\#NodesCompared(n/k) = \log_k \frac{n}{k}$ and substitute this into the recurrence to yield $\#NodesCompared(n) = k(\log_k \frac{n}{k}) + k = k(\log_k n - 1) + k = k\log_k n$. □

The complexity analysis for more general term lattices $T_{\langle k_1,...,k_n \rangle}$ is not so straightforward. We would expect the complexity in this case to be $O(k\log_k n)$ where $k = \frac{k_1+...+k_n}{n}$; in any case it is bounded by $O(k\log_k n)$ where $k = max(k_1,...,k_n)$.

The $factored\_path$ algorithm at least improves on the depth-first $member\_path$ algorithm in terms of nodes visited. In the simple algorithm $member\_path$ this would also translate into a reduction in the number of objects compared. As noted before $member\_path$ can be improved by propagating non-subsumer information. The overhead involved will be significant, whereas the $factored\_path$ avoids most of this overhead at run-time by compiling this information into the program and the recorded structure in φ (the scale structure of the order P).

4 Summary

Order and lattice theoretic analysis of knowledge bases is proving fruitful. Here we have shown how embeddings of generalization hierarchies of conceptual structures into simpler term semilattices can improve retrieval of knowledge from these structures. The real benefit of these techniques will only be evaluated from the implementation which is underway, but we expect they will lead to further scalability results.

Acknowledgments

The authors wish to thank The Defense Science Technology Organization of Australia who have supported this research in a Peirce Conceptual Graph Database research contract, as well as University of Queensland and Royal Melbourne Institute of Technology, who have supported the research of the first author. RMIT also supports the second author with a research scholarship.

References

1. Hassan Aït-Kaci, Robert Boyer, Patrick Lincoln, and Roger Nasr. Efficient implementation of lattice operations. *ACM Transactions on Programming Languages and Systems*, 11(1):115–146, January 1989.
2. Franz Baader, Bernhard Hollunder, Bernhard Nebel, Hans-Jürgen Profitlich, and Enrico Franconi. An empirical analysis of optimization techniques for terminological representation systems. In B. Nebel, C. Rich, and W. Swartout, editors, *Proceedings of the 3rd International Conference on Principles of Knowledge Representation and Reasoning*, pages 270–281, Cambridge, MA, USA, October 1992. Morgan Kaufmann.
3. Yves Caseau. Efficient handling of multiple inheritance hierarchies. In *Proceedings of OOPSLA'93*, pages 271–287, Washington, DC, USA, September 1993.
4. B.A. Davey and H.A. Priestley. *Introduction to Lattices and Order*. Cambridge University Press, Cambridge [England] New York, 1990.
5. Gerard Ellis. Efficient retrieval from hierarchies of objects using lattice operations. In Guy W. Mineau, Bernard Moulin, and John F. Sowa, editors, *Conceptual Graphs for Knowledge Representation*, number 699 in Lecture Notes in Artificial Intelligence, pages 274–293, Berlin, 1993. Springer-Verlag.
6. Gerard Ellis. *Managing Complex Objects*. PhD thesis, Computer Science Department, The University of Queensland, 4072, Queensland, Australia, February 1995.
7. Andrew Fall. Heterogeneous encoding. In *Proceedings of International KRUSE'95 Conference : Knowledge Use, Retrieval and Storage for Efficiency*, pages 162–167. University of California, Santa Cruz, August 11-13 1995.
8. Michel Habib, Marianne Huchard, and Lhouari Nourine. Embedding partially ordered sets into chain-products. In *Proceedings of International KRUSE'95 Conference : Knowledge Use, Retrieval and Storage for Efficiency*, pages 147–161. University of California, Santa Cruz, August 11-13 1995.
9. Robert A. Levinson. A self-organizing retrieval system for graphs. In *The 3rd National Conference of the American Association for Artificial Intelligence*, pages 203–206, Austin, Texas, 1984. AAAI Press, Menlo Park.
10. Thomas A. Lipkis. A KL-ONE classifier. In James G. Schmolze and Ronald J. Brachman, editors, *1981 KL-ONE Workshop*, pages 128–145, Cambridge, MA, June 1982. Published as Bolt Beranek and Newmann Inc. BBN Report No. 4842 and Fairchild Technical Report No. 618.
11. Robert MacGregor. A deductive pattern matcher. In *The 7th National Conference of the American Association for Artificial Intelligence*, pages 403–408, Saint Paul, MI, 1988. AAAI Press, Menlo Park.
12. John F. Sowa. *Conceptual Structures: Information Processing in Mind and Machine*. Addison-Wesley, Reading, MA, 1984.
13. Bernhard Ganter & Rudolf Wille. Conceptual scaling. Technical Report 1174, Technische Hochschule Darmstadt, D-6100 Darmstadt, Schlossgartenstrasse 7, 1988.

C. S. Peirce and the Quest for Gamma Graphs

Peter Øhrstrøm
Department of Communication
Aalborg University, Langagervej 8
9220 Aalborg Øst, Denmark

Abstract: This paper deals with some aspects of the history of C. S. Peirce's Existential Graphs. In his construction of this graphical method during 1896-1897 Peirce was motivated by some interesting considerations regarding diagrammatical reasoning. In the present paper this motivation will be briefly discussed. Whereas Peirce managed to bring the graphical systems of Alpha and Beta Graphs to a high degree of perfection, his treatment of the Gamma Graphs remained tentative and unfinished. Some of his suggestions can also be shown to be mistaken. It is, however, clear that Peirce with his Gamma Graphs was aiming at a complicated system in which one can deal with a number of interesting problems regarding various kinds of modality. Peirce, himself, was well aware of the shortcomings of his treatment of the Gamma Graphs, and he mainly concentrated on the formulation of a Gamma agenda for his followers.

1. Introduction

As Mary Keeler and Christian Kloesel [7] have pointed out Charles Sanders Peirce struggled for over twenty years with his system of graphical logic called Existential Graphs. The question of dating the invention is nevertheless interesting. At least, Peirce himself paid considerable attention to it. He stated that he had invented his Existential Graphs in January 1897, although he did not publish these new ideas until October 1906 [CP 4.618] in The Monist. It should be mentioned that this exact dating of the invention of Existential Graphs is a bit questionable. In a lecture on "The Logic of Relatives" given on February 17, 1898, he stated: "Finally about two years ago, I developed two intimately connected graphical methods which I call Entitative and Existential Graphs." [10, p.151] Given that the former statement is from a paper published in 1908, it seems reasonable to see the latter statement as the more reliable in the sense that he probably was working with the problems already in 1896 (and perhaps even earlier). One very likely scenario appears to be as follows: During 1896 Peirce was very interested in diagrammatical reasoning, which led him to the formulation of the basic ideas involved in Existential Graphs. A first version of the new theory may have been ready by January 1897. Actually, he wanted the system to be called "the Existential System of 1897", since he in this year wrote an account of the system and offered it for publication to the editor of The Monist, who nevertheless did not want to publish it in the form in which it appeared at that time [CP 4.422].

It is obvious that the invention of the Existential Graphs can be seen as a natural continuation of Peirce's work with Venn Diagrams and Euler Circles. His interesting improvements of these classical methods have been carefully studied by

Eric Hammer, who has convincingly emphasised the importance of the fact that Peirce provided "syntactic diagram-to-digram rules of transformation for reasoning with diagrams" [4]. It is likely that it was these efforts which made him aware of the great power of diagrammatical reasoning.

Peirce had earlier worked intensively with the establishment of an algebraic approach to logic. But during 1897 he came to prefer the diagrammatical approach as clearly superior when compared with the algebraic approach to logic [CP 3.456]. He later came to consider diagrammatical reasoning as "the only really fertile reasoning", from which not only logic but every science could benefit [CP 4.571]. According to Peirce the use of diagrams in logic can be compared with the use of experiments in chemistry. Just as experimentation in chemistry can be described as "the putting of questions to Nature", the experiments upon diagrams may be understood as "questions put to the Nature of the relations concerned". [CP 4.530] In this way diagrammatical reasoning may be seen as some sort of game. In fact, as Robert W. Burch [3] has argued, Peirce regarded the system of Existential Graphs as inseparable from a rather game-like activity which is carried out by two fictitious persons, the Graphist and the Grapheus. The two persons are very different. The Graphist is an ordinary logician, and Grapheus is the creator of the universe of discourse, who also makes continuous additions to it from time to time [CP 4.431].

Working with his "Application to the Carnegie Institution" for support for his research in logic (dated July 15, 1902) Peirce established the following interesting defintion of diagrammatical reasoning:

> By diagrammatic reasoning, I mean reasoning which constructs a diagram according to a precept expressed in general terms, performs experiments upon this diagram, notes their results, assures itself that similar experiments performed upon any diagram constructed according to the same precept would have same results, and expresses this in general terms. This was a discovery of no little importance, showing, as it does, that all knowledge without exception comes from observation. [From Draft C (90-102)]

In the same draft Peirce maintained that "all necessary reasoning is diagrammatic". However, Peirce was not quite clear on the question of generality of his diagrammatical approach. In a letter to Lady Welby, dated March 9, 1906, Peirce made clear that there is some limitation to the system of Existential Graphs. He found it hard to see how, for instance, a piece of music or a command from a military officer could be represented in terms of graphs. On the other hand, in the introducing statement in his Monist paper (1906), he was very firm on the question of generality:

> Come on, my Reader, and let us construct a diagram to illustrate the general course of thought; I mean a system of diagrammatization by means of which any course of thought can be represented with exactitude. [CP 4.530]

One possible way of explaining this tension is that although Peirce felt sure that all kinds of human reasoning can be represented in terms of Existential Graphs, he understood that no representation can be perfect or complete. It cannot "directly exhibit all the dimensions of its object, be this physical or psychic." [MS 291; 1905] Every representation will show its object only in a certain light, i.e. from a certain perspective. For this reason, Peirce maintained that no sign can be "perfectly determinate" [CP 4.583]. But on the other hand he stressed that the system of Existential Graphs is "a rough and generalized diagram of the Mind" [CP 4.582].

For the understanding of the Peircean position the notion of a diagram obviously becomes fundamental. According to Peirce a diagram should mainly be understood as "an Icon of intelligible relations" [CP 4.531]. It is, however, very interesting that Peirce appears to have related diagrams of Existential Graphs to the passage of time. He pointed out that Existential Graphs can represent "propositions, on a single sheet, and arguments on a succession of sheets, presented in temporal succession" [14, p.662], and that the system of Existential Graphs may "be characterized with great truth as presenting before our eyes a moving picture of thought." [MS 291; 1905]

Roberta Kevelson has argued that Peirce's reference to time plays an important rôle as "intensification in his explanation of modality in the Existential Graphs" [8, p.102]. This is true for all Existential Graphs. As Mary Keeler [6] has argued, Peirce's Existential Graphs can naturally be viewed as an instrument for investigating semiotic continuity. Like Robert W. Burch's analysis [3] of Existential Graphs in terms of game-theoretical semantics, Mary Keeler's historical investigations emphasise that the Existential Graphs should be understood in relation to time, human experience and communication. These temporal aspects are particularly relevant for the Gamma Graphs, which can in many cases be interpreted in terms of temporal logic. After all, in the Peircean context time should be viewed as one of the most important sorts of modality.

2. The Rules of the Gamma Graphs

Peirce's so-called Alpha graphs correspond to the ordinary propositional calculus, whereas his Beta graphs correspond to first-order predicate calculus. In what he called 'The Gamma Part of Existential Graphs' [CP 4.510 ff.], he put forth some interesting suggestions regarding modal logic. He wanted to apply his logical graphs to modality in general - that is, to use them for representing any kind of modality. However, he was aware of the great complexity in which a full-fledged logic involving modal and temporal modifications would result. This is probably one of the reasons why Peirce's presentations of the Gamma graphs remained tentative and unfinished. In the following I intend to explain some of the problems he was facing and suggest some ideas regarding the possible continuation of his project.

According to Don. D. Roberts, Peirce began working with the Gamma Graphs already in 1898, and both in 1903 and 1906 he dealt intensively with them [14]. As Peirce himself pointed out, the system of Gamma Graphs is "characterized by a great wealth of new signs; but it has no sign of an essentially different kind from those of the alpha and beta part." [CP 4.512] The most important new graphical elements in the Gamma Graphs are the 'broken cuts' and 'tinctures' corresponding to various kinds of modality. However, as long as only one kind of modality is involved we can do with just one new graphical element, the broken cut. This is in fact what Peirce himself did in the main parts of his "Apology for Pragmaticism" in the Monist (1906). Here he presented a logic based on four rules (or "Permissions") for Gamma Graphs, that involve exactly one kind of modality (corresponding to one tincture). In the following I shall discuss these Gamma rules.

Peirce considered a "Phemic Sheet", which refers to a universe of discourse. The two sides are recto and verso, respectively. Graphs on the recto are posited affirmatively and graphs that are "negatived" are scribed on the verso [CP 4.555]. It is important to note that writing something on the verso is not only a matter of negating the graph in question, it may also involve some kind of modality corresponding to one of the tinctures. Peirce did not specify what kind of modality he had in mind in his formulation of the rules. I shall use readings like "it is possible that not p" and "it must be that p" corresponding to the modal expressions M~p and Np and corresponding to the following graphs:

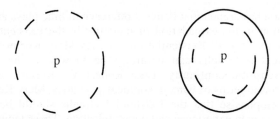

Peirce formulated the first two permissions (rules) for Gamma Graphs in this way:

> *The rule of Deletion and Insertion: Any Graph-Instance can be deleted from any recto Area (including the serving of any Line of Identity), and any Graph-Instance can be inserted in any verso Area (including as a Graph-Instance the juncture of any two Lines of Identity or Points of Teridentity). [CP 4.565]*
> *The rule of Iteration and Deiteration: Any Graph scribed on any Area may be iterated in or (if already Iterated) may be Deiterated by a deletion from that Area or from any other Area included within that. This involves the Permission to distort a line of Identity, at will. [CP 4.566]*

Peirce argued that the rule of Deletion and Insertion is evident, and that the rule of Iteration and Deiteration "will be seen instantly by students of any form of Logical Algebra" [CP 4.566]. As an illustration Peirce showed how one can deduce the proposition "Every catholic must adore some woman" from the proposition "There is a woman, whom every catholic must adore". His deduction is carried out as follows: the premise corresponds to this diagram:

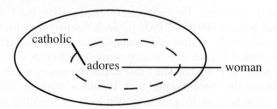

Using iteration one may deduce:

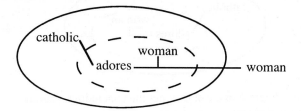

According to Peirce, this may by deletion be transformed into the diagram, which is to be proved:

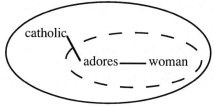

One may wonder how the last step is carried out. It appears that it does in fact involve two steps. First, a deletion of a part of a line of identity according to the rule. This leads to the diagram:

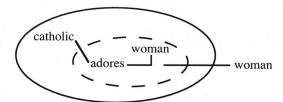

In order to arrive at the above conclusion, it might seem that the following extra principle is needed:

> The rule of retraction of ligatures: Any loose end of a ligature may be retracted outwards through cuts.

This principle is well known from the logic of Beta Graphs, i.e. for unbroken cuts (See [16]). However, one can establish a generalisation of this rule to broken cuts. In the above example the retraction of the ligature may be proved from the above diagram by deleting parts of the ligature from the recto area and by adding the unattached line of identity to the diagram. It should be noted that diagram containing the unattached line of identity is just one of the two axioms in the system of Beta Graphs. The result of these operations is the diagram:

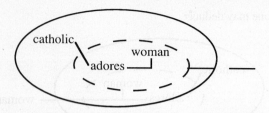

From the above diagram one can then get the concluding diagram by deiteration and deletion of the unattached line of identity.

It is very likely that Peirce had this combination of the rules in mind, but the series of operations is not as simple and straightforward as one may imagine from the reading of Peirce's text.

Expressed in modern modal logic Peirce's deduction takes us from

$$\exists y: (woman(y) \wedge \sim(\exists x: catholic(x) \wedge M\sim adores(x,y))$$

to the conclusion

$$\sim(\exists x: catholic(x) \wedge M\sim(\exists y: woman(y) \wedge adores(x,y))$$

The crucial step in this deduction appears to be based on the truth of the following theorem:

$$M(\forall y: a(y)) \supset \forall y: M\, a(y)$$

or equivalently

$$\exists y: N\, a(y) \supset N(\exists y: a(y))$$

where N is short for ~M~, and 'a' is an arbitrary predicate. As explained in [Øhrstrøm et al. 1994], this theorem is provable in any standard modal logic with standard quantification theory. This may also be expected from the fact that the rule of retraction of ligatures can be proved from the two graphical rules mentioned above, taken together with the rules of the Beta Graphs.

The third rule mentioned by Peirce in his 1906 paper is the following one:

> The Rule of the Double Cut: Two Cuts one within another, with nothing between them, unless it be ligatures passing from outside the outer Cut to inside the inner one, may be made or abolished on any Area. [CP 4.567]

Peirce demonstrated how this rule together with the two other rules mentioned above can be used in order to establish a proof of the modal syllogism:

Any man must be an animal.
Any animal must be mortal.
Ergo:

Any man must be mortal.

The premises correspond to the following two diagrams:

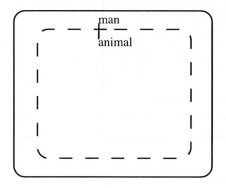 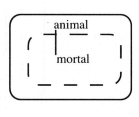

By iteration we easily find:

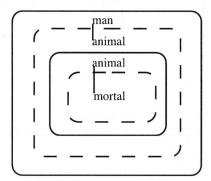

By the rule of insertion:

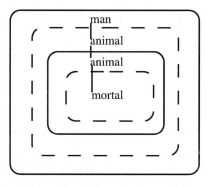

By the rule of deiteration we find:

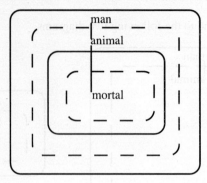

By collapse of the two cuts:

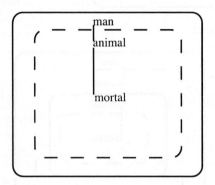

By the rule of deletion, we find:

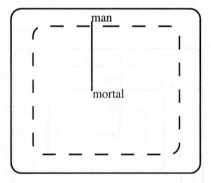

The rule of the double cut is, however, very problematic. If the double cuts in question may be any combination of two broken or unbroken cuts, then a number of rather unattractive propositions can be proved. One may, for instance, prove the theorem:

$$q \supset Nq$$

The reason is that given the unrestricted rule of the double cut one can from the diagram:

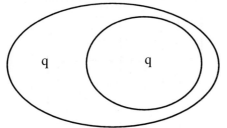

deduce the diagram:

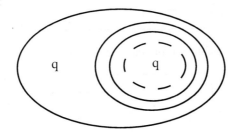

It is likely, judging from his formulations in [CP 4.567], that Peirce was considering the possibility of excluding the rule of the double cut from his system, or at least giving it some secondary status. It is understandable that he gave the rule a place in his system, since it may appear difficult to imagine any usable alternative. But he is not likely to have realised that the use of an unrestricted rule of the double cut would in fact undermine the modal distinctions themselves in the Gamma Graph System. For this reason it is obvious that in order to save the Peircean project we have to restrict the rule of the double cut to unbroken cuts only, and construct some additional rules which can account for reasonable deductions such as the one in the above syllogism. Harmen van den Berg [2] has argued that it would be natural within the Peircean context to introduce the following rule:

> *The rule of modal conversion. An evenly enclosed unbroken cut may be replaced by a broken cut. An oddly enclosed broken cut may be replaced by an unbroken cut.*

Using this rule one can transform the diagram

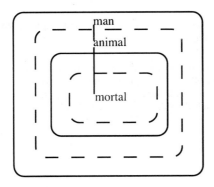

into the diagram

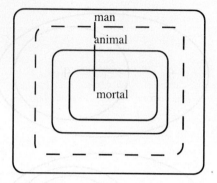

Using the rule of the double cut (on unbroken cuts) this leads to the diagram:

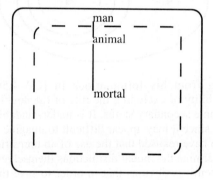

This means that the rule of modal conversion together with the restricted rule of the double cut can in fact do the job of establishing a proof of Peirce's modal syllogism.

The fourth rule which Peirce introduced in his 1906 paper is the following one:

> *Fourth Permission: If the smallest Cut which wholly contains a Ligature connecting two Graphs in different Provinces has its Area on the side of the Leaf opposite to that of the Area of the smallest Cut that contains those two Graphs, then such Ligature may be made or broken at pleasure, as far as these two Graphs are concerned. [CP 4.569]*

This rule seems to be based on the following observation, which may be called 'the existential disjunction theorem':

> The logical principle is that to say that there is some one individual of which one or the other of two predicates is true is no more than to say that there either is some individual of which one is true or else there is some individual of which the other is true. [CP 4.569]

Peirce illustrated the fourth rule by means of the implication from the proposition "There is a man x and a man y, such that if x is bankrupt, then y must commit suicide" to "There is a man x, such that if x is bankrupt, then x must commit suicide". Stated graphically, it follows from the rule applied to this example that the proposition corresponding to the diagram:

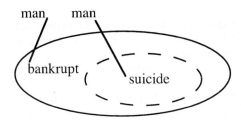

implies the proposition corresponding to the diagram:

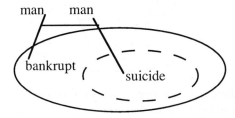

According to Charles Hartshorne and Paul Weiss, Peirce in a letter to F.A. Woods in 1913 expressed scepticism as to the universal validity of the fourth permission [CP 4.569, note]. In the National Academy of Science meeting in Washington, April 1906, he had even called this permission "quite out of place and unacceptable" [CP 4.580] On the other hand, in the same context he also stated that he found himself unable to refute the rule, but he suggested that we may have to reject the idea that "every conditional proposition whose antecedent does not happen to be realized is true." [CP 4.580].

From a modern point of view, Peirce's reaction regarding the fourth rule appears to be quite understandable. The motivating theorem mentioned above (from [CP 4.569]) can be formulated in terms of symbolic (and non-modal) logic in the following way:

$$\exists x \exists y: (\sim b(x) \lor s(y)) \quad \begin{aligned} &\equiv \quad (\exists x: \sim b(x) \lor \exists y: s(y)) \\ &\equiv \quad (\exists x: \sim b(x) \lor \exists x: s(x)) \\ &\equiv \quad \exists x: (\sim b(x) \lor s(x)) \end{aligned}$$

It is interesting that a similar disjunction theorem holds in modal logic with any modal operator N:

$$\exists x \exists y: (\sim b(x) \lor Ns(y)) \quad \begin{aligned} &\equiv \quad (\exists x: \sim b(x) \lor \exists y: Ns(y)) \\ &\equiv \quad \exists x: (\sim b(x) \lor Ns(x)) \end{aligned}$$

This means that the fourth rule holds provided that the cuts mentioned in the above formulation are all unbroken as indeed they are in Peirce's own example. A generalisation to broken cuts would correspond to claims like the following equivalence:

$$\exists x \exists y: M_1(b(x) \lor N_2s(y)) \quad \equiv \quad \exists x: M_1(b(x) \lor N_2s(x))$$

It is easy to see that this equivalence would presuppose something like Barcan's formula, i.e. $\exists x: M_1p \equiv M_1(\exists x:p)$. It is, however, well known that this equivalence can not in general can be assumed as valid, although it is valid in some systems, for instance the system S5 (with standard quantification theory).

Peirce pointed out that an implication from the proposition "there is a man every dollar of whose indebtedness will be paid by one man", corresponding to the diagram

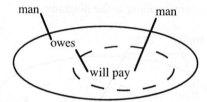

to the proposition "there is a man every dollar of whose indebtedness will be paid by the same man", corresponding to the diagram

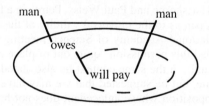

is invalid. With respect to such an implication the fourth rule cannot be applied, since the smallest cut which wholly contains a ligature connecting the two graphs 'owes' and 'will pay' is the same as the smallest cut that contains those two graphs. In terms of modern symbolic logic this difference between the two propositions comes out as

$$\exists x \exists y: man(x) \land man(y) \land (\forall z: owes(x,z) \supset N \; will\_pay(y,z))$$
$$\exists x: man(x) \land \forall z:(owes(x,z) \supset N \; will\_pay(x,z))$$

It is obvious that because of the occurrence of the universal quantification the existential disjunction theorem cannot be used here. (Incidentally, the preoccupation with pecuniary examples may well stem from Peirce's financial plight at this time!)

3. Towards a Logic for Gamma Graphs

The logic of the Gamma Graphs is the logic of modality. Peirce suggested a multi-modal approach to the Gramma Graphs. In fact, he proposed a system involving 12 different tinctures in the diagrams. It is not clear why he suggested exactly that number, but it is important to stress the fact that he wanted the tinctures to stand for various kinds of possibility, intention, and actuality. The specific understanding of a diagram involving tinctures will depend on the relevant interpretation.

Although Peirce suggested a multi-modal system, there is - as far as can be seen from the printed sources - no indication in his works of a study of the ways in which various kinds of modality may interact with each other. It seems that this is just one of the many themes which he left for his followers to pursue. I have shown elsewhere [18] how the Peircean ideas of tinctures and graphs can give rise to a tense logic with two basic tense operators.

Peirce himself realised that there is a lot to do in order to bring the logical system of the Gamma Graphs into a satisfactory form:

> But the gamma part is still in its infancy. It will be many years before my successors will be able to bring it to the perfection to which the alpha and beta parts have been brought. For logical investigation is very slow, involving as it does the taking up of a confused mass of ordinary ideas, embracing we know not what and going through with a great quantity of analyses and generalizations and experiments before one can so much as get a new branch fairly inaugurated. [CP 4.511]

Although Peirce's account of the logic of Gamma Graphs was incomplete, and although it contained some inaccuracies and mistakes, it did in fact define a paradigm for further research. Towards the end of his life he was very much aware of this perspective. Regarding his Existential Graphs he wrote: "I am now working desperately to get written before I die a book on Logic that shall attract some good minds through whom I may do some real good." (Semiotic and Significs; quoted from [6]).

Based on the Peircean ideas it is rather obvious what kind of rules one should study in order to extend a modern system of Existential Graphs, such as the one developed at Loughborough University [5] [9], to cover not only Alpha and Beta Graphs, but also Gamma Graphs.

According to Sun-Joo Shin [15], one of the main reasons why the Existential Graphs never gained great popularity among logicians is the fact that the method is rather complicated. Taking for instance the above mentioned 4th rule for Gamma Graphs into consideration, Sun-Joo Shin's view seems appears reasonable. On the other hand, it is likely that the valid version of this rule as well as other Peircean rules can be reformulated in simpler ways that can make them easier to use. There may even be hints of how to do this in the still unpublished papers by C. S. Peirce. This kind of work seems to be very relevant in the modern context of computer science. As John Sowa has argued for many years, and as clearly demonstrated in a new book edited by Gerard Allwein and Jon Barwise [1], the increasing use of visual displays suggests the introduction of tools for logical reasoning with diagrams.

References

1.   Allwein, G. and Barwise, J. (ed.), *Logical Reasoning with Diagrams*, Oxford University Press, 1996.
2.   Berg, Harmen van den, "Modal Logic for Conceptual Graphs", in Mineau, Guy W.; Moulin, Bernard; Sowa, John (editors), *Conceptual Graphs for Knowledge RepresentationConceptual Graphs for Knowledge Representation*, Springer-Verlag 1993, pp.411-429.
3.   Burch, R. W., "Game-Theoretical Semantics for Peirce's Existential Graphs", *Synthese* 99: 361-375,1994.
4.   Hammer, Eric, "Peirce on Logical Diagrams", *Transactions of the Charles S. Peirce Society*, Fall, 1995, Vol. XXXI, No. 4, 1995, pp. 807-827.
5.   Heaton, J.E., *Goal Driven Theorem Proving Using Conceptual Graphs and Peirce Logic*, Ph.D. Thesis, Loughborough University, UK, 1994.
6.   Keeler, M., "The Philosophical Context of Peirce's Existential Graphs", *Proceedings of the International Conceptual Structures Conference*, University of California, Santa Cruz, 1995, pp. 150-165.
7.   Keeler,M & Kloesel, C., "Communication, Semiotic Continuity, and the Margins of the Peircean Text", in *Margins of the Text*, edited by David C. Greetham, Ann Arbor: University of Michigan Press, 1996 (Can be obtained from http://accord.iupui.edu/accord/margins.txt).
8.   Kevelson, Roberta, *Charles S. Peirce's Method of Methods*, John Benjamins Publishing Company, 1987.
9.   Kocura, P., "Conceptual Graph Canonicity and Semantic Constraints", in Eklund, P., Ellis, G., Mann, G. (ed.), *Conceptual Structures: Knowledge Representation as Interlingua, Auxiliary Proceedings*, 1996, p.133-145.
10. Peirce, C.S., *Reasoning and the Logic of Things*, (edited by Kenneth Laine Ketner), Harvard University Press, 1992.
11. Peirce, C.S., *Collected Papers*, 8 volumes (eds. P. Weiss, A. Burks, C. Hartshorne), Cambridge: Harward University Press (CP). 1931-1958.
12. Peirce, C.S., *Application to the Carnegie Institution* , July 15, 1902, Manuscript L75, Analytical reconstruction and editorial work by Joseph Ransdell, Indiana University, 1994.
13. Roberts, Don D., *The Existential Graphs of Charles S. Peirce*, Mouton, 1973.
14. Roberts, Don. D., "The Existential Graphs", *Computers Math Applic.* Vol.23 (1992), No. 6-9, 1992, pp. 639-663
15. Sun-Joo Shin, "Peirce and the Logical Status of Digrams", *History and Philosophy of Logic*, 15, 1994, pp.45-58
16. Øhrstrøm, P., van den Berg, H., Schmidt, J., "Some Peircean Problems Regarding Graphs for Time and Modality", *Second International Conference on Conceptual Structures,* University of Maryland, 1994, p.78-92.
17. Øhrstrøm, P. & Hasle, P.F.V., *Temporal Logic. From Ancient Ideas to Artificial Intelligence*, Studies in Linguistics and Philosophy 57, Kluwer Academic Publishers, 1995.
18. Øhrstrøm, P., "Existential Graphs and Tense Logic", in Eklund, P., Ellis, G., Mann, G. (ed.), *Conceptual Structures: Knowledge Representation as Interlingua*, Springer-Verlag, 1996, p.203-217.

A Sound and Complete CG Proof Procedure Combining Projections with Analytic Tableaux

Gwen Kerdiles and Eric Salvat

L.I.R.M.M. *(U.M.R. 9928 Université Montpellier II / C.N.R.S.)*
161 rue Ada, 34392 Montpellier cedex 5 - France
tel.: (33) 4 67 41 85 45   fax: (33) 4 67 41 85 00
e-mail: {kerdiles,salvat} @lirmm.fr

Abstract. Conceptual Graphs offer an attractive and intuitive formalism for knowledge representation in Artificial Intelligence. The formalism calls for efficient systems of reasoning. Projection is one such tool for a language limited to conjunction and existential quantification (Simple Conceptual Graphs). Projection is very efficient for certain classes of Conceptual Graphs and offers an original approach to deduction: the perspective of graph matching. The aim of this paper is twofold: Propose an efficient analytic deduction system that combines analytic tableaux with projection for a language of Conceptual Graphs extended to all non functional First-Order Logic formulae and compare this method with the one introduced in [1] for Simple Conceptual Graph rules.

1   Introduction

Conceptual Graphs are a simple and expressive knowledge representation formalism. It combines in a natural graphical presentation Semantic Networks of Artificial Intelligence with the logical graphs of Peirce and Order-Sorted Logic. Conceptual Graphs adopt from Order-Sorted Logic a hierarchy of concept types. The possibility of partitioning the universe of a formal language has been fruitfully investigated since the pioneering work of Herbrand. Some work in Automated Deduction has highlighted the benefits brought by sorting in terms of limitation of search space. For example, [2] studies some computational aspects of different Order-Sorted Logics. Conceptual Graphs also introduce an order over relations of common arity. Projection is an inference mechanism for the language of Simple (non nested) Conceptual Graphs ([3], [4]). Projection presents deduction from the perspective of graph morphisms. It is the cornerstone of the proof procedure in this paper.

Sect. 2 recalls some basic definitions in Simple Conceptual Graph theory. A notion of falsity is introduced in the language and projection is adapted. Sect. 3 presents an extension of the Simple Conceptual Graph language which includes implication and therefore negation: Contrary to some presentations in Peirce's style of an equivalent language where negation is represented by *negative contexts* (for instance, [3], [5] or [6]), in this paper, it is represented by the implication of falsity. In Sect. 4, a tableau procedure for the extended language is defined.

It decomposes the sentences with analytic tableau rules[1] and concludes proofs by using projection. [9] and [1] study Conceptual Graph rules of the form "IF a Simple Conceptual Graph THEN a Simple Conceptual Graph". These rules can be associated with Conceptual Graph bases to generate Simple Conceptual Graphs which follow from the bases and rules (Forward Chaining) or prove that some Simple Conceptual Graphs can be derived (Backward Chaining). Sect. 5 recalls these two mechanisms presented in [1] and compares them to the deduction process performed in tableaux.

2 Simple Conceptual Graphs

We adopt some slightly modified basic definitions which are fully developed, for instance in [3] or [10].

A **support**, $\Sigma = (T_C, T_R, \sigma, M, \tau)$, represents an ontology of a specific application domain. T_C is a poset of concept types with a supremum $\top$ and an infimum $\bot$ (the absurd type). T_R is a partition of posets of relation types of common arity. σ associates to any relation, its signature. The set of markers, M, is partitioned into an infinite set of individual markers I and an infinite set of variable markers V. A special individual marker, $false$, is defined. τ associates to any individual marker the type of the individual represented which cannot be $\bot$ except for $\tau(false) = \bot$. Furthermore, we require for every type of concept that there be at least one individual marker of that type.

A **Simple Conceptual Graph**(SCG) $G = (R, C, U, label)$ related to a support Σ, is a bipartite multi-graph not necessarily connected such that R and C denote respectively the classes of relation and concept vertices, and U is the set of edges. $label$ is a mapping which associates to a relation vertex $r \in R$ its type, and to a concept vertex $c \in C$ a pair $(type(c), ref(c))$ such that $ref(c) \in (M \cup \{*x/x \in V\})$ and if $ref(c) \neq false$ then $type(c) \in T_C \setminus \{\bot\}$. $label$ respects the constraints given by σ and τ and furthermore: *For any two distinct concept nodes $c1$ and $c2$ in C, it holds that $marker(c1) \neq marker(c2)$* (this property corresponds to the so called notion of normality: for instance in [9] or [10]). A concept node with $ref = false$ cannot be connected to an edge. Any SCG containing a such node represents the notion of **falsity**.

Notation: For a concept vertex $c \in C$, if $ref(c) = *x$ then c is called *a declaration of x* and we define $marker(c) = x$, otherwise $marker(c) = ref(c)$. If $ref(c)$ is a marker, m, different from $false$, c is said to be a *free occurrence of m*. We note $Dec(G)$ (respect. $Free(G)$) the sets of pairs $(type(c), marker(c))$ such that c is a declaration in G (respect. c is a free occurrence in G). For any support, two SCGs have a special feature: $False = (\emptyset, \{c\}, \emptyset, label(c) = (\bot, false))$ and $Emptygraph = (\emptyset, \emptyset, \emptyset, \emptyset)$.

Like First-Order formulae and contrary to what is generally called a Conceptual Graph, this definition enables variable markers to occur free in Simple

---

[1] For example, [7] offers a clear presentation of analytic tableaux for Classical First-Order Logic and [8] for Order-Sorted Logic.

Conceptual Graphs, in order to let the definition of the extended language in the next Section be simple and recursive.

A SCG is **closed** if and only if every marker that occurs free in it, is an individual marker (closed SCGs meet the properties of the so called normal Conceptual Graphs). In closed SCGs, variable markers are not necessary; Indeed in the following definition of projection, only the prefix $*$ is primordial. Even in the extended language in the next Section, names of variables could be replaced in closed Conceptual Graphs by so called *coreference links*. Nevertheless, variable markers are useful for a linear notation.

We adapt the notion of **projection** to the marker $false$. It is a property of classical First Order Logic that any graph follows from (can be projected into) a graph representing falsity; This is represented by an empty projection. Projection is only defined on closed Simple Conceptual Graphs: a projection, π, from a closed SCG, G, into a closed SCG, H, is defined if and only if either there exist a concept node c in C_H such that $label(c) = (\bot, false)$, then π is the *empty function*, or there exists a projection as defined in [3], and π is this projection.

3 Conceptual Graphs

We will enrich the language of Simple Conceptual Graphs with a binary symbol ($\Rightarrow$) for implication. As we can notice in Fig. 1 (a representation of "there is somebody who knows a solution to every problem"), the field 'ref' of conceptual nodes will provide the key for the binding of free occurrences of variable markers by declarations and the interlocking structure will define the scope of those declarations. *Coreference links* could as well have been used.

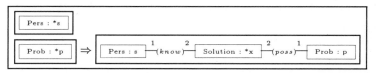

Fig. 1. A Conceptual Graph.

Definition 3.1 A **Conceptual Graph**(CG), G, is a set of entries such that G contains exactly **one**[2] atomic entry where an **entry** is either

- atomic if it is a SCG,
- or complex: $G1 \Rightarrow G2$ where $G1$ and $G2$ are CGs.

Notation: A Conceptual Graph which is a singleton (its only entry is a Simple Conceptual Graph) is called atomic and if this entry is a closed SCG, then the CG is called closed atomic. We represent a set of entries by juxtaposing them inside a frame (see Fig. 1). Note that contrary to Peirce's notation, frames do not represent negative contexts, but are used around Simple Conceptual Graphs and sets of entries. To lighten the representation in the rest

---

[2] Having exactly one atomic entry is not a limitation because this entry may be the *Emptygraph* or a Simple Conceptual Graph which may itself contain many connected compounds, and furthermore, it simplifies notations such as in Def. 3.2.

of the paper, we will draw a single external frame for a Conceptual Graph which is composed of a single Simple Conceptual Graph. Because there will be no ambiguity, we shall as well omit to draw the *Emptygraph* and a frame around the Simple Conceptual Graph *False*. For example, we will draw the

following: 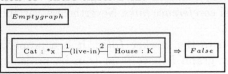 as a representation of the nega-

tion of "there exists a cat who lives in the house K", whereas we should draw:

Definition 3.2 For any Conceptual Graph G, we define (or extend some definitions of Sect. 2):

1. **atomic** is a function which takes a CG and returns its atomic entry and **complex** is a function which returns the set of complex entries of a CG.
2. $Dec(G) = Dec(atomic(G))$ is the set of **declared** variable markers in the atomic entry of G.
3. $Free(G)$ is the set of **free** markers in G, i.e. $Free(G) = (\bigcup_{E \in G} Free(E)) \setminus Dec(G)$ where for a complex entry E of the form $G1 \Rightarrow G2$, $Free(E) = Free(G1) \cup Free(G2) \setminus Dec(G1)$. For instance, let G be the Conceptual Graph represented in Fig. 1, $Dec(G) = \{(Pers, s)\}$ and $Free(G) = \emptyset$.
4. A CG G is **closed** if and only if the markers in $Free(G)$ are all individual markers.
5. A CG is **pure** if it does not contain a variable marker declared twice and if none of its free variable markers occurs in a declaration. A CG which is not pure can easily be transformed into an equivalent pure CG by proceeding outside-in and renaming re-declared variable markers.
   In the rest of the paper, we will only consider pure Conceptual Graphs.
6. The **merging** of two CGs, G and H, is the CG: $G \uplus H = \{atomic(G) \uplus atomic(H)\} \cup complex(G) \cup complex(H)$ where the merging (disjoint sum and internal join of concept nodes with common marker) of two SCGs is defined in for instance [4].

Definition 3.3 Model: A model, $M_\Sigma = ((Dom, \subseteq), F)$ with respect to a support $\Sigma = (T_C, T_R, \sigma, M, \tau)$ consists of

- a poset $(Dom, \subseteq)$ such that for all concept types t_1 and t_2 in T_C which are not the absurd type, it holds that Dom_{t_1} and Dom_{t_2} are elements of Dom and if $t_1 <_{T_C} t_2$ then $Dom_{t_1} \subseteq Dom_{t_2}$.
  These sub-domains of interpretation respect the property of non-empty universe(s): $\forall Dom_t \in Dom, Dom_t \neq \emptyset$.
- an interpretation function F on $(I \setminus \{false\}) \cup T_R$ (individual markers and relation types of the support) verifies:
  - if $c \in (I \setminus \{false\})$ then $F(c) \in Dom_{\tau(c)}$,

- if $R \in T_R$ and $arity(R) = n$ then $F(R) \subseteq Dom_{\sigma_1(R)} \times \ldots \times Dom_{\sigma_n(R)}$ where $\sigma_i(R)$ is the maximal type of the ith-argument of R,
- $\forall R, R' \in T_R, R <_{T_R} R'$ implies that $F(R) \subseteq F(R')$.

We note F_I the restriction of F to the set of individual markers (without $false$).

Notation: An interpretation function g is an **X-extension** of an interpretation function f where X is a set of pairs (t, m) if and only if $dom(g) = dom(f) \cup \{m/(t,m) \in X\}$ and $g \supseteq f$.

Definition 3.4 Truth of a CG: The **truth** of a CG, G, in $M_\Sigma = ((Dom, \subseteq), F)$ under an interpretation f(a partial function from $M \setminus \{false\}$ to Dom_T), noted $(M_\Sigma \models_f G)$ is defined by:

1. $M_\Sigma \models_f G$ if and only if $\exists g$, $Dec(G)$-extension of f, such that for every entry E in G, it holds that $M_\Sigma \models_g E$
2. $M_\Sigma \models_g G1 \Rightarrow G2$ if and only if $\forall h$, $Dec(G1)$-extension of g, $M_\Sigma \models_h G1$ implies $M_\Sigma \models_h G2$
3. for the case of the atomic entry E: $M_\Sigma \models_g E$ if and only if
   - there is no concept node c in E such that $marker(c) = false$ and
   - for every concept node c in E, it holds that $g(marker(c)) \in Dom_{type(c)}$
   - and for every relation node R of arity n in E, if c_i is the ith-neighbour of R in E then $< g(marker(c_1)), \ldots, g(marker(c_n)) > \in F(R)$.

- A CG, G, related to a support Σ, is **satisfiable** if and only if there exists a model M_Σ and an interpretation function f, which is a $Free(G)$-extension of F_I, and such that $M_\Sigma \models_f G$. We also say that G is satisfiable in M_Σ if the model is given.
- A CG, G, related to a support Σ, is **valid** if and only if for every model M_Σ and for every interpretation function, f, which is a $Free(G)$-extension of F_I, it holds that $M_\Sigma \models_f G$.

Theorem 3.5 Soundness and Completeness of projection:
From [3] and [10], it follows that projection is sound and complete: Given a support Σ and two closed SCGs, G and H, there is a projection from G into H if and only if for any model $M_\Sigma = ((Dom, \subseteq), F)$, $M_\Sigma \models_{F_I} \boxed{H}$ implies that $M_\Sigma \models_{F_I} \boxed{G}$.

Definition 3.6 Translation of a CG related to a support Σ into L_Σ, a language of First-Order Logic which has a common vocabulary with Σ (an extension of the usual Φ operator defined on Simple Conceptual Graphs).

1. For a CG, G, $\Phi(G) = \exists x1, \ldots, xn(\Phi(E_1) \wedge \ldots \wedge \Phi(E_m))$ where $< E_1, \ldots, E_m >$ is any ordering of the entries of G and $< x1, \ldots, xn >$ is any ordering of the declared variable markers in G (i.e. the variable markers in $Dec(G)$). If $Dec(G)$ is empty then $n = 0$.
2. Let E be an entry, $\Phi(E)$ is defined by:

(a) E is atomic:

    i. To each concept node c, if $ref(c) = false$, associate an atom $\neg X$ where X is some tautology, otherwise associate an atom $P(marker(c))$ where P is the predicate corresponding to $type(c)$ in L_Σ.

    ii. To each relation node r in E associate an atom $r(marker(c_1), \ldots, marker(c_n))$ where n is the arity of r and c_i the ith-neighbour of r in E

    iii. $\Phi(E)$ is the conjunction of these atoms.

(b) E has the form $G1 \Rightarrow False$ then $\Phi(E) = \neg(\Phi(G1))$.

(c) E has the form $G1 \Rightarrow G2$ with $G2 \neq False$,

let $< x1, \ldots, xn >$ be any ordering of the variable markers in $Dec(G1)$ ($n = 0$ if empty),

$\Phi(E) = \forall x1, \ldots, xn((\Phi(E_1') \wedge \ldots \wedge \Phi(E_m')) \rightarrow \Phi(G2))$

where $< E_1', \ldots, E_m' >$ is any ordering of the entries of $G1$.

For example, if G is the CG in Fig. 1 then $\Phi(G) = \exists s(Pers(s) \wedge (\forall p(Prob(p) \rightarrow \exists x(Solution(x) \wedge Pers(s) \wedge Prob(p) \wedge know(s, x) \wedge poss(p, x)))))$.

We introduce the notion of *signed Conceptual Graphs*, which is a very convenient short-cut in our language of Conceptual Graphs, because negation is represented by the implication of falsity.

Definition 3.7 Signed Conceptual Graphs: Under any interpretation, the truth value of a signed CG, $+G$, is the same as that of G. The one of $-G$ is the same as that of $\boxed{G \Rightarrow False}$. We define $Dec(\pm G) = Dec(G)$, $Free(\pm G) = Free(G)$ and $atomic(\pm G) = atomic(G)$.

Definition 3.8 Let X and Y be two sets of pairs (t, m) such that t is a concept type, m is a marker different from $false$, and if m is an individual marker, then $t = \tau(m)$. $f : X \rightarrow Y$ is a **substitution function** if and only if $f((t, m)) = (t', m')$ implies: (i) $t' \leq_{T_C} t$ and (ii) If m is an individual marker then $t' = t$ and $m' = m$.

Abusing notations, we will also apply substitution functions to Conceptual Graphs, sets of Conceptual Graphs or labels of concept nodes. To any projection π from G into H, we associate an obvious substitution function, π', from the set of pairs (t, m) stemming from $label_G$ to the set of pairs (t', m') stemming from $label_H$. It holds that for every concept node c in G, $\pi'(label_G(c)) = label_H(\pi(c))$.

4 Combining Tableaux with Projections

In [7], Smullyan qualifies analytic tableaux for First-Order Logic as "an extremely elegant and efficient proof procedure". The whole idea for proving that a Conceptual Graph, G, is valid consists in, first asserting the negation of G ($-G$) and then trying to find a refutation of this assertion. For this purpose, $-G$ is systematically decomposed into "smaller" Conceptual Graphs (property of being an analytic method) until reaching a decomposition into atomic Conceptual

Graphs. These graphs are placed as nodes of a (proof) tree with root $-G$, called an analytic tableau for G. A branch of a tableau is read as the conjunction of the graphs occurring on that branch. If it corresponds to a contradiction then the branch is called *closed*. If every branch of a tableau for G is closed then the root itself $(-G)$ is unsatisfiable (because the construction rules preserve satisfiability). Therefore, G is valid. An essential difference between this procedure and the one described in [9] is that atomic Conceptual Graphs obtained in the decomposition are processed as a whole (using projection for those which are closed), whereas in [9], atomic CGs are themselves decomposed into *anti-normal graphs* (atomic CGs containing at most one relation vertex).

We shall distinguish five disjoint classes of signed Conceptual Graphs.

1. We call α a signed CG of the form $+\boxed{\begin{array}{c} \alpha_1 \\ \ldots \\ \alpha_n \end{array}}$, $1 \le n$ such that if $n = 2$ then $atomic(\alpha) \ne Emptygraph$ and if $n = 1$ then $Dec(\alpha) \ne \emptyset$.

2. We call β a signed CG of the form $-\boxed{\begin{array}{c} \beta_1 = atomic(\beta) \\ \ldots \\ \beta_n \end{array}}$, $1 < n$ such that if $n = 2$ then $atomic(\beta) \ne Emptygraph$.

3. We call γ a signed CG of the form $+\boxed{\gamma_1 \Rightarrow \gamma_2}$.

4. We call δ a signed CG of the form $-\boxed{\delta_1 \Rightarrow \delta_2}$.

5. We call χ a negative atomic CG which does not contain any relation node. For convenience, we will talk about $\alpha_1, \ldots, \alpha_n$ as the entries of a node α (Idem for β, γ or δ).

Definition 4.1 Tableaux: A tableau is a pair, $(\mathcal{T}, \Pi)$, composed of an ordered (the successors of a node are ordered) tree $\mathcal{T}$ whose points are occurrences of signed CGs and *a substitution function* Π. $(\mathcal{T}_0, \Pi_0)$ is a tableau for a CG, G, where $\mathcal{T}_0$ is a single node $-G$ and Π_0 is the identity function on the set of all pairs $(type(c), ref(c))$ where c is a concept node in G. Suppose $(\mathcal{T}, \Pi)$ is a tableau for G, let H be a leaf of $\mathcal{T}$, then we may extend the tableau by either of the following rules:

1. **Rule for α's:** If an α occurs on the path P_H, then let Θ_α be a substitution bijection which associates to every referent in $Dec(\alpha)$ **a new individual marker** of the same type ("new" in the sense that it does not appear so far in $\mathcal{T}$) and define $\alpha'_{1 \le i \le n} = +\Theta_\alpha(\boxed{\alpha_i})$. We may adjoin successively $\alpha'_1, \ldots, \alpha'_n$ such that α'_1 is the sole successor of H and $\alpha'_{2 \le i \le n}$ is the sole successor of α'_{i-1} (if $n = 1$ then α'_1 is the sole successor of H).

   Rule for β's: If a β occurs on the path P_H, then let Θ_β be a substitution bijection which associates to every referent in $Dec(\beta)$, **a new variable marker** of the same type and define $\beta'_{1 \le i \le n} = -\Theta_\beta(\boxed{\beta_i})$. We may simultaneously adjoin β'_1 (as leftmost successor of H) to β'_n as successors of H.

Rule for γ's: If a γ occurs on the path P_H, then let Θ_γ be a substitution bijection which associates to every referent in $Dec(\gamma_1)$,**a new variable marker** of the same type, and to every referent in $Dec(\gamma_2)$, **a new individual marker** of the same type. $\gamma_1' = -\Theta_\gamma(\gamma_1)$ and $\gamma_2' = +\Theta_\gamma(\gamma_2)$. We may simultaneously adjoin γ_1' as the left successor of H and γ_2' as the right successor of H.

Rule for δ's: If a δ occurs on the path P_H, then let Θ_δ be a substitution bijection which associates to every referent in $Dec(\delta_1)$,**a new individual marker** of the same type, and to every referent in $Dec(\delta_2)$, **a new variable marker** of the same type. $\delta_1' = +\Theta_\delta(\delta_1)$ and $\delta_2' = -\Theta_\delta(\delta_2)$. We may adjoin successively δ_1' as the sole successor of H and δ_2' as the sole successor of δ_1'.

Rule for χ's: If some χ occurs on the path P_H, then let Θ_χ be a substitution function, which associates to every referent in $Dec(\chi)$, **any individual marker** of a subtype and define $\chi' = +\Theta_\chi(\chi)$. We may adjoin χ' as the sole successor of H.

2. If one of the preceding rules, $x \in \{\alpha, \beta, \gamma, \delta, \chi\}$, is applied, then Π must be adapted to the newly introduced markers: Π becomes $Id_{codom(\Theta_x)} \cup (\Theta_x \bullet \Pi)$.

Example 4.2 Suppose that we have the following representation of "if somebody knows a solution to every problem then this person is a divinity":

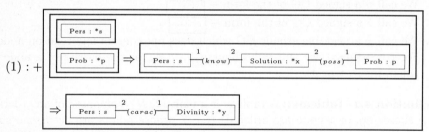

The signed CG (1) is a γ and applying the rule for γ's on it leads to the two signed CGs, (2) and (3), where s' is a new variable marker and d is a new individual marker of type divinity. A γ corresponds to the assertion of a rule "γ_1 implies γ_2". Roughly speaking, we can add the classically equivalent assertion "(NOT γ_1) OR γ_2", the disjunction corresponding to two branches. The replacement of $*y$ by a new individual marker d corresponds to the following reasoning. We have asserted that there exists a divinity. We can call it d because d is new in the tableau, and therefore, was not previously associated to some properties (except being a divinity).

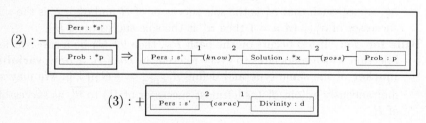

Definition 4.3 A **proof** of G is a tableau for G whose all branches are closed. A branch B of a tableau, $(\mathcal{T},\Pi)$, **closes** if there are in B a closed atomic node $-\boxed{H}$, and some closed atomic nodes $+\boxed{G_1}$, ... , $+\boxed{G_n}$ such that there exists a projection from $\Pi(H)$ into $(\uplus_{1 \leq i \leq n} \Pi(G_i))$ with associated substitution function π. After closing the branch B, Π becomes $\pi \bullet \Pi$.

Example 4.4 A closed tableau: Suppose that we have a knowledge base representing the information "The cat, Poenga, is white", "Poenga lives in the house, K" and "Every feline who lives in a house learns to hunt mice". These two facts and the rule may be represented by three disjoint CGs:

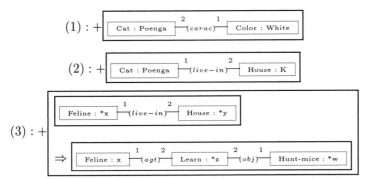

If we want to prove the request "there is a cat of a certain color who learns to hunt mice", then we may start a tableau, $(\mathcal{T},\Pi)$, where $\mathcal{T}$ is a single branch composed of (1), (2), (3) and (4). Π is the identity function on the set of pairs $(type(c), marker(c))$ such that c is a concept node in $\mathcal{T}$ and

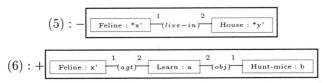

(3) is a δ, thus we may extend the tableau by adding two direct successors to (4): (5) and (6) where x' and y' are new variable markers, a and b are new individual markers of the respective types Learn and Hunt-mice. Π remains the identity function but extended to the newly introduced markers.

Note that (6) cannot yet play a role in closing a branch because we defined projection (in Sect. 2) on closed graphs and x' occurs free in (6). The underlying notion is Skolem dependencies. We adopt Reyle and Gabbay's propositions (in appendix of [11]) of having the advantages of Skolemisation (in terms of restriction of search spaces) without representing Skolem functions. Indeed, in our language of Conceptual Graphs, we do not have functional markers. Therefore, the dependencies are given by the structure of the tableau: x' could have been replaced in another language by a new Skolem function, f_0, on the variable markers occurring free in the branch of (5) (none in this case). Later, when a substitution would have been adopted for f_0 in (5), its value would have been passed on the occurrence in(6). We obtain the same result because we have to wait for a projection of (5) to associate an individual marker to x' before being able to use (6) in another projection.

There is a projection from (5) into (2) which closes the left branch and Π remains the identity function except for x' and y': $\Pi((Feline, x')) = (Cat, Poenga)$ and $\Pi((House, y')) = (House, K)$. There is now another projection from the request (4) into the merging of $\Pi(1)$ and $\Pi(6)$ which closes the tableau.

$$\Pi(1) \uplus \Pi(6) :$$

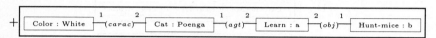

The request is proved and the final value of Π tells that "the white cat, Poenga" is a solution to the request.

Theorem 4.5 Soundness: If there is a closed tableau for a closed and pure CG, G, then G is valid.

Sketch of proof (detailed in [12]): We first show that the five construction rules preserve satisfiability. Let B be a satisfiable branch in a tableau. If a rule is applied on a node in B, then at least one of the extended branches is satisfiable. If the root $-G$ is satisfiable, then at least one branch is satisfiable (by induction on the number of rules applied and due to the preservation of satisfiability). One such branch remains open because of the conditions for closing a branch and completeness of projection. Thus, if the tableau is closed then the root must be unsatisfiable. Therefore, G is valid.

Theorem 4.6 Completeness: If a CG, G, related to a support Σ is valid then there exists a closed tableau for G.

Sketch of proof(detailed in [12]): A branch B of a tableau is called exhausted if the following holds:
- If a β or a γ occurs in B, then the corresponding rule has been applied infinitely many times for each occurrence.
- If an α, a δ or a χ occurs in B, then the corresponding rule has been applied at least once for each occurrence.

We will assume that we have a systematic procedure which guarantees that if the process of extending a branch does not terminate, then the resulting infinite branch will be exhausted (an adaptation of the systematic procedure in [7]).

We first show that every exhausted open branch B, in a tableau $(\mathcal{T}, \Pi)$, is satisfiable. We construct a model M_Σ consisting of the Herbrand universe of all ground atomic CGs related to the support Σ that can be built from atomic nodes appearing in $\Pi(B)$. We then show that every CG occurring as a node of $\Pi(B)$, is satisfiable in our model M_Σ. Finally, we show that if there is no closed tableau for G, then $-G$ is satisfiable: We first note that if there is no closed tableau for G, then there is an exhausted open branch, B, in any tableau constructed systematically for G. Thus, $\Pi(B)$ is satisfiable and therefore is the root, $\Pi(-G) = -G$.

5 Analytic tableaux vs. Graph rules

5.1 Graph rules

In [1], Conceptual Graph rules are introduced. They are inference rules of the form $G_1 \Rightarrow G_2$, where G_1 and G_2 are Simple CGs. G_1 is called hypothesis of the rule, and G_2 conclusion. There may be coreference links between concepts of G_1 and G_2. Such concepts are called *connection points*. Two mechanisms, forward and backward chainings, are proposed for processing these rules. Both are sound and complete procedures with respect to First-Order Logic deduction.

Forward chaining: Forward chaining is typically used to explicitly enrich a knowledge base with information which is implicitly present in rules. A rule $R : G_1 \Rightarrow G_2$ can be applied onto a Simple Conceptual Graph, G, if G fulfills the hypothesis of R, i.e. if there is a projection from G_1 into G. Then G_2 is "added" to G (G and G_2 are joined on connection points of G_2 and images of connection points of G_1 in G).

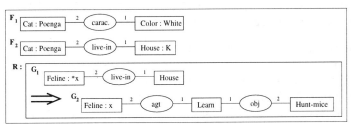

Fig. 2. A knowledge base with a graph rule.

Fig. 2 represents in graph rule formalism the same knowledge base as in Ex. 4.4. The rule R, can be applied to the graph F_2, since there exists a projection from G_1 into F_2. R has only one connection point, materialized by the variable $*x$. The result of applying R to F_2, is the graph $R(F_2)$ of Fig. 3 obtained by joining the concept $\boxed{\text{Feline : x}}$ of G_2 with the concept $\boxed{\text{Cat : Poenga}}$ of F_2.

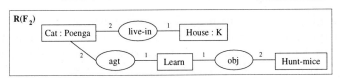

Fig. 3. Result of applying R onto F_2.

Backward chaining: [9] proposes a sound and complete mechanism in backward chaining to prove a request on a knowledge base. The process of the request is similar to the process of a goal in Prolog. Indeed, the request is first split into trivial subgraphs (composed of a relation and its arguments). In [1] the request is decomposed into *pieces* (subgraphs as large as possible that can be processed as a whole). A piece of the rule conclusion is a "unit of information" that a rule would bring when applied in forward chaining. Pieces are defined according to cut points of a graph. A *cut point* of the rule conclusion is either a connection point or an individual concept. Two concepts belong to the same piece if and only if there exists a path from one to the other that does not go through a cut point. In Fig. 2, G_2 has only one cut point and one piece.

To prove a request Q on a knowledge base, we try to unify Q with the conclusion of a rule. Unifying corresponds in forward chaining to obtaining an answer to part of the request. Then, a new request is built from the part of the old request on which the unification has not been done (i.e. the part of the request which has not yet been answered) by joining this part and the hypothesis of the rule on the cut points used in the unification. Indeed, in forward chaining, the rule can provide an answer to part of the request only if its hypothesis can be fulfilled. The initial request is proved when we have built an empty request by a sequence of unifications. For example, to prove the request in Fig. 4, on the

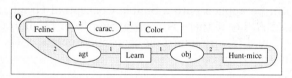

Fig. 4. A request on the knowledge base.

knowledge base in Fig. 2, we can first unify the grey part of Q with G_2. In this unification the vertex $\boxed{\text{Feline}}$ of Q is unified with the cut point $\boxed{\text{Feline : x}}$ of G_2. To build the new request Q' (Fig. 5-left), the unified subgraph of Q is deleted, and G_1 is joined to the remaining part. This join is performed on vertices of Q that were unified with cut points of G_2 and the corresponding vertices of G_1 (in this case, $\boxed{\text{Feline}}$ of Q and $\boxed{\text{Feline : *x}}$ of G_1). Then a new unification is done

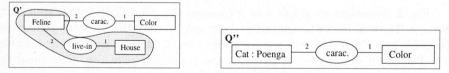

Fig. 5. New requests.

between the grey part of Q' with F_2. The new request Q'' (Fig. 5-right) is built just by deleting the unified part of Q', since a fact can be seen as a rule with the empty graph as hypothesis. In Q'', the vertex $\boxed{\text{Feline}}$ of Q' is specialized in $\boxed{\text{Cat : Poenga}}$ because it is unified with $\boxed{\text{Cat : Poenga}}$ of F_2, which is a cut point of this graph (it is an individual vertex). Finally, Q'' is unified with F_1, and the resulting request is the empty graph.

By keeping a trace of previous unifications, a specialization of the initial request can be built.

5.2 Tableaux vs. Graph rules

The language defined in the first part of this article is more expressive than the one of graph rules, since all First-Order Logic formulae without function can be represented and used in proofs. Formulae corresponding to simple CG rules are of type $\forall x_1...\forall x_n(\Psi(G_1) \rightarrow \exists y_1...\exists y_p\Psi(G_2))$, where :

- G_1 and G_2 are respectively the hypothesis and the conclusion of the rule,
- $x_1, ..., x_n$ are the variables associated to generic concepts of G_1,
- $y_1, ..., y_p$ are the variables associated to generic concepts of G_2 which are not connection points.

Nevertheless, some CGs introduced in the previous sections can be equivalently represented by sets of Simple CG rules. Of course, it is the case of CGs whose all complex entries are of the form $G_1 \Rightarrow G_2$, where G_1 and G_2 are SCGs, and such that all coreference links are internal to each entry (i.e. there is no variable of an entry declared outside the entry).

Another form easy to represent in CG rules is a CG which is a *complex rule*. A complex rule is a CG of type $[G_1 \Rightarrow G_2]$ where G_1 is a SCG, and G_2 is either a SCG or a complex rule. More generally, a complex rule is a CG of the form $[G_1 \Rightarrow [G_2 \Rightarrow [...[G_{n-1} \Rightarrow G_n]...]]$. Such complex rule is equivalent to the graph rule $(G_1 \wedge G_2 \wedge ... \wedge G_{n-1}) \Rightarrow G_n$, where $G_1 \wedge G_2 \wedge ... \wedge G_{n-1}$ is the graph resulting from the join of G_1, G_2,... and G_{n-1} on common variables. Indeed, the formulae associated to these two graphs are equivalent.

Similarly to the language of Sect. 2, in the formalism of graph rules , the introduction of a marker $false$ enables the representation of negative facts. Let G be a CG, if we want to represent the negation of G as a fact of a knowledge base, KB, we insert the CG $G \Rightarrow \boxed{\perp : false}$ into KB. Nevertheless, these type of rules must be processed by an alternative method to unification. For instance, a request Q can be compared to a negative fact $G \Rightarrow \boxed{\perp : false}$. If Q is a specialization of G then the answer to Q is negative.

Let us now compare the mechanisms (tableau method and backward chaining) on a knowledge base consisting of Simple CG rules. The tableau method can be seen as a goal directed computation of forward chaining. In the tableau proof in Ex. 4.4, after assuming the two facts (nodes (1) and (2)) and the rule (node (3)) of the knowledge base, the rule is applied onto a fact of the base. Indeed, the node corresponding to the rule is a γ, so the extension of the tableau produces as left successor the negation of the hypothesis of the rule (node (5)), and as right successor the conclusion of the rule (node (6)). To close the left branch of the tableau, we search for a projection from the hypothesis of the rule into the merging of all facts of the left branch (there is projection from the graph (5) into (2)). This step corresponds to the search of a fact which fulfills the hypothesis of the rule in forward chaining. In Fig 2, the rule R is applied to the graph

F_2 with the same projection as the one which closes the branch of (5) in the tableau. When the branch is closed, a substitution associated to the projection is applied on the conclusion of the rule. This new graph is a specialization of the conclusion of the rule (only variables common with the hypothesis of the rule are specialized).

A CG is in *normal form* if all concept vertices have different markers. A knowledge base is said to be in normal form if the set of Simple CG considered as one global graph is in normal form, and if for each rule $G_1 \Rightarrow G_2$, G_1 and G_2 are in normal form. Finding a projection from the request into the merging of facts in the right branch, is the same as finding this projection in the knowledge base in normal form after the application of the graph rule in forward chaining. Indeed, the graph obtained by the merging of graphs $\Pi(1)$, $\Pi(2)$ and $\Pi(6)$ is equivalent to the graph resulting from the normalization of the base in Fig. 2 with the graph $R(F_2)$: the graphs F_1, F_2 and $R(F_2)$ are joined on the individual concepts with marker *Poenga*.

Some interesting aspects of backward chaining are the following. First, the knowledge base does not need to be in normal form. Indeed, when a unification is performed on a fact, individuals markers are the only cut points of the graph. Unifications with different graphs in which these individual markers appear, add information on these individuals. E.g. the request Q' in Fig. 5 is unified with the graph F_2, then the new request Q'' is unified with F_1, instead of first computing the normal form of the base and then projecting Q' into the normal base. Furthermore, the knowledge base is kept untouched. Only the request changes. Finally, a trace of the reasoning is easy to build and to understand. Indeed a resolution is simply a sequence of graph operations (unifications). For each step, the unified part of the request (which is a subgraph of the request) and the construction of the new request may be visualized (Fig. 4 and 5).

An in-depth comparison of the two methods from the point of view of efficiency calls for implementation.

6   Conclusion and Further Work

In this paper, we presented a sound and complete method of deduction for an extended language of Conceptual Graphs in which all First-Order formulae (without functional terms) can be represented. This method combines analytic tableaux with projection. The tableaux rules decompose complex graphs into atomic ones, and projection is used as the mechanism of closing branches. We compared tableaux for Conceptual Graphs with the inference mechanisms for graph rules (forward and backward chainings defined in [1]).

Tableaux for Conceptual Graphs will be implemented on CoGITo [13], a development environment for knowledge based applications using exclusively Conceptual Graphs. Contrary to other complete systems for Conceptual Graphs ([3], [5] or [6] propose adaptations of Peirce's deduction rules), our deduction system has an essential property in the perspective of an implementation: it is analytic, i.e. in a tableau the eventual successors of a graph are "smaller" graphs.

Further work will study the inclusion of Nested Conceptual Graphs (i.e. graphs with vertices that contain graphs, see [14] [15]) in the proposed systems.

An interesting point on backward chaining is that the knowledge base does not need to be in normal form. It would be interesting to adapt backward chaining to tableaux in order to compute in a branch only partial projections from a request into facts (without the first step of normalization).

Interlacing operations in Graph Theory (projection) and a classical method of Logic (Analytic Tableaux) offers a novel perspective on deduction. It also provides a way to combine results from these two intensively studied fields.

References

1. E. Salvat and M.L. Mugnier. Sound and complete forward and backward chaining of graph rules. In *proceedings of ICCS'96*, volume 1115 of *LNAI*, pages 248–262. Springer-Verlag, 1996.
2. M. Schmidt-Schauß. *Computational Aspects of an Order-Sorted Logic with Term Declarations*, volume 395 of *LNAI*. Springer-Verlag, 1989.
3. J.F. Sowa. *Conceptual Structures, Information Processing in Mind and Machine*. Addison Wesley, 1984.
4. M. Chein and M.L. Mugnier. Conceptual graphs, fundamental notions. *RIA*, 6.4:365–406, 1992.
5. H. v.d. Berg. *Knowledge Graphs and Logic: One of two kinds*. PhD thesis, Universiteit Twente, September 1993.
6. M. Wermelinger. Conceptual graphs and first-order logic. In *proceedings of ICCS'95, Santa Cruz, USA*, volume 954 of *LNAI*, pages 323–337. Springer-Verlag, 1995.
7. R.M. Smullyan. *First-Order Logic*. Springer-Verlag, 1968.
8. P.H. Schmitt and W. Wernecke. Tableau calculus for order sorted logic. In *Sorts and Types in Artificial Intelligence*, volume 418 of *LNAI*, pages 49–60. Springer-Verlag, 1990.
9. B.C. Ghosh. *Conceptual Graph Language: A Language of Logic and Information in Conceptual Structures*. PhD thesis, Asian Institute of Technology, Bangkok, Thailand, February 1996.
10. M.L. Mugnier and M. Chein. Représenter des connaissances et raisonner avec des graphes. *RIA*, 10.1:7–56, 1996.
11. U. Reyle and D.M. Gabbay. Direct deductive computation on discourse representation structures. *Linguistics and Philosophy*, 17:343–390, 1994.
12. G. Kerdiles. Analytic tableaux for an extended language of conceptual graphs. *RR LIRMM*, 97002, 1997.
13. O. Haemmerle. *CoGITo : une plate-forme de développement de logiciel sur les graphes conceptuels*. PhD thesis, Université Montpellier II, France, January 1995.
14. A. Preller, M.L. Mugnier, and M. Chein. Logic for nested graphs. *Computational Intelligence Journal*, 95-02-558, 1995.
15. M.L. Mugnier and M. Chein. Quelques classes de graphes emboîtés équivalentes. *RR LIRMM*, 96063, 1996.

Fuzzy Unification and Resolution Proof Procedure for Fuzzy Conceptual Graph Programs

Tru H. Cao[1], Peter N. Creasy[1], Vilas Wuwongse[2]

School of Information Technology[1]
University of Queensland
Australia 4072
{tru, peter}@it.uq.edu.au

Computer Science and Information Management Program[2]
Asian Institute of Technology
P.O. Box 2754, Bangkok 10501, Thailand
vw@cs.ait.ac.th

Abstract. Fuzzy conceptual graph programs are order-sorted fuzzy logic programs based on fuzzy conceptual graphs (FCGs). In this paper, we develop fuzzy unification and resolution proof procedure, taking into account fuzziness of FCGs and properties of fuzzy reasoning, for FCG programs. General issues of both CG and FCG unifications and resolution procedures are also analysed and solutions to them are proposed. The resolution procedure is proved to be sound with respect to the declarative semantics of FCG programs.

1. Introduction

The notion of fuzzy conceptual graph programs was introduced in [15], based on a new formulation of fuzzy conceptual graphs (FCGs) overcoming shortcomings of previous works ([9, 14]) on FCGs, a system of fuzzy logic ([18]) based on conceptual graphs (CGs) ([12]). The corresponding FCG projection, the declarative semantics of FCG programs and FCG modus ponens were also presented in [15]. The present work deals with the procedural semantics of FCG programs.

In this paper, our attention is focused on the resolution procedure, a typical backward proof procedure of logic programs, in which unification is a key operation ([7]). CG resolution proof procedures have been also developed for CG programs, based on the notion of CG unification ([5, 6, 11]). The main purpose of this paper is to develop FCG unification and the corresponding resolution proof procedure for FCG programs. Although there are differences between CG and FCG programs due to fuzziness and properties of fuzzy reasoning in the latter, there are common issues of both CG and FCG unification operations and resolution procedures, because both CG and FCG programs share common features of order-sorted logic and CG notation.

The paper is organized as follows. Section 2 presents fundamentals of FCG programs. Section 3 discusses general issues relating to unification operations, the normal form, resolution procedures and their completeness of CG/FCG programs. In Section 4, we define FCG unification, taking into account fuzziness of FCGs. In Section 5,

FCG resolution procedure, reflecting properties of fuzzy reasoning, is defined and proved to be sound with respect to (w.r.t.) the declarative semantics of FCG programs. Section 6 is for conclusions and suggestions of further research.

2. Fundamentals of FCG Programs

This section presents the fundamentals of FCG programs. More details can be found in [15, 4].

2.1. Fuzzy Conceptual Graphs

An FCG is defined to be a (not necessarily connected) graph the nodes of which are fuzzy concepts and fuzzy conceptual relations, and the directed edges of which link the conceptual relation nodes to their neighbor concept nodes. Fuzzy concept nodes are possibly joined by coreference links indicating that the respective referents are identical.

Fuzzy values in FCGs are fuzzy sets ([16]) taken from a *fuzzy truth value* set and a *fuzzy measure* set. The fuzzy truth value set is the union of the sets of TRUE, FALSE and UNKNOWN characteristic values, which are denoted by **T**, **F** and **U**, respectively. For the sake of readability, we use linguistic labels to denote fuzzy set values.

A fuzzy conceptual relation is a conceptual relation fuzzified by a *compatibility degree* representing the compatibility of its neighbor concepts in the relation. Fuzzy concepts can be fuzzy entity concepts or fuzzy attribute concepts. A fuzzy entity concept consists of a conceptual type, a referent and a compatibility degree representing the compatibility of the referent to the type. Compatibility degrees are values in the fuzzy truth value set. A fuzzy attribute concept consists of a measure type, a referent and a fuzzy measure, which is a value in the fuzzy measure set.

2.2. FCG Projection

Like CG projection, FCG projection is also a key operation that relates graph operations to logical interpretations and is used to define the semantics of FCGs and the subsumption relation on them. Unlike CG projections, FCG projections may have *mismatching degrees* due to mismatching between fuzzy values in FCGs.

A projection Π of an FCG g_1 onto an FCG g_2 with mismatching degree $\varepsilon_\Pi \in [0,1]$ is denoted by $g_2 \leq_{\varepsilon_\Pi} g_1$, where ε_Π is the maximum of the mismatching degrees between fuzzy values in all concept and conceptual relation projections in Π. When $\varepsilon_\Pi = 0$, we simply write $g_2 \leq g_1$.

In an FCG projection, if a generic concept is projected onto an individual concept, the generic referent is bound to the individual referent. Furthermore, when two or more concepts are projected onto concepts that have a coreference relation, they will also have a coreference relation due to the projection. In other words, an FCG projection Π produces two related operators, a *referent specialization* operator, denoted by ρ_Π, and a *coreference partition* operator, denoted by φ_Π.

2.3. FCG Programs

An FCG program is a set of *FCG clauses*, each of which is either an FCG fact or an FCG rule of the form **if** u **then** v, where u and v are FCGs. For example, the FCG program in Figure 2.1 consists of one FCG fact, saying "My apple is *fairly red*", and one FCG rule, saying "If an apple is *red*, then it is *ripe*".

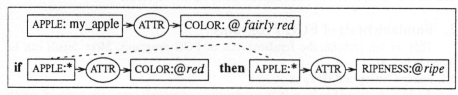

Fig. 2.1. An FCG program

FCG modus ponens states that, if u^+ is an FCG fact and $u^+ \leq_{\varepsilon_\Pi} u$, for some projection Π, then {if u then v, u^+} $\vDash v^+$, where $v^+ = \rho_\Pi \varphi_\Pi(v + \varepsilon_\Pi)$. For example, from the FCG program in Figure 2.1, we can derive the FCG [APPLE: my_apple]$\rightarrow$(ATTR)$\rightarrow$ [RIPENESS: @*ripe+md(red / fairly red)*], where *md(red / fairly red)* denotes the mismatching degree of *red* to *fairly red*.

Similar to a type being subsumed by its super-types, a fuzzy set value A is subsumed by a fuzzy set value B if $A \subseteq B$, on the basis of the entailment principle ([17]). So, $A+\varepsilon$, where $\varepsilon \in [0,1]$, subsumes A, for every fuzzy set value A.

Fuzzy modus ponens ([8]) is consistent with classical modus ponens, that is, when the body of a rule fully matches a fact, then the head of the rule can be derived. Moreover, it also allows one to reason when the body mismatches the fact by some mismatching degree. In this case, we have a degree of indetermination in reasoning, and the conclusion should be more ambiguous and less informational than it is when there is no mismatching. That is, the conclusion should subsume the head, and it is obtained by adding the mismatching degree to fuzzy values in the head.

Throughout this paper, we use the term *query* for *original goal* to distinguish it from *intermediate goals* created during a resolution process. When such a distinction is not needed, we use the term *goal* to mean either an original or intermediate goal.

3. General Issues of CG/FCG Unification and Resolution

CG/FCG unification and resolution proof procedure are closely related to the way a type hierarchy is interpreted, which can be *lattice-theoretic* or *order-theoretic*, and to the way the type hierarchy and the axiomatic part of a knowledge base are coupled, which can be *loose* or *close*. However, these have not been discussed in the previous works [4], [6] and [11] on CG/FCG programs. Another issue is *early type resolution* that is necessary not only for the efficiency, but also for the completeness of a CG/FCG resolution procedure.

3.1. Lattice-Theoretic and Order-Theoretic Interpretations

With the lattice-theoretic interpretation, the greatest lower bound (glb), or the maximal common sub-type, of two types is interpreted by the intersection of their interpretation sets; with the order-theoretic interpretation, it is interpreted by only a sub-set of this intersection ([3, 13]). An example of the order-theoretic interpretation in [13] is that, historic landmarks are the glb of churches and old buildings:

$$\underbrace{\text{CHURCH} \qquad \text{OLD-BUILDING}}_{\text{HISTORIC-LANDMARK}}$$

but an old church (an element of both CHURCH and OLD-BUILDING) is not necessarily a historic landmark (an element of HISTORIC-LANDMARK).

Both [6] and [11] implicitly assumed the lattice-theoretic interpretation. However, under this interpretation, program facts have to be in the *normal form*[1] so that a resolution procedure can be complete. Otherwise, for example, the query G=[HISTORIC-LANDMARK: $*x$] can never be resolved by the program P={[CHURCH: St. Lucia], [OLD-BUILDING: St. Lucia]}[2], when G is actually satisfiable by P with the solution x=St. Lucia. In [6], but not in [11], the set of CG facts of a program was required to be in the normal form for the resolution procedure. In contrast, if the interpretation is order-theoretic, then G is not a logical consequence of P and P cannot be normalized as {[HISTORIC-LANDMARK: St. Lucia]}.

3.2. Loose and Close Couplings

With the loose coupling, the entire taxonomic information is provided by the type hierarchy of a knowledge base, and no rule about relations between types is allowed in its axiomatic part; with the close coupling, there is no such restriction ([3]). [11] implicitly assumed the loose coupling, when requiring that a concept in the head of a rule had the same type as its coreferenced concept in the body of the rule. Without this restriction, [6] implicitly assumed the close coupling.

An example of rules about relations between types is the rule **if** [CHURCH: $*x$]$\rightarrow$(p) **then** [HISTORIC-LANDMARK: $*x$]$\rightarrow$(r), which says "If a church has property p, then it is a historic landmark with property r".

The lattice-theoretic interpretation and the loose coupling are more restrictive than the order-theoretic interpretation and the close coupling, respectively. The lattice-theoretic interpretaion is not quite a reasonable assumption to make in practice, because it would demand an exponential number of type intersection labels when the size of a basic type set grows ([1]). The order-theoretic interpretation is more appropriate for applications in Artificial Intelligence ([3, 13]). On the other hand, the close coupling is essential for a natural language understanding system ([3]). A type hierarchy only says that an object belongs to all super-types of its type, but the close coupling allows one to express in the axiomatic part of a knowledge base rules stating when an object belongs to a sub-type of its type.

3.3. Early Type Resolution

In a CG/FCG goal, not only conceptual relations among objects need to be satisfied, but also types of objects do too. When a sub-graph of a goal is resolved, it is erased from the goal. What needs to be considered is the common concepts in both the sub-graph and the remainder of the goal. In fact, the types of the referents of these con-

---

[1] In the normal form, there is no coreference link and no individual marker occurring in more than one concept in a set of CGs ([6]).

[2] Individual markers represented by the same character string are assumed to be identical.

cepts are already resolved, only the conceptual relations involving these referents in the remainder of the goal are not. In [6] and [11], the types of such connection concepts were kept unchanged in goal remainders. By early type resolution we mean the change of the types of connection concepts as early as possible. Specifically, in the CG case, the type of a connection concept is replaced by the universal type 'T' (at the top of a type hierarchy).

Early type resolution is necessary not only for the efficiency, but also for the completeness of a resolution procedure when either the interpretation is order-theoretic, or the coupling is close, or sets of CG/FCG facts are not in the normal form. Let us consider the two following examples.

Example 3.1 In Figure 3.1 G is clearly satisfiable by P with the solution x=St. Lucia. However, following the procedures in [6] and [11], where early type resolution was not applied, the resolvent of G with the first fact would be [CHURCH: St. Lucia]→(r) that could never be resolved. With early type resolution, G can be first resolved by the first fact, then the new goal [T: St. Lucia]→(r) is resolvable further by the second fact.

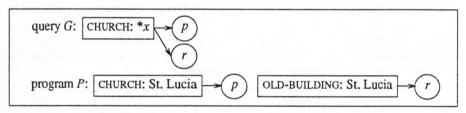

Fig. 3.1. CG program and query to exemplify early type resolution

Example 3.2 Under the close coupling, early type resolution is necessary for completeness even when program facts are in the normal form. In Figure 3.2, it is also clear that G is satisfiable by P with the solution x=St. Lucia. Following the procedure in [6], G was first unified with the head of the rule and the new goal would be [HISTORIC-LANDMARK: *]→(p), →(r) that could not be resolved further. With early type resolution, the result of the first step is [T: *]→(r). Then, it is joined with the body of the rule to produce the new goal [CHURCH: *]→(p), →(r) that is resolvable by the fact in P.

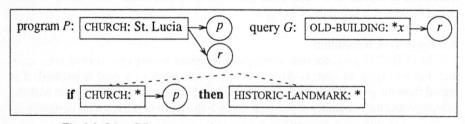

Fig. 3.2. Other CG program and query to exemplify early type resolution

Therefore, we apply early type resolution to FCG unification and FCG resolution procedure. We also generally assume the order-theoretic interpretation and the close coupling in FCG programs, and do not require FCGs forming facts and rules to be in the normal form. Details as well as issues of FCG unification and resolution procedure relating to fuzziness and fuzzy reasoning in FCG programs are presented in the following sections.

4. Fuzzy Unification

In this section, FCG unification and its related operators are defined. For calculation illustration, throughout this paper we suppose the following relations between fuzzy set values denoted by some linguistic labels:

$$red \subseteq fairly\ red = red + \varepsilon_0,\ ripe \subseteq fairly\ ripe = ripe + \varepsilon_0$$

which imply $md(red / fairly\ red) = md(ripe / fairly\ ripe) = \varepsilon_0$.

4.1. Tolerance Degree

As mentioned, a fuzzy rule allows us to draw conclusions not only when its body fully matches a fact, but also when there are mismatching degrees. Let us consider the FCG program P and the query G in Figure 4.1. Applying FCG modus ponens, we obtain the FCG [APPLE:#1]→(ATTR)→[RIPENESS:@*fairly ripe*] which is a projection of, and consequently implies G. In other words, G is a logical consequence of P, so it should be resolvable by P.

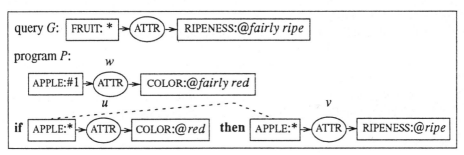

Fig. 4.1. FCG program and query to exemplify the notion of tolerance degrees

However, if we used the conventional resolution mechanism, whereby the resolvent would be the FCG u, which could not be projected onto the FCG w, then G would not be satisfied by P. In fact, there is a *tolerance degree* ([4]) of G to the FCG v, which is of *fairly ripe* to *ripe* and value ε_0 in this case. This tolerance degree should be propagated back to u to produce the resolvent [APPLE:*]→(ATTR)→ [COLOR:@*red*+ε_0], which is resolvable by w.

Therefore, for completeness, the tolerance degree has to be a parameter of FCG unification and taken into account in FCG resolution procedure. Actually, the tolerance degree, used in fuzzy backward chainings, is dual to the mismatching degree, used in fuzzy forward chainings.

Definition 4.1 (Tolerance Degree) Let A and B be two fuzzy set values on a universe U and $A \subseteq B$. The tolerance degree of B to A, denoted by $td(B/A)$, is defined to be

$\text{Inf}\{\mu_B(u) - \mu_A(u) \mid u \in U \text{ and } \mu_B(u) \neq 1\}$. If $\mu_B(u)=1$ for every u in U, i.e., B is an absolutely unknown fuzzy value on U, then $td(B/A)$ is defined to be 1.

In fact, $td(B/A)$ is the greatest value in the real interval $[0,1]$ such that when it is added to A, the result is still subsumed by B.

Property 4.1 Let A and B be two fuzzy set values on a universe U and $A \subseteq B$. Then, $td(B/A)=\text{Sup}\{\varepsilon \mid \varepsilon \in [0,1] \text{ and } A+\varepsilon \subseteq B\}$.
Proof. Let I be the absolutely unknown fuzzy value on U ($\mu_I(u)=1$ for every u in U).
1. $B=I$: $td(B/A)=1$ and $A+td(B/A)=A+1=I=B$.
2. $B \subset I$: $\varepsilon > td(B/A) \Rightarrow \exists u \in U$: $\mu_B(u) \neq 1$ and $\mu_B(u) - \mu_A(u) < \varepsilon \Rightarrow A+\varepsilon \nsubseteq B$.

4.2. FCG Unification

A unification of an FCG onto another one includes unifications of their conceptual referents. In referent unifications, it is necessary to distinguish two kinds of generic markers in the head of a rule: those whose concepts are coreferenced with ones in the body of the rule, and the others whose concepts are not. In fact, generic markers of different kinds are treated differently in backward chainings: a generic marker of the former can be unified with an individual marker, i.e., treated as a variable, but the latter cannot ([5, 4, 6, 11]). This is the reason why only concepts with generic markers of the former were defined to be *cut points* in [11].

Specifically, in FCG programs, we distinguish between ([4]):
1. VAR generic markers, the concepts of which occur in an FCG query, or in the body of an FCG rule, or in the head of an FCG rule and coreferenced with concepts in the body of the rule.
2. NON-VAR generic markers, the concepts of which occur in an FCG fact, or in the head of an FCG rule but are not coreferenced with any concept in the body of the rule.

The attribute of a generic marker in an intermediate goal depends on whether it is bound to a VAR generic marker or a NON-VAR one through unifications during a resolution process.

So, in backward chainings of FCG programs, we have three kinds of referent markers distinguished: individual markers, VAR generic markers, and NON-VAR generic markers. Only a VAR generic marker can be unified with any other referent marker.

Definition 4.2 (Unifiable Referents) Two conceptual referents are said to be unifiable if and only if (iff) either of the following conditions holds:
1. They are the same individual marker, or are identical by coreference, or
2. Either of them is a VAR generic marker.

Two concepts are said to be *referent-unifiable* iff their referents are unifiable.

In [6], a unification of two CGs is defined as a pair of projections onto their *immediate common specialization*. In [11], a unification of a CG, e.g. a goal, with another one, e.g. the head of a rule, has two consecutive steps: (i) a sub-set of the cut points of the latter is unified with a set of concepts of the former, then (ii) the unifiable pieces of the former are projected onto the latter. Here, we define a unification as a 'single' matching operation of two FCGs, similarly to CG/FCG projection, whence all CG type and relation labels and graph structures could be exploited early at the beginning to guide the pattern matching. Before defining FCG unification, we define unifications of FCG components, i.e., concepts and conceptual relations.

Definition 4.3 (Fuzzy Entity Concept Unification) A concept b:$[T_b$: *referent*(b) | $c_b]$ is said to be totally unifiable onto a concept a:$[T_a$: *referent*(a) | $c_a]$ with tolerance degree ε $\in [0,1]$ iff:
   1. *referent*(b) and *referent*(a) are unifiable, and
   2. $[T_a$: _ | $c_a] \le [T_b$: _ | $c_b]$, and
   3. (i)   if $c_a \in \mathbf{T}$ and $T_a \subseteq T_b$ then $\varepsilon = td(c_b/c_a)$, else
      (ii) if $c_a \in \mathbf{T}$ and $T_a \subseteq \mathrm{T}\backslash T_b$ then $\varepsilon = td(\neg c_b/c_a)$, else
      (iii) if $c_a \in \mathbf{F}$ and $T_a \supseteq T_b$ then $\varepsilon = td(c_b/c_a)$, else
      (iv) if $c_a \in \mathbf{F}$ and $T_a \supseteq \mathrm{T}\backslash T_b$ then $\varepsilon = td(\neg c_b/c_a)$, else
      (v)   if $c_a \in \mathbf{U}$ and $T_a = T_b$ then $\varepsilon = td(c_b/c_a)$, else
      (vi) if $c_a \in \mathbf{U}$ and $T_a = \mathrm{T}\backslash T_b$ then $\varepsilon = td(\neg c_b/c_a)$.

In condition 2, the use of underscores in the referent fields means that the referents of the two concepts are ignored (or equivalently, substituted by the same referent) in the FCG projection relation. Like in FCG projection, cases (ii), (iv) and (vi) take into consideration the equivalence of *dual concepts* ([15]). Note that two concepts can be referent-unifiable when neither of them is totally unifiable onto the other.

Definition 4.4 (Fuzzy Attribute Concept Unification) A concept b:$[T_b$: *referent*(b) @ $m_b]$ is said to be totally unifiable onto a concept a:$[T_a$: *referent*(a) @ $m_a]$ with tolerance degree $\varepsilon \in [0,1]$ iff:
   1. *referent*(b) and *referent*(a) are unifiable, and
   2. $[T_a$: _ @ $m_a] \le [T_b$: _ @ $m_b]$, and
   3. $\varepsilon = td(m_b/m_a)$.

Definition 4.5 (Fuzzy Conceptual Relation Unification) A conceptual relation b:$(T_b$ | $c_b)$ is said to be totally unifiable onto a conceptual relation a:$(T_a$ | $c_a)$ with tolerance degree $\varepsilon \in [0,1]$ iff:
   1. $a \le b$, and
   2. $\varepsilon = td(c_b/c_a)$.

In the following, we use *referent-unified*(a, b) to denote that the referents of concept a and concept b are unified (only if they are unifiable referents), or coreferenced (inclusively, a and b have the same individual referent, or they coincide). We also use $td(b/a)$ to denote the tolerance degree of the (total) unification of b onto a, where both a and b are either concepts or conceptual relations.

Definition 4.6 (FCG unification) An FCG g_1 is said to be unifiable onto an FCG g_2 with tolerance degree $\varepsilon_\theta \in [0,1]$ iff there exists a mapping $\theta: g_1 \to g_2$ such that:

1. $\forall c \in V_c$: *referent-unified*(c, θc), and

2. $\forall a, a' \in V_c$:

(a) *referent-unified*(θa, $\theta a'$) if (i) *referent-unified*(a, a'), or (ii) $\exists b, b' \in V_c$: *referent-unified*(a, b), *referent-unified*(a', b') and *referent-unified*(θb, $\theta b'$), and

(b) *referent-unified*(a, a') if (i) *referent-unified*(θa, $\theta a'$), or (ii) $\exists b, b' \in V_c$: *referent-unified*(b, b'), *referent-unified*(θa, θb), *referent-unified*($\theta a'$, $\theta b'$), and

3. $\forall r \in V_r$: r is totally unifiable onto θr, and

$\forall i \in \{1,2, ..., degree(r)\}$: *referent-unified*(*neighbor*(θr, i), θ*neighbor*(r, i)), and

4. No VAR generic maker is unified with different individual markers or non-coreferenced NON-VAR generic markers, and

5. $\varepsilon_\theta = Min\{\varepsilon \mid \varepsilon = td(c/\theta c)$ where $c \in V_c$ and c is totally unifiable onto θc, or $\varepsilon = td(r/\theta r)$ where $r \in V_r\}$,

where V_c and V_r are respectively the sets of concept and conceptual relation nodes of g_1.

For example, Figure 4.2 illustrates a unification of the sub-graph g_1 of the FCG G onto the (unconnected) FCG g_2, supposing that both the generic markers in g_2 are VAR ones but their concepts are not necessarily coreferenced. In this case, both the generic markers in g_2 are unified with the individual marker 'John', and the tolerance degree of the unification is $\varepsilon = Min\{td(fairly\ true\ /\ true), td(young\ /\ very\ young)\}$.

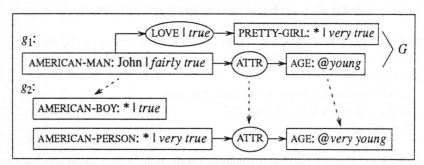

Fig. 4.2. An FCG unification

The main difference between FCG unification and FCG projection is that, in FCG unification there are referent unifications and tolerance degrees are computed instead of mismatching degrees. Efficient matching techniques of CG/FCG projection could also be applied for FCG unification. Now, we introduce the notion of *satisfiable nodes* ([4]), which is a basic notion in FCG resolution procedure.

Definition 4.7 (Satisfiable Nodes) The satisfiable nodes of g_1 by a unification $\theta: g_1 \to g_2$ are the following:

1. All conceptual relations, and
2. Every concept c that is totally unifiable onto θc.

4.3. FCG Unification-Related Operators

Like FCG projection, FCG unification also has related operators, which are the referent specialization and coreference partition operators. Besides these operators, it has the *resolution operator* to erase from an FCG the nodes that are satisfiable through a unification.

Definition 4.8 (Referent Specialization and Coreference Partition Operators) Given an FCG unification θ, the referent specialization and coreference partition operators, respectively denoted by ρ_θ and ϕ_θ, are defined as follows. For every pair of concepts c and c' that are defined to be referent-unified in θ:

1. if *referent(c)* is a VAR generic marker, and *referent(c')* is an individual marker, then $(c, referent(c')) \in \rho_\theta$, and

2. if both *referent(c)* and *referent(c')* are generic markers, then $\exists p \in \phi_\theta: c, c' \in p$.

   When a generic marker occurs in a coreference partition with a NON-VAR one, its attribute is bound to NON-VAR.

For example, if θ is the FCG unification in Figure 4.2, we have:
$\rho_\theta = \{([AMERICAN-BOY: * | true], John)\}$
$\phi_\theta = \{\{[AMERICAN-BOY: * | true], [AMERICAN-PERSON: * | very true]\}\}$

Definition 4.9 (Resolution Operator) Given a unification $\theta: g_1 \rightarrow g_2$, the resolution operator w.r.t. θ, denoted by σ_θ, is defined as the set of all the satisfiable nodes of g_1 by θ.

The result of application of σ_θ to an FCG G is a new FCG $\sigma_\theta G$ derived from G as follows:

1. Erase from G all conceptual relation nodes occurring in σ_θ, and

2. Erase from G all concept nodes that occur in σ_θ but are not neighbors of any conceptual relation not occurring in σ_θ.

3. If a concept node of G occurs in σ_θ but is a neighbor of a conceptual relation that does not occur in σ_θ, then (virtually) erase from this concept its type and fuzzy value fields. It remains in $\sigma_\theta G$ as a *bare concept* ([4]).

For example, given the unification θ in Figure 4.2, we have:
$\sigma_\theta = \{[AMERICAN-MAN: John | fairly true], (ATTR), [AGE: @young]\}$
$\sigma_\theta G = [John] \rightarrow (LOVE | true) \rightarrow [PRETTY-GIRL: * | very true]$

In fact, the creation of bare concepts is a realization of early type resolution introduced in sub-section 3.3, and only their referents need to be further matched in a resolution process.

5. Fuzzy Resolution Proof Procedure

In each step of a CG/FCG resolution process, a sub-graph of a goal is unified onto a fact or the head of a rule in a program. This unifiable sub-graph is then erased from the goal and the remainder needs to be resolved further. Due to unifications, the referents of the common concepts between the sub-graph and the remainder may become NON-VAR generic markers, although in a query they are not. In this case, the remainder can never be satisfied, because NON-VAR generic markers cannot be unified with any other referent markers. So, for efficiency, NON-VAR generic markers should not occur in resolvents. This explains the significance of cut points and the *piece* resolution mechanism introduced in [11]: when piece sub-graphs, rather than arbitrary sub-graphs, are resolved in each resolution step, NON-VAR generic markers will not occur in resolvents. In fact, avoidance of NON-VAR generic markers in resolvents just effects the efficiency, rather than the soundness or completeness of a CG/FCG resolution procedure.

5.1. Resolvents of FCG Clauses

In [4], a resolvable part of an FCG goal could be a subgraph as large as possible, like in [11]. In addition, a resolvable FCG sub-graph could be single concepts. [11] appeared to miss the case that a piece could be a single concept, without which the resolution procedure would not be complete, as given the program in Figure 3.1 for instance. Moreover, a resolvable sub-graph as single concepts can only be realized with early type resolution, which was not applied in [11].

However, in [4] NON-VAR generic markers were not restricted in FCG resolvents. So, for efficiency, we now modify the definitions of FCG resolvents given in [4] by adding conditions preventing occurrences of NON-VAR generic markers in FCG resolvents.

Definition 5.1 (Resolvent of an FCG goal and an FCG fact) Let G be an FCG goal and u an FCG fact. If there exists a unification θ of a sub-graph g of G onto u such that:
1. For every concept c of g, if c is referent-unified with a concept of a NON-VAR generic marker, then c has to be totally unifiable onto θc, and
2. No common concept between g and the remainder without g of G is referent-unified with a concept of a NON-VAR generic marker,

then the resolvent w.r.t. θ of G and u is $R_\theta(G, u) = \rho_\theta \varphi_\theta \sigma_\theta G$.

Definition 5.2 (Resolvent of an FCG goal and an FCG rule) Let G be an FCG goal and **if** u **then** v an FCG rule. If there exists a unification θ of a sub-graph g of G onto v such that:
1. For every concept c of g, if c is referent-unified with a concept of a NON-VAR generic marker, then c has to be totally unifiable onto θc, and
2. No common concept between g and the remainder without g of G is referent-unified with a concept of a NON-VAR generic marker, and
3. No concept of a VAR generic marker in v is referent-unified with a concept of a NON-VAR generic marker,

then the resolvent w.r.t. θ of G and **if** u **then** v is $R_\theta(G, $ **if** u **then** $v) = \rho_\theta \varphi_\theta [\sigma_\theta G \ (u + \varepsilon_\theta)]$.

Figure 5.1 shows an FCG query, an FCG rule and their resolvent. The rule says "If a church is *very old* and located at the place where a president was born, then it is *true* that the church is a historic landmark and it is *very true* that the president (has) visited it". The resolvent is computed as follows:

$\theta: G \rightarrow v$

$\varphi_\theta = \{\{[\text{DEMOCRAT: *}], [\text{PRESIDENT: *}]\}\}$

$\rho_\theta = \{([\text{HISTORIC-LANDMARK: * | } true], \text{St. Lucia})\}$

$\sigma_\theta = \{(\text{VISIT | } true), [\text{OLD-BUILDING: St. Lucia | } fairly\ true]\}$

$\varepsilon_\theta = \text{Min}\{td(true\ /\ very\ true), td(fairly\ true\ /\ true)\}$.

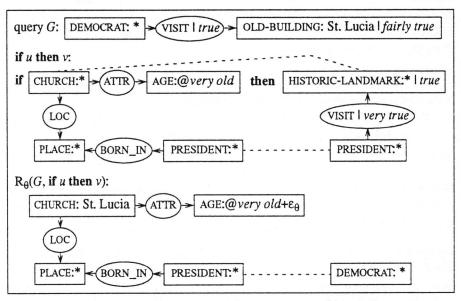

Fig. 5.1. The resolvent of an FCG query and an FCG rule

In fact, a resolvable FCG sub-graph corresponds to a set of resolvable CG pieces defined in [11], whereby NON-VAR generic makers do not occur in resolvents.

Property 5.1 Conditions 1, 2 and 3 in Definitions 5.1 and 5.2 are necessary and sufficient so that NON-VAR generic makers do not occur in FCG resolvents.
Proof. Sufficient: conditions 1 and 2 guarantee that NON-VAR generic makers do not occur in $\rho_\theta\varphi_\theta\sigma_\theta G$, if they do not occur in G before. Condition 3 guarantees that NON-VAR generic makers do not occur in $\rho_\theta\varphi_\theta u$. Necessary: if any of conditions 1, 2 and 3 does not hold, then NON-VAR generic makers will occur in $\rho_\theta\varphi_\theta\sigma_\theta G$ or in $\rho_\theta\varphi_\theta u$.

5.2. FCG Resolution Procedure
Similar to any resolution proof procedure of logic programs, FCG resolution procedure starts with a given FCG goal, resolves it step by step with facts and rules of a given FCG program until the empty FCG is obtained, i.e., success, or until failure.

Definition 5.3 (FCG Resolution Procedure) Let P be an FCG program and G be an FCG goal. At each step i, the current goal G_{i-1} is resolved with a selected clause C_i in P by an FCG unification θ_i, where $G_0=G$, $G_i=R_{\theta_i}(G_{i-1}, C_i)$. Each specific consequence $\{\theta_1, \theta_2,..., \theta_n\}$ defines a specific resolution path of G and P, where n is the length of the path. A resolution path of length n is said to be successful if G_n is an empty FCG. The resolution procedure is said to be successful if there exists a successful resolution path.

Definition 5.4 (FCG-Computed Answer) Let P be an FCG program and G be an FCG goal. If there exists a successful resolution path $\{\theta_1, \theta_2,..., \theta_n\}$ of G and P, then the FCG-computed answer w.r.t. this resolution path is defined to be $\rho\varphi G$, where $\rho=\rho_{\theta_n}\rho_{\theta_{n-1}} \cdots \rho_{\theta_1}$ and $\varphi=\varphi_{\theta_n}\varphi_{\theta_{n-1}} \cdots \varphi_{\theta_1}$.

For example, let us revisit the FCG program P and the FCG query in Figure 4.1. A successful resolution path of G and P is specified by:

$\theta_1: G \to v$, $\varphi_{\theta_1}=\{\{[\text{FRUIT}:*], [\text{APPLE}:*]\}\}$, $\rho_{\theta_1}=\{\}$, $\varepsilon_{\theta_1}=td(\textit{fairly ripe} / \textit{ripe})=\varepsilon_0$

$\theta_2: \rho_{\theta_1}\varphi_{\theta_1}(u+\varepsilon_{\theta_1}) \to w$, $\varphi_{\theta_2}=\{\}$, $\rho_{\theta_2}=\{([\text{APPLE}:*], \#1)\}$, $\varepsilon_{\theta_2}=0$

So, the FCG-computed answer to G is: $[\text{FRUIT}:\#1]\to(\text{ATTR})\to[\text{RIPENESS}:@\textit{fairly ripe}]$.

Definition 5.5 (FCG-Correct Answer) Let P be an FCG program and G be an FCG goal. An FCG-computed answer to G is said to be an FCG-correct answer if it is a logical consequence of P.

Lemma 5.1 Let G be an FCG goal and u an FCG fact. If there exists an FCG unification θ from G onto u such that $R_\theta(G, u)$ is empty, then θ is also a projection from G onto u with mismatching degree 0.

Proof. First, note that two concepts in u are referent-unifiable only if they are coreferenced, because all generic markers in an FCG fact are defined to be NON-VAR generic makers. Also note that all concept and conceptual relation nodes in G are involved in, and satisfiable by, θ because $R_\theta(G, u)$ is empty. We have that θ satisfies the conditions of FCG unification (Def. 4.6). We prove in [4] that these conditions make θ satisfy the conditions of FCG projection, that is, θ itself is an FCG projection from G onto u. Since $R_\theta(G, u)$ is empty, the mismatching degree of this projection is 0.

Theorem 5.1 (Soundness of FCG Resolution Procedure) Let P be an FCG program and G be an FCG goal. Then every FCG-computed answer to G is an FCG-correct answer.

Proof. We prove that every FCG-computed answer to G is a logical consequence of P, by induction on resolution path lengths.

1. $n=1$: The resolvent of G and a selected clause C of P is empty, so C must be an FCG fact. Lemma 5.1 gives $C \leq G$ and Property 3.8 in [15] gives $C \leq \rho\varphi G$, where ρ and φ are respectively the referent specialization and coreference partition operators of the FCG unification/projection from G onto C. Theorem 4.1 in [15] gives $\rho\varphi G$ is a logical consequence of C, and consequently, of P.

2. Induction Hypothesis: Suppose that it is true in the case of resolution paths of length n-1.

3. Consider a resolution path of length n. Let **if** u **then** v be a clause of P that takes part in the first step of the resolution path (The proof is similar if this first selected clause is an FCG fact). Let $\rho'=\rho_{\theta_n}\rho_{\theta_{n-1}} \cdots \rho_{\theta_2}$ and $\varphi'=\varphi_{\theta_n}\varphi_{\theta_{n-1}} \cdots \varphi_{\theta_2}$. On the basis of the induction hypothesis:

$$\rho'\varphi'\rho_{\theta_1}\varphi_{\theta_1}[\sigma_{\theta_1}G\,(u+\varepsilon_{\theta_1})]=\rho'\rho_{\theta_1}\varphi'\varphi_{\theta_1}[\sigma_{\theta_1}G\,(u+\varepsilon_{\theta_1})]=\rho\varphi[\sigma_{\theta_1}G\,(u+\varepsilon_{\theta_1})]$$

is a logical consequence of P (Note that concepts that are specialized by ρ_{θ_1} are not specialized further by φ', so the positions of ρ_{θ_1} and φ' can be exchanged).

It is obvious that $\rho\varphi[\sigma_{\theta_1}G\,(u+\varepsilon_{\theta_1})]\leq_{\varepsilon_{\theta_1}}u$, so on the basis of FCG modus ponens: $\{$**if** u **then** v, $\rho\varphi[\sigma_{\theta_1}G\,(u+\varepsilon_{\theta_1})]\}\vDash\rho\varphi(v+\varepsilon_{\theta_1})$. At this point, we have:

$$P\vDash\{\rho\varphi[\sigma_{\theta_1}G\,(u+\varepsilon_{\theta_1})],\rho\varphi(v+\varepsilon_{\theta_1})\}\vDash\rho\varphi[\sigma_{\theta_1}G\,(v+\varepsilon_{\theta_1})]$$

What remains is to prove that $\rho\varphi[\sigma_{\theta_1}G\,(v+\varepsilon_{\theta_1})]\leq\rho\varphi G$. Intuitively, nodes in $\rho\varphi G$ that remain after the first step of the resolution path can be projected onto themselves in $\rho\varphi(\sigma_{\theta_1}G)$, and those that are satisfiable by θ_1 can be projected onto nodes in $\rho\varphi(v+\varepsilon_{\theta_1})$, onto which they are unified by θ_1. Clearly, the mismatching degrees of all these FCG projections are 0. The actual FCG projection from $\rho\varphi G$ onto $\rho\varphi[\sigma_{\theta_1}G\,(v+\varepsilon_{\theta_1})]$ with mismatching degree 0 is specified in [4].

6. Conclusions

General issues of CG and FCG unifications and resolution proof procedures have been discussed, and early type resolution has been proved to be necessary not only for the efficiency but also for the completeness of a CG/FCG resolution procedure. FCG unification and resolution procedure have been defined for FCG programs, on a general assumption of the order-theoretic type hierarchy interpretation and the close coupling between the type hierarchy and the axiomatic part of a knowledge base. On the basis of fuzzy modus ponens, tolerance degrees have been introduced in FCG unification for the completeness of FCG resolution procedure. The procedure has been proved to be sound w.r.t. the declarative semantics of FCG programs.

When only crisp measures and *absolutely true* truth value are used, FCG programs are reduced to CG programs. Furthermore, by using *absolutely false* truth value one can express negations. Generally, truth qualifications with different degrees of both TRUE and FALSE characteristics are expressible in FCG programs.

Compared to the fuzzy logic programming systems of [2] and [10], FCG programs have the following main differences. First, they are based on FCGs, a kind of order-sorted fuzzy logic which employs type inheritance under uncertainty, while the two mentioned systems are unsorted. Second, FCG programs inherit the graphical feature and other advantages of CGs. Third, the subsumption relations of both CGs and fuzzy values are exploited in reasoning. Finally, FCG programs have a Prolog-like resolution proof procedure with inheritance integrated directly into the fuzzy unification process. With these features, FCG programs could be applied to automated fuzzy reasoning systems with inheritance and taxonomic information.

We are currently studying and proving the completeness of FCG resolution procedure. Furthermore, at the present, fuzzy truth qualifications are only local at concepts and conceptual relations of FCGs. In order to increase the expressive power of FCG programs, global fuzzy truth qualifications on entire FCGs are desirable. This is among the topics suggested for further research.

References

1. Aït-Kaci, H. & Nasr, R. (1986), *Login: A Logic Programming Language with Built-In Inheritance*. J. of Logic Programming, **3**: 185-215.
2. Baldwin, J.F. & Martin, T.P & Pilsworth, B.W. (1995), *Fril-Fuzzy and Evidential Reasoning in Artificial Intelligence*. John Wiley&Sons, New York.
3. Beierle, C. & Hedtstuck, U. & Pletat, U. & Schmitt, P.H. & Siekmann, J. (1992), *An Order-Sorted Logic for Knowledge Representation Systems*. J. of Artificial Intelligence, **55**: 149-191.
4. Cao, T.H. (1995), *Fuzzy Conceptual Graph Programs*. Master's Thesis, Asian Institute of Technology, December 1995.
5. Ghosh, B.C. & Wuwongse, V. (1995), *A Direct Proof Procedure for Definite Conceptual Graph Programs*. In Ellis & Levinson & Rich & Sowa (eds.): Conceptual Structures: Applications, Implementation and Theory, LNAI No. 954, Springer-Verlag, pp. 158-172.
6. Ghosh, B.C. (1996), *Conceptual Graph Language-A Language of Logic and Information in Conceptual Structures*. PhD Thesis, Asian Institute of Technology.
7. Lloyd, J.W. (1987), Foundations of Logic Programming. Springer-Verlag.
8. Magrez, P. & Smets, P. (1989), *Fuzzy Modus Ponens: A New Model Suitable for Applications in Knowledge-Based Systems*. Int. J. of Intelligent Systems, **4**: 181-200.
9. Morton, S. (1987), *Conceptual Graphs and Fuzziness in Artificial Intelligence*. PhD Thesis, University of Bristol.
10. Mukaidono, M. & Shen, Z. & Ding, L. (1989), *Fundamentals of Fuzzy Prolog*. Int. J. of Approximate Reasoning, **3**:179-194.
11. Salvat, E. & Mugnier, M.L. (1996), *Sound and Complete Forward and Backward Chainings of Graph Rules*. In Eklund & Ellis & Mann (eds.): Conceptual Structures: Knowledge Representation as Interlingua, LNAI No. 1115, Springer-Verlag, pp. 248-262.
12. Sowa, J.F. (1984), *Conceptual Structures: Information Processing in Mind and Machine*. Addison Wesley, Massachusetts.
13. Wermelinger, M. & Lopes, J. G. (1994), Basic Conceptual Structures Theory. In Tepfenhart & Dick & Sowa (eds.): Conceptual Structures: Current Practices, LNAI 835, Springer-Verlag, pp. 144-159.
14. Wuwongse, V. & Manzano, M. (1993), *Fuzzy Conceptual Graphs*. In Mineau & Moulin (eds.): Conceptual Graphs for Knowledge Representation, LNAI No.699, Springer-Verlag, pp. 430-449.
15. Wuwongse, V. & Cao, T.H. (1996), *Towards Fuzzy Conceptual Graph Programs*. In Eklund & Ellis & Mann (eds.): Conceptual Structures-Knowledge Representation as Interlingua, LNAI No.1115, Springer-Verlag, pp. 263-276.
16. Zadeh, L.A. (1965), *Fuzzy Sets*. J. of Information and Control, **8**: 338-353.
17. Zadeh, L.A. (1979), *A Theory of Approximate Reasoning*. Machine Intelligence, **9**: 149-194.
18. Zadeh, L.A. (1990), *The Birth and Evolution of Fuzzy Logic*. Int. J. of General Systems, **17**: 95-105.

Reasoning with Type Definitions

M. Leclère

IRIN – IUT de Nantes
2, rue de la Houssinière - BP 92208
44 322 Nantes cedex 03 - France
Ph.: +33 2 40 37 49 01   Fax: +33 2 40 37 49 70
E-mail: leclere@irin.univ-nantes.fr

Abstract. This article presents an extension of the basic model of conceptual graphs : the introduction of type definitions. We choose to consider definitions as sufficient and necessary conditions to belong to a type. We extend the specialization/generalization relation on conceptual graphs to take advantage of these definitions. Type contractions and type expansions are clearly defined. We establish the correspondence with projection by use of the atomic form of conceptual graphs. Finally, we give a logical interpretation of type definitions and prove than the correspondence between logical deduction and generalization relation is maintained.

Keywords: type definitions, contraction, expansion, atomic form, projection, logical interpretation.

1   Introduction

The aim of knowledge representation is to provide formal model which allow to modelize different kinds of knowledge and allow to reason with this knowledge. In conceptual graphs (CGs), the knowledge is split into several levels. The terminological level defines the conceptual vocabulary. It contains the concept type lattice and the relation type poset. These two taxonomies are simply composed of strings, representing concept types or relation types, which are ordered by a specific/generic relation. At the assertional level, one describes some facts by CGs constructed with the conceptual vocabulary.

However, we often dispose of general knowledge which is relevant to terminological level but one can't represent it with a simple specific/generic relation. J. Sowa proposes the formalism of abstractions to answer this need. In this paper, we study a kind of complex terminological knowledge: the type definition.

We start from the basic model of CGs, introduced by J. Sowa in [Sow84] and clarified by M. Chein and M.L. Mugnier in [CM92, MC96], and extend it to take into account type definitions when we are reasoning at assertional level. This extension includes contraction and expansion operations in the specialization rules. The connection with the projection is established once again. After giving a logical interpretation for definition, one demonstrates that the specialization relation always corresponds to a complete and sound set of inference rules.

In section 2, we present the different semantics that we can give to type definitions and we remind the type definition syntax. Then we choose to consider a definition as a set of sufficient and necessary conditions for belonging to a type. Section 3 introduces the four new specialization rules allowing to manipulate the type definitions at assertional level: contraction and expansion of concept types or relation types. Then we show in section 4 how projection allows to compute the new specialization relation. The section 5 is devoted to logical interpretation of type definitions and contraction/expansion rules.

2 Type definition formalism

In the basic model of CGs, the meaning of a type is given by position which hold into the taxonomy of types. The specific/generic relation is thus the only definitional mechanism. Such a representation of meaning for concept types and concept relations is very poor. Often one has some generic information on types which are not expressible by this single mechanism. The mechanism of type definition address this problem by associating a formal description to a type. This description is constructed with atomic types (a type without description) and already defined types. The semantics given to the type/description association determine the reasoning technics to use with this form of knowledge representation.

In this work, we consider that descriptions represent a set of characteristics, properties, attributes which can be shared by objects of the domain of representation. A description is thus the intensional representation of a set of objects. It remains to define the semantics given to the link between a type and its description. In general, two semantics are used:

- the description represents a set of sufficient and necessary conditions to belong to its type. Any object recognized by the description must belong to the type and any instance of type owns the attributes of the description. The description is thus the intensional representation of type. These semantics are given to defined concepts in the KL-ONE derived systems [WS92]. Such semantics are called *definition*;
- the description only represents a set of necessary conditions to belong to its type. These are semantics that is used for natural kinds. In KL-ONE systems, they are given to primitive concepts. These semantics are sometimes called *partial definition*.

In CGs, the descriptions are represented by abstractions (cf. [Sow84] section 3.6). They are conceptual graphs which one or several generic concepts are considered as formal parameters. In the following, we study description mechanism to do some definitions and not to do partial definitions.

Definition 1 *A concept type definition asserts an equivalence between a concept type and a monadic abstraction. We denote $t_c(x) \overset{def}{\Leftrightarrow} D(x)$ the definition of type t_c with x the variable of formal parameter. The formal parameter concept vertex* of $D(x)$ is called the **head** of t_c. In the aristotelician approach, the genus of new type is the type of the head concept and $D(x)$ represents the differentiae from t_c to its genus.

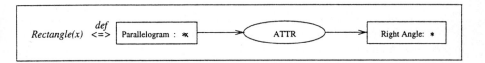

Fig. 1. Definition of a rectangle as a parallelogram with right angles..

Definition 2 *A relation type definition asserts an equivalence between a relation type and a n-ary abstraction. We denote* $t_r(x_1, x_2...x_n) \overset{def}{\Leftrightarrow} D(x_1, x_2...x_n)$ *the definition of type* t_r *with* $x_1, x_2...x_n$ *the variables of formal parameters. The formal parameter concept vertices of* $D(x_1, x_2...x_n)$ *are called the* **arguments** *of* t_r.

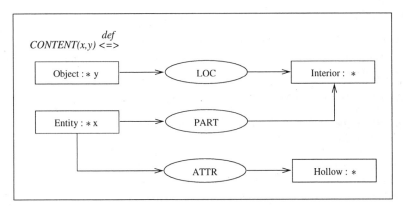

Fig. 2. Definition of content relation as a link between an object and an entity which contains the object.

Restriction: In the following, we assume there is neither direct nor undirect recursive type definition.

3   Extension of the model

The mechanism of type definition grows expressivity of the CG model. It remains to extend the specialization/generalization relation to take advantage of these definitions. We believe that in order to keep the specificity of the CG model, such an extension must keep the triple correspondence: specialization rules / projection / logical inference. We need to modify some parts of CG model to perform that.

3.1 The canon

The canon contains the conceptual vocabulary of the domain of representation. It represents the terminological component of the CG model. The canon is shared in two taxonomies: T_c the concept type taxonomy and T_r the relation type taxonomy.

These two type posets are composed of atomic types or defined types. A type definition is attached to each defined type. For atomic types, the order relations $\leq_c$ and $\leq_r$ are simply given by the user (who is supposed to be a specialist of the domain of representation).

The concept type definitions extend the $\leq_c$ relation. Since a concept type is introduced by definition as subtype of its genus, the $\leq_c$ relation is modified:

Definition 3 *Let t_c be a defined concept type, we have $t_c \leq_c genus(t_c)$.*

The genus of a defined concept type may be atomic or defined. If the genus is defined then, by transitivity of t_c, the new type is also a subtype of the genus of its genus. As no recursive definition is permitted, one can define the smallest atomic super-type of a defined concept type that we call *atomic genus*.

Definition 4 *Let t_c be a defined concept type, the atomic genus of t_c denoted by $AG(t_c)$ is the atomic type obtained from t_c by successive applications of the genus function.*

We immediately get the following property:

Property 1 *Let t_c be a defined concept type and t'_c an atomic concept type, $t_c \leq_c t'_c$ if and only if $AG(t_c) \leq_c t'_c$.*

On the contrary, the relation type definitions does not extend the $\leq_r$ relation. This is because such a kind of definition does not derive from Aristotle's method. Each relation type owns a signature which fixes the arity and the concept authorized as maximal types for arguments of a relation of this type. The signature of atomic relation types is given by the user. For defined relation types, the signature is deduced from type definition in the following way.

Definition 5 *Let t_r be a defined relation type:*

- *the arity of t_r is equal to the arity of the abstraction which defines t_r;*
- *the concept maximal type of the i-th argument of t_r is the type of the i-th formal parameter of the abstraction.*

A conceptual graph must verify the canonicity property and the conformity property. The first property enforces the neighbour concepts of a relation vertex to accord with the signature of the relation type. The second property enforces the concepts to have a referent according with the type.

3.2 Extension of specialization relation

A specialization relation is defined on CGs by a set of specialization rules. The reverse relation, the generalization, corresponds to logical deduction. First part of this theorem (the soundness of the CG model) has been demonstrated by J. Sowa [Sow84]. The reciprocal part (the completeness) has been demonstrated by M. Chein and M.L. Mugnier [CM92, MC96] with CGs in *normal form* (normal form does not allow CGs to have several individual concepts with the same referent).

Using type definitions allows to represent knowledge to different levels of abstraction. But generalization relation forgets some inferences. In figure 3 for instance, there is no possibility to derive H from G (or G from H) though these two CGs represent the same knowledge (according to the type definition of figure 1).

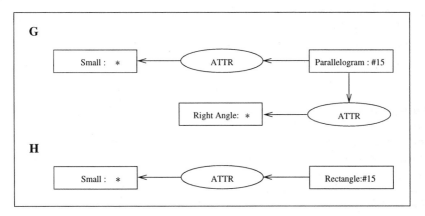

Fig. 3. Two equivalent graphs according to the type definition of Rectangle.

In order to preserve the truth of the generalization relation, we must add four new specialization rules: type contractions and type expansions. These operations are close to J. Sowa's operations [Sow84] but not identical. First we introduce these operations in order to extend the formal model of CGs when J. Sowa presents these operations as tools to simplify CGs. So we search to give a well-defined framework to manipulate type definitions rather than an algorithm of graph reduction. Secondly, J. Sowa's contraction (cf. section 3.6.5 of [Sow84]) does not compute an equivalent graph but a more general one than the original (e.g. some informations can be lost). As for his expansions, either it does not preserve truth, or the defined type is not replaced by its genus.

We state that our following operations preserve truth. We define them as reversible operations which replace a defined type with its definition and conversely replace a graph corresponding to a type definition with a defined type.

Definition 6 (concept type expansion) *Let G be a conceptual graph containing a concept c with a defined type t_c. Let $t_c(x) \overset{def}{\Leftrightarrow} D(x)$ be the definition. H is obtained by expansion in the following way:*

1. *replace t_c the type of c by its genus;*
2. *restrict the head of t_c by changing its referent to the referent of c;*
3. *join G and D on c and the head of t_c;*

This operation is permitted only if the resulting graph is canonical (e.g. c does not violate signature of neighbouring relation types).

Definition 7 (relation type expansion) *Let G be a conceptual graph containing a relation r with a defined type t_r. Let $t_r(x_1, ...x_n) \overset{def}{\Leftrightarrow} D(x_1, ...x_n)$ be the definition. H is obtained by expansion in the following way:*

1. *restrict each argument c_i of t_r by changing its type and referent to type and referent of the i-th neighbour of r;*
2. *perform n joins to identify the i-th neighbouring concept of r with the i-th argument of t_r;*
3. *delete r and all its edges from G.*

Definition 8 (concept type contraction) *Let G be a CG containing a subgraph G' isomorphic to the definition $D(x)^1$ of a defined concept type t_c (concept c in G' corresponding to formal parameter of $D(x)$ in isomorphism can have an individual referent yet). If moreover c is a cut point between G' and the other part of G (i.e. all the paths from one vertex of G' to one vertex of the rest of G go through the vertex c), H is obtained by contraction in the following way:*

1. *replace the type of c with t_c;*
2. *delete subgraph $G' \setminus \{c\}$ from G.*

This operation is permitted only if c in the resulting graph is conform.

Definition 9 (relation type contraction) *Let G be a CG containing a subgraph G' isomorphic to the definition $D(x_1, ...x_n)$ of a defined relation type t_r (concepts c_i ($i \in [1..n]$) in G' corresponding to formal parameters of $D(x_1, ...x_n)$ can have yet individual referents and types lesser than types of corresponding arguments of t_r). If moreover the concepts c_i ($i \in [1..n]$) allow to disconnect G' from G, H is obtained by contraction in the following way:*

1. *insert a relation having type t_r to G;*
2. *$\forall i \in [1..n]$ link the i-th edge of r to concept c_i of G';*
3. *delete subgraph $G' \setminus \{c_1, ...c_n\}$ from G.*

---

[1] The variable x in the formal parameter concept is though not considered for isomorphism.

Observe that concept type expansion and concept type contraction are not necessarily permissible due to the respect constraint of canonicity and conformity. There is no problem with relation contraction and relation expansion.

Let us recall the basic specialization rules before defining the specialization relation on CG model extended to type definitions. We consider the following four rules:

- concept restriction which allows to replace a type with a subtype or generic referent with an individual one (conformity must be preserved);
- relation restriction which allows to replace a type with a subtype (canonicity must be preserved);
- join which allows to merge two identical concepts (perform from two graphs or only one);
- simplification which allows to delete a duplicated relation.

Definition 10 *H is a specialization of G, written $H \leq_D G$, if and only if one can derive H from G using both the basic specialization rules and the new specialization rules. Respectively G is thus a generalization of H.*

Note that the new specialization relation produces equivalent classes greater than the equivalent classes of the basic specialization relation. This is due to the semantics of definition which enforce to consider the resulting graph of a contraction (or expansion) operation equivalent to the original graph.

4 Computing the specialization relation

In the basic model, the projection operator is used to detect the specialization relation between CGs. In our extended model, projection can still exhibit the specialization relation but it is not a complete operator to detect the specialization relation between CGs. For instance, there is no projection between conceptual graphs in figure 3 though there exists a mutual specialization relation between these two graphs.

Using projection operator to detect again specialization relation needs to define a transformation on conceptual graphs on which one computes projection. For that, we reuse the *atomic form* introduced in [CL94]. Indeed, we extend this form to take into account the defined relation types.

Definition 11 *Let G be a CG, we call atomic form of G, written $AF(G)$, the graph without defined types obtained by a succession of type expansions.*

Property 2 *Without recursive type definitions, one can compute the atomic form of a CG in a finite number of type expansions.*

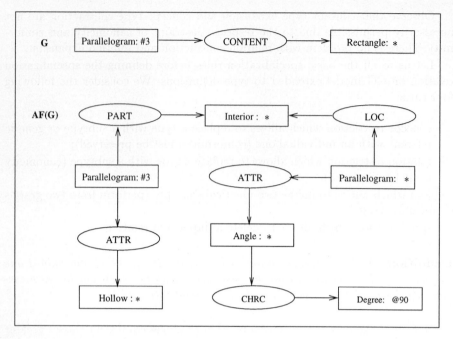

Fig. 4. G and its atomic form according to Rectangle definition, CONTENT definition and supposing that Right Angle is defined as an Angle of 90 degrees.

Proof

Let G be a CG containing vertices with defined types. As no constraint exists to perform relation type expansions, one can compute from G a graph G' without any relation having a defined type (we have just to perform relation type expansions while it remains relations with defined types). Since relation type definitions are not recursive, one can do it in a finite number of relation type expansions.

Let's prove that we can now perform concept type expansion from any concept c of G' having a defined type (e.g. the resulting graph is canonical). Since G' is canonical, any neighbouring relation r of c is such as: $type(c) \leq_c Sig_i(type(r))$ with c being the i-th neighbour of r and Sig_i representing the maximal concept type of the i-th argument of a relation. Moreover, the type of r is atomic according to the construction of G'. With property 1, we conclude that $AG(type(c)) \leq_c Sig_i(type(r))$. Thus $genre(type(c)) \leq_c Sig_i(type(r))$ and so the resulting graph will be canonical. Consequently, we can perform the concept type expansions from all the concepts with defined types of G'. Let G'' the resulting graph. The concept type definitions may contain vertices with defined types, we must redo the two preceding phases. But these definitions being no recursive, we will obtain a graph without defined types in a finite number of repetitions. □

Theorem 1. *The atomic form of a CG is unique.*

Proof

We must demonstrate that whatever succession order chosen to perform the type expansions, we obtain isomorphic graphs. We use the proof schemes proposed by G. Chaty and M. Chein in [CC81] for studying the graph reductions. These proof schemes based on the Church-Rosser's systems take advantage of properties on the confluent reductions demonstrated by G. Huet [Hue80].

Firstly, we define the $\mathcal{R}$ relation on set of the isomorphic CG classes by: $C_g \mathcal{R} C_h$ *iff* $\exists G \in C_g$ and $\exists H \in C_h$ *such as H is obtained from G by a type expansion.*

Secondly, we demonstrate $\mathcal{R}$ is locally confluent that is $\quad \forall C_g, C_{g_1}, C_{g_2} \in \mathcal{G},$ *if* $C_g \mathcal{R} C_{g_1}$ *and* $C_g \mathcal{R} C_{g_2}$ *then* $\exists C_h \in \mathcal{G}$ *such as* $C_{g_1} \mathcal{R}^* C_h$ *and* $C_{g_2} \mathcal{R}^* C_h$ *with* $\mathcal{R}^*$ *is the transitive closure of* $\mathcal{R}$.

With the different properties on confluent reductions, we conclude that atomic form of a graph is unique. A detailed proof is given in [Lec95]. $\quad\square$

Atomic form of CG being obtained by type expansions, we immediately have the following property:

Property 3 *Atomic form of a graph is equivalent to the original graph according to the specialization relation. We thus have* $AF(G) \leq_D G$ *and* $G \leq_D AF(G)$.

In order to establish the link between projection operator and specialization relation in the extended model, we first demonstrate that two CGs belong to specialization extended relation if and only if their atomic forms belong to specialization basic relation. In the following, specialization basic relation is written $\leq$ while the specialization extended relation is written $\leq_D$.

Theorem 2. *If* $AF(H) \leq AF(G)$ *then* $H \leq_D G$.

Proof

Since $AF(H) \leq AF(G)$ then we have $AF(H) \leq_D AF(G)$ because $\leq_D$ is defined as an extension from $\leq$. With the property 3 and considering the transitivity of $\leq_D$ we immediately conclude that $H \leq_D G$. $\quad\square$

In order to demonstrate the reciprocal lemma, we show that to each specialization rule allowing to derive H from G in the extended model, there corresponds a sequence of specialization elementary rules allowing to derive AF(H) from AF(G) (a detailed correspondence is established in [Lec96]):

- for a join on concepts with atomic types there corresponds a similar join on atomic forms;
- for a join on concepts with defined types there corresponds a sequence of joins and simplifications;
- for a simplification with relations having atomic types there corresponds a similar simplification;

- for a simplification with relations having defined types there corresponds a sequence of joins and simplifications;
- for a relation restriction which necessarily replaces an atomic type with an atomic subtype there corresponds a similar restriction;
- for a concept restriction which replaces a generic referent with an individual referent there corresponds a similar restriction;
- for a concept restriction which replaces an atomic type with an atomic subtype there corresponds a similar restriction;
- for a concept restriction which replaces an atomic type with a defined subtype there corresponds a concept restriction and an external join (to connect the atomic form of the subtype definition);
- for a concept restriction which replaces a defined type with a defined subtype there corresponds an external join and a sequence of internal joins and simplifications (to merge the two type definitions);
- for a contraction or expansion there corresponds the empty sequence of specialization rules on atomic forms.

Theorem 3. *If $H \leq_D G$ then $AF(H) \leq AF(G)$.*

Proof

For each rules of specialization extended relation allowing to derive H from G, we have exhibited a sequence of rules of specialization basic relation allowing to derive $AF(H)$ from $AF(G)$. Since $H \leq_D G$, we obtain, by composition of specialization rules, $AF(H) \leq AF(G)$. □

Consequently, any sequence of specialization rules allowing to derive H from G in the extended model can be reorganized in:

1. a sequence of expansions to derive $AF(G)$ from G;
2. a sequence of specialization elementary rules to derive $AF(H)$ from $AF(G)$;
3. a sequence of contractions to derive H from $AF(H)$.

Let us recall the complete correspondence between projection and specialization basic relation established by M. Chein and M.L. Mugnier in [CM92]:

Theorem 4. $H \leq G$ *iff there exists a projection from G to H.*

The three previous theorems yield:

Theorem 5. $H \leq_D G$ *if and only if there exists a projection from $AF(G)$ to $AF(H)$.*

5 Logical interpretation

We extend the operator ϕ to take into account type definitions in the logical interpretation of CG model. A type definition represents an equivalence between a type and its description, we must preserve these semantics in logical interpretation. Let $t(x_1, ...x_n) \overset{def}{\Leftrightarrow} D(x_1, ...x_n)$ be a relation type definition or a concept type definition (in this case $n = 1$). The logical formula associated with this definition, written $\phi(Def(t))$, is determined by the following construction:

- with the type t, associate a n-adic predicate having the same name;
- with the left hand of the definition, associate the atomic formula $t(x_1, ...x_n)$;
- with the right hand of the definition, associate the formula $\phi(D(x_1, ...x_n))$ computed in the same way as for a normal CG but without existential closure of variables $x_1, ...x_n$;
- connect the two previous formula with an equivalence logical operator;
- universally quantify the variables $x_1, ...x_n$ of the constructed formula.

For instance, the formulae associated with type definitions in figures 1 and 2 are:

$$\forall x (Rectangle(x) \leftrightarrow \exists y (Parallelogram(x) \land RightAngle(y) \land Attr(x,y)))$$

$$\forall x \forall y (Content(x,y) \leftrightarrow \exists w \exists z (Object(y) \land Interior(w) \land Entity(x) \land Hollow(z)$$
$$\land Loc(y,w) \land Part(x,w) \land Attr(x,z)))$$

With this extension of ϕ, we can show that type expansions and type contractions preserve the semantics of the formulae associated with CGs before and after the operation.

Property 4 *Let H be a CG derived from G by a type expansion (or G derived from H by a type contraction) of a defined type t, we have $\phi(Def(t)) \vdash \phi(G) \leftrightarrow \phi(H)$.*

Proof

Suppose that H is obtained from G by a concept type expansion on c. Let $t(x) \overset{def}{\Leftrightarrow} D(x)$ be the type definition of the type of c. We have $\phi(Def(t)) = \forall x (t(x) \leftrightarrow F_D[x])$ ($F_D[x]$ is the formula associated with $D(x)$).
If c is a generic concept then $\phi(G) = EC(t(x) \land A[x])$ and $\phi(H) = EC(F_D[x] \land A[x])$ (EC represents the existential closure). Furthermore $A[x]$ and $F_D[x]$ do not contain any common variable except x. Since $\phi(Def(t))$ is valid, we also have $\phi(G) \leftrightarrow \phi(H)$ is valid.
If c owns an individual referent m then $\phi(G) = t(m) \land EC(A)$ and $\phi(H) = EC(F_D[m] \land A)$ with $EC(F_D[m]) = Subst_{[x/m]}(EC(F_D[x]))$. Furthermore $A[x]$ and $F_D[m]$ do not contain any common variable. Since $\phi(Déf(t))$ is valid, we also have $\phi(G) \leftrightarrow \phi(H)$.

Suppose that H is obtained from G by a relation type expansion on r. Let $t(x_1, ...x_n) \overset{def}{\Leftrightarrow} D(x_1, ...x_n)$ be the type definition of type of r. We have $\phi(Def(t)) = \forall x_1...\forall x_n (t(x_1, ...x_n) \leftrightarrow F_D[x_1, ...x_n])$ ($F_D[x_1, ...x_n]$ is the formula associated with $D(x_1, ...x_n)$).

$\phi(G) = EC(t(t_1, ...t_n) \wedge A[t_1, ...t_n])$ and $\phi(H) = EC(F_D[t_1, ...t_n] \wedge A[t_1, ...t_n])$ and $t_1, ...t_n$ are constants or variables according to referents of neighbouring concepts of r.

$EC(F_D[t_1, ...t_n]) = Subst_{[x_1/t_1, ...x_n/t_n]}(EC(F_D[x_1, ...x_n]))$. Furthermore, $A[t_1, ...t_n]$ and $F_D[t_1, ...t_n]$ do not contain any common variable except possible variables in $t_1, ...t_n$. Since $\phi(Def(t))$ is valid, we also have $\phi(G) \leftrightarrow \phi(H)$ is valid. □

We denote $\phi(\mathcal{S})$ the set of formulae associated with the relations $\leq_c$ and $\leq_r$ defined on the atomic type sets. Let us recall that with each pair (concept or relation types) such as $t \leq t'$ is associated the formula $\forall x_1...\forall x_n(t(x_1, ...x_n) \rightarrow t'(x_1, ...x_n))$. We denote $\phi(\mathcal{D})$ the set of formulae associated with type definition set. We now demonstrate that generalization rules (the reverse specialization rules) still corresponds to a set of inference rules.

Theorem 6. *If* $G \leq_D H$ *then* $\phi(\mathcal{S}), \phi(\mathcal{D}), \phi(G) \vdash \phi(H)$.

Proof

Using $\phi(\mathcal{S})$, M. Chein and M.L. Mugnier demonstrated that elementary rules of generalization correspond to inference rules [CM92]. Using $\phi(\mathcal{D})$, I have demonstrated that type contractions and type expansions correspond to equivalence rules (cf. previous property). By composition of different rules, we obtain $\phi(G) \rightarrow \phi(H)$ is valid. □

Moreover, we have immediately:

Property 5 *Let* G *be a CG*, $\phi(\mathcal{D}) \vdash \phi(G) \leftrightarrow \phi(AF(G))$.

The normal form restriction is not still sufficient to demonstrate the completeness of the CG model. Suppose that a normal form graph G contains a concept with the individual referent m and another concept with a defined type t. And suppose that the type definition of t contains a concept with the same individual referent m. Then the atomic form of G will not be in a normal form and so the specialization relation can miss some deductions. For instance, the graph G in figure 5 can not be derived from H but the logical formula associated with G implies the logical formula associated with H.

Theorem 7. *Let* G *and* H *be two CGs with atomic forms in normal form.* $\phi(\mathcal{S}), \phi(\mathcal{D}), \phi(G) \vdash \phi(H)$ *if and only if* $G \leq_D H$.

Proof

Since $\phi(\mathcal{S}), \phi(\mathcal{D}), \phi(G) \vdash \phi(H)$, property 5 entails $\phi(\mathcal{S}), \phi(\mathcal{D}), \phi(AF(G)) \vdash \phi(AF(H))$, and then the set $\{\phi(\mathcal{S}), \phi(\mathcal{D}), \phi(AF(G)), \neg\phi(AF(H))\}$ is inconsistent. The resolution method allows to derive the empty clause from clauses generated by this set.

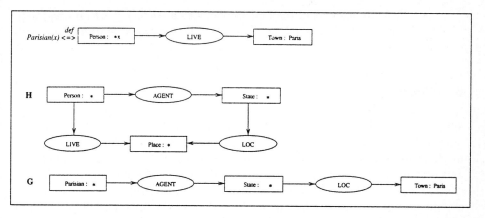

Fig. 5. An example with normal form graphs such as $G \not\leq_D H$ while $\phi(\mathcal{S}), \phi(\mathcal{D}), \phi(G) \vdash \phi(H)$.

Each formula in $\phi(\mathcal{S})$ is of type $\forall x_1...\forall x_n(t(x_1,...x_n) \rightarrow t'(x_1,...x_n))$ and generates a clause of type: $\neg t(x_1,...x_n) \vee t'(x_1,...x_n)$.

Each formula in $\phi(\mathcal{D})$ is of type:
$$\forall x_1...\forall x_n(t_d(x_1,...x_n) \leftrightarrow \exists y_1...\exists y_m(P_1[x_1,...x_n,y_1,...y_m]\wedge...P_k[x_1,...x_n,y_1,...y_m]))$$
with t_d the defined type predicate and P_i ($i \in [1..k]$) predicates associated with other vertices of the definition. The argument terms are either defined type variables $x_1,...x_n$, or variables $y_1,...y_m$ and constants associated with other concepts. Each formula generates k clauses of type:
$$\neg t_d(x_1,...x_n) \vee P_i[x_1,...x_n,f_1(x_1,...x_n),...f_m(x_1,...x_n)]$$
(f_i functions result from the skolemisation of variables y_i), and a clause of type: $t_d(x_1,...x_n) \vee \neg P_1[x_1,...x_n,y_1,...y_m] \vee ...\neg P_k[x_1,...x_n,y_1,...y_m]$.

The formula $\phi(AF(G))$ is of type: $\exists x_1...\exists x_n(Q_1[x_1,...x_n] \wedge ...Q_m[x_1,...x_n])$ with Q_i ($i \in [1..m]$) predicates associated with vertices of $AF(G)$. It generates m clauses of type: $Q_i[a_1,...a_n]$ (with a_i the Skolem constants).

The formula $\neg\phi(AF(H))$ is of type: $\neg\exists x_1...\exists x_n(R_1[x_1,...x_n]\wedge...R_l[x_1,...x_n])$ with R_i ($i \in [1..l]$) predicates associated with vertices of $AF(H)$. It generates a clause of type: $\neg R_1[x_1,...x_n] \vee ...\neg R_l[x_1,...x_n]$.

The empty clause is necessarily produced in emptying the clause associated with $\neg\phi(AF(H))$. The clauses generated by the type definitions are needless in the resolution process. Using these clauses will introduce some new predicates (the predicate associated with defined type). Since to delete these predicates one should have to re-introduce previous deleted predicates. Thus the empty clause can be produced from the other clauses and thus $\{\phi(\mathcal{S}), \phi(AF(G)), \neg\phi(AF(H))\}$ is inconsistent.

Then the completeness of the basic model yields $AF(G) \leq AF(H)$ and with the theorem 2 we can conclude $G \leq_D H$. The reciprocal is immediate with theorem 6.   □

6 Conclusion

We have defined an extension of the CG model allowing to enrich the terminological knowledge representation. This extension clarify the type definition mechanism proposed by J. Sowa in stating precisely the semantics of definitions and providing a framework to exploit type definitions. We have introduced type contractions and type expansions as new specialization rules and consequently extend the specialization relation. We have showed the correspondence between the new specialization relation and projection operator on atomic forms of CGs. Finally we have demonstrated the soundness and completeness of the extended model.

This extension allows to define a type classifier in a KL-ONE-style [Lip82] which was introduced in [CL94]. We are currently working to take into account partial definitions in order to provide terminological knowledge representations for natural types.

This theoretical work must be completed by a study of good algorithms for computing specialization relation. Totally compute the atomic forms to test projection does not seem to be the best solution !

References

[CC81] G. Chaty and M. Chein. Réduction de graphes et systèmes de Church-Rosser. *R.A.I.R.O.*, 15(2):109–117, 1981.

[CL94] M. Chein and M. Leclère. A cooperative program for the construction of a concept type lattice. In *Supplement Proceedings of the Second International Conference on Conceptual Structures*, pages 16–30, College Park, Maryland, USA, 1994.

[CM92] M. Chein and M.L. Mugnier. Conceptual graphs : fundamental notions. *Revue d'Intelligence Artificielle*, 6(4):365–406, 1992.

[Hue80] G. Huet. Confluent reductions : Abstract properties and applications to term rewriting systems. *J.A.C.M.*, 27:797–821, 1980.

[Lec95] M. Leclère. *Le niveau terminologique du modèle des graphes conceptuels : construction et exploitation.* PhD thesis, Université Montpellier 2, 1995.

[Lec96] M. Leclère. Raisonner avec des définitions de types dans le modèle des graphes conceptuels. Rapport de recherche 143, IRIN, Nantes, France, Novembre 1996.

[Lip82] T. Lipkis. A KL-ONE classifier. In J.G. Schmolze and R.J. Brachman, editors, *Proceedings of the 1981 KL-ONE Workshop*, pages 126–143, Jackson, New Hampshire, 1982. The proceedings have been published as BBN Report No. 4842 and Fairchild Technical Report No. 618.

[MC96] M.-L. Mugnier and M. Chein. Représenter des connaissances et raisonner avec des graphes. *Revue d'Intelligence Artificielle*, 10(1):7–56, 1996.

[Sow84] J.F. Sowa. *Conceptual Structures - Information Processing in Mind and Machine*. Addison-Wesley, Reading, Massachusetts, 1984.

[WS92] W.A. Woods and J.G. Schmolze. The kl-one family. In F. Lehmann, editor, *Semantic Networks in Artificial Intelligence*, pages 133–177, Pergamon Press, Oxford, 1992.

Universal Marker and Functional Relation: Semantics and Operations

Tru H. Cao and Peter N. Creasy

School of Information Technology
University of Queensland
Australia 4072
{tru, peter}@it.uq.edu.au

Abstract. The universal marker (i.e., universal quantifier) and the functional relation are two useful notations that make Conceptual Graph (CG[1]) representations more concise in expressing universally quantified facts and functional dependencies, which are commonly used in knowledge bases, logic programs and data conceptual schemas. We introduce an expansion rule that formally defines the semantics of CGs containing universal markers and/or functional relations. On the basis of this formal semantics, we define two reasoning operations that are performed directly on CGs with these two notations to make them more useful. One operation is the universal CG projection defining the subsumption relation on the extended CGs. The other operation is the universal concept join performing universal instantiations and inheritances simultaneously in one graph operation. Both the operations are proved to be sound with respect to their described interpretations.

1. Introduction

A strength of CG is that it is a graphical language having a logic formalism, which most other graphical languages do not have. This not only gives CG representations precise meanings, but also allows direct reasoning on these representations. CGs themselves are logical formulas and reasoning can be performed on them as logical operations. For example, given two CGs, one can determine whether one subsumes the other by a CG projection performed directly on the two CGs.

On the other hand, basic CG[2] is a fine granularity language in the sense that CG constructs mainly consist of only existentially quantified concepts and conceptual relations connected by simple arcs. As a result, many new notations and constructs have been proposed to make CGs express more concisely information in certain application areas.

---

[1] We use 'CG' to mean the Conceptual Graph language or theory, and use 'CGs', 'a CG' or 'the CG' to mean particular conceptual graphs.

[2] By 'basic CG' we mean the first-order subset of CG, where conceptual referents are only generic or individual referents, conceptual relations are not functional, and contexts are only negative contexts.

Among the extended CG notations and constructs that have been used so far, the-universal marker and the functional relation are two useful ones. They make CG representations more concise in expressing universally quantified facts and functional dependencies, which are commonly used in knowledge bases, logic programs and data conceptual schemas ([10, 6, 1, 8, 7]). The universal marker was introduced early in [8], and functional dependencies were expressed through *actors* there. In [7], the author noticed the lack of the functional relation notion in CG and integrated functions into it by fully revising the canonical formation operations, taking into account functional dependency constraints imposed by functional relations in the extended CGs.

However, an important issue of a formal integration of these two notations into CG has not been adequately resolved yet. That is a definition of the formal semantics of CGs containing universal markers and/or functional relations. Such a formal semantics is necessary to give the extended CGs precise meanings, and provide a formal basis for the development of operations on them and the proof of the operations' soundness.

The meanings of universally quantified CGs were defined by the expansion rule given by Assumption 4.2.7 in [8], but those of CGs containing functional relations and/ or universal markers have not been defined yet in terms of basic CGs. This is also a shortcoming of [7] in an effort to formally integrate functions into CG. Without a formal semantics nothing can be proved. In this paper, we introduce an expansion rule defining the meanings of the extended CGs.

On the basis of this formal semantics, we define two reasoning operations that are performed directly on CGs with the two notations. So far, CGs containing universal markers and/or functional relations still have not been fully manipulable on the basis of formally defined and proved sound operations performed directly on them. It is desirable that these CGs could be stored and processed directly, instead of indirectly through their defining expansions, which are less concise and more complicated. It would be more natural and efficient in both storage and computation, and make them more useful.

The projection and join are two essential operations of CG. The existential CG projection defines the subsumption relation on CGs, and existentially quantified facts in a CG knowledge base can be organized in a subsumption hierarchy that considerably speeds up searching and retrieval operations ([3]). However, when facts are universally quantified and/or contain functional relations, the existential CG projection is not applicable for extended CGs representing them. We define the *universal CG projection* performed directly on CGs with universal markers and/or functional relations, as an extension of the existential CG projection.

The conventional join is a canonical formation rule, allowing construction of larger canonical CGs from smaller ones, but not a rule of inference ([8]). However, given universally quantified CGs, possibly with functional dependency constraints, reasoning can be performed on them by join operations. We define the *universal concept join* as a realization of universal instantiations and inheritances, performed simultaneously in one graph operation, for the extended CGs.

The paper is organized as follow. In Section 2, we define the universal CG projection and the universal concept join operations for universally quantified CGs. In Section 3, we present the expansion rule for CGs that are universally quantified and/or

418

contain functional relations, then the two operations are extended further for these CGs. Both the operations are proved to be sound with respect to their described interpretations. Finally, Section 4 draws conclusions.

2. Universal Markers

We call a concept with the universal marker ('∀') a *universal concept*, and a concept with the generic marker ('*') or an individual marker an *existential concept*. A *plain CG* is one that does not have any context, but is not necessarily connected and may have coreference links. A plain CG is called a *universal CG* if it contains universal concepts, and an *existential CG* otherwise.

According to Assumption 4.2.7 in [8] (see also [9]), a universal CG g with n universal concepts $c_1, c_2, \ldots, c_n$ is defined by $g_e = \neg[\ c_{1\exists}\ c_{2\exists} \ldots c_{n\exists}\ \neg[g_\exists]]$, where each $c_{i\exists}$ is the concept derived from c_i by replacing the universal marker by the generic marker, $g_\exists$ is derived from g by replacing every c_i by $c_{i\exists}$, and every $c_{i\exists}$ is coreference-linked to its copy in $g_\exists$.

For illustration, Figure 2.1 shows a universal CG and its defining expansion, saying "Every course is offered by a lecturer who assigns a PhD student to be its tutor".

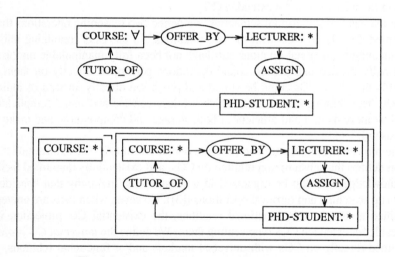

Fig. 2.1. A universal CG and its defining expansion

2.1. Universal CG Projection

If an existential CG v has a projection onto an existential CG u, then by definition v subsumes u ([3]), and on the other hand, v is a logical consequence of u, i.e., v is derivable from u by the CG first-order inference rules (Assumption 4.3.5 in [8])[3]. Such a projection operation can also be defined on universal CGs.

---

[3] So, we consider the subsumption relation between two CGs as their logical consequence relation.

First, we extend the subsumption partial order on existential concepts for both existential and universal concepts. We use the same notation '$\leq$' to denote this partial order, the conventional partial orders of type and conceptual relation hierarchies, and the projection partial order. It will be clear in a specific context which partial order the notation '$\leq$' applies to.

Definition 2.1 A concept a: $[T_a: referent(a)]$ is said to be subsumed by a concept b: $[T_b: referent(b)]$, denoted by $a \leq b$, if and only if (iff) either of the following holds:

1. *referent(a)* and *referent(b)* are identical individual markers, and $T_a \leq T_b$
2. *referent(a)* is either an individual marker or '*', *referent(b)* is '*', and $T_a \leq T_b$
3. *referent(a)* and *referent(b)* are both '$\forall$', and $T_b \leq T_a$
4. *referent(a)* is '$\forall$' and *referent(b)* is '*', and $T_b \cap T_a \neq \perp$[4].

The first two cases are the same as defined in basic CG. It is clear that case 3 also satisfies the intended meaning of the subsumption relation, where '$T_b \leq T_a$' is the condition instead of '$T_a \leq T_b$'. Here, the intuition is, for example, the CG $[T_b:\forall]\rightarrow(p)$, saying "Every instance of T_b has property p", is a logical consequence of the CG $[T_a:\forall]\rightarrow(p)$, saying "Every instance of T_a has property p", given $T_b \leq T_a$. Case 4 does not always satisfy the intended meaning, but it is included for possibilities in which generic concepts are projected onto universal concepts in universal CG projections.

Definition 2.2 is an extension of the definition of the existential CG projection, for both existential and universal CGs. In a universal CG projection Π: $v \rightarrow u$, one concept in v can be mapped to one or more concepts in u. So, for every concept c in v, we denote Πc as the set of the concepts in u to which c is mapped. We also denote $\mathbf{C}_{\exists g}$, $\mathbf{C}_{\forall g}$, and $\mathbf{R}_g$ respectively as the sets of existential concepts, universal concepts and conceptual relations in a CG g.

Definition 2.2 Let u and v be two plain CGs with no coreference links. Then, v is said to have a projection onto u, denoted by $u \leq v$, iff there exists a mapping $\Pi: v \rightarrow u$, such that:

1. $\forall r \in \mathbf{R}_v$: $\Pi r \leq r$, and

   $\forall i \in \{1,2, \dots, degree(r)\}$: $neighbour(\Pi r, i) \in \Pi neighbour(r, i)$,
2. $\forall c \in \mathbf{C}_{\forall v}$: $\Pi c = \{c' \mid c' \in \mathbf{C}_{\forall u} \text{ and } c' \leq c\}$,
3. $\forall c \in \mathbf{C}_{\exists v}$: (i) $\Pi c = \{(\text{only one}) \ c'\}$ where $c' \in \mathbf{C}_{\exists u}$ and $c' \leq c$, or

   (ii) $\Pi c = \{c' \mid c' \in \mathbf{C}_{\forall u} \text{ and } c' \leq c\}$ where $referent(c) = \text{'*'}$,
4. There exists a mapping Π^{-1}: $\mathbf{C}_{\forall u} \rightarrow \mathbf{C}_{\forall v}$, such that:

   (i) $\forall c' \in \mathbf{C}_{\forall u} \exists! c \in \mathbf{C}_{\forall v}$: $c = \Pi^{-1}c'$ where $c' \leq c$,

   (ii) if $c' \in \Pi c$ where $c \in \mathbf{C}_{\forall v}$, then $\Pi^{-1}c' = c$,

   (iii) if $\{c'_1, c'_2, \dots, c'_k\} = \Pi c$ where $c'_i \in \mathbf{C}_{\forall u}$ and $c \in \mathbf{C}_{\exists v}$, then

---

[4] $T_b \cap T_a$ is the greatest lower bound, i.e., maximal common sub-type, of T_b and T_a. $\perp$ is the absurd type, at the bottom of a type hierarchy.

$$\Pi^{-1}c'_1 = \Pi^{-1}c'_2 = \ldots = \Pi^{-1}c'_k \text{ and}$$

$$type(\Pi^{-1}c'_i) \le type(c), type(c'_1), type(c'_2), \ldots, type(c'_k).$$

Here, 'Π^{-1}' is just a mnemonic notation to denote a mapping from $C_{\forall u}$ back to $C_{\forall v}$, but Π^{-1} is not the inverse function of Π. In fact, Π^{-1} is partially defined by Π, and it is mainly a checking step. Conditions 4.(i) and 4.(ii) imply that two or more universal concepts in v cannot be mapped to the same universal concept in u. Condition 4.(iii) is the constraint on projections of generic concepts in v onto universal concepts in u.

For example, the CG "Every student has a name" can be projected onto the CG "Every person has a name and age", as shown in Figure 2.2.

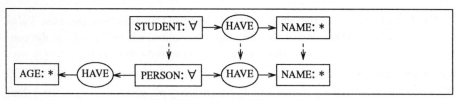

Fig. 2.2. A universal CG projection

Theorem 2.1 states the soundness of the universal CG projection with respect to the subsumption relation. That is, if v has a projection onto u according to Definition 2.2, then v subsumes u, or in other words, v is a logical consequence of u.

Theorem 2.1 Let u and v be two plain CGs with no coreference links. If $u \le v$, then v is a logical consequence of u.
Proof. When both u and v are existential CGs, the theorem holds on the basis of Theorem 4.3.7 in [8] (Note that, when CGs are existentially quantified, the universal CG projection performs as the existential one). Now, we assume that either u or v is universally quantified. On the basis of Definition 2.2, if either u or v is universally quantified and $u \le v$, then so is the other. Let u_e and v_e be respectively the defining expansions of u and v:

$$v_e = \neg[\, c_{1\exists}\, c_{2\exists} \ldots c_{m\exists}\, \neg[v_\exists]\,] \quad u_e = \neg[\, c'_{1\exists}\, c'_{2\exists} \ldots c'_{m\exists} \ldots c'_{n\exists}\, \neg[u_\exists]\,]$$

On the basis of Definition 2.2, we must have $n \ge m$, and each of universal concepts c_is in v can be projected onto one or more universal concepts c'_js in u. Then, v_e is derivable from u_e by the CG first-order inference rules[5] as follows:

$$u_e$$
$$\Rightarrow \neg[\, c_{1\exists}\, c_{2\exists} \ldots c_{m\exists}\, \neg[u'_\exists]\,] \qquad \qquad (Insertion\ \&\ Coreferent\ Join)$$

Every c'_j is type-restricted to $\Pi^{-1}c'_j$ ($\Pi: v \to u$), and

c'_js that have the same $\Pi^{-1}c'_j$ are coreferent-joined, then

---

[5] Throughout this paper, the names of these rules are written with their initial characters in capitals.

the corresponding concepts in $u_\exists$ are type-restricted and
coreferent-joined accordingly, whence $u_\exists$ becomes $u'_\exists$ satisfying $u'_\exists \le v_\exists$

$$\Rightarrow \neg[\, c_{1\exists}\, c_{2\exists} \dots c_{m\exists}\, \neg[v_\exists]] \equiv v_e \qquad\qquad (u'_\exists \le v_\exists \ \& \text{ Theorem 4.3.7 in [8]})$$

$u'_\exists$ is replaced by $v_\exists$, and coreference links attached to concepts in $u'_\exists$
are transferred to the corresponding concepts in $v_\exists$.

Figure 2.3 shows subsumption relations between some example universal CGs, on the basis of Definition 2.2 and Theorem 2.1.

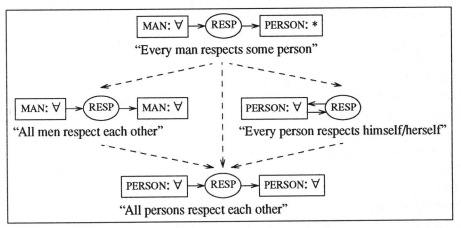

Fig. 2.3. Subsumption relations between universal CGs

Using the universal CG projection, we can determine subsumption relations between universal CGs directly on these CGs, without the need to expand them to less concise and more complicated existential CGs with nested negations. This subsumption relation can also be exploited to organize universally quantified facts in hierarchies, which help to speed up searching and retrieval operations, as it was done for existential CGs ([3]). Since the universal CG projection is also based on graph structures and concept and conceptual relation partial orders, efficient matching techniques of the existential CG projection could be applied to it.

2.2. Universal Concept Join

In a universal concept join, only universal concepts in a CG can be joined with concepts in another CG.

Definition 2.3 Let u be a universal CG and v be any plain CG, with no coreference links in u and v. One or more universal concepts in u can be joined with concepts in v as follows: a concept a: $[T_a: \forall]$ in u can be joined with a concept b: $[T_b: referent(b)]$ in v, where $T_b \le T_a$, into the concept $[T_b: referent(b)]$. We denote the resulting CG by $J_\forall(u, v)$.

Note that, in Definition 2.3, u can be joined with v at several universal concepts in u in one operation, and the order of the two arguments of $J_\forall()$ is significant. After a join, duplicate conceptual relations in the resulting CG can be eliminated. Before proving the soundness of the universal concept join, we prove the following lemma.

Lemma 2.1 If $\neg[u \neg[v]]$ is true, where u and v are existential CGs and there may be coreference links between them, then the following inference rules are sound:

1. In an evenly enclosed context, u is replaced by v and coreference links attached to concepts in u are transferred to the corresponding concepts in v.
2. In an oddly enclosed context, v is replaced by u and coreference links attached to concepts in v are transferred to the corresponding concepts in u.

Proof. 1. The proof is similar to that of a case in Theorem 4.3.7 in [8].

2. This is dual to case 1.

Theorem 2.2 Let u be a universal CG and v be any plain CG, with no coreference links in u and v. Then, any $J_\forall(u, v)$ is a logical consequence of $\{u, v\}$.

Proof. Assume that v is also a universal CG. The proof is similar otherwise. Let u_e and v_e be respectively the defining expansions of u and v:

$$u_e = \neg[\, c_{1\exists} \cdots c_{i\exists}\, c_{i+1\exists} \cdots c_{j\exists}\, c_{j+1\exists} \cdots c_{n\exists}\, \neg[u_\exists]] \quad v_e = \neg[\, c'_{1\exists} \cdots c'_{i\exists} \cdots c'_{m\exists}\, \neg[v_\exists]]$$

supposing that $c_1, \ldots, c_i$ are the universal concepts in u that are joined with universal concepts $c'_1, \ldots, c'_i$ in v, and $c_{i+1}, \ldots, c_j$ are the universal concepts in u that are joined with existential concepts in v. So, the defining expansion of $J_\forall(u, v)$ is:

$$J_\forall(u, v)_e = \neg[\, c'_{1\exists} \cdots c'_{i\exists} \cdots c'_{m\exists}\, c_{j+1\exists} \cdots c_{n\exists}\, \neg[J_\exists(u_\exists, v_\exists)]]$$

where $J_\exists(u_\exists, v_\exists)$ is the conventional join of $u_\exists$ and $v_\exists$ that corresponds to $J_\forall(u, v)$.

$J_\forall(u, v)_e$ is provable from u_e and v_e as follows:

$$\{u_e, v_e\}$$

$\Rightarrow \neg[\, v_\exists\, c_{1\exists} \cdots c_{i\exists}\, c_{i+1\exists} \cdots c_{j\exists}\, c_{j+1\exists} \cdots c_{n\exists}\, \neg[u_\exists]]$ *(Insertion)*

$v_\exists$ is inserted into the oddly enclosed context of u_e

$\Rightarrow \neg[\, v_\exists\, c_{1\exists} \cdots c_{i\exists}\, c_{i+1\exists} \cdots c_{j\exists}\, c_{j+1\exists} \cdots c_{n\exists}\, \neg[u_\exists\, v_\exists]]$ *(Iteration)*

$v_\exists$ is copied into $\neg[u_\exists]$

$\Rightarrow \neg[\, v_\exists\, c_{j+1\exists} \cdots c_{n\exists}\, \neg[u'_\exists\, v_\exists]]$ *(Coreferent Join)*

$c_{1\exists} \cdots c_{i\exists}$ are type-restricted to and joined with $c'_{1\exists} \cdots c'_{i\exists}$ in $v_\exists$,

$c_{i+1\exists} \cdots c_{j\exists}$ are type-restricted to and joined with the corresponding existential concepts in $v_\exists$, then $u_\exists$ is type-restricted accordingly to become $u'_\exists$

$\Rightarrow \neg[\, v_\exists\, c_{j+1\exists} \cdots c_{n\exists}\, \neg[J_\exists(u_\exists, v_\exists)]]$ *(Coreferent Join)*

$u'_\exists$ and $v_\exists$ are coreferent-joined accordingly

$\Rightarrow \neg[\, c'_{1\exists} \cdots c'_{i\exists} \cdots c'_{m\exists}\, c_{j+1\exists} \cdots c_{n\exists}\, \neg[J_\exists(u_\exists, v_\exists)]] \equiv J_\forall(u, v)_e$ *(Lemma 2.1)*

$v_\exists$ is replaced by $c'_{1\exists} \cdots c'_{i\exists} \cdots c'_{m\exists}$, and coreference links attached to concepts in $v_\exists$ are transferred accordingly to $c'_{1\exists} \cdots c'_{i\exists} \cdots c'_{m\exists}$.

For example, in Figure 2.4, u says "For every man and every woman, there is a color that he likes but she hates", and v says "Every American man loves some woman". Applying Theorem 2.2, we can derive the shown $J_\forall(u, v)$ from u and v by joining [MAN: $\forall$] with [AMERICAN-MAN: $\forall$] and [WOMAN: $\forall$] with [WOMAN: *].

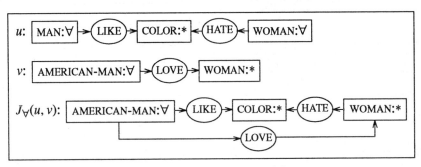

Fig. 2.4. A universal concept join

We could see that the universal concept join is actually a realization of the universal instantiation and inheritance rules, which are performed simultaneously in one graph operation.

3. Functional Relations

Functional relations are special and essential cases of conceptual relations. Sowa initially used actors to express functional dependencies among concepts ([8]). However, the use of actors loses the declarative feature and the concept-relation uniformity of CGs. In [7] and [9], the authors proposed the use of a double-lined arrow for an arc connecting to a functionally dependent concept, but did not formally define the semantics of CGs containing functional relations in terms of basic CGs.

Syntactically, each functional relation has one *dependent concept* linked to the relation by a double-lined arc, and one or more *determining concepts* linked to the relation by single-lined arcs. In Figure 3.1, F is a functional relation, c_n is its dependent concepts and $c_1, c_2, ..., c_{n-1}$ determining concepts. Determining concepts can be existential or universal concepts, but dependent concepts are only existential ones. We exclude universal concepts being dependent concepts because they are not usual.

3.1. Defining Expansion Rule

We now present the expansion rule for CGs that contain universal markers and/or functional relations. This expansion rule formally defines the meanings of these extended CGs in terms of basic CGs, and provides a formal basis for the development of direct operations on them and for the proof of the operations' soundness.

For illustration, if the CG in Figure 3.1 is an existential CG, i.e., every *referent*(c_i) is only either generic or individual, then its defining expansion is the CG in Figure 3.2, where '$\neq$' is the dyadic inequality relation (see [8]) stating that the two respective concepts have different referents. The general definition is presented in Definition 3.1.

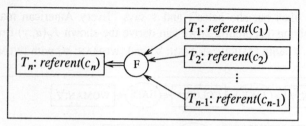

Fig. 3.1. A functional relation

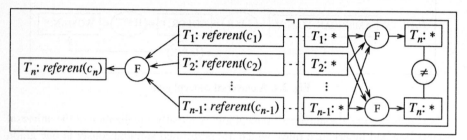

Fig. 3.2. The defining expansion of a CG with a functional relation

Definition 3.1 (a) The defining expansion of an existential CG g with functional relations is [$g_0 \neg[g_1] \neg[g_2]$ ... $\neg[g_m]$], where g_0 is the same as g except that all double-lined arcs in g are replaced by single-lined arcs in g_0, each $\neg[g_i]$ corresponds to a double-lined arc in g, and there are coreference links between occurrences of determining concepts in g_0 and their copies in $\neg[g_i]$s (see Fig. 3.2); (b) If g is a universal CG, then its universal markers are expanded prior to the expansion of functional relations.

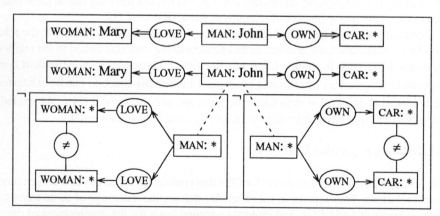

Fig. 3.3. An existential CG with functional relations and its defining expansion

For example, Figure 3.3 shows a CG with two functional relations and its defining expansion, saying "Man John loves only one woman, who is Mary, and owns only one car". Although in this paper (also in [7]) functional relations are defined to have only one dependent concept, the expansion rule and operations introduced in this paper could be generalized straightforwardly for functional relations having one or more dependent concepts.

With the expansion order of universal markers and functional relations defined in case (b) of Definition 3.1, the meanings of plain CGs containing universal markers and functional relations are as natural as those we usually want to express with universally quantified statements conveying functional dependency constraints. Figure 3.4 is an example of a universal CG with a functional relation saying "Every person has only one date of birth" together with its defining expansion.

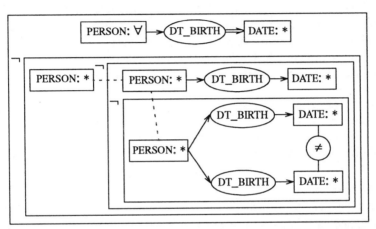

Fig. 3.4. A universal CG with a functional relation and its defining expansion

In [7], the CG canonical formation operations were revised for CGs containing functional relations. An algorithm to join these extended CGs was defined, taking into account functional dependencies so as not to produce inconsistent CGs. Here, we develop operations of a different kind for reasoning. The universal CG projection and the universal concept join defined in Section 2 are now extended further for CGs that may also contain functional relations.

3.2. Extended Universal CG Projection

Given two plain CGs u and v, possibly containing functional relations, Definition 2.2 is extended, so that, if v has a projection onto u then v is a logical consequence of u.

Definition 3.2 Let u and v be two plain CGs (possibly containing functional relations) with no coreference links. Then, v is said to have a projection onto u, denoted by $u \leq v$, iff there exists a mapping $\Pi: v \rightarrow u$ satisfying Definition 2.2 and the following condition: If the i-th arc of a relation r in v is double-lined, then the i-th arc of Πr in u must also be double-lined, and $type(neighbour(r, i)) = type(neighbour(\Pi r, i))$.

For example, the CG "Man John owns some car" can be projected onto the CG "Man John owns only one car", but not vice versa, as shown in Figure 3.5.

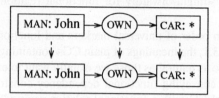

Fig. 3.5. An extended universal CG projection

The added condition '$type(neighbour(r, i)) = type(neighbour(\Pi r, i))$' in Definition 3.2 is necessary, because, for instance, the fact "Man John owns only one vehicle" is not a logical consequence of "Man John owns only one car", although CAR is defined to be a sub-type of VEHICLE. That is, the CG [MAN: John]→(OWN)⇒[VEHICLE: *] cannot be projected onto the CG [MAN: John]→(OWN)⇒[CAR: *].

Theorem 3.1 Let u and v be two plain CGs (possibly containing functional relations) with no coreference links. If $u \leq v$, then v is a logical consequence of u.

Proof. The proof is first carried out for the case when u and v are not universally quantified. Then it is similar to the proof of Theorem 2.1, when u and v are universally quantified. Let u_e and v_e be respectively the defining expansions of u and v:

$$v_e = v_0 \neg[v_1] \neg[v_2] \ldots \neg[v_m] \quad u_e = u_0 \neg[u_1] \neg[u_2] \ldots \neg[u_n]$$

where v_0 is the same as v except that all double-lined arcs in v are replaced by single-lined arcs, and each $\neg[v_i]$ corresponds to a double-lined arc in v; u_0 and $\neg[u_j]$s are defined similarly (see Def. 3.1).

On the basis of Definition 3.2, $u_0 \leq v_0$ and each $\neg[v_i]$ corresponds to some $\neg[u_j]$, where $\neg[v_i]$ is the same as $\neg[u_j]$ except that the types of the determining concepts in $\neg[v_i]$ are super-types of (inclusively, the same as) those of the corresponding determining concepts in $\neg[u_j]$. If $m < n$, then some $\neg[u_j]$s can be deleted from u_e. If $m > n$, which implies that two or more functional relations in v are projected onto one functional relation in u, then some $\neg[u_j]$s can be copied. So, we can derive v_e from u_e by the following sound steps:

1. Make the number of $\neg[u_j]$s be the same as the number of $\neg[v_i]$s by deletions or copies as explained above. If $\neg[u_k]$ is copied from $\neg[u_j]$, then the functional relation and its neighbour concepts that correspond to $\neg[u_j]$ are copied accordingly for $\neg[u_k]$. This makes u_0 become u'_0. Note that we still have $u'_0 \leq v_0$.
2. Replace the types of determining concepts in u'_0, which are involved in the projection, and their copies in $\neg[u_j]$s by the types of the corresponding concepts in v_0. This makes $\neg[u_j]$s become $\neg[v_i]$s and u'_0 become u''_0 that still satisfies $u''_0 \leq v_0$.
3. Applying Theorem 4.3.7 in [8], replace u''_0 by v_0 and transfer coreference links attached to the concepts in u''_0 to the corresponding concepts in v_0. Then v_e is obtained.

In the proof of Theorem 3.1, note that for each dependent concept c in u_0 there are two corresponding concepts of the same type as c in some $\neg[u_j]$, which are connected by a '$\neq$' relation but are not coreferenced with c outside (see Fig. 3.2). Since c occurs in an evenly enclosed context and its two corresponding concepts occur in an oddly enclosed context, without being coreferenced, neither type specialization nor generalization can be performed on them for derivation, so that their types would be still identical. This explains the added condition '$type(neighbour(r, i)) = type(neighbour(\Pi r, i))$' in Definition 3.2.

In [5], the authors attempted to define the projection operation performed on arbitrary negation-nested CGs, but did not give a justification for the soundness of the projection algorithm sketched there. In contrast, the sound universal CG projection is performed directly on CGs containing universal markers and/or functional relations, instead of their defining expansions that are negation-nested CGs.

3.3. Extended Universal Concept Join

Definition 2.3 and Theorem 2.2 of the universal concept join are still applicable to CGs with functional relations.

Theorem 3.2 Let u be a universal CG and v be any plain CG, with no coreference links in u and v (possibly containing functional relations). Then, any $J_\forall(u, v)$ is a logical consequence of $\{u, v\}$.

Proof. This proof is similar to that of Theorem 2.2. The only difference is that we have to take into account dependent concepts in Coreferent Join steps as carried out in the proof of Theorem 2.2. As we just explained after the proof of Theorem 3.1, each dependent concept, after expansion, has two corresponding concepts in a negative context. We need to prove that the types of a dependent concept and its two corresponding concepts are still identical after Coreferent Join steps.

Only coreferent joins of universal concepts in u with dependent concepts in v demand attention. If a dependent concept c in v is joined with some universal concept in u, then the type of c is unchanged, on the basis of Definition 2.3. That is, the types of c and its two corresponding concepts are still the same after the join.

For example, it is sound to join the CG "Every id is a string of characters" $[ID:\forall] \rightarrow (ISA) \rightarrow [CHAR\_STR:*]$ with the CG "Every student has only one student id" $[STUDENT:\forall] \rightarrow (HAVE) \Rightarrow [STUD-ID:*]$, at the concepts $[ID: \forall]$ and $[STUD-ID: *]$, to derive the CG $[STUDENT:\forall] \rightarrow (HAVE) \Rightarrow [STUD-ID:*] \rightarrow (ISA) \rightarrow [CHAR\_STR:*]$, supposing that STUD-ID is a sub-type of ID.

In order to check and produce consistent CGs with respect to functional dependency constraints, the extended universal concept join can be equipped with the join algorithm defined in [7]. That is, when the determining concepts of two identical functional relations are joined, their dependent concepts are joined accordingly also. Since the extended universal concept join is sound, the fact that inconsistency occurs during this chained join procedure implies inconsistency in the participating CGs.

Figure 3.6 is adapted from [8] (Section 6.5-Database Inference), using functional relations instead of actors and unique existential quantifiers. Graph u says that, for every person and every date, the person's age on the given date is uniquely determined by the difference between this date and his/her date of birth. Graph v gives the fact that employee Lee was born on 3/1/74 and hired on some date. Then a logical consequence of u and v is the shown $J_\forall(u, v)$, from which Lee's age on the date he was hired can be computed.

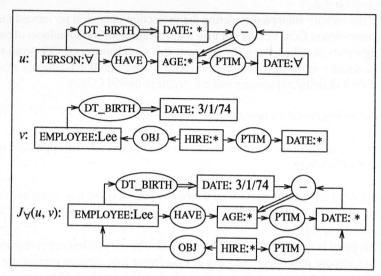

Fig. 3.6. Universal concept join equipped with chained joins

In [8], a special join operation called *schematic join* was used for database inference, joining conceptual *schemas*, *type definitions* and query graphs in deriving answers. Unlike type definitions, which are universally true and can be represented by universal CGs, a schema may not be true for every use of the type. Moreover, a schematic join is a *maximal join* that joins not only concepts of a graph with their corresponding formal parameters of a schema, but also other concepts of the graph and the schema. Therefore, the schematic join is a *plausible reasoning* operation rather than an *exact reasoning* operation like the universal concept join.

4. Conclusions

We have revisited the universal marker and the functional relation. The expansion rule formally defines the semantics of CGs with these two notations and provides a formal basis for the development of direct and sound operations on them. The universal CG projection, which is an extension of the existential CG projection, and the universal concept join bring a number of advantages for the use of the extended CGs in CG knowledge bases, CG programs and CG data conceptual schemas.

A query to a CG knowledge base is answered by comparing the query CG with the

CGs in the knowledge base to see if it is present among or implied by these CGs ([3]). The operation that defines this subsumption relation is the existential CG projection, and as is well-known, the search is speeded up considerably when the CGs are organized in a subsumption hierarchy in the knowledge base. The extended CGs, representing universally quantified facts and functional dependencies, could have the same advantage with the universal CG projection.

In CG programs so far ([4]), universally quantified facts have been assumed to be expressed by rules instead naturally by facts as in predicate calculus programs. It is desirable that universally quantified facts are expressed directly in CG programs and proof procedures manipulate them directly. It would be more natural and efficient in both storage and computation. On the other hand, CGs with functional relations can be used to represent terms in logic programs as *feature term graphs* ([11]). The defined formal semantics and operations of the extended CGs could be useful for these purposes.

CGs with universal markers and functional relations could represent data conceptual schemas. The universal concept join provides a formal sound reasoning operation for the database inference mechanism described in [8]. The universal concept join and universal CG projection could be used for manipulation of CG data conceptual schemas in their integration process, which would involve finding common generalizations, defined through projections, of the participating schemas ([2]).

This work adds to the effort of the formal integration of the universal marker and functional relation into CG, which was initiated in [8] and [7], and makes the CG extension with these two notations more complete and useful.

References

1. ANSI (1993), *IRDS Conceptual Schema*. X3H4/93-196.
2. Creasy, P.N. & Ellis, G. (1993), *A Conceptual Graphs Approach to Conceptual Schema Integration*. In Mineau & Moulin & Sowa (eds): Conceptual Graphs for Knowledge Representation, LNAI No. 699, Springer-Verlag, pp. 126-141.
3. Ellis, G. & Lehmann, F. (1994), *Exploiting the Induced Order on Type-Labelled Graphs for Fast Knowledge Retrieval*. In Tepfenhart & Dick & Sowa (eds): Conceptual Structures: Current Practices, LNAI No. 835, Springer-Verlag, pp. 293-310.
4. Ghosh, B.C. (1996), Conceptual Graph Language - A Language of Logic and Information in Conceptual Structures. PhD Dissertation, Asian Institute of Technology.
5. Heaton, J.E. & Kocura, P. (1991), *Negation and Projection of Conceptual Graphs*. Proc. of the Sixth Annual Workshop on Conceptual Structures, 1991.
6. Lloyd, J.W. (1987), Foundations of Logic Programming. Springer-Verlag.
7. Mineau, G.W. (1994), *Views, Mappings and Functions: Essential Definitions to The Conceptual Graph Theory*. In Tepfenhart & Dick & Sowa (eds): Conceptual Structures: Current Practices, LNAI No. 835, Springer-Verlag, pp. 160-174.
8. Sowa, J.F. (1984), *Conceptual Structures: Information Processing in Mind and Machine*. Addison Wesley, Massachusetts.
9. Sowa, J.F. (1994), *Knowledge Representation: Logical, Philosophical and Computational Foundations*. A draft book in preparation.

10. Ullman, J.D. (1988), *Principles of Database and Knowledge-Base Systems*, Vol. 1. Computer Science Press, Rockville, Maryland.

11. Willems, M. (1995), Projection and Unification for Conceptual Graphs. In Ellis & Levinson & Rich & Sowa (eds): Conceptual Structures: Applications, Implementation and Theory, LNAI No. 954, Springer-Verlag, pp. 278-292.

Animating Conceptual Graphs

Ryszard Raban

School of Computing Sciences
University of Technology, Sydney
P.O. Box 123
Broadway, NSW 2007, Australia
richard@socs.uts.edu.au

Harry S. Delugach

Computer Science Department
Univ. of Alabama in Huntsville
Huntsville, AL 35899 U.S.A.
delugach@cs.uah.edu

Abstract. This paper addresses operational aspects of conceptual graph systems. This paper is an attempt to formalize operations within a conceptual graph system by using conceptual graphs themselves to describe the mechanism. We outline a unifying approach that can integrate the notions of a fact base, type definitions, actor definitions, messages, and the assertion and retraction of graphs. Our approach formalizes the notion of type expansion and actor definitions, and in the process also formalizes the notion for any sort of formal assertion in a conceptual graph system. We introduce definitions as concept types called assertional types which are animated through a concept type called an assertional event. We illustrate the assertion of a type definition, a nested definition and an actor definition, using one extended example. We believe this mechanism has immediate and far-reaching value in offering a self-contained, yet animate conceptual graph system architecture.

1. Introduction

This paper addresses operational aspects of conceptual graph systems, as distinct from the theory. The conceptual graph theory itself is well-developed, as evidenced by books [8] [3], conference proceedings [7] [10], [5] [6], and a forthcoming book [9]. Conceptual graph systems, however, have not been nearly as well developed. This paper is an attempt to formalize operations within a conceptual graph system by using conceptual graphs themselves to describe the mechanism.

As a first-order logic vehicle, conceptual graphs are excellent and powerful in describing static facts, constraints and other relationships that exist in the world. Yet there is more to the world of interest than facts and relationships; the world also contains dynamic behaviors and ever-changing relationships. In order to use conceptual graphs in a practical information system, we must therefore have some mechanism for inserting, updating and deleting information, along with a way to representing the operations of making such changes. An early attempt was shown in [1] [2].

At present, most conceptual graph systems maintain a stored collection of conceptual graphs to represent their knowledge base, yet the operations on these graphs are "stored" in the form of external programs that manipulate the graphs. Their architectures resemble something like Fig. 1(a). Our approach calls for including operations as bona fide graphs, so that they are a part of the conceptual graph collection. We call the collection of graphs, definitions, and their operations a *conceptual graph system* as suggested by Fig. 1(b).

In this paper we outline a unifying approach that integrates the notions of a fact base, type definitions, actor definitions, messages, and the assertion and retraction of graphs. Our approach formalizes the notion of type expansion, and in the process also formalizes the notion for any sort of formal assertion in a conceptual graph system. This has been achieved by introducing a very few conceptual graph concept types and relations, and attaching an operational semantics to them.

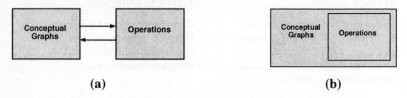

(a)   **(b)**

Fig. 1. Graph Operation Architectures

We first summarize how graph operations have previously been described [8] [9] [6] and evaluate their limitations. We then introduce the dual notions of an *assertional type* and an *assertion event*, which are necessary for animating conceptual graphs, explain assertion semantics of the types using a simple example, and finally summarize our results and point to future work.

2. CGs Behavior in Its Current Form

In [Sowa1984] there is the following use of actors, shown in Fig. 2. Thus Sowa's original concept of actors allows:
  • calculation of referent values for concepts of type MEASURE,
  • selection   between   different   paths   of   processing   similar   to
    if c **then do** A **else do** B,
and therefore with sequence and recursion allows for expressing any algorithm, but there is no way to specify changes to concepts and relations and make graphs evolve.

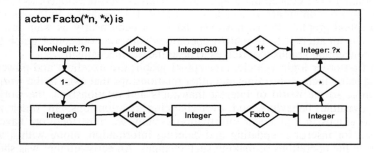

Fig. 2. Actors As Originally Described

Fig. 3 shows how in [9] actors have been replaced with functions and actors are represented as relations.
  Now the notion of functional relations allows:
  • calculation of referent values for measure type concepts, and
  • selection between different values in assigning values to a referent similarly to
    $x :=$ **if** c **then** a **else** b.
  This conditional assignment statement is not sufficient to express all algorithms. Therefore in this paper we adopt the original notation for actors [8] and its construct for branching control flow. At the same time we do accept function cond as useful in some algorithm definitions.

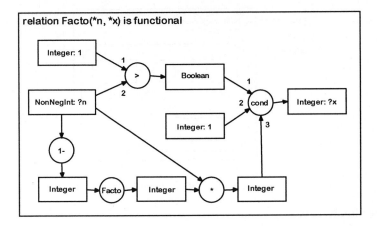

Fig. 3. Functional Relations

Modeling behavior of CGs models has been addressed in [9] in the context of object-oriented modeling. This work introduces operations on messages, objects and process descriptions. There is however a great deal of confusion about general rules of applying these operations. In particular, it is not clear how one state of an object can be retracted and replaced by a new one. The state modification problem was addressed in [6] with the use of a state based view of conceptual graphs. But in this case the result has been achieved at the expense of simplicity and readability of the original CGs formalism. The solution to CGs behavior modeling proposed in this paper uses unmodified CGs notation to capture dynamic aspects of CGs models related to concept creation and actor evaluation.

3. Assertion Semantics

This main section describes assertion semantics, including how definitions turn into conceptual graph facts and how nested definitions are handled as well. Our purpose here is to suggest the flavor of assertional semantics and illustrate their capabilities, not to provide detailed definitions. We describe the types needed in assertional semantics, then introduce the assertion event which controls the process, and follow through a detailed example to show how assertion semantics works on type definitions, nested definitions, and actor definitions.

3.1. Assertional Types

Graphs on the assertion sheet are created and evolve according to some predefined patterns constituting a system definition. We introduce three assertional types to handle the notion of definition. The new types are Definition, Actor, and Type. (We can easily allow for extensions to these types for queries, messages, etc.) We say that

> Definition < Proposition,
> Type < Definition,
> Actor < Definition.

We illustrate this as follows in Fig. 4. In our descriptions of rules, we also use functions *type*, *referent* and *graph* to refer to the three elements of an assertional concept.

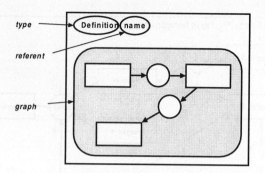

Fig. 4. Format Of An Assertional Type

3.2. Assertion Events

Assertion events are the sources of concepts appearing on the assertion sheet as well as the agents of change as the asserted graphs need to adopt to changing circumstances. The assertion events are represented by the additional type the Assertion_Event. It is a subtype of ACT and similar to the assertional types contains three elements: a type label which is Assertion_Event, a referent which is a unique sequence number, and a graph. Its context forms an operational block. It turns an enclosed graph into one or more asserted concepts, with origination links. The process of asserting concepts through assertion events always happens in three stages:

Initiation when an assertion event is brought to existence by an appearance of or change to a concept which is connected some way with an assertional type definition. Such a concept is called the *initiator*. At this stage the assertion event is related to the two concepts: the initiator *i (there can be one or more of these) via relation (a_init), and the definition *d via relation (a_agnt), as shown in Fig. 5.

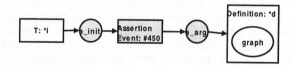

Fig. 5. Initiation Of Assertion Event

Evaluation when the definition graph is copied into the assertion event's context and acted upon according to the rules adopted for the definition type until all concepts in the graph are instantiated and ready to be asserted as shown in Fig. 6.

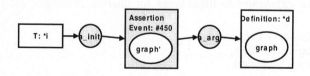

Fig. 6. Evaluation Of An Assertion Event

Exporting when concepts *p_i called *product* concepts, of which there may be arbitrary number, are now exported together with relations between them out of the assertion event context into the assertion sheet. Every concept exported that does not have a coreference link with a concept already on the assertion sheet is connected with the assertion event via relation (a_obj). In this way the assertion event does not contain a graph anymore. The final product of activating an assertion event is a new graph which has to be maximally joined with graphs already on the assertion sheet. The product concepts are shown in Fig. 7:

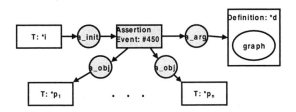

Fig. 7. Asserting Concepts Using An Assertion Event

The additional types and relations are shaded merely for convenience in presenting the material here; otherwise they represent bona fide conceptual relations in the ordinary way. It is important to understand that the result of assertion is a set of ordinary graphs; the only additions are these few concept types and the (a_obj), (a_init), and (a_agnt) relations.

An assertion event is inserted in a graph as a result of type or actor definition activation. When inserted, it represents an operational prescription to be performed. That is, when a type's assertion event is inserted, that type's definition fills the role of definition graph *d; the operational prescription is that the definition graph *d is to be instantiated and exported from the Assertion_Event context, with appropriate values for the definition's parameters inserted. For an actor's assertion event, again the definition fills the role of graph *d; the operational prescription is that its constraints are evaluated (see below) and any resulting concepts are exported from the context. In either case, whatever is exported is linked to its original Assertion_Event concept via (a_obj) relation. These operational prescriptions are defined as assertional semantics of assertional types. In general, animating an assertion event means first asserting its definition graph into the assertion event context, and then activating its attached operational semantics

As we describe later, a definition may contain concepts or actors that themselves have definitions. In the case of nested definitions (e.g., the appearance of an actor in a type or actor definition), an Assertion_Event may also be spawned by another Assertion_Event; in that case, the new Assertion_Event will be linked via an (a_obj) relation to the original Assertion_Event. We show an example of it in Section 0 below.

3.3. Assertional Semantics of Type Definition

This section describes the semantics of a type definition, and how an instance of a concept invokes an assertion event on the definition of that concept's type. We show the steps in creating an assertion event which results in new concepts and relations appearing in the graph.

3.3.1. Type Initiation

The initiator for a type definition is some asserted instance of a concept of that type. For example, consider the type definition for Cat shown in Fig. 8. If we want to insert the concept [Cat: #293] then we would actually initiate a concept Assertion_Event as shown in Fig. 9 below.

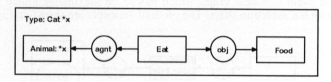

Fig. 8. Example Type Definition

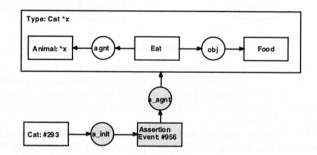

Fig. 9. Example Assertion Event Initiation

3.3.2. Type Evaluation

The definition graph of the Cat type is now copied into the assertion event context, a coreference link is drawn between the [Cat: #293] concept and the genus, and all the other concepts are instantiated which leads to the graph in Fig. 10.

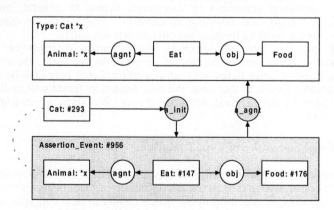

Fig. 10. Example Assertion Event Evaluation

3.3.3. Type Exporting

All concepts required by the type definition have now been instantiated. These concepts (and the relations between them) can now be exported outside the assertional event context as ordinary graph elements. Concepts which do not have coreference links with the outside of the context are linked back to the assertional event via relation (a_obj) to denote their origin on the assertion sheet. The final product of asserting an instance of Cat is shown in Fig. 11. Note that these are the exported products of the Cat type definition assertion only. We will later modify the example assuming that an additional type definition exists for type Food.

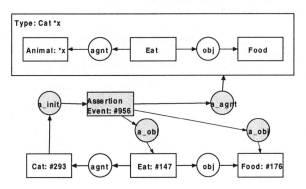

Fig. 11. Example Of Completed Type Assertion

3.4. Nested Assertions

Let us suppose now that Food also has a type definition on the assertion sheet as shown in Fig. 12. Then Fig. 10 cannot be turned into Fig. 11 immediately; we first need to activate the Food type definition via its assertion event.

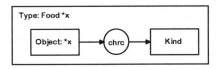

Fig. 12. Second Example Type Definition

In this case, before concepts can be exported from the assertion event context, another (nested) type instantiation process must be initiated by creation of the concept [Food: #176]. This after the processing leads to the graph in Fig. 13.

Exporting graphs from such a nested assertion event context is performed from the outermost event (Assertion_Event: #956) to the innermost (Assertion_Event: #491), following the already outlined rule. Export of the concepts from the outer context leads to Fig. 14. The next step of exporting from the inner event context (#491) produces the final result in Fig. 15.

One note on the assertion semantics is called for here. Updating and deleting are important operations that must also be supported by any animation mechanism. We have developed semantics for updates and deletions, based essentially on the reverse of exporting. This ensures that all graph elements that were the result of creating a concept are withdrawn (imported back to the assertion event context) before the concept can be deleted or updated. The only difference between deletion and update is that when deleting a concept also its assertion event is deleted, while change is followed by a subsequent export of modified graph. Although slightly more complex than the ones shown here, we believe that the creation semantics and their resulting products clearly offer the support needed for updates and deletions. We have omitted the details in the interests of space.

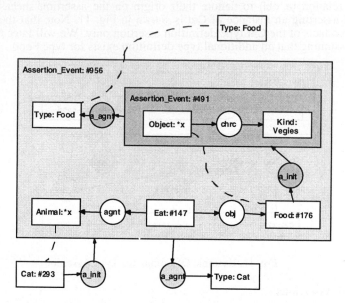

Fig. 13. Example Of Nested Assertion Evaluation

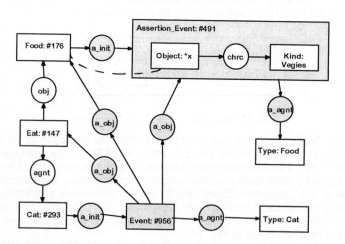

Fig. 14. Nested Assertion Event After Exporting

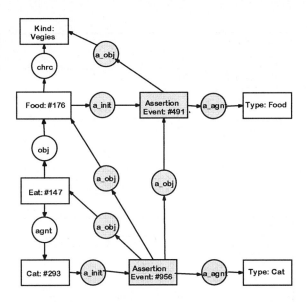

Fig. 15. Completed Nested Assertion Results

3.5. Assertional Semantics of Actor Definition

In order to illustrate the assertional semantics of an actor definition, let us expand our example with some procedures associated with type Cat. Suppose when instantiating a Cat we want to establish what kind of cat it is depending on the type of food it eats. If a cat eats meatless food like vegetables it should be created as a vegetarian animal, if, on the other hand it eats food containing meat it should be created as a non vegetarian animal.

In Fig. 16d a new type definition for the Food type has been created for this purpose. It differs from the one in Fig. 12 by having actor CatType, shown in a diamond shaped box, initiated by the creation of concept [Kind] or by a change of its referent value. The definition of the actor is shown in Fig. 16a. The definition graph of an actor contains concepts with "?*" referents, generic or individual concepts as well as other actors or functions. The concepts marked with the "?" sign are all input concepts for the actor. In fact, as they are related to each other they specify an *input graph* for the actor's evaluation. These concepts will not be asserted by when the actor is evaluated.

The actor CatType definition contains also function Ident and the selection construct as described in [8]. So, if the food kind turns out to be of VegType (vegetarian type) then we initiate actor VCat (Fig. 16b) which asserts that a cat is a vegetarian animal; otherwise actor NVCat (Fig. 16c) is initiated and asserts that it is a non vegetarian cat. The rest of this section uses this expanded example to describe the assertional semantics of actors for initiation, evaluation and exporting.

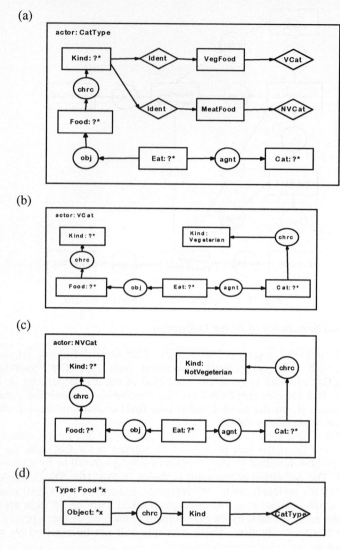

Fig. 16. Expanded Definitions For A Cat's Food

3.5.1. Actor Initiation

An actor is initiated when its initiator (a concept with an arrow pointing to the actor's box) is created or changes its value. To illustrate the process let us assume that [Cat: #293] which was created before in Fig. 9 is now instantiated with the modified Food type definition. Initiation of the Cat type is not affected by the change so we will not show it; we start from the point where the new Food type is initiated as shown in Fig. 17. In a similar way to type definition instantiation, this causes a new assertion event (Assertion_Event: #401 in this case) to be created and linked with the initiator via relation a_init and with the actor definition via relation a_agnt. Note here again that the actor definition is coreferent with its original assertion on the assertion sheet.

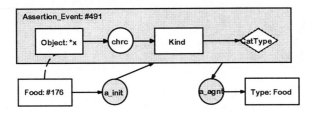

Fig. 17. Initiation of the New Food Type Assertion

3.5.2. Actor Evaluation

Once an actor is initiated its definition is copied into the assertion event context. But before the actor's definition can be evaluated all its input concepts (the ones with ?* referents) must be available. The input itself always comes from the enclosing context, whether it is another assertion event's context or the assertion sheet. The input is available if there is a fully instantiated projection of the input graph in a graph of the enclosing context. As shown in Fig. 18 the input graph in **Assertion Event #401** does not have all inputs determined in its enclosing **Assertion Event #491** context. In fact only two concepts of type **Food** and **Kind** have established values, what is shown in Fig. 18 by the coreference links.

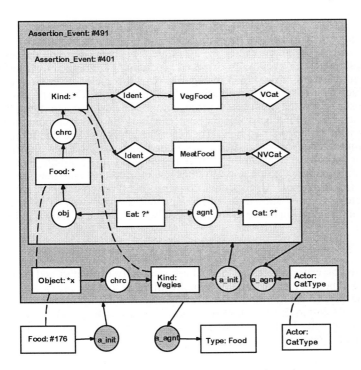

Fig. 18. Initiation of CatType Actor Assertion

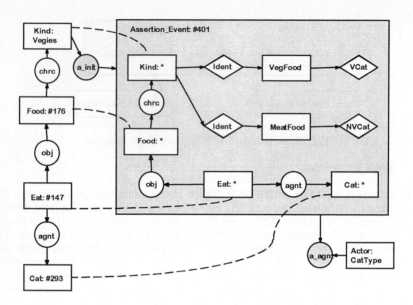

Fig. 19. Exporting of Cat Type Assertion Products

At this stage, actor CatType stays initiated and is ready to be further evaluated as soon as the remaining input values are available. This happens when the results of Assertion Event #491 are eventually exported producing the graph shown in Fig. 15 whose relevant parts are now shown in Fig. 18. Now the enclosing context contains all input values required for the evaluation of actor CatType. The coreference links in the figure show how the input graph projects in the enclosing context graph.

The value of food kind activates the top branch of the selection construct since actor Vegies is of actor VegFood type. The other branch is blocked due to an output value type mismatch. Now actor VCat is initiated and goes through the initiation process described above. The graph which is produced in Assertion_Event: #401 context contains the input graph and all other instantiated concepts that are directly or indirectly linked with it. All the other concepts and blocked actors are discarded from the context. Actor VCat can be fully evaluated straight away. The final effect of initiation and evaluation of the two actors CatType and VCat is shown in Fig. 20.

3.5.3. Actor Exporting

The two nested assertion events in Fig. 20 are now fully evaluated and hence their graphs can be exported and maximally joined with the rest of the previously asserted concepts as in Fig. 15 leading to the final result as shown in Fig. 21.

Note that Assertion_Event: #489 does not have a relation a_init to any of the concepts in the graph. That simply means that it can be only initiated through the evaluation of CatType.

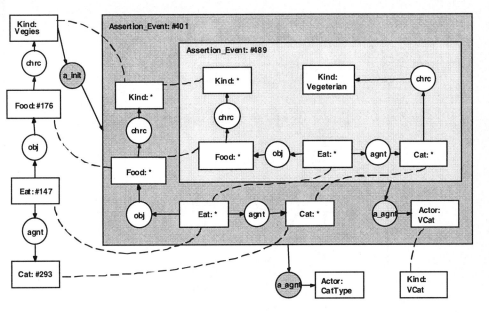

Fig. 20. Evaluation Of Food Actor Assertion

3.5.4. Actor Reevaluation

An actor that was already evaluated can be initialized again by a change to its initiator's referent value. This leads to a process very similar to the updating of type defined concepts as briefly described in section 3.4.

To illustrate it, let's suppose that the Cat: #293 changed its eating habits and started eating meat. This fact would be reflected in a change to its food kind value from Vegies to Meat. The change of the food kind value initiates actor CatType which is the agent of Assertion_Event: #401. Before the usual actor initiation, evaluation and exporting can be performed, the result of the last CatType evaluation has to be retracted. This is done by importing into Assertion_Event: #401 all elements that resulted from its previous activation. All concepts generated by the last evaluation of Assertion_Event: #401 can be easily found by following all a_obj and a_init links coming out of Assertion_Event: #401. Once these concepts are retrieved and placed back into Assertion_Event: #401 context they are erased, and the evaluation process starts exactly the same way as described earlier. Now the situation is exactly as shown in Fig. 19 except that the food type is not Vegies anymore but Meat so the evaluation proceeds appropriately activating the lower branch of the selection construct. In effect actor NVCat is initialized and the final result will have the cat characterized as a non vegetarian.

The process of actor reevaluation allows for dynamic changes to a CGs model as it allows us to retract previous results and to assert new concepts. The example showed a very simple modification of a concept value, but the two actors involved, VCat and NVCat, might generate completely different graphs describing the eating habits of a cat and lead to more substantial changes to the asserted graph than those shown in the example.

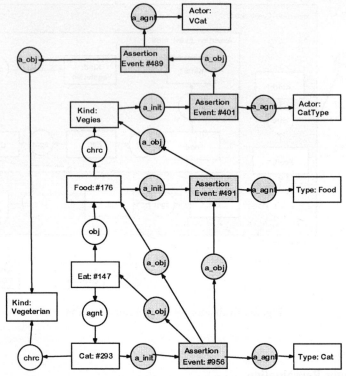

Fig. 21. Exported Products Of Food Actor Assertion

4. Conclusion

We have developed a representation and semantics for animating conceptual graphs. We have described the informal algorithmic procedures that have been used to operate on conceptual graph fact bases in practical systems. Since the representation is expressed in conceptual graphs themselves, all operations are explicit and visible instead of being hidden in one's (possibly erroneous) implementation.

An important result of this approach is that changes to a fact base are explicitly recorded and displayed, along with the source events which caused the change. Although such traceability is not strictly required for logic purposes, it is a great benefit for practical information systems where information must be verified, updated and deleted on a regular basis.

We believe this mechanism has immediate and far-reaching value in offering a self-contained, yet animate conceptual graph system architecture. Our future work will be focused on a detailed discussion of the rules for initiation/evaluation/exporting and implementing these mechanisms, then applying them to new assertional types such as queries, inference rules, messages, etc. It is even possible that some assertional types could include instructions for making new assertional types automatically, thereby offering the capability for a self-modifying and self-adapting fact base that has been hitherto unavailable in conceptual graphs.

5. References

[1]   Delugach, Harry S., "Specifying Multiple-Viewed Software Requirements With Conceptual Graphs," *Jour. Systems and Software*, vol. 19, pp. 207-224, 1992.

[2]   Delugach, H.S. "Analyzing Multiple Views Of Software Requirements" in *Conceptual Structures: Current Research and Practice*, Eklund, P., Nagle, T., Nagle, J. & Gerholz, L. eds., 1992, 391-410, Ellis Horwood.

[3]   Eklund, P., Nagle, T., Nagle, J. & Gerholz, L. [Eds.], *Conceptual Structures: Current Research and Practice*, Ellis Horwood, 1992.

[4]   P.W. Eklund, G. Ellis and G. Mann, eds., *Conceptual Structures: Knowledge Representation as Interlingua, Lecture Notes on Artificial Intelligence,* LNAI 1115, Springer Verlag, 1996.

[5]   Ellis, Gerard, Levinson, Robert, Rich, William, and Sowa, John F., eds., *Conceptual Structures: Applications, Implementations and Theory, Lecture Notes on Artificial Intelligence,* LNAI 954, Springer Verlag, 1995.

[6]   Ellis, Gerard. "Object-Oriented Conceptual Graphsî in *Conceptual Structures: Applications, Implementations and Theory, Lecture Notes on Artificial Intelligence,* LNAI 954, Springer Verlag, 1995, pp. 144-157.

[7]   Mineau, Guy W., Moulin, Bernard, and Sowa, John F., eds., *Conceptual Graphs for Knowledge Representation, Lecture Notes on Artificial Intelligence,* LNAI 699, Springer Verlag, 1993.

[8]   Sowa, John F., *Information Processing in Mind and Machine*, Addison-Wesley Publ., Reading, MA, 1984.

[9]   Sowa, John F., *Knowledge Representation: Logical, Philosophical and Computational Foundations*, (in press), 1997.

[10] Tepfenhart, William M., Dick, Judith P., and Sowa, John F., eds., *Conceptual Structures: Current Practices, Lecture Notes on Artificial Intelligence,* LNAI 835, Springer Verlag, 1994.

Accounting for Domain Knowledge in the Construction of a Generalization Space

Isabelle Bournaud and Jean-Gabriel Ganascia

LAFORIA-LIP6, University of Paris 6
4, place Jussieu 75252 Paris Cedex 05 FRANCE
email : {bournaud,ganascia}@laforia.ibp.fr

Abstract. Our study registers in the framework of the automatic construction of classifications. We tackle an issue which has been less explored, that of the discovery of classifications. To tackle this problem we have chosen to pursue and develop the works of Mineau in the domain of the organization of knowledge bases using generalization [20]. We propose an original approach, called COING, to the discovery of classifications of structured objects represented using conceptual graphs. This approach consists in building a space of concepts which generalize the objects descriptions, called the *Generalization Space*, and then exploring this space so as to iteratively extract one or several conceptual classifications [2]. In this paper, we describe the method of construction of the Generalization Space that has been implemented in COING. This method is an extension of the MSG proposed by Mineau [19], which enables to account for knowledge about the types during the construction of the Generalization Space. However, we propose a formalization of the use of constraints in the construction of the Generalization Space. We present empirical results of the method proposed that show its effiency.

1 Introduction

In Artificial Intelligence, the problem of the automatic construction of classifications has been the subject of much research for about fifteen years [18]. Historically, this study has taken place in the field of inductive learning. Most of the developed approaches have defined this task as the search for a classification that would best predict unknown features of new objects. The developed methods have proved successful through a number of applications [18] [7] [10].

Our study registers in the framework of the automatic construction of classifications but we tackle an issue which has been less explored, that of the *discovery of classifications*. In this prospect, the predictive capacity of a classification becomes much less important than its intrinsic power of representation on the one hand, than what it reveals of the hypotheses of those who built it on the other hand. To tackle this problem which had been somewhat abandoned [8], we have chosen to pursue and develop the works of Mineau in the domain of the organization of knowledge bases using generalization [19]. Mineau proposes to construct an *objective* organization of objects, reflecting all the existing similarities between these objects without emphasizing any point of view about the objects considered [20]. This organization is called the "Generalization Space" (GS). More precisely, given a set of objects being described by conceptual graphs, the Generalization Space is a set of concepts which generalizes those descriptions. This space is partially ordered by a generality relation between concepts. In general the construction of the GS brings some complexity problems. The objects descriptions under the form of conceptual graphs correspond to what is called *structured descriptions*. In order to generalize such descriptions, one must first be able to match them. The problem of matching structured descriptions is related to the problem of the sub-graphs isomorphism which is a NP-complete problem [13]. The solution proposed by Mineau in the MSG to counter this matching problem consists in considering the generalization of arcs of the graphs taken individually rather than generalizing whole graphs [19]. This restriction enables to limit drastically the complexity of the graphs generalization.

We propose an original approach to the discovery of classifications (named COING for COnceptual clusterING) which consists in exploring the GS so as to iteratively extract one or several conceptual classifications [3]. Knowledge of the domain play a key role in the construction of classifications [4] [11]. Yet they are not used very often in structured domains because they bring an adding complexity factor. Indeed, they may cause the size of the GS and its construction time to increase considerably. In order to use our method to build classifications when domain knowledge is available, we have proposed an approach enabling to integrate domain knowledge in the GS construction. This approach is an extension of the original method proposed by Mineau in the MSG [19].

In part 2, we describe the descriptions generalization process and the principle of the Generalization Space construction. In part 3, we present a method which enables to account for knowledge about the types during the GS construction while limiting the underlying complexity. In the following parts, we formalize the use of semantic constraints in the GS construction. We present in the last part an empirical evaluation of the GS construction method that we proposed.

2 Constructing the Generalization Space

Given a set of objects being described by conceptual graphs, the Generalization Space is a set of concepts which generalize those descriptions. This space is partially ordered by a generality relation between concepts. This partial order supplies the GS with a pruned lattice structure which may be represented by an inheritance network [11]. Numerous works have been carried on knowledge retrieval from structures similar to the GS [6] [12] [24]. Levinson proposed a representation of such structures supporting efficient retrieval [16]. However, the problem we are interested in is not to find an efficient representation of the Generalization Space but to construct it. In order to construct the GS, we have adopted an ascending method. The method builds upon the objects descriptions to construct generalizations of these descriptions. The operation of generalizing descriptions is thus central to this method.

2.1 Generalizing the objects descriptions

Building up the GS requires generalizing the objects descriptions, or under the circumstances, generalizing conceptual graphs. In order to generalize such descriptions, one must first be able to match them. The problem of matching structured descriptions is related to the problem of the sub-graphs isomorphism which is a NP-complete problem [13]. In effect, the complexity of the generalization increases exponentially with the structuration degree of the objects [27]. It is thus necessary to introduce a restriction (a *bias*) which enables to limit the complexity of the generalization of such descriptions so that algorithms within the power of machines may be found.

The solution we have adopted to counter the problem of matching conceptual graphs is the one proposed by Mineau in the MSG [19]. It consists in restricting oneself to a limited number of matchings between graphs. Instead of trying to match a graph G1 with a graph G2, one restricts oneself to searching partial matchings of an arc of G1 with an arc of G2. In order to look for these matchings, it is sufficient to consider a representation of the graphs as a set of *independent* arcs. The matchings sought are those of an arc from the set of arcs of G1 with an arc from the set of arcs of G2. The arcs being oriented, they are totally matched. Indeed, given two conceptual arcs $[Cs1] \rightarrow (r1) \rightarrow [Cd1]$ and $[Cs2] \rightarrow (r2) \rightarrow [Cd2]$, the only possible matching between these arcs is the type $Cs1$ with the type $Cs2$, the type $r1$ with the type $r2$ and the type $Cd1$ with the type $Cd2$. From an algorithmic point of view, given two graphs G1 and G2 having respectively $n1$ and $n2$ arcs, this decomposition limits the set of possible matchings of G1 with G2 $((n2)^{n1})$ to $n2 \times n1$ matchings. The problem of generalizing the objects descriptions then reduces to the problem of the generalization of arcs.

2.2 Principle of the GS construction

Once the graphs describing the objects have been decomposed into a set of arcs, the principle of construction of the GS nodes is the following. One first generates all the generalizations of each of the arcs describing the objects. The second step consists in clustering the generalized arcs covering the same objects (i.e. the arcs describing the similarities between these objects). Finally, the generalized arcs are filtered. In effect, for a given matching there exists numerous possible generalizations and this last step consists in choosing, for each set of arcs, those which will make up the description of the node. This principle is illustrated on Figure 1. One still need to link the nodes by using an inclusion relation based on their covering.

In the GS construction, and more classically in conceptual clustering, one limits the number of generalizations by considering only the most specific ones. The most specific generalizations contain the largest amount of information about the objects whereas the most general generalizations being those which retain the least information about the objects [17]. The filtering step (3, Fig. 1) consists in storing only the most specific arcs. We designate as *useless* an arc which is not maximally specific [2].

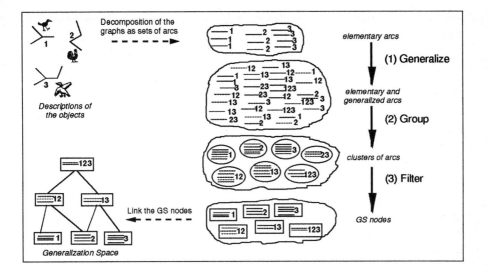

Fig. 1. Principle of the GS construction

In the classical MSG, the generalization step (1, Fig. 1) consists in systematically transforming the types of an arc into a generic type, denoted "?" [19]. From any arc one constructs at the most $7(= 2^3 - 1)$ generalizations. Indeed, given 3 types (i.e. two types of concept and one type of relation), each of them may or not be generalized, the number of possible arcs is thus 8. In the following section, we bring a refinement to this generalization process which consists in accounting for knowledge about the types, without enduring any important loss of efficiency.

3   Accounting for knowledge about the types in the GS construction

Introducing knowledge in order to improve the generalization step has been the subject of numerous studies in machine learning [15] [22]. One of the aspects we have been interested in our approach is the taking into account of some knowledge expressing relations of generality between the different types occurring in the arcs. This knowledge is represented in the conceptual graphs model by the type lattice. Such knowledge enables to express, for instance in the domain of food, the fact that any corn is a seed. As is emphasized by several studies, accounting for such knowledge may improve the relevance of the nodes of the GS [26] [4] [11].

In machine learning, accounting for domain knowledge is often performed by using a preliminary stage of *saturation* [15] [1]. The saturation consists in an exhaustive application of a set of rules onto the objects descriptions. These rules allow to improve objects descriptions by adding all the knowledge which may be deduced from domain knowledge and their descriptions. A rule which is frequently used is the rule called "ascension in a concept lattice" [17]. In the next section we define this operation in terms of conceptual graphs and present in section 3.2 its use in the GS construction.

3.1 The arc generalization

We define the *arc generalization* as an operation of generalization allowing the generation of an arc *ag* from an arc *a* by accounting for a type lattice. This operation relies upon the classical operations of concept augmentation and relation augmentation of the conceptual graphs model [5]. For homogeneity with the typology of the conceptual graphs, a generalization of arc will also be called an augmentation of arc. The arc generalization is defined in the following way :

Definition 1 (Arc generalization) *The generalization of arc* $a = [Cs] \rightarrow (r) \rightarrow [Cd]$ *consists in augmenting at least one of the types of the arc a, that is Cs, r or Cd. The arc obtained is called "augmented arc"; it is a generalization of a. We call degree of an arc generalization the number of types of the arc which are augmented.*

In order to obtain all the generalizations of an arc, it is sufficient to proceed to all possible augmentations of this arc. A classical approach to take into account a type lattice in constructing the GS would consist in proceeding to all the possible augmentations of the arcs (1 Fig. 1) before clustering them and possibly filter them. In the case of the MSG this approach is easily applied. Indeed, each type having a single more general type (the type lattice is said to be "flat"), there exists at the most 7 distinct generalizations of any arc [19]. On the other hand, in the general case where the type lattice is deep and of maximum depth D_{max}, this approach may hardly be considered. Indeed, any type of arcs has at the most D_{max} more general types ; the maximum number of arcs to be stored if we consider N objects each described by $kmax$ arcs is thus $N \times kmax \times (D_{max} + 1)^3$. Let us take as an example the case of the Chinese characters. If we consider 400 Chinese characters, each being described by 12 arcs at the most and a lattice of depth 4, the maximum number of arcs to be stored is 600,000 ($400 \times 12 \times 125$). Of course this number corresponds to an upper limit since on the one hand the lattice of relations is generally flat, and on the other hand the generated arcs often generalize several arcs. However, it remains that the number of arcs to be stored is considerable. In other words, the use of purely declarative knowledge during a saturation stage creates a large number of arcs and may therefore require a large amount of memory space. We present in the following section an efficient method which accounts for a type lattice in the Generalization Space construction. This method consists in anticipating the filtering step (3 Fig. 1) by generalizing the descriptions "by layers".

3.2   The generalization by layers

The method which we propose for accounting for knowledge about the types in the generalization relies, as in CHARADE [9] or KBG [1], upon a saturation stage. Nevertheless, we propose a way to counter the problem of the size of the required memory space. Our method consists in anticipating the filtering step (3 Fig. 1) by eliminating as soon as possible the useless arcs added (i.e. those which are not maximally specific). This method, which is based on the notion of *depth* of an arc (the depth of an arc is equal to the sum of the depths of its types in the type lattice [2]), reduces to performing the descriptions generalization in an iterative manner. It consists in partitioning the set of arcs which may be created in successive layers, a layer corresponding to the set of arcs of same depth. The principle is then to generalize the descriptions one layer after another, starting with the deepest ones. In other words, this reduces to fix the order in which the arcs are generalized. One of the interests of this approach is that it enables to eliminate the useless arcs generated as soon as possible. As soon as the arcs of a particular layer have been generalized, it is indeed possible to remove the useless arcs. The steps 1 and 3 of Figure 1 are, in fact, no longer sequential. Figure 2 below summarizes the principle of the generalization by layers.

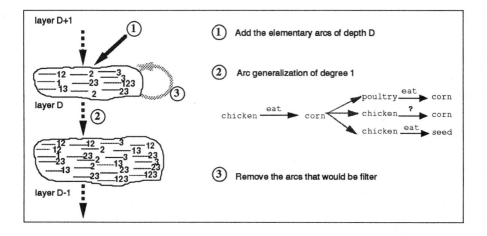

Fig. 2. The generalization by layers

Let us notice that the arc generalizations performed are all of degree 1 (cf. Def. 1). The depth of the layers corresponding to the depth of the arcs, itself calculated as a function of the depth of the three types of the arc, the passage from a layer to the next layer is achieved by augmenting only one of the three types of the arc. The set of arcs obtained when the elimination of the useless arcs is achieved by layers (Fig. 2) is the same as if one had performed filtering of arcs after the generalization ("classical" approach, Fig. 1). However, as is shown in section 5, time and memory space required are much smaller when the filtering is anticipated.

In our approach, the Generalization Space construction is constrained with different kinds of constraints. We present in the next section two kinds of constraints which are used : the first one relates to the generalization language and the second one to the gathering of arcs. We also formalize the use of these constraints in the GS construction.

4 Constraints on the GS construction

4.1 Constraints on the generalization of arcs

One of the underlying problems in the methods based on generalization is the very large number of generalizations which one may build from a set of descriptions. As a matter of fact, the use of constraints on the generalization language has been the object for several years of numerous studies in machine learning. It is particularly the case in inductive logic programming (ILP) [21]. In this domain, the constraints on the generalization language are explicitly defined and are usually called "bias of language" [25]. In ILP, the basic formalism used is first-order logic. The bias of language considered are usually restrictions of Horn clauses. Examples of restrictions are the ij-determined clauses [21], the DATALOG clauses (i.e. without functions) [22] [23], the models of rules [14], the schemes of literals [25]. Restricting the generalization language to a set of unconnected arcs (cf. Part 2.1) corresponds to a syntactic constraint on the generalization language, as is also the case for the previous different constraints. We shall focus in this section on the use of semantic constraints on the generalization language.

Arc generalization under constraints Among the $(D_{max} + 1)^3$ generalized arcs which may be generated from any arc, it is possible that some be undesirable or non relevant with respect to the domain considered. We have proposed to constrain the generalization process so as not to generate such arcs. To formalize the process of generalization under constraints, we have defined the operation of *arc generalization under constraints*. First of all, let us introduce a formal representation of a set of constraints C. We let the predicate satisfy_C take for argument an arc a and be true if and only if this arc satisfies the whole set of constraints C. We then introduce the operation of arc generalization under constraints in the following way :

Definition 2 (Arc generalization under constraints C) *An arc generalization under constraints C is an arc generalization such that for the predicate satisfy_C the generalized arc is evaluated to true.*

The mechanism of generalization of arcs (1 Fig. 2) set in the COING approach relies upon the arc generalization under constraints. By construction, the generalized arcs obtained satisfy the constraints defined. Naturally the definition of the predicate satisfy_C governs the implementation of the constraints considered. In the next section, we present a formalization of the constraints which we have been using in COING. The formalization proposed allows us to characterize the semantic constraints which we impose on the generalization language used for building the Generalization Space.

An example of semantic constraint A kind of semantic constraints which aims at limiting the number of possible generalizations of arcs consists in giving a definition of the arcs *relevant* to the domain considered. Such knowledge is defined in the conceptual graphs model by the *star graphs* of the relations [5]. The star graphs of the relations correspond to the most general and acceptable augmented arcs. In other words, the star graphs define the "limits" (in terms of generalization) which the augmented arcs must not exceed. Such constraints correspond to semantic constraints on the augmentation of the types of arcs containing a given relation r.

In the case where the considered constraints correspond to star graphs of relations, the predicate satisfy_C can be defined as follows. Let the arc representing the star graph of the relation r be $([Cs] \rightarrow (r) \rightarrow [Cd])$. satisfy_C is true if and only if $Cs \triangleleft Css$ and $r \triangleleft rs$ and $Cd \triangleleft Cds$ ($C1 \triangleleft C2$ means that C2 is more general than C1). In a logic programming syntax, this is denoted by :

$$\text{satisfy\_C} \ ([Cs, r, Cd]) \ :- Cs \triangleleft Css, \ r \triangleleft rs, \ Cd \triangleleft Cds$$

It is precisely this implementation of the predicate satisfy_C which has been retained in our approach to the GS construction.

4.2 Constraints on the construction of the GS nodes

The constraints which we have presented above are imposed on the arcs and their generalizations. However, once the set of the generalized arcs is generated, the GS itself remains to be built (i.e. the groups corresponding to the nodes of the GS must be formed). As we have seen, all the arcs having the same covering on the objects (step 2, Fig. 1) must first be gathered, then it may be chosen among all these generalized arcs those which will constitute the description of the GS node (step 3, Fig. 1). A first approach consists in storing all the generalized arcs in the node description. The drawback is that the descriptions of the GS nodes may become very large. Rather, we propose to formalize the constraints which can be imposed on the descriptions of the GS nodes, as well as those which we use in the COING approach.

Let $\mathcal{O}$ be the set of objects to be classified and $\mathcal{P}(\mathcal{O})$ the set of parts of $\mathcal{O}$. For improved clarity in this formalization, we introduce the notion of *aggregate* as follows :

Definition 3 (Aggregate) *An aggregate $\mathcal{A}_E$ is a set of generalized arcs having the set $E \in \mathcal{P}(\mathcal{O})$ for covering.*

The inclusion relation between the parts of $\mathcal{O}$ allows to define a generality relation between aggregates. An aggregate $\mathcal{A}_{E'}$ is more general than an aggregate $\mathcal{A}_E$ (denoted $\mathcal{A}_E \lhd \mathcal{A}_{E'}$) if and only if $E \subset E'$. We define a Generalization Space from the notion of aggregate as follows :

Definition 4 (Generalization Space) *The Generalization Space is the set of the non-empty aggregates* $filter(\mathcal{A}_E)$, *where* $filter$ *is a function which associates to a set of arcs a sub-part of this set.*

We shall designate as node of the GS one of the aggregates which constitute it. Let us notice that if the `filter` function is the identity function (i.e. there are no constraints on the choice of the generalized arcs), the GS is exactly the set of the non-empty aggregates $\mathcal{A}_E$. In this case, the descriptions of the GS nodes contain all the generalized arcs of the arcs describing the objects of $\mathcal{O}$. In the case where the function `filter` is not the identity, it can be interpreted as the expression of constraints on the gathering of arcs.

One may consider defining different types of `filter` functions. It is thereby important to notice that any aggregate being a set of generalized arcs, the arcs constituting it are partially ordered by the generality relation between arcs. One can thus define `filter` functions which make the most of this relation between arcs of a same aggregate. In our approach, we have proposed to use the function `filter`=MSA which associates to a set of arcs those which are maximally specific (for the generality relation between arcs). The GS built by using the function MSA is such that the descriptions of its nodes contain only the most specific arcs.

It is possible to define some `filter` functions based on semantic constraints, unlike the MSA function which is founded on the generality relation between arcs. As Mineau stresses (personal communication), one may want to account for constraints of functional dependence between types, or also for constraints of exclusivity between instances. Such constraints, which allow to constrain the construction of the GS nodes, may easily be represented in the framework proposed above by choosing an appropriate `filter` function. The method of generalization by layer relies upon an anticipation of the filtering step. Naturally, its efficiency strongly depends on the `filter` functions used.

5 Empirical Evaluation

The method of construction of the GS presented here is used in the COING system [2]. It is implemented in LeLisp and runs on Sun Sparc 10. In this section, we present an empirical evaluation of our approach to constructing the GS. After an evaluation of the impact of domain knowledge on the GS in terms of the number of nodes, we show the gains of the generalization by layers in comparison with a classical approach to generalization. The experiments presented have been carried out on databases of Chinese characters possessing at the most 4 components and being each described by 12 arcs at the most (see [2]). The considered type lattice contains 813 types, has a mean depth of 2.18 and a maximum depth of 4.

5.1   Impact of domain knowledge on the GS

In order to evaluate the impact of the type lattice on the number of nodes of the GS, we have built Generalization Space for sets of 10, 25, 50, ..., 200 Chinese characters with and without domain knowledge. It may be noted that we are counting the objects in the GS nodes (i.e. the nodes of the GS correspond to the generalized nodes and to the nodes representing the objects considered). The results are presented on Figure 3.

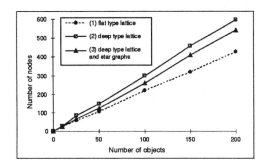

Fig. 3. Impact of domain knowledge on the GS

It clearly appears on Figure 3 that the number of nodes of GS constructed by using a deep type lattice is not so much larger than in the case where the lattice used is flat. This has been confirmed by experiments in other domains which show that the multiplicative factor is in practice much smaller than the theoretical factor $(\frac{D_{max}+1}{2}^3)$ [2]. We see that accounting for a type lattice in the generalization process does not considerably increase the number of nodes of the GS and may thus be also considered for large knowledge bases.

The curve (3) of Figure 3 represents the evolution of the number of nodes as a function of the number of characters when GS are builded by using a deep lattice and star graphs of relations. One can see here that the use of star graphs allows to decrease the number of nodes of the GS. This can be explained by the fact that the constrained generalization (Def. 2) aims precisely at limiting the allowed generalizations to those which verify a set of constraints. Yet this is not sufficient to explain the fact that accounting for a deep lattice in the generalization results in a small increase in the number of nodes of the GS. Further experiments on various data are still needed if we want to understand better the reasons for this small increase.

5.2   Evaluation of the generalization by layers

If accounting for a deep type lattice in the generalization does not considerably increase the size of the GS, on the other hand it requires a large memory space (cf. Part 3.1). We evaluate in this section the gains in memory and time of our method of generalization by layers.

Gain in space The Figure 4(a) presents the evolution of the number of arcs in memory as a function of the number of processed layers. This experiment has been achieved on a database of 200 Chinese characters while accounting for a deep lattice and for star graphs of relations. The curve (1) of Figure 4(a) presents the evolution of the total number of useful arcs (present in the GS nodes); the curve (2) presents the evolution of the total number of arcs in memory when the arcs are filtered by layer (method used in COING); at last the curve (3) presents the evolution of the total number of arcs in memory when the filtering of arcs takes place at the end of the generalization.

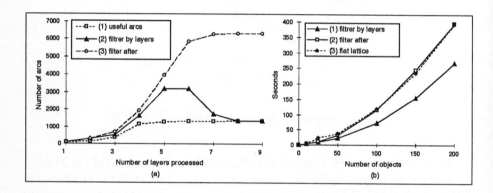

Fig. 4. Memory and time required for the GS construction

The curves (2) and (3) of Figure 4(a) emphasize the gain obtained when the filtering of the arcs is performed by layer. One notices that eliminating the useless arcs as soon as they have been generalized leads to a better memory management. The gain obtained is equal to the ratio between the maximum of curve (2) and the maximum of curve (3). The gain here is 1.98. In other words, the method of generalization by layers makes it possible, in this particular case, to divide by 2 the memory space required.

Gain in time The Figure 4(b) presents the evolution of the generalization time required as a function of the number of characters considered in the cases where the arcs are filtered by layers or at the end.

One can see on this figure that the time required for the generalization is shorter if the elimination of the useless arcs is carried out by layers (curve (1)). This is explained by the fact that the number of arcs upon which one tries to apply the arc generalization decreases when layers are being used. Indeed, the operation is successively applied to every layer, but each time the number of arcs upon which it applies is only constituted by the arcs contained in a layer.

The curve (2) of Figure 4(b) represents the construction time of the GS when the lattice used for the generalization is flat. In theory, constructing the GS by using a flat lattice requires less time than when the lattice is deep. However, the use of the notion of "layer" allows to eliminate as soon as possible the useless arcs and thus to minimize the processes. Now, the use of a flat lattice does not allow the partitioning of the elementary arcs into layers (they are all of depth 3). Hence the useless arcs are not progressively removed during the generalization. This explains that the construction time of the GS with a flat lattice is perceptibly the same as when a deep lattice is used. Let us emphasize that the use of star graphs of relations in the generalization decreases the construction time of the GS (curve (2), Fig. 4(b)). This can be understood by the fact that the number of arcs obtained with constrained generalization is smaller than with unconstrained generalization. The number of arcs to process at each layer is thus smaller and so is the time necessary to deal with all the layers.

6 Conclusion

This paper presents a method for constructing a Generalization Space. This method is an extension of the MSG [19] since it allows to account for knowledge about the types in the generalization process while limiting the underlying complexity. We have developed an algorithm which is efficient in terms of computing time and memory space and which enables the application of this method to the construction of GS for large knowledge bases. The experiments in the domain of Chinese characters presented here illustrate the gain of the method of generalization by layers in comparaison with a "classical" method of GS construction using domain knowledge. Furthermore, we have proposed to formalize the use of domain knowledge in the construction of the Generalization Space.

The formalization of the constraints on the generalization language relies upon the definition of the operations of arc generalization with or without constraints. In the case where the constraints considered are star graphs of relations, the formalization gives a very concise representation of the constraints on the generalization. Accounting for the star graphs of relations, which define some restrictions on the generalizations independently of the objects yet with respect to domain knowledge, is to our knowledge an original contribution to the construction of a space of concepts by generalization. One aspect we are working on consists in enriching the generalization language by tackling the issue of the matching between conceptual graphs. This would allow genuine treatment of structured descriptions in the construction of the GS. We are considering exploring the approach used in the learning system of concepts REMO [27] which consists in progressively relaxing the restrictions on the graph matching in order to enrich the generalization language. The formalization of the constraints on the construction of the GS nodes allows to characterize the presented approach as a way among others to constrain the construction of the GS nodes.

These formalizations aim at offering a more general framework for future studies in the domain of the use of semantic constraints for restricting the generalization language used for building the Generalization Space as well as the language of the nodes of the Generalization Space.

Acknowledgements

We would like to thank Guy William Mineau for his suggestions in formalizing the ideas presented in this paper, and Jean-Daniel Zucker for providing valuable comments on this work.

References

1. G. Bisson. Learning in fol with a similarity measure. In *Proc. of AAAI*, pages 82–87, San Jose, CA, 1992.
2. I. Bournaud. *Regroupement conceptuel pour l'organisation de connaissances.* Thèse de Doctorat, Université Paris VI, France, 1996.
3. I. Bournaud and J.G. Ganascia. Conceptual clustering of complex objects: A generalization space based approach. In *Lectures Notes in AI n-954. Proc. of the third International Conference on Conceptual Structures*, pages 173–187. Springer-Verlag, 1995.
4. C. Carpineto and G. Romano. GALOIS: An order-theoretic approach to conceptual clustering. In *Proceedings of the Tenth International Conference on Machine Learning*, pages 33–40, 1993.
5. M. Chein and M. L. Mugnier. Conceptual graphs: Fundamental notions. In *Revue d'intelligence artificielle 6(4)*, pages 365–406. 1992.
6. G. Ellis. Efficient retrieval from hierarchies of objects using lattice operations. In *Lecture Notes in AI n-699. Proc. First International Conference on Conceptual Structures*, pages 274–293. 1993.
7. D. Fisher. Knowledge acquisition via incremental conceptual clustering. In *Machine Learning 2*, pages 139–172. 1987.
8. D. Fisher. Iterative optimization and simplification of hierarchical clusterings. In *Journal of Artificial Intelligence Research, Vol. 4*, pages 147–179. 1996.
9. J.G. Ganascia. AGAPE et CHARADE, *deux techniques d'apprentissage symbolique appliquées à la construction de bases de connaissances - Doctorat d'Etat.* Thèse d'Etat, Université de Paris XI - Orsay, France, 1987.
10. J.H. Gennari, P. Langley, and D. Fisher. Models of incremental concept formation. In *Artificial Intelligence 40 - 1(3)*, pages 11–61. 1989.
11. R. Godin, G. Mineau, R. Missaoui, and H. Mili. Méthodes de classification conceptuelle basées sur les treillis de galois et applications. In *Revue d'Intelligence Artificielle, 9(2)*, pages 105–137. 1995.
12. O. Guinaldo. Techniques d'indexation pour aider à la classification dans le modèle des graphes conceptuels. In *Actes de la conférence Langages et Modèles á Objets, LMO'95*, Nancy, France, 1995.
13. D. Haussler. Learning conjunctive concepts in structural domains. In *Machine Learning 4*, pages 7–40. 1989.

14. J. U. Kietz and S. Wrobel. Controlling the complexity of learning in logic through syntactic and task-oriented models. In S. Muggleton, editor, *Inductive Logic Programming*, pages 335–359. Academic Press, 1992.

15. Y. Kodratoff and J. G. Ganascia. Improving the generalization step in learning. In J.G. Carbonell and T.M. Mitchell, editors, *Machine Learning: an Artificial Intelligence Approach*, pages 215–244. San Mateo, CA: Morgan Kaufmann, 1986.

16. R. Levinson. Uds : A universal data structure. In *Lectures Notes in AI n-835, Proc. Second International Conference on Conceptual Structures, ICCS'94*, pages 230–250. 1994.

17. R.S. Michalski. A theory and methodology of inductive learning. In *Machine Learning, An Artificial Intelligence Approach, Volume I*, pages 83–129. Morgan Kaufmann, 1983.

18. R.S. Michalski and R.E. Stepp. An application of AI techniques to structuring objects into an optimal conceptual hierarchy. In *Proceedings of the Seventh International Joint Conference on Artificial Intelligence*, pages 460–465, 1981.

19. G. Mineau, J. Gecsei, and R. Godin. Structuring knowledge bases using automatic learning. In *Proc. Sixth International Conference on Data Engineering*, pages 274–280, Los Angeles, USA, 1990.

20. G. Mineau, R. Godin, and J. Gecsei. La classification symbolique: une approche non subjective. In *Actes des Journées Françaises d'Apprentissage*, Lannion, France, Avril 1990.

21. S. Muggleton and C. Feng. Efficient induction of logic programs. In S. Muggleton, editor, *Inductive Logic Programming*, pages 281–298. Academic Press, 1992.

22. M. Pazzani and D. Kibler. The utility of knowledge in inductive logic. In *Machine Learning (9)*, pages 57–94. 1992.

23. R. Quinlan. Learning logical definitions from relations. In *Machine Learning (5)*, pages 239–266. 1990.

24. J. D. Roberts. A new parallelization of subgraph isomorphism refinement for classification and retrieval of conceptual structures. In *Lectures Notes in AI n-954. Proc. of the third International Conference on Conceptual Structures*, pages 202–216. Springer-Verlag, 1995.

25. B. Tausend. Representing biases for inductive logic programming. In Lecture Notes in Artificial Intelligence 784 F. Bergadano and L. De Raedt, editors, *Proc. of the European Conference on Machine Learning, ECML-94*, pages 427–430, 1994.

26. K. Thompson and P. Langley. Concept formation in structured domains. In *Concept Formation: Knowledge and Experience in Unsupervised Learning*. Morgan Kaufmann, San-Mateo, California, 1991.

27. J.D. Zucker and J.G. Ganascia. Changes of representation for efficient learning in structural domains. In *Proc. International Conference on Machine Learning*, Bari, Italy, 1996.

Rational and Affective Linking Across Conceptual Cases - without Rules

Graham A. Mann

Artificial Intelligence Laboratory
School of Computer Science & Engineering
University of New South Wales
Sydney, NSW 2052, Australia.
mann@cse.unsw.edu.au

Abstract. Human reasoning across experiential cases in episodic memory seems quite different from conventional artificial reasoning with conceptual representations by systematically manipulating them according to logical rules. One difference is that in humans linkages between particular experiences can apparently be made in a number of qualitatively different ways, forming recollective chains along different dimensions. For example, watching one movie may recall another which had a similar ending, cinematography, or common actors. It may also recall an otherwise unrelated movie which produced the same emotional impact. These linkages do not appear to be economically or simply described by rules. Yet case-based reasoning systems could benefit from sequential indexing of this kind. A conceptual-graph-based FGP (Fetch, Generalise, Project) machine using a small database of intellectual property law cases could enable such "memory-walks" to be computed without rules.

This research was supported by Australian Research Council grant A49600961 "Providing Sophisticated Access to Legal Information" (Gedeon, Greenleaf & Mowbray, 1996-1998).

1 Introduction

Inference engines that manipulate expressions by means of deductive logic can accomplish useful work inside expert systems and knowledge based systems, but operations of this sort seem qualitatively different from a certain kind thinking which seem valuable to humans. A good part of our problem-solving evidently draws on our store of experiences in episodic memory, through which our attention makes excursions that are sometimes called "memory-walks". Whether in well-organised, goal-oriented, rational trains of thought or in free associations with one idea recalling another apparently at random, these memory-walks are a familiar part of mental life.

The mechanisms by which one experiential pattern may summon another, somehow similar pattern has also been of interest since the origins of case-based reasoning (CBR) in the late 70s. "Reminding" has been a recurring topic in memory-oriented views of conceptual processing [10]. More recently, Gelernter has articulated a theory of "musing" to account for the memory walk phenomenon [9]. The theory conjectures that human minds link ideas together in different ways according to a continuum of mental focus, ranging from strictly constrained and quite logical operations on specifically selected ideas at the high end, through looser, more idiosyncratic and general connections between groups of related ideas in the middle range, to affect-linked or random "daydreaming" across the entire episodic memory at the low end.

To ratify the theory, a data-driven algorithm, called the FGP (Fetch, Generalise, Project) machine, was written. It can find high-end relationships between cases [7]. In one experiment, the machine is tested on a series of room descriptions encoded as simple attribute pairs (e.g. ((oven yes) (computer no) (coffee-machine yes) (sink yes) (sofa no)). The program is to describe the kind of room which has a particular attribute or attributes, and is given an appropriate probe vector (e.g. (oven yes)). The FGP machine first fetches all the cases which closely match the probe attributes. It places these into a "memory sandwich" and examines this longitudinally for common attributes. If all cases in the sandwich have a common attribute, the program generalises, guessing that all cases with the probe attribute will also have the common attribute. If many of the cases have some attribute, the program "projects", or speculates that this attribute could be characteristic, and recursively uses this attribute as a probe. If the memory sandwich returned by this recursive probe is a good match to the characteristic pattern of attributes built up so far, the program accepts the putative attribute; if not, the attribute is discarded and the next attribute considered. The process continues this *fetch-generalise-project* cycle until all attributes have been accounted for, returning a composite room description.

An FGP machine may be described formally as follows:

Let T be a feature tuple, M be an unordered collection of feature tuples and L be a list of Ts, ordered by a suitably defined proximity metric to some arbitrary T. Then let the following functions be defined:

- *fetch* $(T, M) \longrightarrow L$ Given a single pattern T, returns an ordered list L of patterns from M which are closer than some threshold to T in the problem space.

- *generalise* $(L) \longrightarrow T$ Given a list L of patterns, generate a new pattern T, which captures general features of all patterns in L. The contribution of each element of L to the new generalised pattern depends on its ordinal position in L and on the element's status as either a prototype or an ordinary case.

- *project* $(T) \longrightarrow T'$ Given a single pattern T, returns a new pattern T', which contains a subset of the most "evocative" features of T, and which thereby shifts subsequent processing into new regions of the problem place.

The aim is to answer a query input by the user. Queries can be either a pair $(T_0, a?)$ consisting of a test feature tuple and a single attribute, or else a single tuple T_0. An answer to the first query will be value of a for T_0, while the second query examines the cases for a prototypical redescription of T_0. In effect, it summons the memory walk process, asking "what does T_0 bring to mind?"

From the initial T_0 the following two-step executive cycle calls the functions:

repeat $(T_i, \mathcal{M})$
     1) Extend: *generalise (fetch $(T_i, \mathcal{M})$)* $\longrightarrow T$
     2) Refocus: *project (T)* $\longrightarrow T'$
         for each feature of T'
               generalise (fetch $(T', \mathcal{M})$) $\longrightarrow T_{i+1}$

The cycle iterates until either a) a value for a is discovered or b) an iteration produces no new conclusions. Various experimental variations on the above basic principle are possible. For example, the global variable which represents Gelernter's degree of focus can be made to influence the behaviour of the functions in different ways. Tests of the FGP machine on two databases [7] show a predictive or diagnostic performance level close to or better than human domain experts in the same task (65% vs. a human expert's 53% correct predictions of the nationality of folkdances from a database of descriptive features, and 70% vs a medical specialist's 60% correct differential diagnoses from descriptions of mammograms). Qualitatively, the behaviour of the program is interesting, as it interacts with the user, reporting its progress through the cycle of reminding, generalisation and speculation in a very lifelike fashion.

The ability to find logical relationships in data without any explicit rules or principles is valuable enough, but Gelernter wants to improve the FGP machine in two ways. First, he argues that more and better relationships could be found if the cases were better representations of real situations i.e. at a higher resolution of detail and with more structure. This might also allow interactions with the machine to be carried out in natural language, making it easier to use. Second, he requires a method by which affective states - emotions - can be represented in cases. These could be used to perform a different kind of low-focus linking. In affect, the machine could speculate about situations which had the same emotional connotations, finding commonalities across its entire experience with happy, sad or funny situations. In Gelernter's hypothetical dialogues about such linking [9, p.145-146], a low-focussed FGP machine recognises the displeased state of a user, is reminded of a literary character by the plaintiff of a legal case study, and refuses to obey a request for financial help because the subject matter is boring! The point of this is to imagine how affective associations might be used to make computer reasoning more lifelike, more creative, and better grounded in emotions.

Could an improved representation language, such as conceptual graphs, overcome the first problem? How can the design of the FGP machine be updated so that it functions with structured propositional forms instead of unstructured collections of features? And could the improved representations also incorporate emotional patterns, thus enabling affect-linking? Is there a way to for the device itself to generate affective patterns without copying them from a human emoter? What other improvements could a conceptual-graph-based FTG machine (hereafter referred to as a CG-FGP machine) enjoy? In this paper, I answer these questions in the course of modifying the design of the FGP machine.

2   Encoding of Legal Cases

A CG-FGP machine requires a rich database of conceptual graphs describing experiences in which regularities may be found. An example would be a series of cases presented to an Appeals Court, within a subdivision of the law such as intellectual property. Such cases are often summarised into brief paragraphs for rapid assimilation into briefs or student study e.g. [11]. Legal information is notoriously difficult to represent, but fortunately the exact semantics used to encode the cases is not crucial to the operation of the CG-FGP machine. Good efforts to represent complex legal information using conceptual graphs have been described by Dick [4,5]. Dick adapted Toulmin's [16] six-aspect model of argument representation, composing a schema from a set of distinct contexts for claim, grounds, reasons, backing, modalities and rebuttal. Within each context was nested a set of conceptual graphs or subcontexts, encoding the facts, events, or acts pertinent to this context, expressed in a rich semantics combing Somer's case grid and Sowa's methods for situations [14]. This method covers the dispute which motivates the legal action, but, from a case-based reasoning point of view, needs to be extended to cover the resolution of the dispute

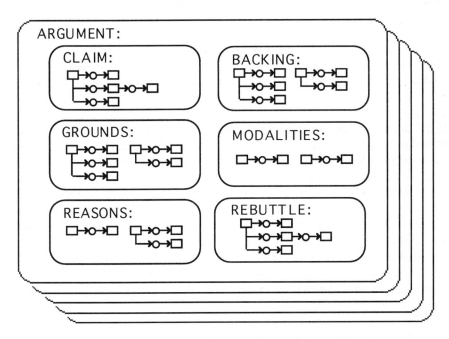

Fig. 1. Database of case arguments encoded according to Dick's scheme. Each case is an argument including a maximum of six contexts. CLAIM is the goal or conclusion sought. The GROUNDS are factual assertions used to support the CLAIM, while REASONS are logical connections relating the GROUNDS to the CLAIM. The BACKING is the law authorising the REASONS and its validity, subject to scoping limits specified in MODALITIES, and to possible failures of applicability due to assertions in REBUTTLE.

(findings, judgements, dismissal) and its final outcome (penalties, appeals, settlements). It should not be difficult to provide two more contexts into which this information could be placed.

Other difficulties are encountered in the construction of such a database, centred around the cost and difficulty of encoding case briefs into conceptual graphs that properly capture the essences of each aspect of the model. The chief problem appears not to be that of providing a useful ontology - the legal profession already has established precoded indices of specialised terms from which conceptual hierarchies may be derived [1]. Rather, it is the complexity and open-textual nature of the content of the briefs. These often presuppose a great deal of commonsense knowledge as well as a specialised knowledge of the law. Dick experienced particular difficulty dealing with multiple views or perspectives on the same situation and with expressing continuity and other temporal aspects of cognition in a natural way.

Suppose that, for any of these reasons, it proved difficult to write graphs to properly represent a state of affairs for one of the model's contexts. Perhaps suitable specific concepts or relations did not exist in the catalogues, or it was unclear which of a number of alternatives applied. There might be no alternative but to choose overly general concepts or relations for some parts of the graph. As we shall see in the next section, there is reason to believe that such underspecificity would weaken, but not fully invalidate, the rational linkages which the CG-FGP machine must make in high-focus mode.

Perhaps, too, the correct structure for the graphs would prove elusive. If the pattern of relationships was especially troublesome, a possible non-optimal solution would be to identify each graph's principal, or head, concept and attach the remaining concepts to it directly by links of the highest generality (e.g. an "(ASOC)" relation or even the non-committal (LINK) relation), forming a loose cluster which is nevertheless a syntactically well-formed graph. In the extreme case, one could be left with a structureless "degenerate graph" consisting of an unordered set of concepts and relations. Yet even this worst case need not render the database useless for the CG-FGP machine. It is possible that the three functions could be modified so that they use structure as a constraint if available, yet tolerate poor or missing structure by treating the concepts and relations as an unordered collection of features, with which the simpler FGP machine was designed to deal.

Regardless of these problems and details of encoding, for the purpose at hand let us assume the existence of a database $\mathcal{M}$ encoded as shown in Figure 1.

3 Rational Linking

What modifications need to be made to *fetch*, *generalise* and *project* to make them work in the CG-FGP machine? To begin with, they should be renamed *CG-fetch*, *CG-generalise* and *CG-project*, to distinguish them from the existing functions. We must deal with each function in turn to make it perform the equivalent function over the legal database, taking into account both the effect of the focus setting and the possibility of poor structure as raised in the last section. Before that, however, it is necessary to discuss some general obstacles to this development.

One obstacle is that the original T feature tuples contained numerical weightings indicating their frequency of occurrence in generalisations, or their importance in prototypes, which information was needed by the *generalise* function. For example, if in a barrel of 100 apples half were red and the other half green, this would be represented as

((name apple 100) (type fruit 100) (colour (red 50) (green 50))).

How may this be done in a conceptual graph i.e. what is the equivalent of an attribute-value pair, and where should the weight be stored? The position advocated here is that since the numbers could mean relative importance as well as frequency counts, it is better not to try to explicitly represent them in the graph. Instead, the weights will be invisibly annotated to the concepts, as in the following graph, thus avoiding explicit commitment about frequency or importance.

[APPLE: {*}] -
> (CHRC) -> [COLOUR:Red]
> (CHRC) -> [COLOUR:Green]

With this kind of representation (related to philosophical Realism), the "scope" of the weights can be restricted to the concepts themselves, which seem to encompass both attribute and value. With some representations, relation-concept pairs are attached to a main stem as cases. To correspond to attribute-value weighting, a weight would have to apply to the linkage of relation and concept only. But the picture is complicated by the realisation that the colours of apples should not necessarily be confused with the colours of other objects, though their relation-concept pairs appear the same. So perhaps the weights should cover triples. Yet this argument applies recursively, threatening to demand that weights must only apply to entire graphs, which is clearly going too far. Moderation here seems appropriate: triples or pairs may have enough structure to avoid great semantic confusion during weight summation, yet not so much as to blur necessary distinctions.

The richer, more complex conceptual structures pose other problems for finding commonalities across cases, compared to the simple feature tuples in the original representations. For one, the legal cases of Figure 1 are partitioned into separate contexts. Should longitudinal comparisons across cases be made only through corresponding contexts, or should relationships to different contexts be sought? It could be argued that the value of reminding lies in its power to discover unexpected relationships across cases, so that restricting the possibilities is defeating the purpose of the exercise. On the other hand, the point of contexts is to establish boundaries of relevance, and restricting processing to only corresponding contexts reduces computation by at least a factor of five. So if only for feasibility's sake, the corresponding-context-only policy may be forced.

Furthermore, each context may contain one or more graphs. How can corresponding contexts be directly compared if they have different numbers of graphs? A strict policy would forbid any such matching as potentially invalid. A more liberal approach seems preferable here. If total structure is not regarded as essential for a match, pairs or triples from any graph within a context should still probably be permitted to match pairs or triples from any graph in the corresponding context of another case. The more structure there is in a match probe, the less likely this is to lead to an invalid result, and not to permit within-context graph confusion would be to risk missing important relationships.

Better opportunities to find commonalities across cases could be had if their graphs were entered into the database in standard forms. This would, however, create an additional expressive constraint, making the already difficult acquisition task harder. Mineau [13] has suggested that excess variability in graphs - a range of ontological choices and structural compositions beyond that which is strictly necessary to encode different meanings - might be reduced by automatically *normalising* the graphs. He outlines eight basic methods for normalisation, including checking against a graph grammar, restriction of choice to "privileged" relations and the use of rewrite rules. If methods like these were applied to a database immediately after input, runtime performance would not be compromised.

Consider now the modifications required for the three basic functions.

• *CG-fetch* must now take a conceptual graph T and return a short list of graphs L from M which are semantically close to T and to each other. The semantic distance measure used must take account of similarity in, but not depend on, graphical structure, and must take into account the "evocativeness", or ability to add informative value, of the features within the graphs with respect to the query. The query, incidentally, would be a request to instantiate T with one or more concepts.

Since this mapping is expected to be far more computationally expensive than that of *fetch*, an effort must be made to define an efficiently computable single measure. This should begin with the work of others to refine the orthodox index of semantic distance between conceptual graphs - the sum over corresponding concepts in the two graphs of the distances on the type hierarchy between each concept's type and the most specific type which subsumes them both (minimal common generalisation) [8] - into a measure which exhibits more of the characteristics of human similarity judgements, such as salience, context and perceiver effects [3]. Wuwongse and Niyomthai [17] have suggested that as concepts are not regarded as equally important in human intuitions of similarity, so the existing semantic distance (SD) sum should be influenced by salience values associated with the concepts. In their CBR matching score

$$\sum_{i=1}^{n} (IV_i - SD_i)$$

the summed importance values (IV) of matching concepts are discounted by the corresponding semantic distances. One question that may be raised about this formula is the validity of subtracting terms with what may be dissimilar units (importance and relatedness). Despite their name, the IVs are really only arbitrary numbers that have no natural units, so it would seem they may be compared with distance measures in any convenient but systematic way. The arrangement was apparently satisfactory for case matching in [17].

The main difficulty with the IV term is that of supplying a number for each concept in the system. Here, we have the opportunity to experiment with the use of the weights discussed above. It will be interesting to see whether or not, in this domain, the importance/frequency weights are context-dependent enough to need to be somehow corrected according to which of the eight contexts encloses the concepts they appraise. One might also ask whether the frequency interpretation of this number adversely affects the behaviour of the measure. The answers to these questions are probably best determined empirically. Notice that, although strictly numbered with concepts, the use of weights that apply to relation-concept pairs (or triples) brings some structure into the measure.

fetch originally also took into account the evocativeness of $\mathcal{T}$ with respect to a query. This means how strongly and clearly $\mathcal{T}$ brings to mind a value for the query. Fertig and Gelernter [7] use an information-theoretic formula in which the information gain (negative entropy) on the query from $\mathcal{T}$ is defined as:

$$ S = \frac{1}{\ln N} \left(- \frac{\Sigma_i \, n_i \, \ln n_i}{T} + \ln T \right) $$

where

N is the total number of possible values for the goal in $\mathcal{M}$
n_i is the number of times goal value i appears in the top-cluster
T is the total number of goal values found in the top-cluster

Roughly speaking, S ranges from a 0 (maximum evocativeness) to 1 (minimum evocativeness). The equivalent computation must be made over conceptual graphs to find the correct concept for a relation slot, and possibly to find the correct referent for a concept. Evocativeness can then be included as a factor in our graph-matching measure:

$$ \sum_{i=1}^{n} (IV_i - SD_i) \, (1 - S) $$

It might be necessary to bias the evocativeness factor to avoid zero values. We might also wish for some way of ensuring that the graphs in $\mathcal{L}$ form a tight cluster. However, the computation is already intensive enough, so this will be neglected. It is hoped that the conceptual structures used to encode cases will, by virtue of their composition from a common pool of concepts and relations under selectional constraints, naturally group themselves into semantic clusters in the problem space.

• *CG-generalise* The requirement to collapse the ordered list L of Ts into one which captures essential elements of each presents a real problem if multiple graphs are allowed in case contexts. While it is possible to imagine techniques which find the correct correspondences for longitudinal coalescence across multi-graph contexts, the problem is formidable, and when taken together with the other problems this causes, a good case emerges for requiring contexts to be standardised (or normalised) single graphs, even if that reduces expressivity. (Dick's example cases in [4] are written this way). With that constraint, the problem simplifies to the extraction of high-frequency or weighty features and the elimination of contradictory features. These features would then be instantiated into a template for this context's standard graph form. Since Ts which are ranked higher on L should exert a greater influence than later Ts, this had best be organised as an iterative process.

1) Begin T as a new copy of the template for the current context
2) Find all features with a weight $> \varepsilon$ in T_1, and instantiate into T
3) For i=2 to #(L):

   For each feature f with a weight $> \varepsilon+i$ in T_i

   If f agrees with a corresponding feature in T

   Then increment weight of corresponding feature in T

   Else decrement weight of corresponding feature in T

   If f has no corresponding feature in T instantiate f into T

4) Eliminate all features of T with a weight $<\theta$

This adds new features to the evolving generalisation, but records agreements and disagreements in later graphs, allowing features with contradictory values to eliminated at step 4. ε and θ are arbitrary thresholds which are set to correctly detect commonality or contradiction, respectively, across small lengths of L.

• *CG-project* This function must take a single graph and return a subgraph containing its most evocative and distinctive features. Evocation has been discussed already; the feature's attribute must be evocative with respect to the query. To be distinctive, a feature's possible range of values over M must be "small". For a given relation-concept pair, the possible different concept attachments could be simply counted as D.

A feature may thus be excluded from T if both

$$(1 - S) < \alpha \quad \text{and} \quad D < \beta$$

where

α and β are arbitrary thresholds set for a given database.

The structure of the returned subgraph should probably resemble something of the original. With stem-and-case structures, this is simply a matter of preserving the stem, and pruning some of the relation-concept cases away. Theoretically, any substructure may contain regularities for future processing, so it is not essential; but it would help with higher level processing, such as natural language description.

For this kind of high-focus linking, the two-step cycle remains the same as for the FGP machine.

4 Affective Linking

Another aspect of human cognition which might be brought into the picture is emotional reaction, which humans commonly experience in response to a situation like a legal case. Emotions are hypothesised to play a number of important roles in human cognition, including evaluation of otherwise unclassifiable objects, events and situations, distinctive signalling of important bodily and mental states for behaviour control, and positive and negative reinforcement for and learning. Here, we focus on the Gelernter's hypothesised affective linking. The basic idea is that at low focus, the connections between one episode and the next could be chiefly that they made the experiencer "feel the same".

Emotional states could be simulated in our CG-FGP machine in two basic ways. Given a model of what emotional states exist and how they relate to each other, a human modeller could read each legal case in the database, and try to model his or her own reactions, creating a new conceptual graph and attaching it to the situation as the machine's reaction. Given a good model, this has the advantage of simplicity, and the machine's reactions should be at least recognisable to another human. But it requires extra human effort at the point of input of all new cases, and is essentially derivative. Alternatively, one could wish for a method of having the machine generate its own emotional graphs in response to cases. To do this, the model would need a set of attitudes, which map objects, events and situations onto a set of affective states. The mapping could come from arbitrary policy or systematic design based on the "survival-value" or other machine-oriented teleological principle. This could be difficult to create, and the resultant reactions are not guaranteed to be recognisable to a person, but it requires no extra effort at input. Tepfenhart has pointed to the advantages of explicitly represented attitudes for problem solving, and has suggested practical methods for representing and computing with them [15].

The reader might wonder what justifies any effort at including affective states at all. Are not these affective states simply more features of input for the CG-FGP machine to link on? If the affective states are simply input by a human emotive trainer, then in a way, that is all they are. But affective patterns do not originate in the external state of affairs represented in the cases. They arise in a classificatory mechanism, whether in the human emoter or inherent in the system itself. They add important information about the relationship between the cognitive agent and a particular aspect of the world.

If they are systematically assigned, affective commonalities might be discoverable between objects, events and situations which were otherwise unrelated.

For example, suppose an attitudinal set could detect situations in the outcome context where the severity of a penalty was disproportionate to the seriousness of the offence (according to, say, a "natural justice" table listing classes of misdemeanour and their corresponding fair types of punishment). If the system could add a conceptual graph representation of "unfairness" into the reaction context, one unfair outcome on a case could recall others. This kind of linkage could prove extremely valuable in legal reasoning.

Imagine a very simple affect model, in which situations are experienced only as either pleasant or unpleasant. The affective state is characterised by three features: an arousal level, indicating the strength of reaction; a valency, representing its direction, and an object, in this case, the situation:

[AFFECTIVE-STATE] -

$$(ALEV) \rightarrow [CARDINAL:*a] \qquad 0 \leq a \leq 10$$
$$(VALC) \rightarrow [INTEGER:*v] \qquad -10 \leq v \leq +10$$
$$(OBJ) \rightarrow [SITUATION:]$$

The arousal concept is simply a non-negative integer indicating excitement or interest, while the valency is a signed integer that allows positive and negative aspects to be summed. Only the sign of the final computed valency is interpreted. To evaluate a situation, its representation is examined with respect to the table in Figure 2.

Pattern Detected in Input	ΔArousal	ΔValence	Object
Violent Act (physical)	+9	-8	situation
Violent Act (verbal)	+4	-3	situation
Theft of Property	+5	-7	situation
Loss of Amenity	-1	-6	situation
Disproportionate punishment	+4	-8	situation

Fig. 2. Attitudinal policy table indicates changes to arousal and valency for specified features of a legal situation. Each pattern would be implemented as a conceptual graph, which would be projected onto the situation. If a match was found, the specified changes are made to the evolving affect graph. Realistic reactions should also allow separate reactions to different objects within the overall situation as extra entries in the table, e.g. feelings about an attacker and a victim could differ.

The situation is tested against each entry in the left hand column of the table. If the specified pattern is detected, then the values of variables *a and *v are altered accordingly. When every test has been performed, the summed values represent the sign and magnitude of the overall reaction to the situation. These could be attached directly to the context, but for modularity, it is probably better to partition all reactions into one separate affective context and tie them to their situations by coreferential links. Far more sophisticated emotional models have been advanced e.g.[6], but this is sufficient to explore the principle.

Now let a new function be defined:

- $CG\text{-}affect$ (T) $\longrightarrow$ A Given a single pattern T, returns a new context A, representing the output of the machine's attitudinal policy.

To make the CG-FGP machine register its own reactions, the existing two-phase executive cycle requires an extra step:

repeat (T_i, M)
      1) React: $CG\text{-}affect$ (T_i) $\longrightarrow$ A
      2) Extend: $CG\text{-}generalise$ $(CG\text{-}fetch$ $(T_i, M))$ $\longrightarrow$ T
      3) Refocus: $CG\text{-}project$ (T) $\longrightarrow$ T'
              for each feature of T'
              $CG\text{-}generalise$ $(CG\text{-}fetch$ $(T', M))$ $\longrightarrow$ T_{i+1}

Between steps 1 and 2, A is incorporated into T_i , thus becoming available for linking using the existing mechanism.

5   Implementation

Currently, only part of the CG-FGP machine is implemented. A few test cases have been encoded into the database, and the basic Lisp executive and primary functions are being modified as described in Section 3. The first step is to demonstrate basic high-focus relational linking. At present the function *affect* has not been perfected, so that kind of linking will have to wait. Testing of the CG-FGP machine will additionally require an expanded library of legal cases, and writing these, and extending the conceptual and relational catalogues is expected to be difficult and time consuming.

The final stage of this work should be to bring the machine's interaction with the user alive by adding a natural language, thus substantiating Gelernter's dream. A faster version of the conceptual parser used by BEELINE [12], could be used to accept case inputs in natural language, while a generator such as [2] could describe cases and keep the user informed about the state of the machine, its simulated speculation and any conclusions reached. Both components depend on considerable amounts of hand-encoded knowledge - enough to force the project toward the creation of automatic or semi-automatic knowledge tools. A relatively simple dialogue manager would also be required, which dealt with the limited pragmatics of the CG-FGP's operation.

References

1. D.R. Corkery: *Hallsbury's Laws of Australia*. Vol. 29, Index. Canberra: Butterworth, 1993.
2. S. Dogru & J.R. Slagle. A System that Translates Conceptual Structures into English. In H.D. Pfeiffer & T.E. Nagle (Eds.), *Conceptual Structures: Theory and Implementation*, Lecture Notes in AI 754, Springer-Verlag, Berlin,1992, 283-292.
3. H.S. Delugach: An Exploration into Semantic Distance. In H.D. Pfeiffer & T.E. Nagle (Eds.), *Conceptual Structures: Conceptual. Structures: Theory and Implementation*, Lecture Notes in AI 754, Springer-Verlag, Berlin,1992, 119-124.
4. J. Dick: A Conceptual, Case-Relation Representation of Text for Intelligent Retrieval.*Technical Report CSRI-265*, Computer Systems Research Institute, University of Toronto, 1992.
5. J. Dick: Using Contexts to Represent Text. In W. M. Tepfenhart, J. P. Dick & J. F. Sowa, (Eds.), *Conceptual Structures: Current Practices*, Lecture Notes in AI 835, Springer-Verlag, Berlin, 1994, 196-213.
6. M.G. Dyer: Emotions and Their Computations: Three Computer Models. *Cognition and Emotion*, 1978,1,1, 223-234.
7. S. Fertig & D.H. Gelernter: The Design, Implementation and Performance of a Database-driven Expert System. *Technical Report #851*, Department of Computer Science, Yale University, 1991.
8. N. Foo, B.J. Garner, A. Rao, & E. Tsui: Semantic Distance in Conceptual Graphs. In T.E. Nagle et. al. (Eds.), *Conceptual Structures: Current Research and Practice*. Chichester: Ellis Horwood, 1992, 150-154.
9. D. Gelernter: *The Muse in the Machine*. London: Fourth Estate, 1994.
10. J. Kolodner: *Case-Based Reasoning*. San Mateo, CA: Morgan Kaufmann Publishers, 1993. Chapter 4.
11. J.P. McKeough: *Intellectual Property*. Canberra: Butterworth's Student Companions, 2nd Edition, 1994.
12. G.A. Mann: BEELINE - A Situated, Bounded Conceptual Knowledge System. *International Journal of Systems Research & Information Science*, 1995, 7, 37-53.
13. G.W. Mineau: Normalizing Conceptual Graphs. In T.E. Nagle et. al. (Eds.), *Conceptual Structures: Current Research and Practice*. Chichester: Ellis Horwood, 1992, 339-348.
14. J.F. Sowa: *Conceptual structures*. Menlo Park, California: Addison-Wesley Publishing Company, 1984.
15. Tepfenhart, W.M. Attitudes: Keys to Problem Identification. In W. M. Tepfenhart, J. P. Dick & J. F. Sowa, (Eds.), *Conceptual Structures: Current Practices*, Lecture Notes in AI 835, Springer-Verlag, Berlin, 1994, 127-143.

16. Toulmin, S.E. *The Uses of Argument*, Cambridge: Cambridge University Press, 1958.
17. V. Wuwongse, S. Niyomthai: Conceptual Graphs as a Framework for Case-based Reasoning. *Proceedings of the 6th Annual Workshop on Conceptual Graphs*, 1990, 119-133.

Conceptual Graphs for
Corporate Knowledge Repositories

Olivier Gerbé

DMR Consulting Group Inc.
1200 McGill College, Montréal, Québec, Canada H3B 4G7
e-mail: Olivier.Gerbe@dmr.ca

Abstract. The challenge companies will have to meet when making the leap from the industrial era to the knowledge era is the memorization of corporate knowledge and its dissemination to employees throughout the organization. Developing a corporate memory is the means chosen by DMR Consulting Group to capitalize on and manage its expertise in information technology. This paper presents a study conducted to choose a formalism to represent the know-how and methodologies – processes, techniques and learning materials – in corporate memory. It compares modeling formalisms against specific requirements and demonstrates that conceptual graphs are well suited to implement corporate memories. More specifically, we show that conceptual graphs support: (i) classification and partial knowledge, (ii) category or instance in relationship and (iii) category or instance in metamodel.

1 Introduction

Nowadays there is consensus on the value of corporate knowledge. The knowledge is at the center; an employee who must know a work process; a manager who must anticipate market trends; a researcher who must know about the state of the art. Corporate knowledge is made up of strategies, visions, rules, procedures, policies, traditions and people. The knowledge assets and the learning capacity of an organization are seen as the main source of a competitive advantage [3], and the challenge the management of this corporate knowledge [12].

The challenge companies will have to meet is the memorization of knowledge, its storage and, its dissemination to employees throughout the organization. Knowledge may be capitalized on and managed in corporate memories in order to ensure standardization, consistency and coherence. Knowledge management requires the acquisition, storage, evolution and dissemination of knowledge acquired by the organization [15] and computer systems are certainly the only way to realize corporate memories [16] which meet these objectives.

DMR Consulting Group is one of the largest service providers in the information technology (IT) in the world. Mastering the evolution and management of IT is a challenge that requires methods, processes, software tools and training programs, as well as a systematic and consistent approach to implement them.

Several products describing methods and processes, such as guides, tools, lecture materials, self-learning courses, reference texts, templates and videos, have been developed [11]. These products are stored in a corporate memory whose data structure and functionality allow for easy consultation, adaptation to the particular needs of an organization, and evolution at acceptable levels of cost [8]. This corporate memory, called the Method Repository, plays a fundamental role. It captures, stores [9], retrieves and disseminates [10]throughout the organization all the consulting and software engineering processes and the corresponding knowledge produced by the experts in the IT domain. During the early stage of the development, the choice of a knowledge representation formalism was identified as a key issue of the development of the Method Repository. That leaded us to define specific requirements for corporate memories, to identify suitable knowledge representation formalisms and to compare them in order to choose the most appropriate formalism.

This paper presents the study we conducted to choose a formalism to represent the know-how and methodologies – processes, techniques and learning materials – for the Method Repository. It compares five modeling formalisms, extended entity-relationship, object-oriented (UML), relational model, Classic – a KL-One-like formalism and conceptual graphs, against our specific requirements. This study demonstrates that conceptual graphs are particularly well suited to implement corporate memories since they support: (i) classification and partial knowledge, (ii) category or instance in relationship and (iii) category or instance in metamodel.

The paper is organized as follows. Section 2 defines specific requirements for corporate memories. Section 3 compares both traditional formalisms used in data modeling and formalisms used in knowledge representation and demonstrates that conceptual graphs support our predefined requirements. Finally, Section 4 concludes and provides some technical information about the Method Repository we have developed.

2 Requirements

Modeling techniques aim at defining simplified, computerized models of real or hypothetical worlds, in order to gather and store information about them. Defining model starts with the identification and definition of categories of things and of relationships between these things. They help describe the application domain. Things are interrelated and organized into categories according to established similarity criteria. Models must reflect the structure and present these categories and their interrelationships.

In order to compare modeling formalisms, we looked at the three specific concerns, encountered when we began to study how to represent methods, procedures and techniques, and which were not obvious.

- Before gathering and storing information about things, do we need to define all the possible categories of things that exist in the application domain?

- Is it possible to represent associations between categories and things?
- If we define categories of categories, is it possible to integrate this higher-order information in the same knowledge base?

As we shall see later in this paper above questions may be translated into three requirements:

- Classification and partial knowledge (2.2);
- Category and/or instance in relationship (2.3); and
- Category or instance in metamodel (2.4).

This section is organized to first introduce briefly the basic notions and terminology used in modeling techniques, and then to detail each of our three requirements.

2.1 Basic Notions

The main elements used by modeling techniques are categories or types, instances and relationships. A *category* represents a set of things that share the same properties: attributes, relationships and behavior. An *instance* of a category is a thing that conforms to the definition of the category. A *relationship* is an association between things or categories.

Let Employee be a category that defines what an employee is. An employee may have attributes like employee number, name, etc. An employee also has a relationship with a company he or she is working for. Figure 1 illustrates this example. EMPLOYEE and ORGANIZATION are categories, and instances of these types should be John, Paul, IEEE, UNU (United Nations University), etc. 'Works for' is a relationship between the two categories.

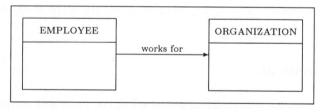

Fig. 1. Model containing two categories and one relationship.

Inheritance and classification are two other important notions used in modeling techniques. *Inheritance* is a kind-of relationship between categories. A category inherits the properties of a higher category. *Classification* is the hierarchy of categories built upon the inherited properties. For example, the mammal classification is a hierarchy that defines types of species.

2.2 Classification and Partial Knowledge

Categories are defined based on common properties of their instances. Therefore, defining categories corresponds to defining partitioning criteria that help to distinguish one instance from another. If the number of partitioning criteria increases, the number of categories may increase dramatically and rapidly become unmanageable.

In most modeling formalisms, a thing is an instance of one and only one category. This limitation forces the definition of all the possible or conceivable categories where a thing may be potentially classified. For example, a car dealer wants to classify cars to be sold. Considering the number of seats, three categories can be defined: coupe, sedan and van. Considering the number of wheels that provide propulsion, there are two categories: two-wheel drive (2WD) and four-wheel drive (4WD). Considering the origin of cars, three categories are possible: American, European and Asian. Figure 2 illustrates the different hierarchies resulting from these three criteria.

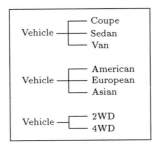

Fig. 2. Different vehicle hierarchies according to three criteria.

If the car dealer wants to classify all the cars according to these three criteria, all possible combinations must be considered. In this example there are 18 possibilities, as shown in Figure 3.

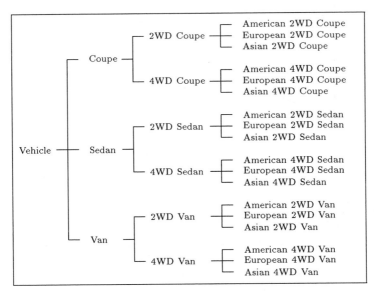

Fig. 3. A unified vehicle hierarchy comprising all three criteria.

Another aspect of classification but rarely connected with classification is partial knowledge. How can we classify a thing if all its properties are partially known? For instance, how is the person Brown classified if we have only two categories Male and Female and if Brown's sex is unknown? Let us assume that our car dealer receives two new cars. The first one is a 2WD Sedan. However the dealer does not know in which category to classify it because the car has been assembled in Europe from Asian parts. The car may be included in the 2WD Sedan category. But if the dealer can later describe the car's origin, what will happen? The second car he receives is an American one. In which category may it be included?

To fulfill this classification and partial knowledge requirement, the formalism should support multi-classification; i.e. an object may be an instance of more than one category, or the system should dynamically migrate instances from one category to another more specialized category.

2.3 Category and/or instance in relationship

In most modeling techniques, relationships are established between categories and are applicable to their instances; however, this is often not sufficient.

In the previous example concerning employees and organizations, we want to distinguish UNU employees that work for the United Nations University from other employees. Let UNU-EMPLOYEE be a category that specializes UNU employees. Since employees of UNU have the same properties, same attributes and same kinds of relationships as employees, UNU-EMPLOYEE is defined as a sub-category of the category EMPLOYEE (see Figure 4).

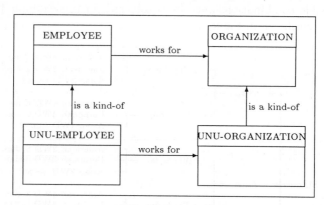

Fig. 4. Complete Model.

In most modeling techniques, it is very difficult to express that any UNU employee has a relationship with the UNU organization. Modeling techniques formulate knowledge at the category level. Relationships link categories and are applicable at the instance level. In our example, a category UNU-ORGANIZATION, that has only one instance (UNU), has to be defined to be able to link UNU-EMPLOYEE and UNU-ORGANIZATION as shown in Figure 4.

However, what we want to express is: "Each instance of UNU Employee has a relationship 'works for' with UNU, instance of the type Organization." This is stated by the following predicate:

$$\forall x, \text{UNU-EMPLOYEE}(x) \wedge \text{ORGANIZATION}(UNU) \wedge \text{works-for}(x, UNU)$$

In most modeling techniques, this is stated by the following two expressions where the first expression states that a UNU employee works for a UNU organization and the second expression states that any UNU organization is the UNU:

$$\forall x, \exists y, \text{UNU-EMPLOYEE}(x) \Rightarrow \text{UNU-ORGANIZATION}(y) \wedge \text{works-for}(x, y)$$

$$\forall z, UNU - ORGANIZATION(z) \Rightarrow z = UNU$$

These two formulations are equivalent. However, the first formulation is simpler and therefore preferable.

To fulfill this requirement, the formalism should support category and/or instance in relationship; i.e. a category may be linked to an instance, or relationships may be established at the instance level.

2.4 Category or instance in Metamodel

We seek to develop models that are formal descriptions of objects or notions in order to make sound and complete inferences from this model. This formal description uses different kinds of components and different kinds of relationships. A metamodel describes these components and their relationships. Metamodels deal with categories and categories of categories. In most cases, it is easy to make a distinction between categories and instances. Categories give information about instances, but categories may be seen as instances in a metamodel that gives information about the categories themselves.

In the previous example, let FEDERAL-ORGANIZATION be a kind-of ORGANIZATION. This means that FEDERAL-ORGANIZATION is a category that is a subcategory of ORGANIZATION. FEDERAL-ORGANIZATION is a category therefore FEDERAL-ORGANIZATION is an instance of a model component CATEGORY. Figure 5 illustrates this situation where FEDERAL-ORGANIZATION may be seen as a category that carries its associated semantics, or as an instance of the category CATEGORY, depending on the perspective.

From the category perspective, FEDERAL-ORGANIZATION is a subtype of the category ORGANIZATION and it inherits its attributes. The attributes of ORGANIZATION are: label that names the organization, owner and sub-organizations. The inherited attributes of FEDERAL-ORGANIZATION are: label, owner that specializes in government, and sub-organizations.

From the instance perspective, FEDERAL-ORGANIZATION is an instance of CATEGORY. The attributes of CATEGORY are: super category that establishes the position of the category in the classification, category label that names the category, and attributes that list the attributes of the category. These attributes have the following values for FEDERAL-ORGANIZATION:

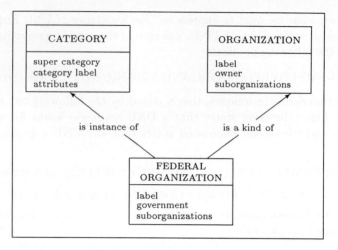

Fig. 5. Category or instance.

- *super category*: ORGANIZATION
- *category label*: FEDERAL-ORGANIZATION
- *attributes*: label, government, sub-organizations.

To fulfill this requirement, the formalism should support category or instance in metamodel; i.e. an element should allow to be viewed as a category or an instance.

3   Considered Formalisms

This section examines five formalisms we believe the most promising: three well-known formalisms frequently used in the domain of information technology, Extended Entity-Relationship [4], Object-Oriented [1], and Relational Model [5] formalisms; and two formalisms frequently used in the domain of knowledge representation, Classic: a KL-One-like language [2] and Conceptual Graphs [14]. For each of them, we introduce pertinent notions and discuss to which extent they fulfill our requirements.

3.1   Extended Entity-Relationship Formalism

The Extended Entity-Relationship formalism was originally developed by Peter Chen [4] in 1976 and was later extended [6, 7].

The Extended Entity-Relationship formalism supports categories through entities and relationships; there is no explicit component in the formalism to represent instances. All the knowledge is specified at the entity level. Entities are represented by boxes with the entity's name at the top, and relationships by ovals with two lines that link the entities concerned. The Extended E-R diagram corresponding to "employee works for organization" is presented in Figure 6

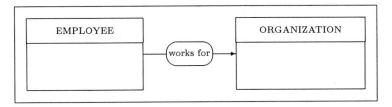

Fig. 6. Extended E-R: Employee works for organization.

Classification and Partial Knowledge. There is no means to represent knowledge about instances. In the Extended E-R formalism, an instance is implicitly an instance of one and only one entity and there is no support for migration of instances and partial knowledge.

Category and/or instance in relationship. As instances are not represented in the formalism, there is no possibility to specify constraints at the instance level. The commonly used solution in this case is to define an entity that has only one instance, to express the relationship at the entity level (Figure 7), and to add an explanatory note.

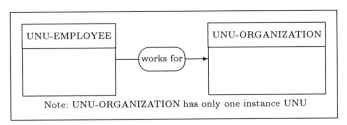

Fig. 7. Extended E-R: Category and/or instance in relationship.

Category or instance in Metamodel. The Extended E-R formalism is applicable at one and only one level. The only way is to define two diagrams, one for each level, and to link them by an explanatory note (Figure 8).

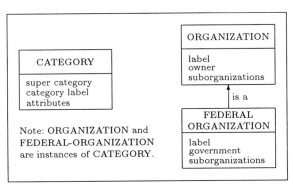

Fig. 8. Extended E-R: Category or instance in Metamodel.

3.2 Object Oriented Formalism

As representative of object oriented formalisms, we chose the Unified Modeling Language [1], a language for specifying, visualizing, and constructing the artifacts of software systems, as well as for business modeling, that was developed by Grady Booch, Jim Rumbaugh and Ivar Jacobson from the unification of the Booch, OMT and OOSE methods.

The Unified Modeling Language defines instances called objects and categories called classes, and supports classification. In Unified Modeling Language notation, classes are represented by rectangular boxes, objects are represented as classes with an underlined label and relationships are represented by lines. "Employees work for organization" is represented as in Figure 9.

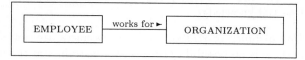

Fig. 9. UML: Employees works for Organization.

Classification and Partial Knowledge. The ordinary UML semantics assume multiple inheritance, no multiple classification, and no dynamic classification. But different semantics can be permitted by identifying semantic variation points that users and tools could understand.

Category and/or instance in relationship. Relationships are defined at the class level. Similar to the Extended E-R formalism, the solution is to define a class with only one instance as shown in Figure 10 to represent singleton classes.

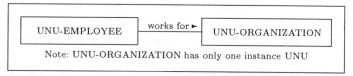

Fig. 10. UML: Category and/or instance in relationship.

Category or instance in Metamodel. Formalisms are different to represent classes and objects and there is no means to state that an instance may also be seen as a class. The solution is to define a class and an object with the same label as shown in Figure 11.

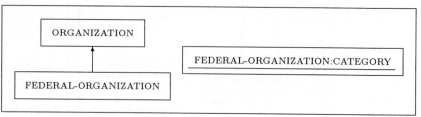

Fig. 11. UML: Category or instance in Metamodel.

3.3 Relational Model Formalism

The relational model [5] is probably the simplest and most flexible formalism of the compared formalisms, but unfortunately its current implementations in database management systems reduce its power.

The relational model defines tuples (instances) and tables (categories). Most relational database management systems use the same formalism to describe tuples and tables, and the table descriptions are stored in a set of tables generally identified as system tables or metabase. Relationships are expressed by using join attributes that link tables. "Employees work for organizations" is represented in Figure 12 by two tables, and in the Employee table the attribute OrgId is a foreign key that establishes the relationship between an employee and the organization he or she works for provided that no null values are allowed to ensure referential integrity.

Organization	OrgId	Name	...	Employee	EmpId	Name	OrgId

Fig. 12. Relational Model: Tables and Join Attributes.

Classification and Partial Knowledge. A tuple is a row of one and only one table and there is no specific means to implement partial knowledge, but the relational view mechanism and the use of null values may be used to implement partial knowledge management though they introduce other problems.

Category and/or instance in relationship. Relationships are defined at the instance level and implemented using join attributes. For the UNU example, the solution is to define a UNU Employee table with a join attribute, OrgId, always equal to the organization identifier of UNU as shown in Figure 13.

Organization	OrgId	Name	...	UNU Employee		OrgId
	123	UNU				123
						123

Fig. 13. Relational Model: Category and/or instance in relationship.

Category or instance in Metamodel. The descriptions of tables created in a relational database management system are normalized and stored in system tables (Figure 14). Like any other tables, system tables may be manipulated using SQL statements, and using a SELECT clause in a FROM clause like in SELECT Government FROM (SELECT TableName FROM Attribute WHERE AttributeName='Government') would be a solution, but current implementations do not authorize such SQL statement.

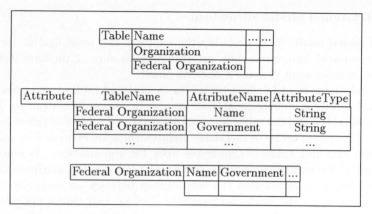

Fig. 14. Relational Model: Category or instance in Metamodel.

3.4 Classic: A KL-One-like Language

Classic [2] is a KL-One-like system; it is a frame-based knowledge representation system. Knowledge is represented by describing objects using frames, as opposed to asserting arbitrary logical sentences. Classic defines instances called individuals and categories called concepts. Relationships between individuals are implemented using attributes, called roles in Classic. Figure 15 shows the definition of the concept EMPLOYEE that represents "employee works for an organization". EMPLOYEE is defined as a subtype of PERSON that has an attribute works-for which represents the relationship with the concept ORGANIZATION.

```
EMPLOYEE ⇔ (AND PERSON(ALL works-for ORGANIZATION))
```

Fig. 15. Classic: Concepts and relationships.

Classification and Partial Knowledge. Classic supports multiple classification. An individual can satisfy more than one concept. Classic also supports partial knowledge through dynamic classification. When a new individual is introduced into the system, classification is invoked to find all the concepts that are satisfied by the individual.

Category and/or instance in relationship. Relationships between individuals are implemented using roles. Operators have been defined to express restrictions on roles. One of these operators, FILLS, specifies that a role is filled by some specified individual. Figure 16 shows the definition of the concept EMPLOYEE as a person who works for a company, and the definition of UNU-EMPLOYEE as an employee that works for UNU.

Category or instance in Metamodel. Classic distinguishes individuals from concepts and does not support the notion of metaconcept. Therefore, the system is not suitable in situations where some individual may be viewed as a class with instances.

```
EMPLOYEE ⇔ (AND PERSON(ALL works-for ORGANIZATION))
UNU-EMPLOYEE ⇔ (AND EMPLOYEE(FILLS works-for UNU))
```

Fig. 16. Classic: Category and/or instance in relationship.

3.5 Conceptual Graphs

Conceptual graphs are a formalism whereby the universe of discourse can be modeled by concepts and conceptual relations. A concept represents an object of interest or knowledge. A conceptual relation makes it possible to associate these concepts. Conceptual graphs were developed by John Sowa in the early 80s [14]. They are a system of logic based on the existential graphs of C.S. Peirce [13] and semantic networks. Conceptual graphs define knowledge both at the type and instance levels. Concepts are represented by boxes and relationships by circles with arrows that link the concepts associated. Figure 17 represents the sentence "There exists an employee that works for an organization"

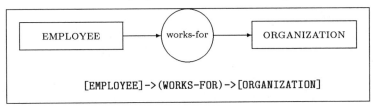

[EMPLOYEE]->(WORKS-FOR)->[ORGANIZATION]

Fig. 17. Conceptual Graphs: Concepts and Relationships.

Concepts may be categorized based on the type of conceptual relations they have with other concepts. Concept types define these categories. A concept type is defined by a definition graph to which any instance of that concept type must comply with. Figure 18 presents the definition graph of EMPLOYEE that means that all employees are persons that work for some organization.

```
Type EMPLOYEE(x) is
   [PERSON:*x]->(WORKS-FOR)->[ORGANIZATION].
```

Fig. 18. Conceptual Graphs: Type Definition.

Classification and Partial Knowledge. Conceptual graph theory defines one type hierarchy; this hierarchy is a lattice with the universal type ⊤ at the top and the absurd type ⊥ at the bottom. Multiple inheritance and multiple classification are supported by conceptual graph theory. Multiple inheritance is often difficult to use in a practical way, especially when dealing with hundreds of concepts. Multiple classification allows multiple perspectives, each perspective with its own vocabulary.

If we go back to the car dealer example, we do not need multiple inheritance. This situation corresponds to multiple perspectives: number of seats, propulsion and origin, and the use of multi-classification is certainly better. For example, if car #123 is an American 4WD Sedan, then we will create three concepts [SEDAN:#123], [4WD:#123], [AMERICAN:#123] each of them corresponding to the perspectives, number of seats, propulsion and origin. For partial knowledge, known information is stated according to its relevant perspective independently of other information.

Category and/or instance in relationship. Instances must conform the definition of the concept type to which they are associated. For instance, in the case of UNU employees, the definition graph of UNU-EMPLOYEE is:

```
Type UNU-EMPLOYEE(x) is
  [EMPLOYEE:*x]->(WORKS-FOR)->[ORGANIZATION:UNU].
```

Fig. 19. Conceptual Graphs: Category and/or instance in relationship.

Category or instance in Metamodel. A concept is the association of two markers: one for type and one for instance. Changing the position of an instance marker from left to right promotes it to a type marker. Figure 20 illustrates how an individual may be seen either as a type or as an instance.

When the marker is on the left side [CONCEPT-TYPE:FEDERAL-ORGANIZATION], it represents an instance of the right side type, and when the marker is on the right side, it represents a category as in [FEDERAL-ORGANIZATION:ENV-AGENCY].

```
[CONCEPT-TYPE:FEDERAL-ORGANIZATION]-
  (SUBTYPE)->[CONCEPT-TYPE:ORGANIZATION]
  (SYMB)<-[CATEGORY-LABEL:'federal-organization']
  (DEFINED-BY)<-[GRAPH: Type FEDERAL-ORGANIZATION(x) is
                  [ORGANIZATION:*x]-
                    (SYMB)<-[LABEL:*]
                    (ATTR)<-[GOVERNMENT:*]
                    (MEMBER)<-[FEDERAL-ORGANIZATION:*] ].

[FEDERAL-ORGANIZATION:ENV-AGENCY]-
  (SYMB)<-[LABEL:'environment agency']
  (ATTR)<-[GOVERNMENT:'Canada'
  (MEMBER)<-[FEDERAL-ORGANIZATION:AIR-AGENCY].
```

Fig. 20. CG Formalism: Category or instance in Metamodel.

3.6 Summary

Table 1 presents a summary of the formalisms compared.

	Classification Partial Knowledge	Category or instance in Relationship	Category or instance in Metamodel
E-R	No	No	No
OO	Yes	Using Composite	Using Same Label
Relational	Using View	Yes	No
Classic	Yes	Yes	No
CG	Yes	Yes	Yes

Table 1. Summary.

4 Conclusion

This paper has compared five knowledge representation formalisms. Our aim is to develop corporate knowledge repositories. This comparison has shown how conceptual graphs are a response to the specific requirements involved in the development of corporate knowledge repositories.

Using this formalism, a corporate knowledge repository [8] is developed and implemented at the Research & Development Department of DMR Consulting Group Inc. in order to memorize the methods, know-how and expertise of its consultants. This corporate knowledge repository, called Method Repository, is a complete authoring environment used to edit, store and display the methods used by the consultants of DMR. The Method Repository has three components: a Method Knowledge Acquisition facility, a Conceptual Graph Knowledge Base and a Knowledge Dissemination engine. Method Knowledge Acquisition is an ad hoc module based on the method metamodel that allows method developers to create, maintain and adapt methods. The CG Knowledge Base is the core of the environment; it is a knowledge engineering system based on conceptual graphs. The Knowledge Dissemination facility provides view mechanism [10] using available technologies such as HTML files, SGML files and RTF files, among others.

In June 1996, five methods were commercially delivered: Information Systems Development, Architecture, Benefits, Technical Infrastructure and Estimating, in hypertext format generated from conceptual graphs. From about 80,000 conceptual graphs, we generated more than 100,000 HTML pages that can be browsed using commercial Web browsers. The power of conceptual graphs in terms of expressiveness and flexibility for corporate knowledge modeling has been demonstrated through the development of the Method Repository now being used in the field by thousands of consultants at DMR.

References

1. G. Booch, J. Rumbaugh, and Jacobson I. *Unified Modeling Language, Version 1.0.* Rational Software Corporation, 1997.
2. R. J. Brachman and al. Living with classic: When and how to use a kl-one-like language. In John Sowa, editor, *Principles of Semantic Networks: Exploration in the Representation of Knowledge*, pages 401–456. Morgan Kaufmann, 1991.

3. Prahalad C. and Hamel G. The core competence of the organization. *Harvard Business Review*, pages 79–91, 1990.
4. P. Chen. The entity-relationship model - toward a unified view of data. *ACM Transactions on Database Systems*, 1(1):9–36, 1976.
5. E. F. Codd. A relational model of data for large shared data banks. *Communications of ACM*, 13(6):377–387, 1970.
6. R. Elmasri and S. Navathe. *Fundamentals of Database Systems.* The Benjamin/Cummings Publishing Company Inc., Redwood City, California, 1989.
7. G. Engels, M. Gogolla, U. Hohenstein, Hülsmann K., Löhr-Richter P., Saake G., and Ehrich H.-D. Conceptual modelling of database applications using an extended er model. *Data & Knowledge Engineering*, 9(2):157–204, 1992.
8. O. Gerbé et al. *Macroscope Architecture: Architecture of DMR Repository.* DMR Group Inc., Montréal, Québec, 1994.
9. O. Gerbé, B. Guay, and M. Perron. Using conceptual graphs for methods modeling. In *Proceedings of the 4th International Conference on Conceptual Structures*, Sydney, 1996.
10. O. Gerbé and M. Perron. Presentation definition language using conceptual graphs. In *Peirce Workshop Proceedings*, Santa Cruz, California, 1995.
11. DMR Consulting Group. *DMR Macroscope.* DMR Consulting Group Inc., 1996.
12. C. Havens. Enter, the chief knowledge officer. *CIO Canada*, 4(10):36–42, 1996.
13. C. S. Peirce. *Collected Papers of C.S. Peirce.* Harvard University Press.
14. J. F. Sowa. *Conceptual Structures: Information Processing in Mind and Machine.* Addison-Wesley, 1984.
15. E. W. Stein. Organizational memory: Review of concepts and recommendations for management. *International Journal of Information Management*, 15(1):17–32, 1995.
16. G. van Heijst, R. van der Spek, and E. Kruizinga. Organizing corporate memories. In *KAW 96 Proceedings*, Banff, 1996.

An Experiment in Document Retrieval Using Conceptual Graphs

D. Genest, M. Chein

LIRMM (CNRS & University Montpellier 2), genest@lirmm.fr

Abstract. In this paper an experiment using conceptual graphs to represent documents and queries is described. This paper is centred on the representation and answering language, a variant of simple CGs, and not on automatic indexing. Two important specificities of the experiment described in this paper are: first, it uses an existing general "ontology", the thesaurus of RAMEAU, which is used in all the french universitary libraries; secondly, the CG indexation has been obtained from the existing indexation by adding a new "relational" level. The main goal of this paper is to show that it is possible to add, to general traditional document retrieval systems, a CG level, and that it is worthwhile to pay an additional cost, because, using CG for indexing documents and for representing queries, leads to significant improvements in the answers obtained.

1 Introduction

In document retrieval — finding documents whose content is relevant to a user information need — difficulties induced by the large size of document databases, by their heterogeneity, or by their distribution on a computer net, are often emphasized. All these problems are very important since efficient real-time document retrieval systems (DRS) for such document databases are needed for more and more present applications. Nevertheless, in traditional DRS, a fundamental limit is due to the weakness of representations used for documents — sktechily, structured sets of weighted key-words — and for queried information (see for instance [1] or [9]). Richer document representation schemes can be obtained by considering not only "words" but also "semantic relations" between words (and this to represent both documents and queries). Different languages and systems have been built, in artificial intelligence, using entities and relations between these entities, under the generic name of semantic nets (for a recent survey see [13]). Sowa [22] introduced such a model, the conceptual graph model, which seems to have interesting properties for developing DRS ([4], [19], [18], [12]). In this paper an experiment using conceptual graphs for representing documents and queries is described. Within this model, the four main steps of the "query-by-content" fundamental problem can be sketchily described as follows:

1 represent the documents of a database B by a set of conceptual graphs: $R(B) = \{G1, G2,..., Gk\}$, i.e. index documents by an indexing language based on CG,
2 represent a query Q by a CG: $R(Q)$,
3 look for documents corresponding to the query: i.e. search in $R(B)$ conceptual graphs which "match" with $R(Q)$ (this is done by computing a kind of projection),
4 modify the query and loop from 2 until the user is satisfied.

The first step, representation of documents, which is considered by Blair [1] as "the central issue of information retrieval", is not directly addressed in this paper. This paper is centred on the representation and answering language, CG, and not on automatic indexing, which is a natural language processing problem (see for instance [14]). Indeed, the meaning of a document — a text for example — is the result of an interpretation done by a reader. This interpretation task needs much more information than the data contained in the document itself. Thus, a unique representation of a document, satisfying different readers with different unknown goals, does not probably exist. Automatic indexation needs important progress in semantic analysis of documents. There are some prototype softwares for indexing general texts by CGs (see for instance [18] for the Wall Street Journal), but, even for very technical and specific texts, automatically indexing by CGs, or by any other kind of semantic nets, is a very difficult task (see for instance the results of the Menelas project dealing with automatic CG indexation of patient discharge summaries [23]). As a matter of fact, today, in french university libraries, indexing is done by hand, and *we believe that manually indexing by CGs is no more difficult than indexing by key-words*. Indeed, a person who is indexing a document must understand the document content, and this understanding involves not only the determination of the main notions but also the determination of semantic relations between these notions. In our experiment, which concerns exclusively texts, we manually enrich the existing indexation by adding general relations between some present terms. Let us mention that a problem much more simple than indexation, the problem of "information extraction" — i.e. to find in a text the information needed to fill a given template — is far from being solved, today: " IE systems deal only with specific types of texts and are only partly accurate" say [6]. Thus, it is not very original to claim that, at the present day, and for general document or text databases, only semi-automatic indexing methods can be considered, and, in this case, indexing by CG is only a little bit more complex than traditional indexation. The main goal of this paper is to show that it is possible to add, to general traditional document retrieval systems (DRS), a CG level, and that it is worthwhile to pay the additional cost induced, because, using CG in the above steps 3 and 4, leads to significant improvements in the answers obtained.

This paper is organized as follows:
- in the second part the french indexing language RAMEAU, and the DRS SIBIL, used in the french university libraries, are presented;
- the third part is devoted to a presentation of the above steps 3 and 4;
- finally, the experiment is presented in the fourth part, and a conclusion ends the paper.

2 The RAMEAU-SIBIL System

2.1 RAMEAU

Rameau is a precoordinated documentary language used by French university libraries in order to index documents. The goal of Rameau is to allow indexation of all-domain documents. An indexation is a *combination* of words. The language is constituted by a *subject heading list* which contains words librarians can use to index documents, and *relations* between these words.

These subject headings (called *"vedettes"*) are terms representing concepts which can be used to build indexations.

There are various types of subject headings:
some represent only one concept   ex: *computers, surgery*
whereas other represent several concepts. These multi-concept subject headings can be classified in two classes:
- non-subdivided subject headings   ex: *children and war*
- subdivided subject headings   ex: *surgery ** patients*

Subdivided subject headings are divided into two types: emancipated subdivisions and non-emancipated subdivisions. Subject headings (SH) built using emancipated subdivisions are not in the subject headings list (subdivision is in the list but the built SH is not), whereas SH built using non-emancipated subdivisions are in this list. Non-emancipated subdivisions can only be applied to a few subject headings. So each regular SH built using non-emancipated subdivisions appears in the list. A Rameau sentence is called a "built SH". These SH are combinations of subdivisions: each subdivision has a role, and this role introduces a semantic link between the different parts of the built SH. The subdivisions must respect an order:
          "Vedette" head ** Topical subd. ** Geographic subd. ** Chronological subd. ** Form subd.

ex: *medicine $w social angle $x France $t 19th century*
          redress (legal) $r thesis

Vedette head represents the theme of the document, and topical subdivision specifies which aspect of the vedette head the document is about. Form subdivision permits to represent the kind of the document (biography, atlas, thesis, ...). Subdivisions are optional. Emancipated subdivisions can be applied to many SH. They appear in the subject headings list. SH librarians can build using them are not in the list: for instance, *insurance* and *France* are in the list, then *insurance ** France* is a regular SH even if this built SH is not in the subject headings list. So, both "real" subject headings and subdivisions figure in this list, then, in order to specify the type of each element, some flags are present, and mark SH uses (usable as topic subdivision, usable as geographic subdivision, ...).

An *authority record* is associated with each SH. This authority record contains the above mentioned flags and relations between this SH and other ones. There are several relations. The *synonymy* relation links several synonymous (or quasi-synonymous) terms (called *excluded terms*) to a particular SH (called *selected term*). The goal of this relation is to homogenize indexations: librarians must use the same (selected) term in order to represent documents about the same theme. So, this relation permits to restrict the terms librarians can use. This limitation is not necessary if a DRS is used.

ex:

Selected term	Excluded term
children and war	*war and children*

The *hierarchical grouping* relation represents generic links between SH. The goal of this relation is to assist librarians to build the most specific indexation.
ex:

Narrower term	Broader term	
occipital bone	*skull*	*"is a part of"*
occipital bone	*bone*	*"is a kind of"*

The *non hierarchical grouping* relation links subject headings which are semantically related.
ex:

Related term	Related term
accident	*catastrophe*

Rameau relations don't always respect their semantic: certain relations are wrongly used (synonymy relation between non-synonymous terms, hierarchical grouping relation between terms which have not a generic connection, ...) and several relations lack: for instance, there is no hierarchical relation between *drug addict* and *humans*. Because this relation is obvious and because Rameau relations are only "manually" used by librarians, Rameau authors didn't create this link.
ex:

Selected term	Excluded term	
treatment of animals *(animaux, traitement des)*	*brutality against animals* *(sévices aux animaux)*	two (excluded) antonymous terms are linked to the same (selected) synonym.
treatment of animals *(animaux, traitement des)*	*protection of animals* *(protection des animaux)*	

	Broader term	
cinema and children	*children*	link between a multi-concept SH and one of its concept.
drug addict	*drug addiction*	non hierarchical link.

2.2 SIBIL

Sibil is a document retrieval system created by the BCU Lausanne, it is one of the most used DRS in the French university libraries since 1976. A document representation is called *"bibliographic record"*. This bibliographic record especially contains book title, authors names, editor name, libraries where the book is available, and, of course, a representation of the document content. The document content is represented by a set of Rameau subject headings. Several kinds of queries are available: a document can be retrieved by its name, author or content.

Nevertheless, multi-criteria queries are impossible: for instance, Sibil can't retrieve documents owned by a library *L* and which respect another criterion (title, content, ...): this is an important problem because Sibil's database contains documents owned by all the libraries of the Sibil Net, then a query result contains several documents to which user can't easily access. There are two kinds of query by title: Using the first type, query is executed on the title beginning (Sibil retrieves documents of which the title beginning is what the user typed). Using the second type, Sibil retrieves documents of which the title contains the word typed by the user. However, user can choose only one word. Multi-criteria queries lack produces silence and queries by title are silent (first type) or noisy (second type). Sibil's search-by-content method retrieves documents of which the content representation is exactly what the user typed: the system does not use Rameau relations. Though document indexation should describe as exactly as possible the document content, a "too narrow" indexed document will never be retrieved even if it is useful for the user (*silence*). Nevertheless, documents have to be precisely indexed because if a document indexation is "too broad", document will be retrieved by a query on a near subject (*noise*). Search-by-content is executed from only *one* built SH: users can't consider several SH. An indexation may be constituted by a *set* of built SH: if a document is about a very specific subject, building a single SH is impossible, then the subject is divided, and each part of the subject is represented by a built SH. But queries are executed from only one SH, then the document will be retrieved by every query which concerns a part of its subject.

For instance, regarding a document called *"commentary and reflection about forensic procedure applied to corporeal damage redress concerning a victim of a road accident that occurred in Australia"*, these concepts seem pertinent: *redress, corporeal damage, road accidents, Australia*. Then it is impossible to build only one SH representing these concepts (according to Rameau rules). The Rameau indexation of this document is:

> road accident **Australia** thesis
> corporeal damage ** thesis
> redress (legal) ** thesis

So, this document is retrieved by queries about *redress* and user can't specify which kind of redress he is interested about. *Redress* appears in indexations of many documents (about various subjects), then the query *redress* produces too many documents.

3. A Documentary Language Based on Conceptual Graphs

This part describes how the simple conceptual graph model (as defined in [22] and [3] can be modified to provide a precoordinated documentary language. A DRS is not able to perfectly determine if a document is relevant for a user but a system has rather to estimate this relevance by calculating the semantic resemblance between the query and the indexation of a document. If simple CGs are used for query and indexation representation, and if projection is used for answer computing, some problems appear:

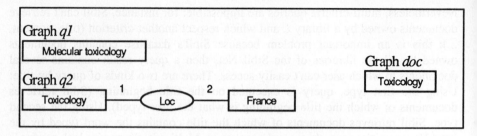

Fig. 1. Example Conceptual Graphs

Considering Figure 1, if the support contains "*molecular toxicology*" < "*toxicology*", there does not exist a projection from *q1* (or *q2*) to *doc*. However, if *q1* and *q2* represent queries and *doc* represents an indexation of a document, this document seems to be relevant for these queries: "classic" projection does not respect the above-mentioned condition. Moreover, support does not allow to represent a thesaurus structure: concept types set is provided with an order representing an "A-Kind-Of" hierarchy, whereas thesaurus relations are more numerous and richer.

3.1 Th-Support: Thesaurus Structure Modeling

Th-support permits a thesaurus structure representation: the order on the concept type set, which corresponds to A-Kind-Of relation between types, is replaced by a *set* of relations representing thesaurus relations.

A *Th-support* is a 5-tuple $S = (TC, OC, TR, I, \tau)$ such that :
   TC is a set of concept types.
   OC is a set of binary relations on TC . $OC = \{OC1, OC2,... OCu\}$
   TR, I and τ are defined in the same way as they are in "classic" support (i.e. respectively set of relation types, set of individual markers, typing of the individual markers, see for instance [3]).

The lack of a partial order representing an AKO hierarchy causes the impossibility to define the maximal type ot i-th argument of a relation. Hence, relations signatures are unusable with a Th-support (there are no constraints on OC relations, so they may not be orders). Definition of a conceptual graph related to such a Th-support is the same as the one of conceptual graph on "classic" support, minus the condition on neighbours of relation vertices (signature) which disappears. In a same way, projection has to be redefined because it is now impossible to determine if a concept vertex can be projected to another one.

Let L be a relation on N

P, called *projection capability relation*, is a binary relation on TC defined as follows :

$\forall c \in TC, \forall c' \in TC,$

$P(c,c')$ *iff* $\quad c = c'$ or

$\quad\quad \exists\, S = (s1, s2,... sk)$ a sequence of integers $\in [1,u]$ (let us recall that $u = |OC|$)

$\quad\quad\quad$ such that: $\quad$ (1) $c\ OCs1\ c1, c1\ OCs2\ c2, ... , ck\text{-}1\ OCsk\ c'$ and

$\quad\quad\quad\quad\quad\quad\quad\quad\quad$ (2) $L(s1, s2, ... , sk)$

Condition *(1)* defines P as the transitive closure of the union of OC relations, and condition *(2)* restricts P to pairs *(c,c')* verifying a condition concerning relations sequence between c and c'. This condition, which may concern the sequence length or relation types used in the sequence, permits to limit projection capabilities of a concept type. If this condition lacks, depending on OC relations usage (modelled thesaurus structure), a concept type may be projected to many other ones. For example, if a thesaurus has an unique root (referring to a relation type) or if it is not constituted by disjoint sets (referring to the union of relations), each concept type may be projected to any other one. So, the role of this relation is projection capabilities limiting (in a rational way). Then, definition of projection between graphs defined on a Th-support is immediately deducted from the one between graphs related to a classic support: Order on TC is replaced by the relation P.

3.2 Evaluated Projection: Relevance Evaluation

Considering two conceptual graphs, projection from one of them to the other one *is* or *is not* possible. This implies that documents which partially answer a query are judged irrelevant (there is no projection). Then, projection has to be adjusted so as not to produce a boolean result but rather a value representing the semantic resemblance between the query and the document.

Let and be two mappings such that:

$\quad \alpha\colon TC\ xI\ xTC\ xI\ \text{-> } R+$ and $\beta\colon TR\ xTR\ \text{-> } R+$

An $\alpha\beta$-*projection* from a conceptual graph $H=(RH, CH, UH, labH)$ to a conceptual graph $G=(RG, CG, UG, labG)$ is an ordered pair $\Pi=(f,g)$ of mappings $f{:}RH\text{->}RG$, $g{:}CH\text{->}CG$ such that:

In the same way as "classic" projection, $\alpha\beta$--projection preserves edges and edges labels.

$\quad r \in RH, \beta(\,labH(r)\,, labG(f(r))\,) \neq 0$

$\quad c \in CH$, given $labH(c)=(t,m)$ and $labG(g(c))=(t',m')$, $\alpha(\,t, m, t', m'\,) \neq 0$

The goal of the mapping α is to value the projection from a label *(type, marker)* of a concept vertex to another one, whereas the purpose of the mapping β is to value the projection from a relation vertex to another one. Partial orders on concept types and relation types are not needed because α and β replace them.

A *valuation pattern* of an $\alpha\beta$-projection $\Pi=(f,g)$ from a conceptual graph H to a conceptual graph G is defined as follows:

$$v(\Pi)=V_H\left(\underset{c\in C_H}{S}\ \alpha\big(lab_H(c),lab_G(g(c))\big),\underset{r\in R_H}{S'}\ \beta\big(lab_H(r),lab_G(f(r))\big)\right)$$

This definition provides a valuation pattern: to obtain a real-valued function, mappings VH, S and S' have to be defined. These mappings may be defined as follows:

$$S = + \qquad\qquad S' = + \qquad\qquad VH(a,b) = (a + b) \,/\, (|RH| + |CH|)$$

These choices result in the following valuation formula:

$$v(\Pi)=\frac{\underset{c\in C_H}{\Sigma}\ \alpha\big(lab_H(c),lab_G(g(c))\big)+\underset{r\in R_H}{\Sigma}\ \beta\big(lab_H(r),lab_G(f(r))\big)}{|C_H|+|R_H|}$$

Then, if α and β are rationally defined, $\alpha\beta$ -projection provides a semantic resemblance calculus, and concerning the above-mentioned problem, a solution is produced: using classic projection, there is no projection from *q1* to *doc*. Now, *toxicology* and *molecular toxicology* are semantically "near", therefore (if α is rationally designed) here is a $\alpha\beta$–projection from *q1* to *doc* even if *doc* does not perfectly answer the query: this will be represented by the valuation which will not be maximum. Nevertheless, even if $\alpha\beta$-projection is used, there is no projection from *q2* to *doc*: the concept vertex labelled *France* and relation vertex *loc* cannot be projected on any vertex of *doc*.

There is an $\alpha\beta$-partial-projection from a conceptual graph H to a conceptual graph
    G, iff there exists an $\alpha\beta$-projection from H', a sub-conceptual graph of H, to G.
Valuation v of such an $\alpha\beta$-partial-projection only depends on H' vertices and is defined like (total) $\alpha\beta$-projection.
See [2] for more details about partial projections.

3.3 Valuation and Th-support

$\alpha\beta$-projection and valuations can easily be used on conceptual graphs defined on a Th-support: we just have to determine α. α may be defined by using the sequence used in the projection capability relation. Such a projection (and valuation) then uses OC relations.
Ex.
$\alpha\,(t,m,t',m') = 0$ if $notP(t,t')$ or if $m \neq *$ and $m' \neq m$
otherwise $\alpha\,(t,m,t',m') = Max\,(M(s1, s2, \dots , sk))$ if $\exists S = (s1, s2,\dots\, sk)$ a sequence of integers $[1,u]$ such that: $t\ OCs1\ t1,\ t1\ OCs2\ t2,\ \dots ,\ tk\text{-}1\ OCsk\ t'$

Mapping M allows to consider the various OC relations used in the sequence. This is useful because the semantic resemblance of two terms linked by an OC relation may vary depending on the used OC relation. For instance, a synonymy relation links terms which are more semantically likeness than terms linked by a hierarchical grouping relation. α depends on the maximum of this calculus because there may exist many sequences " from t to t' " (according to the thesaurus structure). Moreover, α permits to consider the "semantic remoteness" between a generic term and a specific one: using classic projection, if $t1 > t2 > ... > tn$, a concept vertex of type $t1$ can be projected on a vertex of a same type in the same way as a concept vertex of type $t1$ can be projected on a vertex of type tn . Now, projection is used for document retrieval, therefore, if an user chooses a generic term, he surely wants to retrieve documents about this term ($t1$, ex: *medicine*), and not about a specific term (tn, ex: *pediatric pharmacology*, which is a specific term of *medicine*). In spite of differentiation of these two types is impossible with classic projection, $\alpha\beta$ -projection can represent this "semantic remoteness" with $\alpha\,(t1,t1) > \alpha\,(t1,tn)$.

Above described formulas allow valuation of *one* projection, now, the purpose of a searching method is the semantic resemblance calculus (between a query and an indexation) therefore we have to define this calculus. Let q be a query graph and d an indexation of a document, relevance of the document for the query is estimated by $Rel(q,d) = Max(\,v(\Pi)\,/\,\Pi\;\alpha\beta$ -partial-projection from q to d)

3.4 Relation Vertices

A conceptual graph is constituted by concept vertices and relation vertices. When these graphs are used for representing a document subject, concept vertices represent concepts of the thesaurus treated in the document, and relation vertices represent semantic links between these concepts. This is a distinctive feature of conceptual graphs: they allow to represent semantic links between concepts whereas poor representation models cannot. Rameau, as a precoordinated language, represents some semantic links but these links cannot represent many various relations, that's why several built vedettes have to be created to index a document about only one (complex) subject. The use of conceptual graphs allows to index such a document by a connected graph (semantic links are represented by relation vertices, and if a document is about only one subject, there are semantic links between concept vertices, then the graph is connected). For instance, there is no "person" subdivision in the Rameau rules; then representing a document about something concerning a "person category" is difficult. The Rameau indexation of a document called *"A propos de la réparation médico-juridique des séquelles intéressant la tête et le cou chez le jeune enfant victime de sévices corporels"* (*About forensic redress of after-effects on head and neck among young child victim of corporeal damage*) is:

> *mistreated children*
> *head ** lesions and injuries*
> *neck ** lesions and injuries*
> *corporeal damage*

Rameau indexation does not represent the semantic link between *corporeal damage* and *children*. The use of a semantic relation (*"pers"*) with a conceptual graph indexation allows a more precise representation of the topic of this document: it is not about children *and* corporeal damage (two independent topics) but about corporeal damage *among* children. Concerning this document, Rameau indexation allows to deduce the semantic link between *corporeal damage* and *children*, but if a person category is present in an indexation, we can generally not deduce this link. For instance, if *corporeal damage* and a person category are present in an indexation, this may represent *corporeal damage* among the *person category, corporeal damage* caused by the *person category,* or *person category* point of view about *corporeal damage.* Now, different indexations are needed for documents about different subjects to obtain an efficient searching method, therefore, representing semantic links between concepts seems to be necessary.

4. An Experiment

Evaluating a system is difficult even if a large database is available for it, consequently, in order to better understand conceptual graphs contribution, an experiment on real documents seemed to be necessary. However, this experiment is not an evaluation of a DRS: if a prototype has been constructed, we cannot call it a DRS, moreover the number of indexed documents used in this experiment is too small to permit a reliable evaluation. This part describes how the prototype has been designed from above described definitions.

4.1 Domain Choice and Determination of Semantic Relations

To experiment our prototype on documents about a specific subject: *forensic medicine* and particularly *corporeal damage* was chosen. The first step is the selection of the "semantic" relations. These relations have to be chosen by considering the domain. We choose a few relations: most of them are quite generic and no forensic medicine specific (see Table 1). Determination of the amount of relations and relations themselves is empirical and has been done by consulting the Sibil's database. We could surely find some more specific relations but this would require knowledge about forensic medicine. Nevertheless, we think chosen relations produce interesting results.

4.2 Th-Support Determination

Conceptual graphs building requires a support: the subjects headings list of Rameau (excluded terms and selected terms) has been chosen as a base of the support.
Chosen *OC* relations are Rameau relations:
$OC = \{ EP, VOIR, TG, TS, TA \}$
Names of these relations are the abbreviations defined by Rameau:
Synonymy relation: selected term EP excluded term, excluded term VOIR selected term
Hierarchical grouping relation: narrower term TG broader term, broader term TS narrower term
Non hierarchical grouping relation: related term TA related term
Some properties appear: (Let *t1* and *t2* be two subject headings)
t1 EP *t2* iff *t2* VOIR *t1*   *t1* TS *t2* iff *t2* TG *t1*   *t1* TA *t2* iff *t2* TA *t1*

Relation	Meaning	Example
obj	Object.	evaluation —1→ obj —2→ corporeal damage
loc	Localisation of a pathology.	traumatism —1→ loc —2→ orbit
aim	Aim.	evaluation —1→ aim —2→ redress (legal)
pers	Subject *among* a person category.	corporeal damage —1→ pers —2→ motorists
infl	Influence.	airbag —1→ infl —2→ corporeal damage
cause	Cause.	accidents —1→ cause —2→ corporeal damage
care	Medical care.	chemotherapy —1→ care —2→ cancer
circ	Circumstances.	accidents —1→ circ —2→ military service
asp	Aspect, point of view	evaluation —1→ asp —2→ forensic medicine
geo	Geographic localisation.	accidents —1→ geo —2→ Australia
hist	Time, historic precision.	corporeal damage —1→ hist —2→ civil war

Table 1. Semantic Relations

4.3 α and β: Valuation Functions

We have chosen $\alpha : TC \times I \times TC \times I \rightarrow [0,1]$ and $\beta: TR \times TR \rightarrow [0,1]$ for coherence reasons: 1 will represent the equivalence between two types and 0 will represent the lack of any semantic link between them. Expression of these functions has arbitrarily been determined. Even if this determination is arbitrary, particularities of chosen semantic relations imply some constraints:

$\forall r \in TR, \beta(r,r) = 1$

obj is more generic than the other ones: $\quad \forall r \in TR, r \neq obj, \beta(obj,r) > \beta(r,obj)$

infl seems to be more generic than *care*: $\quad \beta(infl,care) > \beta(care,infl)$

circ seems to be more generic than *geo* and *hist* : $\beta(circ,geo) > \beta(geo,circ)$
$\qquad \beta(circ,hist) > \beta(hist,circ)$

The values of β, which have manually been chosen, respect these conditions and allow projection from any relation vertex to any other one ($\forall r \in TR,\ \forall r' \in TR,\ \beta(r,r') \neq 0$). This privileges projection between c and c', two concept vertices linked by a relation r to $c1$ and $c'1$, concept vertices linked by a relation vertex $r1$ even if r and $r1$ are not relations of the same type, in that way, an user may use in a query a relation type which is not exactly appropriate.

The semantic of Rameau relations implies some new constraints:
$\forall t \in TC, (t,*,t,*) = 1$

$\forall t \in TC, \forall t' \in TC, \forall R \in OC,$ if $t\ R\ t'$ then $\alpha(t,*,t',*) \neq 0$

$\forall t1 \in TC, \forall t2 \in TC, \forall t3 \in TC, \forall t4 \in TC,$

if $t1$ EP $t2$ and $t3$ TS $t4$ ($t1$ is the selected term of $t2$ and $t4$ is a specific term of $t3$)

then $\alpha(t1,*,t2,*) > \alpha(t3,*,t4,*)$

> (Synonymy relation links terms which are more semantically likeness than terms linked by hierarchical grouping relation)

if $t1$ TS $t2$ and $t3$ TA $t4$ ($t2$ is a specific term of $t1$ and $t3$ is related to $t4$)

then $\alpha(t1,*,t2,*) > \alpha(t3,*,t4,*)$

> (Hierarchical grouping relation links terms which are more semantically likeness than terms linked by non-hierarchical grouping relation)

if $t1$ TS $t2$ and $t3$ TG $t4$ ($t2$ is a specific term of $t1$ and $t4$ is a generic term of $t3$)

then $\alpha(t1,*,t2,*) > \alpha(t3,*,t4,*)$

> (A document about a specific term better answers a query about one of its generic terms than a document about a generic term answers a query about one of its specific terms)

Semantic resemblance quickly decreases by using Rameau relations, for instance *accidents* is related to *lightning* which is related to *electricity* which is related to *electrons*. So as to respect Rameau semantics, excluded terms usage requires that the valuation between concept types representing subject headings linked by a synonymy relation has to be near to the maximum (these terms are strictly equivalent by considering Rameau semantics):

$\forall t1 \in TC, \forall t2 \in TC, \forall t3 \in TC$

if $t1$ EP $t2$ and $t1$ EP $t3$ ($t1$ is the selected term of $t2$ and $t3$)

then $\alpha(t1,*,t2,*) \cong 1$ and $\alpha(t2,*,t3,*) \cong 1$

Determining manually the values of α is impossible because of the amount of concept types. A calculus which respects the above conditions may be obtained by applying the following:

```
α(c, c')
L.Add(1, c)
while not(L.Empty)
        (xval, xtype) = L.Dequeue
        if (xtype = c') then return xval
        if (xval * val_EP > Threshold)
                for each d such that xtype EP d
                        L.Add(xval * val_EP, d)
        // Repeat this if for VOIR, TG, TS, TA
    return 0
```

L is a priority queue: *Add(p,c)* adds to the queue the item c with the priority p. *Dequeue* returns (and deletes from the queue) an item having the higher priority. *Threshold* is the minimum non-nil value of : concepts c' such that $(c,c') < Threshold$ are regarded as "too semantically distant" concepts and the calculus returns 0. *Val_EP, Val_VOIR, Val_TG, Val_TS* and *Val_TA*, which represent the semantic resemblance between two concept types linked by an *OC* relation, are (arbitrarily) defined like this:

val EP	val VOIR	val TG	val TS	val TA
0.95	0.9	0.7	0.9	0.7

These values and the heuristic calculus which allows to stop the search before considering too dissimilar concepts, permit to quickly obtain a result which respects the above constraints.

Relevance evaluation, which allows answers sorting, is provided by calculating the value of the best $\alpha\beta$-partial-projection from the query graph to indexation graphs, and uses the above-defined formula. Obtaining every partial projection from the query graph to each indexation graph leads to combinatorial explosion. The algorithm used in the prototype is based on the partial projection of *Rock* [2].

4.4 Examples

Query 1

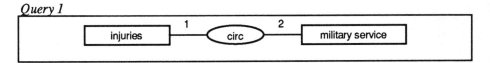

Fig. 2. Query

This query furnishes only one document having a significant relevance ratio (86%), this document is called "*#14: legal redress of corporeal damage among a national service conscript victim of an accident during his service*". Figure 3 shows the indexation of this document and the best projection from the query. The values on dotted lines represent values of α and β for each vertex of the query.

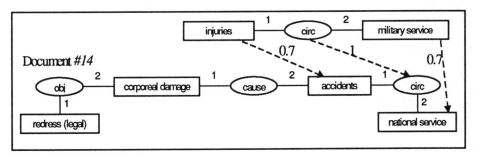

Fig. 3. Indexation

Some other documents are retrieved but their ratio is judged too low and the prototype does not present them in the results list (indexation of these documents contains *injuries* or a near concept). A consultation of the documents list shows there are no other relevant documents than *#14*, then we can conclude the prototype provides good results for this query.

Comparison with the Rameau indexation.
In the Sibil's database, document *#14* is represented by the following Rameau indexation:

*national service ** accidents, corporeal damage ** legal, redress (legal)*

As terms used in the query are different from those used in the indexation, Sibil can not retrieve this document. Moreover, if conceptual graphs are used, the user need not guess which term is the vedette head (*national service*), which term is a topic subdivision (*accident*), or if the indexation is constituted by several vedettes (*national service* and *accidents*). Consequently, an efficient usage of Sibil requires knowledge of Rameau's indexation rules, whereas building a query with conceptual graphs seems easier: the user chooses concepts and links them by using semantic relations.

Query 2

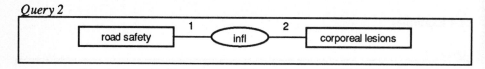

Fig. 4. Query

This query produces two documents: the document *#33* (relevance ratio: 88%) is called *"study about protective effects of an airbag on corporeal damage among motorists"* and the document *#23* (relevance ratio: 80%) *"commentary and reflection about forensic procedure applied to corporeal damage redress concerning a victim of a road accident that occurred in Australia"*.

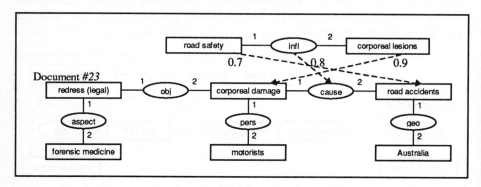

Fig. 5. Projection of the Document #23

The projection to the document *#23* (Figure 5) has an higher value than the one to the document #33 (Figure 6) because the relation vertex labelled "*infl*" is projected on a vertex of another type. The results produced by the prototype are then adequate.

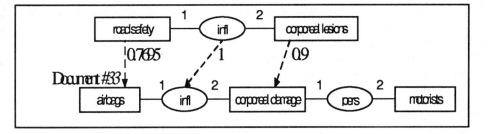

Fig. 6. Projection to the Document #33

5. Conclusion

The kind of approach described in this paper has been used in intelligent information or knowledge based applications. One preliminary task consists in building an "ontology" of the relevant domain — a structured set of concepts and conceptual relations (see [8] for a pioneering work, and [16] for a recent one, both using CG). Two important specificities of the experiment described in this paper are: first, it uses an existing general "ontology", the thesaurus of RAMEAU, which is used in all the french universitary libraries and in most of the french public libraries; secondly, the CG indexation has been obtained from the existing indexation by adding a new "relational" level. The ABES (Agency of french university libraries) found the results sufficiently interesting — with respect to the main criteria for a DRS: precision and recall — to think of more significant tests, in order to see if it is worthwhile to use CG in next generation of SU (the french universitary DRS). We only addressed here the fundamental problem in DR, the query-by-content problem. Different DR problems, such as document or text categorization, document routing, document filtering,..., could be tackled with the same techniques. Furthermore, the simple CG model can be extended in different directions (probabilistic CG [4], nested graphs [17], rules [21], ...). This homogeneous model enables one to represent different levels (parts of document, documents, document databases, libraries,...), and different kinds of documents (bibliographic data, texts, images,...). Moreover, the natural graphical representation of graphs is an interesting feature for building friendly interfaces. The important problem of the management of large graph databases is also studied [7], [11]. Finally, we hope that our work will contribute to show that CGs, or other similar semantic net models (see for instance [15]), will be fruitful for IR.

References

[1] D.C. Blair. *Language and Representation in Information Retrieval*. Elsevier Science Publishers.

[2] B. Carbonneill, O. Haemmerlé. ROCK: un système de Question/Réponse fondé sur le formalisme des graphes conceptuels. *Actes du 9ᵢ Congrès Reconnaissance des Formes et Intelligence Artificielle*, Paris, 1994.

[3] M. Chein, M.L. Mugnier. Conceptual Graphs: Fundamental Notions. *Revue d'Intelligence Artificielle*, 6, 4, 1992, 365-406.

[4] J .P. Chevallet. Le modèle logique de recherche d'informations. PhD, Université de Grenoble, 1992.

[5] W.S. Cooper. A Definition of Relevance for Information Retrieval. *Information Storage and Retrieval*, 7, 19-37.

[6] J. Cowie, W. Lehnert. Information Extraction. *Communications of the ACM*, vol.39, n¡ 1, 80- 91.

[7] G.Ellis. *Managing Complex Objects*. PhD thesis, Computer Science Department, University of Queensland.

[8] J. Fargues, A. Berard-Dugourd, M.C. Landau, L. Catach. Conceptual Graphs for Semantics and Knowledge Processing. IBM Journal of Research and Development, vol.30, n¡1, 1986.

[9] W.B. Frakes, R. Baeza-Yates, editors. *Information Retrieval*. Prentice Hall.

[10] GENEST, D., Une utilisation des graphes conceptuels pour la reherche documentaire. Mémoire de DEA, University of Montpellier (France).

[11] O. Guinaldo. Etude d'un gestionnaire d'ensembles de graphes conceptuels, Ph. D. Thesis, University of Montpellier (France) 1996.

[12] T. Huibers, I. Ounis, J.P. Chevallet. Conceptual Graph Aboutness, in *Conceptual Structures: Knowledge Representation as Interlingua*, Eklund, P.W., Ellis, G., Mann, G., edts, *Lecture Notes in AI*, #1115, 130-144.

[13] F. Lehmann, editor. *Semantic Networks in Artificial Intelligence*. Pergamon Press, 1992.

[14] D.D. Lewis, K. Spark Jones. Natual Language Processing for Information Retrieval. *Communications of the ACM*, vol.39, n¡ 1, 92- 101.

[15] D. Lucarella, A. Zanzi. A Visual Retrieval Environment for Hypermedia Information System. *ACM Transactions on Information Systems*, 14, 1, January 1996, 3-29.

[16] P. Martin, L. Alpay. Conceptual Structures and Structured Documents. in *Conceptual Structures: Knowledge Representation as Interlingua*, Eklund, P.W., Ellis, G., Mann, G., edts, *Lecture Notes in AI*, #1115, 145-159.

[17] M.L. Mugnier, M. Chein. Représenter des connaissances et raisonner avec des graphes. *Revue d'Intelligence Artificielle*, 10, 1, 7-56.

[18] S.H. Myaeng, C. Khoo. Linguistic Processing of Text for a Large-Scale Conceptual Information Retrieval System. in ICCS'94 ?

[19] D. Puget. Aspects sémantiques dans les systèmes de recherche d'informations. Thèse, Université de Toulouse, 1993.

[20] E. Riloff, L. Hollaar. Text databases and Information Retrieval. *ACM Computing Surveys*, Vol.28, No. 1, 133-35.

[21] E. Salvat, M.L. Mugnier. Sound and Complete Forward and Backward Chainings of Graph Rules. in *Conceptual Structures: Knowledge Representation as Interlingua*, Eklund, P.W., Ellis, G., Mann, G., edts, *Lecture Notes in AI*, #1115, 248-262.

[22] J. Sowa. *Conceptual Structures: Information Processing in Mind and Machine*, Addison Wesley.

[23] P. Zweigenbaum, editor. MENELAS: an Experiment in the Processing of Natural Language Medical Reports. Final Report. AIM Project # A2023.

PORT: A Testbed Paradigm for Knowledge Processing in the Humanities

Mary A. Keeler[1], Leroy F. Searle[1], and Christian Kloesel[2]

[1] Center for the Humanities, University of Washington
[2] Indiana University-Purdue University at Indianapolis

Abstract. The Peirce On-line Resource Testbed (PORT) is, materially, a digital resource of primary data: C.S. Peirce's manuscripts, archived in the Houghton Library at Harvard, which have remained largely and essentially inaccessible for 80 years. Building upon this "raw data" archive (as digitized page images, transcribed text, and indexed links), we have the opportunity to increase the contributions of scholars and researchers in a networked, continuing resource development testbed. Conceptually, Peirce's pragmatic philosophy encourages us to treat this form of communication as a continuing argument, with premises, conclusions, and an account of the interpretational procedure by which the result is reached from the evidence. But Peirce's pragmatism also cautions that judgment should proceed heuristically–not algorithmically, since the conceptual basis for judgment established by any group of inquirers may, at any time, be mistaken. With inquiry conceived as an ongoing, sophisticated communicational challenge, the collective "editorial role" in this procedure is to stabilize collective inquiry with respect to the manuscript evidence. This model of an on-line research testbed requires a conceptual map or meta-representation of the work of individual scholars, as well as a means for representing the possible order(s) of the primary materials. In this respect, our work with knowledge processing researchers to develop PORT as a testbed for tool development depends less on producing definitive algorithms for editorial work than on maintaining an intelligible communication pathway for a community of scholars carrying out in practice Peirce's idea of inquiry as a self-correcting and self-directing experiment.

1   Knowledge Processing Testbeds for Digital Resources in the Humanities

At the 1993 (Quebec) ICCS conference, Brian Gaines summarized the development of computer technology, from the rudimentary hardware in the 1940s and 1950s through more recent knowledge-based system advancements, and concluded that this technology is reaching a stage of maturity at which we must begin to address its cumulative impact on application tool development. According to Gaines, we have advanced beyond information processing to knowledge processing and entered a new era of innovation and development that will focus "as much on the content and intentions of computer-based activities as on the underlying technologies of hardware, software, communications and human computer interfaces" [6](39).

At the same time, Princeton's Center for Electronic Texts in the Humanities (CETH) ended its 1993 Summer Seminar with a "wishlist" for humanities computing, envisioning future development of networked digital resources in the humanities that go beyond the scholarly resources available through printed texts. This vision for computing in the humanities emphasized, first of all, the development of clearer perspectives about what we want in computer-based tools. Both hardware and software must be specifically adaptable to advanced work in the humanities, where ambiguity is not a defect but a source of insight, and where the incorporation of sophisticated visual information is not an embellishment but an essential feature of the subject matter. Secondly, the future development of digital resources will require a sophisticated and flexible infrastructure to support interdisciplinary collaboration and the sharing of research materials and results at widely distributed physical sites. We will need non-prejudicial standards for text encoding, for application and interface design, as well as new models for scholarly publishing–and the education of colleagues, deans, and senior administrators.

We concur in the opinion of Willard McCarty (founding editor of the electronic bulletin board Humanist) that the immediate need is for "focus on the actual use of electronic communication," and the development of "a vision or paradigm for its creative development" that takes into account the complexity of the relation "between human subject and technological object" [17] (50, 51). The traditional technology of the printed book is commonly taken for granted, even when it may become an impediment to research. Yuri Rubinsky (President of SoftQuad) has argued that we should "redefine the content to incorporate the most useful raw or eloquent data in a processable form to provide the information in the most useful and economical way" [24] (14). Though computing capacity between 1980 and 1994 increased a thousand-fold, according to John Seely Brown of Xerox PARC, most of us still work, learn, and publish in the same way we did in 1980. But the even more remarkable development of Internet resources since 1994 makes clear that we are at the threshold where (in Brown's words) "personal computing becomes social computing" (Comments made at the Coalition for Networked Information task force meeting, 1994). John Gage (Sun Microsystems) made a similar point (at the same meeting) in urging that we "lift the level of our attention from the mechanism to human interaction."

Electronic publishing must do what print publishing cannot do, by developing the inherent capabilities of the medium–but without compromising editorial and scholarly standards. As Gaines has pointed out, "electronic publications will have to support mechanisms for refereeing, authenticating and time-stamping publications so that quality and priority of publication may be evaluated in ways that can be trusted, and editing of versions of electronic documents can be detected"[6] (49). As traditional publishing increases in cost and as institutions and agencies reduce funding, the preparation of high-quality journals, books, and (particularly) large scale print editions for research and education becomes more difficult to accomplish. Intellectually valuable source material in large manuscript archives becomes more fragile in time, further limiting primary research. (For example, the curator of the Peirce collection in the Houghton Library at Harvard has determined that the papers will survive no more than fifty more years–if no one touches them.) As these conditions worsen, interpretive work based on primary source material suffers, and subsequent inquiry and intellectual growth are the ultimate casualties.

The Lederberg/Uncapher conception of "collaboratory" development in the sciences (such as for mapping the human genome) suggests that the method they call "testbeds" is readily adaptable to many fields in the humanities and social sciences, to create a reliable but flexible observational context in actual working situations. The testbed model could be the means to address the development of computational tools, during the advancement of specific scholarly projects, taking into account the self-awareness of human intellect (along with the desire of humans not to be simply observed but to be involved in creating better ways of doing things). Collaboration between knowledge processing professionals and scholars working with inaccessible and highly complex documentary materials can clarify precisely what tools are suitable to those materials without neglecting the human dimension of software interface design [7]. Those who are observed also observe themselves, as they participate in refining and validating both the tools that are supposed to augment their work and their own habits of behavior– in an interactive mode of development. The testbed method can integrate the subjective and objective conditions of tool development so that those for whom tools are created have more critical control in the process of creation, and can directly monitor possibilities that emerge as research continues.

Knowledge science research has been working to create a conceptual basis for such integration by bringing together theory, features, and experience from different disciplines as a foundation or framework for application development. Many of the presuppositions of artificial intelligence research have been challenged in this effort, particularly its focus on the individual brain rather than the social context in which human intelligence operates. Supporting rather than replicating human knowledge processes has become the objective of knowledge science, based upon the one well-established result of artificial intelligence research: that humans are very good at analogic reasoning while computers are very good at keeping track of details. A more adequate conception of the scope and limits of AI confirms that, within the confines of what is possible, AI can provide methods and tools of immense benefit to all mankind [5](303). Institutions such as the Knowledge Science Institute (KSI) at the University of Calgary, established by Brian Gaines in 1985, exemplify growing efforts to monitor and encourage "new forms of scholarly cooperation mediated through electronic networks." By focusing on modeling the "structure and function of knowledge processes in society" and "the infrastructure of information technology," knowledge processing tools derived from AI research can have a more general social impact [6] (36).

Gaines's research convinces him that Peirce's theoretical framework (semiotic and pragmatism) can provide "secure foundations for significant applications" of mature computer network technology. As the basis for product development in the 1990s, when the rapid decline in the cost of computing and networking is changing the structure of these industries, this foundation can give sufficient support for the growth of user communities, particularly as "the internet itself has evolved as a public domain activity"[6] (42).

The challenge for knowledge science is to determine–in specific application contexts and communities of computer users–which tasks in knowledge creation require human intelligence, and how much of the burden is more appropriate to the quite different capabilities of computers. Testbeds for designing, building, and testing systems in partnerships between technologists and content experts, may be the method required in the new era of knowledge processing tool development and knowledge representation research, to address their "cumulative impact." Peirce's semiotic theory provides an appropriately comprehensive conceptual framework, and the testbed method embodies the pragmatism implied by his theory (see accounts in [10], [11]).

2   The Testbed Method: Peirce's Pragmatism in Operation

The leading idea for research testbeds is that the conduct of research should incorporate the refinement and modification of the research program itself, following what W. Edwards Deming calls the mode of continuous quality improvement. In the testbed, all salient features of a system are possible focal points for learning: the development of new hardware, new software, modified conceptions of "content," and numerous desiderata for the actual users, affecting interface design, institutional infrastructure, and communication throughout a dispersed "telecommunity"[9]. The "testbed" coordinates research with common global objectives, modified by feedback from specific sites, to reconnect interface and system design to users in their actual context of operation [8].

The testbed method is not just another research strategy, but part of a shift in conceptual perspective. Jon Barwise, for example, has identified the following five interrelated assumptions that would be directly affected by a testbed model:

1. Research in the humanities, mathematics, and non-lab sciences is carried out by individuals working more or less in isolation.
2. Research developments are best communicated in medium to large sized chunks, in articles and books.
3. The contribution of any given individual to the advancement of knowledge is easily determined and rewarded.
4. The best way for novices to learn is to have an expert present the information to them in written form.
5. Colleges and universities are physically located at some one geographical place. [9] (18)

In the particular example of testbed method at issue here, there are three concurrent problems. The first is the design and implementation of workstations and software, with particular attention to the interface required for individual users to act, in practice, as communicating users. The second is the institutional environment in which research is carried out; and the third is the conceptual description and representation of the complex tasks the research itself requires. A constant issue for humanists is the transition from studying the archived object (increasingly, not available for physical inspection or handling) to studying a digital representation or reproduction of it, which requires a degree of resolution, scaling, and sheer computing power that has, until recently, been prohibitively expensive and technically intimidating. A digitized image of a single manuscript page from the Peirce papers, for example, can take up to twenty-one megabytes of disk storage, imposes serious workstation display requirements and raises obvious questions about the use of network bandwidth. But the only practical design approach in the long-run is to specify first what working scholars actually need– even if it seems to run ahead of current technology or present budgets. Testbed experiments must aim for the highest possible quality in image capture and reproduction from the start, as a fundamental requirement of the work itself. On the basis of recent experience, no one can know in advance the limits of technological possibility, but it is certain that if the actual requirements of the work are not made explicit, advances in network and imaging technology become practically irrelevant in any case.

The second issue, institutional infrastructure, is the undramatic but essential core of the testbed method. All of the five assumptions identified by Barwise come into play here, since the conditions under which scholars can actually develop new approaches and employ new technology intelligently are distributed through subtle networks of support, ranging from access to technical resources (including expertise, consulting, and collegial advice) to the support of the Provost or Dean or Department Chair. Nothing here can be taken for granted. For a testbed approach to serve a network of communicating users, we must first create the "network," (not just in the physical or electronic sense) as a feature of a complex intellectual-industrial culture in which modern universities are already situated. Work of this kind cannot be accomplished by individuals, working alone, and research developments may be communicated, in some cases, without publication in the traditional sense at all, through the sharing of software, processing strategies, exemplary data structures and so on. The way new members of a group are brought in (and brought up) is very unlikely to be by way of pedagogical relics such as the graduate seminar since the only form of effective learning is by actual, concrete participation in a long-term project. Finally, the success of a testbed project at a single institution, while it must have its own local support infrastructure to survive, must be conceived from the outset as only one node in a dispersed network of other projects, at other sites, all doing work of a similar general kind.

The third ingredient of the testbed method, as we conceive it, is the conceptual description and representation that such collaborative research requires. Peirce's theoretical view of knowledge representation (in semiotic terms) conceives knowing as the provisional result of continuing hypothetical inference, in which meaning is virtual, not merely factual. Any conventional system we create tends to become algorithmic in our habitual application of it, allowing us to believe we can do only what that system can do. Peirce's pragmatic method, implied by his theory, reminds us to maintain a provisional view of our conventions by self-critically examining the actual and possible outcomes of our habitual behavior, using whatever means we can create to do so: observational instruments, modes of expressing and comparing results of these observations, and augmentation of these techniques and powers through the invention of communication media for that purpose.

The knowledge processing problem in such a project must start from a discrimination of semantic contexts, so as not to suppose that the content of Peirce's thought presents the same problem as understanding Peirce's inordinately complicated manuscripts. In the absence of the usual published record, produced during an author's lifetime, the manuscript record, even in disarray, is the primary ground for verifying "Peirce's thought"–not the cumulative decisions of his twentieth century editors concerning what to publish, and in what contextual order. The immediate task is to develop a tractable conceptual representation of the manuscript pages, to model the documentary evidence for CG processing. Following Jay Zeman's metaphor (derived from Peirce himself) that the Graphs "are to be considered 'maps'; they are not pictures of facts–for who ever saw a fact that looked like an existential [or Conceptual] graph–but they are supposed to indicate continuity where there is continuity, and represent discontinuity where there is discontinuity"[29] (24). To put the matter simply, what scholars need now is less a "map" of Peirce's thought than a conceptual map of his manuscripts.

3 The Peirce On-line Resource Testbed and Knowledge Processing

Considering the nature and condition of his manuscripts (which we describe below) it seems entirely fitting that Peirce's corpus had to wait for a publication medium capable of capturing its nature and purpose. Now, after 500 years of the book and printed page determining the conduct and character of inquiry, we are barely beginning to realize just how much (as Marshall McLuhan said) "the medium has been the message." Without the limitations of the paper delivery medium, we must now discover the new capabilities of intellectual resources in digital media-integrated modes of representation under powerful user control (as interactive multimedia of sound, text, and image), affording more freedom of expression but requiring more capabilities for critical management.

Building a networked community based on Peirce's work as a digital resource begins with what we consider the primary data: his manuscripts. Each page of Peirce's manuscripts must be digitized (as raster image, as character-based transcript, and as index record) to be retrieved and viewed or linked to any other page, for the purpose of searching and sorting the collection into compositional order and for selecting manuscripts pages to order in other ways (such as reading or topical orders) for specific purposes, as they arise. The interface must be designed to incorporate any management, retrieval, manipulation, and encoding capabilities determined appropriate in the testbed's operation, including systems of cross-referencing and specialized transcription markup. Network access and communication then make it possible to increase the contributions of scholars to the editorial process which, in Peirce's semiotic terms, will require a regularizing procedure to bring their many views together on some (tentative) basis. In some cases, it will be clear early in the process that a particular body of manuscript material should be developed with the end of producing an authoritative version for conventional publication. In other cases, however, the dissemination of Peirce's work may take other forms, as illustrations of Peirce's mode of authorial work and reasoning, or in on-line or distributable resources for scholars that might not require conventional publication in book form.

Since "publication" is fundamentally a form of scholarly communication, the validation represented by the printing of a physical artifact (the book or article) is an outcome of a process whose integrity is assured; but in the arena of networked digital "publication," the process of validation differs. It is the entire process of scholarly communication that must be treated as an extended argument and an experiment in semiotic terms, with premises (assumptions about the evidence), interpretational procedures or rules of inference applicable to the evidence, and conclusions, as the results of reasoning about or interpreting the evidence. In preparing of material for communication, the first level of validation is tentative acceptance of the material as experienced in and through the communication. For scholars who have never seen one of Peirce's manuscripts and know his work only from the available print record, the entire context of discourse is constrained in ways that the participants may not even recognize. Similarly, the validation that occurs through communication itself constitutes a kind of collaboration even when the contributors to specific discourse may not realize that they are collaborating. In the PORT project, creating effective access to the "primary data of experience" is fundamental for securing reliability in interpretation and validity in collective judgments. Our way of proceeding–our method of inquiry– must entail checking the veracity of our representations with respect to their referents [14] (132). The authentication of results and procedures depends on keeping a clear pathway back to the primary data, and a means to represent in logical, semiotic terms collective decisions made about it, including evaluation of the functional role of technology in validating research and learning practices.

To augment the validation process (in Peirce's sense, as a function of on-going inquiry), the PORT project addresses some of the most basic problems in the development of academic resources: preparing the archival source material, which must serve reliably as the "raw data" to be developed for many resource purposes in research and academic pursuits. Our initial "Transcriber Testbed" focuses on the conversion of selected manuscript pages from Peirce's work on graphical logic (Existential Graphs), with two primary purposes: (1) to ascertain the requisite digital image quality of the source material for effective transcription of the manuscripts by means of electronic display and (2) to determine what augmentation can be developed for the transcriber in the form of specialized application tools.

Peter Robinson's instructive Oxford Humanities Communication reports explain the complexities of this process and raise issues on which technical decisions can be made only in the working transcriber's context, using the testbed method (Robinson 1993). Questions of color quality and resolution need to be answered most carefully in the case of an author like Peirce who made extensive use of graphics, strike-outs and over-writes, and colored pen and pencil in his writings. Several manuscripts with these features selected from the Harvard archive, which have been digitally imaged using a high-resolution CCD camera, will be intensively tested. Only from the point of view of a testbed transcriber who actually uses these images, can we determine what are the image quality requirements (what compression methods produce the best interaction performance and legibility results). No requirement in digital archive development can be considered in isolation or as primary, determining all others. Factors such as image resolution, color accuracy, and sound fidelity for feature detection; file compression for speed of transmission; and storage and management for safe and effective access are all complexly interrelated–and the difficulty of establishing these for any particular collection and user community is further complicated by the fast pace of technology advancement. For the advancement to be genuinely beneficial, however, there must be ongoing communication between technology users and developers in which the digital quality required for any particular resource material must be judged by those who study the material, in the actual context of research. High-quality encoded text is especially difficult to judge and must be developed collaboratively among resource users in communities of common application interest.

4   The Nature and Condition of Peirce's Corpus: Obstacles to Archive Access

The notion of a resource or edition capable of representing the style of Peirce's work and his "intentions" (as critical editions claim to do), defies the limitations of the print medium and will significantly challenge the potential of digital systems in the realm of creating digital editions as resources. Although all of Peirce's scientific writings have been published in print, his most valuable philosophical works–at least 40,000 pages, or nearly half the Harvard collection (including the Existential Graphs)–remain in manuscript (less than 10which has ever been printed), with only monochrome microfilm as a "convenient" means of access. Perhaps the most serious obstacle to print publication is the highly graphical nature of these manuscripts: text enclosed in graphical figures, graphics embedded in text, text contoured around graphics, and entire pages of graphics without any text, with color often used in text and figures to make important distinctions and to key text elements to related graphical elements. Of all his invented symbols, especially the Existential Graphs are pervaded with colors and tinctures.

Editing and publishing only his previously printed writings has produced an entirely misrepresentative sample of about one sixth of his entire work. As Peirce indicates in his unsuccessful 1902 grant application to the Carnegie Institution (to enable him to write 36 memoirs detailing his system of logic): "Those things I have published have been slight and fragmentary, and have dealt little with the more important of my results" (L75: E 33 and 37-38)–and represent only "scattered outcroppings of a rich vein" of unpublished fragments, which "no human being could ever put together" (MS 302). More serious than even he estimated are the difficulties in reassembling, reorganizing, and dating his papers, which arrived at Harvard in a chaotic state of disarray in 1915 and have never regained coherent order. Several thousand pages in the Harvard collection remain unplaced within their original manuscripts–and will so remain until each page has been carefully scrutinized. Unfortunately, certain gaps will always remain, because pages and papers have been lost or stolen over the years (during paper shortages of wartime, some papers were actually given away).

Although today the collection is sorted into folders, pages in one folder are often found to belong to manuscripts in another. Richard Robin's Annotated Catalogue of the Papers of Charles S. Peirce (the most valuable and comprehensive view of the disordered manuscripts and their microfilm edition) conveys a misleading sense of order in a topical arrangement that has proven counterproductive in dealing with a polymath such as Peirce. MS 1043, a five-page untitled note on chemical valency, for example, is listed in the "Chemistry" section and said to belong with MSS 1041 and 1042, both entitled "Valency"; but careful editorial work has shown that it completes a now reassembled excellent twenty-five page version of the 1906 essay on "The Basis of Pragmaticism," consisting of one loose title page from MS 280 (in the "Pragmatism" section), nineteen pages from MS 908 (in the "Metaphysics" section), and the note from MS 1043 (whose first page has Peirce's inscription, in red ink, "Note to be printed in small type at the end of the article"). Peirce editions editors can recount numerous instances of such disarray and misappellation, finding manuscripts under "Mathematics" (or "Pragmatism" or "Metaphysics") that could just as well belong under "Logic"– and visa versa–others under "Astronomy" that could belong under "Physics," still others under "Physics" that equally concern "Psychology."

Foremost among his personal works (not intended for publication) is the so-called Logic Notebook, his journal of logical analyses, casually inscribed over the years between 12 November 1865 and 1 November 1909. This manuscript may be the single most fruitful and important of all of Peirce's archived works, of which Peirce wrote on 23 March 1867: "I cannot explain the deep emotion with which I open this book again. Here I write but never after read what I have written for what I write is done in the process of forming a conception." The notebook is devoted to discussions and developments of formal logic, categoriology, graphical logic, semiotic, mathematics, and metaphysics–but its highly graphical nature poses extreme typesetting and printing difficulties.

More generally, his writings have presented serious obstacles for traditional publication methods. Rather than writing a series of (collatable) drafts that are subsumed in a final version, Peirce's writings generally proceed along a path of thought, return to a "crossroads," and then "blaze new trails"–with the roads not taken becoming at times important stretches of a subsequent exploration. As a result, nearly every manuscript has its own "textual authority" and the copy text is easily established; but selecting a version to print must be a highly arbitrary decision, since each may have equal significance in the formation of his thought. Examples of this difficulty can be seen in the four chapters that make up the 1901-1902 "Minute Logic" (MSS 425–434), an incomplete logic book of about 1,000 pages. Many pages exhibit Peirce's trail-blazing style, which one persevering scholar worked for over a year to map (in several carefully constructed stemmata), revealing complex and organic convolutions. Even if all 2,300 pages of this book (in the main body as well as in the editorial and textual apparatus) were published, the printed page and bound book would keep the reader from viewing the character of its many evolutions.

The traditional print medium makes the notion of producing an authoritative text that represents an author's "final intention" seem reasonable. But Peirce's very concept of pragmatism serves as an example by which to show what happens to concepts in communication.' From its early formulation, even those who greatly respected Peirce's intellect (such as William James and John Dewey) began to interpret his conception of pragmatism in ways that served their own (more limited) purposes. Today, by dictionary definition and philosophical tradition, pragmatism is hardly recognizable as what Peirce's manuscripts indicate it meant to him. Because his theory is so comprehensive, pragmatism's commonly accepted meaning is not contrary to Peirce's use of the term, but it falls far short of what it might mean to us if we had more evidence of the actual process by which his idea was articulated and developed.

As Peirce himself remarked, "[t]he data from which inference sets out and upon which all reasoning depends are the perceptual facts, which are the intellect's fallible record of the percepts," or "evidence of the senses" (2.143) but in the case of his own papers, traditional premises of editorial interpretation have compromised or occluded the "evidence" by establishing a classificatory order that does not reflect the context in which Peirce's writing was done. The primary conceptual problem is to establish a data system by which to track the relations between pages, with co-reference to diverse publications and prior ordering schemes in print and archive collections. Most significantly, Peirce's own account of the production of meaning makes clear that the notion of being able, once and for all, to settle an author's "intention" fails to capture the dynamic quality of all authentic thinking and inquiry. As Peirce describes semiotic evolution, "[E]very symbol is a living thing, . . . its meaning inevitably grows, incorporates new elements and throws off old ones"[18] (2.222). Symbolic meaning has continued to grow since Peirce's writing: "How much more the word electricity means now than it did in the days of Franklin; how much more the term planet means now than it did in the time [of] Hipparchus. These words have acquired information"[18] (7.587). While we may be able to construct a sense of what terms and concepts might have meant at the moment of their introduction, it is at least as important to construct a system of scholarly access and communication that is, by design, open to this process of semiotic evolution–particularly when Peirce's papers provide superb exemplification.

In his 1903 lecture on "Multitude and Continuity" given before the mathematics faculty at Harvard, Peirce says that "No sooner is a paper of mine worked up to a finish and printed than I immediately begin to take a critical attitude toward it and go to work to raise all the objections to it big and little that I can"(MS 316a). "He says repeatedly that he struggled for decades with the veracity of his semiotic; for nearly thirty years with the truth of pragmatism; and for over twenty years with his system of graphical logic that, in January 1897, he came to call Existential Graphs (though he had it in mind for a dozen years)"[12] (303). In one of the 1903 Pragmatism Lectures, referring to his realization that his graphical logic would provide a proof of pragmatism, Peirce recounts:

> When I first got the general algebra of logic into smooth running order [in 1884], by a method that has lain nearly twenty years in manuscript and which I have lately concluded is so impossible to get printed that it had better be burned,–when I first found myself in possession of this machinery I promised myself that I should see the whole working of the mathematical reason unveiled directly. (MS 303)

Peirce's manuscripts record many comments indicating his concern to develop his philosophical ideas further, including many where he poses as his own sharpest critic. For Peirce, meaning was "a continuous process, which we determine, with arbitrary precision (depending on 'different circumstances and desires'), in communities of inquiry." From Peirce's famous slogan–"Do not block the way of inquiry"[18](1.135)–it follows that "no belief is ever ultimate, and no one ever gets the last word" [25](562).

5   Critical Control in Knowledge Processing: A Paradigm to Fulfill Peirce's Dream

Traditionally established structures for organization and control have promoted the notion that knowledge (as the "last word" for the foreseeable future) is located somewhere, such as in critical editions. Now, the development of pervasive high-speed networks for research and learning will require new means of effective critical control for scholarly work, at its foundations. Critical editions have been extremely costly to produce in print form, with critical editorial work painfully slow. Not only does this work need the efficiency to be gained from the new medium, but scholarly work has begun to require the electronic enhancement of such products, in the form of computer-based resources and tools. At a minimum, these tools will automate (make more efficient to use) such features as tables of contents, indices, and bibliographies that are essential to an edition's effective critical purpose. New means of finding and keeping track of information will be even more crucial to research conducted in electronic network media (see [20](25-27) on map makers, filters, and ferrets); but just as in the transition to books from manuscripts, when printers were merely trying to automate the process of manuscript production, we are currently trying to automate the book and have barely begun to understand how the electronic form makes a qualitative and quantitative difference [4].

During the transition from books to electronic media (as we lose the printed page and with it the binding, the bookshelf, and the floors of indexed storage) we will need what Patricia Battin (former Director of the Commission on Preservation and Access) calls a "new set of lifelines" [1] (172). The operations that serve investigation searching, sorting, selecting, and keeping track of where related ideas seem to be going or where they came from–must be provided in the new environment of online network resources. Such functions (already essential in the responsible scholarship of an individual investigator) will take on dramatically new "overhead" dimensions in the context of a telecommunity, calling for largely unprecedented facilities for investigation and communication (or critical control, in terms of Peirce's pragmatism). Cable, cellular, and Internet II development of network capacity and speed–making text, image, audio, and video mergeable into one user controlled digital medium–should continue to improve the capability of researchers and students to meet and work together, depending on how well their media needs can be identified and corresponding technology requirements can be accurately specified.

This new era of digital technology innovation and development must clearly involve knowledge representation and processing (in hypermedia, with intelligent agents), but also our judicious application of this augmentation in our academic discourse on worldwide communication networks. We must become more closely engaged in determining what we want automation technology to do for us. Although much of the technical innovation for so-called "hyper-" or "electronic editions" already exists, critical functions remain to be carefully specified and responded to in digital systems development, if we are to use these resources effectively in scholarship and continue their development in virtual communities of inquiry. Humanists have only begun to experiment with and speculate about the nature of such resources. Charles Faulhaber (Director of the Bancroft Library at UC Berkeley) outlines three aspects of electronic or hypereditions (content, creation, and use), with content as a "top layer" in the "critical text itself, but underlying that . . . are the paleographical transcriptions of all witnesses, in turn underlain by the digitized facsimiles of the MSS themselves" [4] (123).

These possibilities reflect a significant extension of the notion of a critical edition, and move in the direction of what Jerome McGann (Director of the Rossetti Hypermedia Archive) has described as a "critical archive" rather than a "critical edition," going beyond the "exigencies of the book form" to link dispersed subarchives into an extensible scholarly resource, available on the Internet. And Peter Robinson (creator of the electronic Canterbury Tales for Cambridge U. Press) describes, "a resource bank . . . an accumulation of materials without the privileging of any one text at all, . . . [which stores] data in the most complete forms possible (both as logically marked-up etext and as high-resolution digitized images) [and that] must be able to accommodate the collation of pictures and the parts of pictures with each other as well as with all kinds of purely textual materials." He summarizes, "The goal . . . is not to produce a critical text through mechanistic means but rather to present the editor with as much usable evidence as possible, allowing human judgment to operate as efficiently as possible." The hyperedition "must provide facilities to allow users to annotate the text by attaching commentaries to it, which might form the basis for an article or a class discussion, or merely contain a query concerning a puzzling feature of the text. In turn it must be possible to filter out these commentaries on the basis of the author, their date of composition, or their subject matter"[23]. Faulhaber stresses the changes in our intellectual conduct that such hypereditions will require, when scholars "must rely more than ever on cooperation with their peers and with specialists from other disciplines, particularly computing and information science" and that "[f]ar more than in the traditional print environment, advances in electronic scholarship will depend on enlightened collaboration among specialists in widely separated fields"[4] (128).

We do not imagine that electronic editions will supplant print, but they will change profoundly the function and the relative status of the printed edition. Procedures of authentication for reference and archival copies will need to be established as workable standards in digital archives, along with protocols for selecting which particular resources will be incorporated into printed books. Such specific questions as how the authority of particular versions of texts would be established to authenticate reference copies or maintain critical editions need to be carefully studied, without a traditional regime and regiment restricting the development of new methods and means. Toward this investigation, we find many promising tools under development in the ICCS community, which might be integrated and tested in the PORT context for their "cumulative impact." The most obvious applications are:

- concept-based search and retrieval,
- knowledge acquisition,
- interlingua (for both natural language translation and system integration),
- database, document, and knowledge base management,
- communication services support, and
- discourse management.

In 1996, Gerard Ellis listed reasons for the attraction of Conceptual Graphs, in a thriving and diverse community of philosophers, linguists, mathematicians, logicians, computer scientists and others: "interlingua, standards, complexity and expressiveness, speed, human-computer interaction, and cooperation and diversity of the community." He concluded that "conceptual structures have advantages over other knowledge representations because they emphasize natural language as their design motivation" while promoting "a wider human experience than traditional knowledge representation which is too reliant on symbolic logic"[3].

Peirce would surely have agreed, but a fuller exploration of the semiotic and pragmatic benefit of his graphic logic will be essential in the creation of testbeds for the humanities. The Existential Graphs were developed specifically not as a calculus (algorithmic problem-solving system) but as a "topology of logic," or an observational instrument, which Peirce even considered to be a "proof of pragmatism"[29] (CSP-MS 303). These "moving pictures of thought" were intended to be a communication tool (CSP-MS 291). To fulfill his dream (in creating effective human-to-human partnerships that, after all, are the ultimate purpose of human-computer communication), an essential complement to Conceptual Graphs will be the multicontext and conceptual data system development brought to the Conceptual Structures community by Rudolf Wille and Gerd Stumme, as a form of (analogic) "conceptual landscape" (or conceptual mapping) tool to augment coordinated network communication.

If we hope to overcome the de-humanizing, technology-driven pattern of this century, we must have conceptual grasp of the operational potential and the intellectual purpose of a new medium, rather than continue to be led by whatever the established technology and its poorly defined market forces can provide. Considering the current state (as Gaines describes it) of computer and network technology development, we have established the (humanities-based) Peirce On-line Resource Testbed (PORT) project (with participants from the Conceptual Structures community) as a means for developing and integrating knowledge processing tools and testing relevant technology in the actual context of making Peirce's papers available. We cannot think of a more fitting subject for such a testbed experiment.

Note: *All "MS" and "L" references are to Peirce's unpublished manuscripts and letters.*

References

1. Battin, Patricia. "The Library: Center of the Restructured University." *College & Research Libraries* 45 (1984): 170- 76.
2. Deming, W. Edwards. *Out of Crisis*. Cambridge, MA: MIT Center for Advanced Engineering Study, 1982.
3. Ellis, Gerard, Peter Eklund, and Graham Mann. *Conceptual Structures: Knowledge Representation as Interlingua, Auxiliary Proceedings*. Berlin: Springer Verlag, 1996).
4. Faulhaber, Charles B. "Textual Criticism in the 21st Century." *Romance Philology* 45 (August 1991): 123-48.
5. Fetzer, James H. *"Artificial Intelligence: Its Scope and Limits*. Dordrecht: Kluwer, 1990.
6. Gaines, Brian R. "Representation, Discourse, Logic and Truth: Situating Knowledge Technology." In *Conceptual Graphs for Knowledge Representation*, eds. Guy W. Mineau, Bernard Moulin, and John F. Sowa. (Berlin: Springer-Verlag, 1993), pp. 36-63.
7. Heckel, Paul. *The Elements of Friendly Software Design* (2nd edition). San Francisco: Sybex, 1991.
8. Keeler, Mary A. and Susan M. Denning. "The Challenge of Interface Design for Communication Theory." *Interacting with Computers: an International Journal of Human-Computer Interaction* 3:3 (1991): 283-301.
9. Keeler, Mary A. and Christian Kloesel (eds.). *Casting the Net: Towards a Model for Communication and Collaboration on the Electronic Network*. Report of an In-

vitational Symposium at George Washington University, 4-5 June 1992 (Supported by the NSF, Organized by the Electronic Peirce Consortium). 15 October 1992.

10. Keeler, Mary A. "The Philosophical Context of Peirce's Existential Graphs." *Third International Conference on Conceptual Structures: Applications, Implementation and Theory, Proceeding Supplement.* Department of Computer Science, University of California Santa Cruz (1995): 94-107.

11. Keeler, Mary A. "Communication and Conceptual Structuring." *Conceptual Structures: Knowledge Representation as Interlingua (Auxiliary Proceedings).* University of New South Wales, Sydney, Australia (1996): 150-164.

12. Keeler, Mary A. and Christian Kloesel. "Communication, Semiotic Continuity, and the Margins of the Peircean Text," in David Greetham, ed., *The Margins of the Text.* (Ann Arbor: University of Michigan Press, 1997), pp. 269-322.

13. Lederberg, Joshua, and Keith Uncapher. *Towards a National Collaboratory: [NSF] Report of an Invitational Workshop.* Rockefeller University, New York City, 13-15 March 1989.

14. Lukose, Dickson, (et al.). "Conceptual Structures for Knowledge Engineering and Knowledge Modelling." *Third International Conference on Conceptual Structures: Applications, Implementation and Theory, Proceeding Supplement.* Department of Computer Science, University of California Santa Cruz (1995): 126-137.

15. McLuhan, Marshall. *Understanding Media: The Extensions of Man.* New York: McGraw-Hill, 1964.

16. McGann, Jerome. "The Rationale of Hypertext," a World Wide Web document at URL http://Jefferson.village.virginia.edu/public/jjm2f/rationale. html . (See also "The Rossetti Archive and Image-Based Electronic Editing," by substituting "imagebase" for "rationale" in the URL above).

17. McCarty, Willard. "Heraclitus' River: The Humanist and Electronic Communication." In *The New Medium: Proceedings of the ALLC/ACH Annual Conference.* (Siegen, Germany, June 1990), pp. 4951.

18. Peirce, Charles S. *Collected Papers of Charles Sanders Peirce.* Eight volumes. Edited by Arthur W. Burks, Charles Hartshorne, and Paul Weiss. Cambridge: Harvard University Press, 1931-58.

19. Peirce, Charles S. *Writings of Charles S. Peirce: A Chronological Edition.* 5 volumes to date. Edited by Christian J.W. Kloesel et al. Bloomington: Indiana University Press, 1982-93.

20. Rawlins, Gregory J. E. *The New Publishing: Technology's Impact on the Publishing Industry over the Next Decade.* Technical Report no. 340, Computer Science Department, Indiana University. November 1991.

21. Robin, Richard S. *Annotated Catalogue of the Papers of Charles S. Peirce.* Amherst: University of Massachusetts Press, 1967.

22. Robinson, Peter. *The Digitization of Primary Textual Sources.* Oxford: Office for Humanities Communication, 1993. (OHC Publications Number 4).

23. Robinson, Peter. *The Transcription of Primary Textual Sources.* Oxford: Office for Humanities Communication, 1993. (OHC Publications Number 6).

24. Rubinsky, Yuri. "Scholarly Publishing on the Networks," *CETH Newsletter,* September 1993.

25. Searle, Leroy F. "Charles Sanders Peirce." In Michael Groden and Martin Kreiswirth (eds.), *The Johns Hopkins Guide to Literary Theory and Criticism.* (Baltimore and London: Johns Hopkins University Press, 1994): 558-62.

26. Stumme, Gerd. "Knowledge Acquisition by Distributive Concept Exploration." *Third International Conference on Conceptual Structures: Applications, Implemen-*

tation and Theory, Proceeding Supplement. Department of Computer Science, University of California Santa Cruz (1995): 98-111.

27. Stumme, Gerd. "Local Scaling in Conceptual Data Systems." *Proceedings of the 4th International Conference on Conceptual Structures.* Berlin: Springer Verlag, 1996. 308- 320.

28. Wille, Rudolf. "Conceptual Structures of Multicontexts." *Proceedings of the 4th International Conference on Conceptual Structures.* Berlin: Springer Verlag, 1996): 23-39.

29. Zeman, J. Jay. "The Graphical Logic of C.S. Peirce" (Diss. Chicago, 1964), p. 27.

Using Access Paths to Guide Inference with Conceptual Graphs

Peter Clark[1] and Bruce Porter[2]

[1] The Boeing Company, PO Box 3707, Seattle, WA 98124
(clarkp@redwood.rt.cs.boeing.com)
[2] University of Texas at Austin, TX 78712 (porter@cs.utexas.edu)

Abstract. Conceptual Graphs (CGs) are a natural and intuitive notation for expressing first-order logic statements. However, the task of performing inference with a large-scale CG knowledge base remains largely unexplored. Although basic inference operators are defined for CGs, few methods are available for guiding their application during automated reasoning. Given the expressive power of CGs, this can result in inference being intractable.

In this paper we show how a method used elsewhere for achieving tractability — namely the use of *access paths* — can be applied to conceptual graphs. Access paths add to CGs domain-specific information that guides inference by specifying preferred chains of subgoals for each inference goal (and hence, other chains will not be tried). This approach trades logical completeness for focussed inference, and allows incompleteness to be introduced in a controlled way (through the knowledge engineer's choice of which access paths to attach to CGs). The result of this work is an inference algorithm for CGs that significantly improves the efficiency of reasoning.

1 Introduction

Consider asking a large-scale Conceptual Graph (CG) knowledge-base the question "What is the age of the American president?" Answers to this question may be determined in many ways by applying CG inference rules. Some inference chains are more likely to be fruitful than others. For example, a fruitful chain might be: "The age of a person is probably (approximately) the age of his/her spouse", hence "Is the president married?" and "What is the age of his spouse?". A less fruitful chain might be: "The age of a person is the age he/she declared on his/her last job application", hence "What was the last job the president applied for?", hence "A person's last job probably matches his/her interests", hence "What are the president's interests?", hence "A person's interests probably match his/her friends' interests", hence "Who are the president's friends?", etc. Clearly, to answer questions in a timely fashion and to cope with the inherent intractability of inference with expressive languages such as CGs, some mechanism is needed to guide the application of the basic inference rules. Although conceptual graph research has significantly improved the efficiency of basic inference operations (eg. performing joins), there is still a need for methods that guide their application so that questions can be efficiently answered from large CG knowledge-bases.

In this paper, we apply ideas from *Access-Limited Logic* [1] to Conceptual Graphs, as a means of guiding CG inference. Access-Limited Logic specifies *access paths* that (1) relate together the concepts in a knowledge-base, and (2) constrain inference to follow only those paths when answering queries that require navigating the knowledge-base. Although this introduces logical incompleteness, it does so in a controlled way, as the knowledge engineer chooses which access paths to include in the knowledge-base. In addition, access-limited logic retains the property of "Socratic Completeness", which guarantees that no consequence of the knowledge base is inherently unreachable – there will always be some sequence of queries which can be issued to the KB allowing any logical consequence of it to be inferred. Our experience in applying this approach to a large-scale CG knowledge-base is that suitable paths can generally be determined from domain knowledge and encoded within a CG framework. The result of this approach is that complex chains of reasoning become possible with conceptual graphs because the inference process is focused in fruitful ways.

We have used the representation language described here, which adds access paths to standard CGs, to build a large knowledge-base about plant biology [2] and a smaller knowledge-base about distributed computing [3]. We have fully implemented the inference algorithm presented here, and used it in conjunction with these knowledge bases [4].

The paper is organized as follows. Section 2 outlines the CG semantics used in this work. Section 3 describes access paths and how they can be used to guide inference, and Section 4 presents an inference algorithm based on this approach. Sections 5 and 6 discuss some of the benefits and limitations of this approach.

2 Inference with Conceptual Graphs

2.1 The CG Knowledge Base

We consider a CG knowledge-base (KB) to contain two basic types of representational structures, namely *schemata* and *type definitions*, as illustrated in Figure 1. Intuitively, a schema for a concept C denotes "contingent facts" about C, namely those facts that are implied by membership in C, but which are not sufficient to conclude membership. This corresponds to uni-directional logical implication. A type definition for concept C, in contrast, denotes the "definitional properties" of C, ie. those facts that are both implied by membership in C and (together) are sufficient to conclude membership in C for some individual. A type definition corresponds to bi-directional logical implication. In terms of Description Logics (eg. [6]), type definitions express terminological (TBox) knowledge, while schemata express assertional (ABox) knowledge. In a knowledge-base, a concept will typically have both a type definition (describing its definitional properties) and a schema (describing additional implied properties). In addition, the KB contains a type lattice ('hierarchy') asserting generalization/specialization relationships among types.

2.2 Inference Rules

There are different, interchangable vocabularies with which CG inference can be described. At the most primitive level, (Sowa's CG adaptation of) Peirce's beta

Conceptual Graph:	Semantics:

schema for *concept(x)* **is**
   graph of preds using $x,y_1,..,y_n$

$\forall x \; concept(x) \rightarrow$
   $\exists y_1, ..., y_n$ *set of preds using* $x,y_1,..,y_n$

type *concept(x)* **is**
   graph of preds using $x,y_1,..,y_n$

$\forall x \; concept(x) \leftrightarrow$
   $\exists y_1, ..., y_n$ *set of preds using* $x,y_1,..,y_n$

For example:

Conceptual Graph:	Semantics:

schema for PERSON(x) **is**
   [PERSON:*x]-
     (PART)→[HEAD].

$\forall x \; person(x) \rightarrow$
   $\exists y \; part(x,y) \wedge head(y)$

type RED-WINE(x) **is**
   [WINE:*x]-
     (COLOR)→[RED].

$\forall x \; red\text{-}wine(x) \leftrightarrow$
   $\exists y \; wine(x) \wedge color(x,y) \wedge red(y)$

Fig. 1. Semantics used for CG schemata and type definitions.

rules provide a set of graph rewriting rules for CGs, equivalent in power to the rules of first-order predicate calculus [5, page 154]. These rules can be thought of as the "assembly code" for CG inference, in that they are building blocks with which other inference rules ("derived rules" [5, page 151]) can be described. For example, modus ponens is equivalent to applying a particular sequence of four alpha/beta rules. This approach is used in other CG implementations (eg. Prolog+CG [7]), and also here.

Type definitions and schemata express three types of inference rules, corresponding to the three directions of implication in Figure 1:

1. concept membership implies facts (schema expansion)
2. concept membership implies facts (type expansion)
3. facts imply concept membership (type refinement)

These three rules form the basic inference rules for our knowledge-base. They are derived rules, as they can each be implemented as a sequence of Peirce's beta rules. However, for simplicity we treat and implement them as primitives.

In graph theoretic terms, a **schema expansion** is a minimal join of a schema to a conceptual graph, and is equivalent to applying the inference rule which the schema represents. Similarly, we define a **type expansion** as a minimal join of a type definition to a concept in a conceptual graph, which is equivalent to applying the type definition's inference rule in the forward direction ("concept membership implies facts"). Finally, we define a **type refinement** as the recognition that a type's definition is satisfied by a concept in a conceptual graph (ie. the definition subsumes the graph at that node), and the resulting specialization of that node's type in the graph to be that in the type definition. A type refinement is equivalent to type contraction, but without detaching items from the CG, and corresponds to applying the type definition's inference rule in the

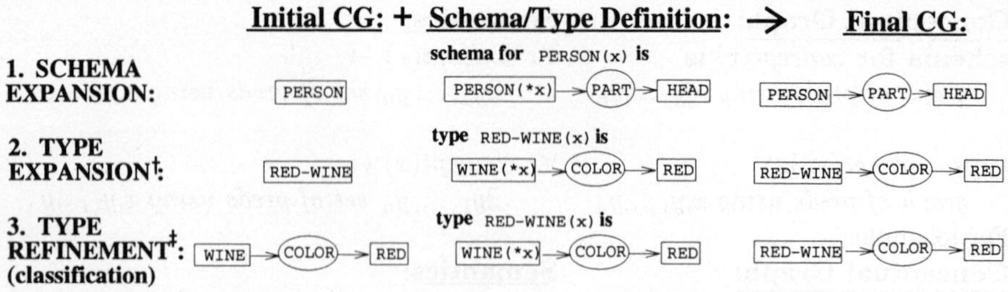

† This definition slightly deviates from that in [5], in that RED-WINE in the initial CG is *not* replaced by WINE in the final graph, but instead is retained. This enables types/schemas for RED-WINE and its generalizations to be located for future expansions.
‡ Type refinement is equivalent to type contraction [5, page 108] without detaching items from the CG.

Fig. 2. The three basic CG inference operators used.

backward direction ("facts imply concept membership"). Expansions correspond to the *elaboration* of initial knowledge, and type refinements correspond to the *classification* of concepts in a conceptual graph. These are illustrated in Figure 2.

2.3 Cardinality Assumptions

In normal CG syntax, relations are assumed to denote a many-to-many mapping unless otherwise specified. To denote functional relations (ie. one whose second argument is unique, given the first), the '@1' notation is normally used (eg. to denote that a person has just one father, the PERSON schema would include [PERSON]→(FATHER)→[MAN:@1]). Functional relations are common in knowledge-bases, and also have particular significance when performing minimal joins: minimal joins will only merge relations which are functional (ie. will only assume coreferentiality when logically implied), and hence their use is essential if a schema/type definition is intended to imply extra properties of objects in a CG (rather than assume objects in the schema are distinct from those in the CG). Thus, as a notational convenience in this paper and our implementation, we assume that relations *are* functional unless otherwise specified (using a @* annotation). As a result, minimal joins will by default merge matching relations together, unless annotated otherwise.

3 The Inference Task and the Use of Access Paths

3.1 Inference with CGs

Our goal is to use a CG knowledge base to answer queries. A query comprises a set of assertions (eg. "there exists a table") and a question ("what is it made of?"). We call the assertions the 'scenario', and represent the scenario as a CG whose nodes are all instances. The scenario is a temporary CG, built just for answering the question, and to be elaborated until it contains the answer. We can thus describe the inference problem as follows:

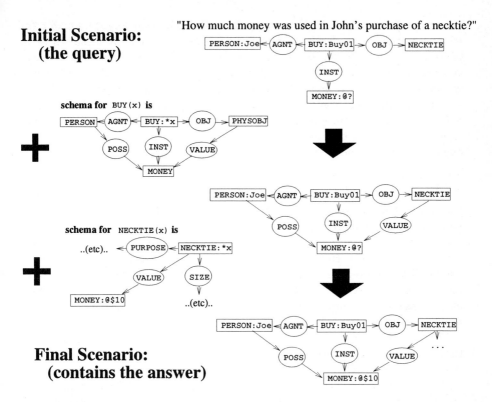

Fig. 3. Inference with schemata only: Expanding the Buy01 and NECKTIE nodes in the initial CG produces an answer to the query.

- **GIVEN:**
  - A CG KB of type definitions and schemata, and a type lattice
  - A query, comprising assertions (the initial scenario CG) and an unknown variable (the target)
- **FIND:**
  - A sequence of expansions/refinements to the initial scenario CG which results in it containing a value for the target.

3.2 The Search Problem

A simple example of this inference task is shown in Figure 3, an adaptation of [5, page 110]. (We ignore issues related to temporal reasoning, as they are not our concern here). In this example, the initial CG describes "Joe buying a necktie", and the query asks "How much money was used in the purchase?" In the scenario graph, there are three nodes that could be expanded by joining each node's schema (from the KB) to the initial graph. (We assume that schema are defined for each node.) Although in this case just two schema expansions will answer the query (as shown in Figure 3), many more could be made; most of these alternatives fail to answer the query. For example, suppose the first expansion joined the BUY(x) schema from the KB with Buy01. As this fails to answer the query, further expansions are needed, and (at least) three are possible: expand PERSON, NECKTIE, or Buy01 (using a schema from some other generalization of BUY). However, there is no information available to suggest which should be tried first: all three relate to the node of interest (MONEY) in structurally identical ways. If PERSON (say) is expanded, then again no answer is found and more concepts will be added to the initial CG, providing yet more choices for the next expansion, and so on. In general, the space of sequences of expansions and refinements is too large to search exhaustively.

3.3 Access Paths

Of course, this problem is not specific to CGs, but occurs whenever a logically complete inference procedure is applied to an expressive representation language. One solution to this problem is to significantly weaken the language's expressive power. This approach is pursued by many within the Description Logic community (eg. [6]). However, the language restrictions required to guarantee tractability are severe and can limit the language's applicability [8]. The alternative we have chosen is to restrict the inference process, thus achieving tractability at the expense of completeness. The challenge is to adopt restrictions that, in practice, have minimal effect on question-answering ability.

The inference control method we present for CGs is based on "access limitation", which is used elsewhere in the knowledge representation community [9]. Access limitation is based on the use of *access paths*, which state how a value can be computed via a chain of inferences. In the context of Conceptual Graphs, an access path describes how the value of a CG node can be determined by traversing a path in the graph. Incorporating access paths into CGs requires a small extension to the CG formalism, whereby a variable at a CG node can be replaced with a path, *at the end of which the value of that variable may be found*. To the inference engine, the path describes a sequence of subgoals that may result in a variable's value being found. As an example of access paths, consider the following schema for BUY(x):

Standard CG Schema: **Schema with access paths:**

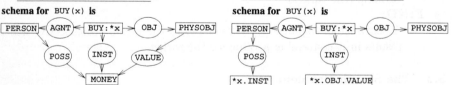

where *x.OBJ.VALUE denotes the path [*x]-(OBJ)→[*]-(VALUE)→[*], or equivalently in logic notation:

$$obj(Self, y), value(y, z)$$

(where *Self* refers to the instance of BUY under consideration). This particular path provides guidance for computing the INSTrument of the BUY. Specifically, it instructs the inference engine to first find the OBJect bought, then find its VALUE (thus, it indirectly instructs the inference engine *not* to consider (say) the PERSON node further for computing this value). In this way, paths encode a preferred sequence of computations for computing the identity of one concept from others, and this knowledge can be used by the inference engine to reduce the search to answer a query. Paths may point to concepts which themselves are described by paths, and hence complex chains of inference may be required to find the concept(s) to which a path refers.

More formally, we define an access path as a chain of binary predicates P_i linking some individual $X0$ to other individuals x_n, such that the second argument of P_i is the same as the first argument of P_{i+1}:

$$P_1(X0, x_1), P_2(x_1, x_2), ..., P_n(x_{n-1}, x_n)$$

where the x_i's are free variables. (This is a slightly more restricted definition than in [9], but one suitable for conceptual graphs.) An access path denotes the set S of values for x_n (the last variable in the path) for which there exists at least one value for all the other variables in the path, ie.

$$\forall x_n (\ x_n \in S \leftrightarrow \exists x_1, ..., x_{n-1}\ P_1(X0, x_1) \wedge ... \wedge P_n(x_{n-1}, x_n)\ )$$

For example, the access path $parent(John, x), sister(x, y)$ denotes "John's parents' sisters". In terms of conceptual graphs, an access path corresponds to the set of paths that start at node $X0$ and traverse arcs labeled $P_1, ..., P_n$. For legibility, we write paths using a 'dot' notation of the form $X0.P_1.\ ...\ .P_n$, for example $parent(John, x), sister(x, y)$ would be written `John.parent.sister`.

A chain of predicates like this is very common in knowledge-based systems (and object-oriented programming) and has many different names, including "attribute paths" in CLASSIC [6, page 425], "relation compositions" in LOOM [10], "role chains" in KRYPTON [11, page 421], and "chains" in [12]. The difference here is that chains are used not only to express coreferentiality of concepts, but also to guide inference: Paths do not just denote the concepts they point to, they also give *a sequence of steps by which the identity of those concepts can be computed*, as shown in the next section. It is thus important to retain paths within the CG knowledge-base, rather than "compile them out" into shared variables (as is done, for example, in the language LIFE [13]). Replacing paths with shared variables loses the control information that paths contain.

4 Inference with Access Paths

4.1 An Algorithm for Inference

Inference is performed when a query is issued to the KB, either from a user or an application system. As described in Section 3.1, a query consists of a set of assertions, plus a target variable of interest related to those assertions. In the CG framework, these assertions are provided as an initial 'scenario' CG, and the target is specified as an access path from a node in that graph to the variable of interest. For example, the query "What is needed to assemble a table?" is specified as a scenario, in which there is an assembly (`Assemble01`, say) of a table, and a path `Assemble01.INST` denoting the variable of interest (namely the instrument used in that assembly):

Initial Scenario CG: **Path to the variable of interest:**

                                      `Assemble01.INST`

The path starts from a node in the initial graph, but it may reference nodes that are not in the graph; consequently, the graph may need to be expanded to add nodes, in order to find the item(s) which that path points to.

We define **evaluating a path** as the computation of the values which that path denotes. Answering a query is thus the task of evaluating the path, specified in the query. To evaluate a path, it is traversed one arc at a time in the scenario CG. If the CG does not contain the arc to traverse, then the graph is first elaborated by doing a join of some CG in the KB with the current node which the inference algorithm is at in its path traversal. If, at the end of the traversal, the resulting node is itself a path, then the algorithm is called recursively to evaluate that path. Finally, a value will be found and returned. Note that the path guides the algorithm by telling it which arcs in the scenario to traverse, and hence which nodes in the scenario to elaborate (via joins) should an arc be missing. This algorithm is in Figure 4, without the "classify" step being used.

If the scenario is missing an arc that the algorithm wishes to traverse, the algorithm searches up the type hierarchy from the current node N in the traversal, looking for type definitions or schemata describing generalizations of that node, and which contain the missing arc. If one is found, it is joined with node N in the scenario, and hence path following can resume. This algorithm is suitable for a CG KB containing schemata only. However, an extra complication is added if the KB also contains type definitions. Recall from Figure 1 that type definitions allow instances to be classified. Now, it could happen that the currently known type in the scenario for node N is not the most specific possible (as some type definition may imply this instance is of a more specific type). Without adding a classification step to check N is currently classified as specifically as possible, the algorithm's search up the type hierarchy will start at too general a type, and hence potentially miss some applicable CGs in the KB. For example, if the algorithm is looking for CGs to join with a node N, whose type is ADHESIVE and which (according to the scenario) is being used to affix WOOD, and a NAIL is defined as an ADHESIVE which joins WOOD, then N could first have its type refined from ADHESIVE to NAIL. It is important that such refinements are performed before searching up the type hierarchy, in order to ensure that search starts at the most specific types. To achieve this, we add a 'classify' step in the algorithm in Figure 4, which first checks that the instance N is classified as specifically as possible before it is elaborated. Again based on access-limited logic, classification is tractable (but incomplete), as it may call the algorithm recursively to compute properties of the individual being classified (in order to test whether it satisfies a type definition) which itself is a path-guided (hence tractable but incomplete) process.

An additional feature of our implementation is that it performs only *partial joins* of CG schema to the scenario CG. We define a partial join of CG_1 to CG_2 as the join of a *subgraph* of CG_1 to CG_2. During elaboration, after finding an applicable schema to join to the scenario (step 1 of Elaborate in Figure 4), the algorithm joins only the part of the schema containing the arc of interest with the scenario CG. This part consists of the arc of interest, plus all nodes reachable

---

Procedure `evaluate_path(Path` $= (P_1(\text{X0},x_1),...,P_n(x_{n-1},x_n)),$ `CG)`

        returning values for x_n:

/* *GOAL: Evaluate the path left-to-right to find the x_n* */

1. For $i = 1$ to n:

    Compute all values of x_i that satisfy $P_i(x_{i-1}, x_i)$ by calling:

        $x_i =$ `evaluate_step(`$P_i(x_{i-1}, x_i)$`,` `CG)`

2. Return the set of values x_n

Procedure `evaluate_step(`$P(\text{X0},y)$`,` `CG)`

        returning values for y:

I: CLASSIFY (type refinement):

    /* *GOAL: Try to refine* X0*'s classification, so as to ensure the subsequent elaboration step (below) finds all relevant info about* X0*.* */

    1. Find the current most specific class(es) $\{C_1, ..., C_n\}$ of X0

    2. For each C_j in $\{C_1, ..., C_n\}$:

        2.1 For each direct specialization $SpecC$ of C_j:

            - Compute whether X0 satisfies the type definition of $SpecC$ (this may involve calling `evaluate_path()` recursively, to query for other facts about X0).

            - **if**   X0 satisfies the type definition of $SpecC$

            - **then**   - refine the class of X0 by replacing C_j with $SpecC$ in CG (= type refinement)

                        - search for further refinements, by computing whether direct specializations of $SpecC$ apply (by recursively applying step 2.1 with $SpecC$ instead of C_j).

II: ELABORATE (type/schema expansion):

    /* *GOAL: Expand* CG *at node* X0 *to include arc(s)* P(X0,y)*.* */

    if   CG already contains arc(s) $P(\text{X0},y)$

    then return the value(s) y (no elaboration needed)

    else 1. Search generalizations $XGen_i$ of X0 for type definitions or schemata with info about $P(\text{X0},y)$, ie. which include the arc $P(XGen_i, Yexpr)$ in their graphs

          2. For the $Yexpr$ found:

             if $Yexpr$ is a named individual

             then $y = Yexpr$

             else if $Yexpr$ is a concept (denoting an unnamed individual)

             then   create $y =$ a new Skolem constant denoting it in CG

             else if $Yexpr$ is a path

             then   compute values for y with `evaluate_path(`$Yexpr$`,CG)`

          3. Add the arc(s) $P(\text{X0},y)$ to CG for all y found (partial join)

          4. Return the value(s) for y.

Fig. 4. The full inference algorithm, describing how access paths guide the choice of schema/type expansions/refinements to perform.

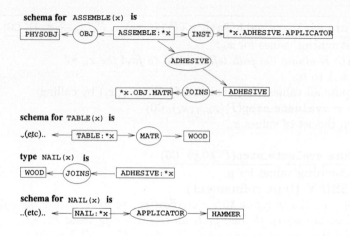

schema for ASSEMBLE(x) **is**

schema for TABLE(x) **is**

type NAIL(x) **is**

schema for NAIL(x) **is**

Fig. 5. Part of a CG KB required for answering the query of Section 4.2

from it in the schema (this will not necessarily be the whole schema). This is done for efficiency, to avoid computing the joins of other parts of the graphs unrelated to answering the current query.

4.2 An Example

We continue with the 'table assembly' example to illustrate the algorithm, where the query was "What tool is needed to assemble a table?" (Answer: a hammer). The query is stated as an assertion ("there is an assembly of a table"), which forms the initial scenario CG, plus a path to the variable of interest ("what is the instrument of the assembly?"):

Initial Scenario CG: **Path to the variable of interest:**

TABLE ← (OBJ) ← ASSEMBLE: Assemble01   Assemble01.INST

Given a CG knowledge-base which includes the CGs shown in Figure 5, the algorithm proceeds as follows:

1. For the first (and only) predicate in the path, inst(Assemble01,y):
   Classify:
   First, the algorithm tries to refine the class of Assemble01 from ASSEMBLE to something more specific. No refinements are found.
   Elaborate:
   As an INST arc from Assemble01 is not yet contained in the scenario CG, it searches for generalizations of Assemble01 whose schema contain information about the INST relation. One such schema is ASSEMBLE(x), shown in Figure 5, and so it adds this arc to the scenario:

   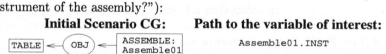

   Now this arc can be traversed in the scenario CG. The algorithm thus finds that the object y at the end of this arc is:

   y = Assemble01.ADHESIVE.APPLICATOR
      = "the thing which applies the adhesive in this assembly event"
      = the path adhesive(Assemble01,x_1),applicator(x_1,x_2)

So it recursively calls `evaluate_path((adhesive(Assemble01,$x_1$),`
`applicator($x_1$,$x_2$)),CG)` to find the individual(s) this path refers to (ie. what the applicator is):

1.1 For the first predicate in this path, `adhesive(Assemble01,y)` ("What is the adhesive used in the assembly?"):

Classify:
First, the algorithm tries to refine the class of `Assemble01` from `ASSEMBLE` to something more specific. No refinements are found.

Elaborate:
As the `ADHESIVE` arc is not yet contained in the scenario, the algorithm searches for generalizations of `Assemble01` whose schema contain information about the `ADHESIVE` relation. One such schema is again `ASSEMBLE(x)`. From this schema, y is found to be an instance of (the concept) `ADHESIVE` which `JOINS` the material of the assembled object. This part of the schema is joined with the scenario CG, and a Skolem individual `Adh01` is created to label this instance:

1.2 For the second predicate in the path, `applicator(Adh01,y)` ("What is used to apply this adhesive?"):

Classify:
First, the algorithm tries to refine the class of `Adh01` from `ADHESIVE` to something more specific. By searching type definitions in the KB, the algorithm finds that one possible specialization of `ADHESIVE` is `NAIL`, defined as an `ADHESIVE` which joins `WOOD`. To test if this applies, it recursively calls the algorithm to find `Adh01.JOINS` ("What does the adhesive join?"). Traversing this arc in the scenario, the algorithm finds the path `Assemble01.OBJ.MATR`, and hence the algorithm is again called recursively to evaluate this path. First, `Assemble01.OBJ` is found (in the scenario CG) to be `TABLE`, hence a Skolem constant `Table01` is created:

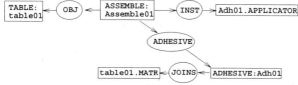

Then, `Table01.MATR` is found by joining in the schema for `TABLE`:

Hence the value Wood01 is returned as the thing which Adh01 joins. Hence the definition is satisfied, and Adh01's type can be refined from ADHESIVE to NAIL.

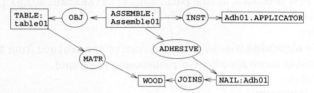

Elaborate:

As the APPLICATOR arc is not yet contained in the scenario, the algorithm searches for generalizations of Adh01 whose schema contain information about the APPLICATOR relation. One such schema is NAIL(x), shown in Figure 5. This is joined with the scenario graph:

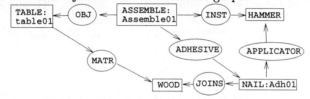

Traversing this arc, the node HAMMER (denoting an instance of hammer) is found. A Skolem individual Hammer01 is again created to label that instance.

  1.3 Hence return y = Hammer01 (an instance of HAMMER)

2. Hence return y = Hammer01 (an instance of HAMMER)

Note that the conclusion (that the APPLICATOR of Adh01 is a HAMMER) results from the refinement of Adh01's type from ADHESIVE to NAIL. Without the classify step in Figure 4, the inference engine would have failed to answer this query.

5 Discussion

5.1 Sources of Incompleteness

Our use of access paths is inspired by the work on Algernon [9], which similarly uses incomplete reasoning and access paths to achieve tractable inference. However, our work has evolved to exploit different properties of Algernon compared with those that Algernon was originally based on. In particular, the original research proved polynomial time inference for 'pure' Algernon, which did not include statements involving existential quantification or restrictions on the cardinality of relations (and thus is similar to Datalog [14]), and achieved tractibility by limiting the number of bindings a variable in a rule could take [1]. In contrast, our interest has been in 'full' Algernon (due to our requirements for these features), and we exploit different sources of incompleteness, and hence efficiency, in its reasoning.

There are three sources of incompleteness in full Algernon which we exploit.

Consider the following schema with access paths:

Standard CG Schema:	Schema with access paths:

schema for X(x) **is**

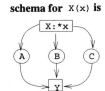

schema for X(x) **is**

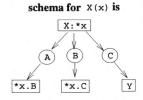

Let $X0, Y0$ be instances of types X, Y respectively. Again assuming all these relations are all functional (Section 2.3), access paths limit inference in the following three ways:

Excluded Paths: The algorithm will try computing $C(X0, y)$ to find $B(X0, y)$, but not try $A(X0, y)$.

Directionality of Coreference: The algorithm will try computing $C(X0, y)$ to find $B(X0, y)$, but not the converse, ie. will not try computing $B(X0, y)$ to find $C(X0, y)$. (This can be viewed as a special case of excluded paths).

Hiding of Inverses: Given the inverse relation of B is B^{-1}, the algorithm will not infer from $C(X0, Y0)$ that $B^{-1}(Y0, X0)$ is also true[3].

In all these cases, these inferences logically follow from the information in initial schema, plus the constraint that the relations are functional. However, the algorithm is (deliberately) not 'clever' enough to draw these conclusions (if it did, then access paths would be merely a notational convenience for indicating coreferential variables). In terms of the inference algorithm, the incompleteness arises due to the query path not passing through the node containing the access path, hence the access path is not evaluated, hence the information it encodes is not seen by the inference engine.

We are exploiting this property to allow the knowlege engineer to focus reasoning with CGs, so that inference follows just those access paths that he/she specifies. It is important to note the knowledge engineer can exploit this incompleteness as much or as little as desired, to hide parts of the knowledge base which would be undesirable to visit when answering a particular question. If such masking is not desired, suitable additional paths can always be added. Our implementation allows multiple alternative paths to be specified at a node (separated by the symbol &&), and similarly for paths and types to be mixed. For example, the above schema for X could be expressed with less incompleteness as:

schema for X(x) **is**

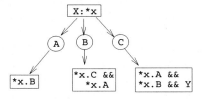

---

[3] eg. the algorithm will not be able to infer [Car1]→(DRIVEN-BY)→[Fred] from [PERSON:*p]→(DRIVES)→[*p.CAR] and [Fred]→(CAR)→[Car1].

This provides additional paths by which the values of *A*, *B* and *C* can be found. With multiple paths at a node, the algorithm tries each in turn until a solution is found[4]. A looping-checker prevents infinite recursion.

Finally, it is important to note that, although inference may be incomplete when answering a particular question, there is no logical consequence of the knowledge base which is inherently unreachable: leading questions can always be asked which touch the required access paths, hence causing them to be evaluated and the information they contain to be added to the scenario. This property, in which all logical consequences are inferable by *some* sequence of questions, is referred to as "Socratic completeness" in access-limited logic [9].

5.2 Extensions

Access paths can be viewed as a special case of defining a CG node's value as a computation. Based on this view, we can extend the path language, and hence the query language also, to include other forms of computation besides chains of relations. The full path language we use is described in [4], and includes additional constructs for equality, negation (as failure), conditionals, arithmetic, and filtering of values. We can then express, for example, that a person's favorite clothes are those which are his/her favorite color by placing an expression in the node at the end of the FAVORITE-CLOTHES arc in the PERSON schema, eg.

schema for PERSON(p) **is**

6 Summary and Conclusions

The most important result of this work is an inference method which makes reasoning with CGs possible in reasonable run-time. On a DEC Alpha 3000/500, our implementation runs at approximately 600 inferences per cpu-second, where an inference consists of making a single step along a path (which itself may require joining schema/type information to the conceptual graph being queried). More important than absolute speed, however, is the fact that inference is channeled down specific chains of subgoals, as specified by the access paths. This allows queries requiring complex composition of CGs to be answered quickly.

The cost of tractability is the loss of guarantee that the inference algorithm will derive all conclusions implied by the knowledge-base (though of course a time-bounded general inference algorithm will similarly lose this guarantee, which is perhaps a fairer comparison to make). The degree to which incompleteness is a practical problem depends crucially on whether an adequate set of access paths can be encoded, such that logical incompleteness has minimal effect on question-answering ability. Our experience is that this is generally possible; however, we are currently lacking adequate theoretical and empirical frameworks to characterize this, in particular to characterize the questions which can be answered, and the degree of speed-up which is thus achieved. Unfortunately, there

---

[4] This is an incompleteness in our implementation: strictly, it should try all paths and then unify the result.

are few theoretical frameworks available which distinguish between the pragmatically different cases we are interested in (namely, how the degree of incompleteness impacts coverage and efficiency of answering questions); many frameworks (eg. worst-case complexity analysis) collapse the distinction that we are trying to clarify. Similarly, empirical analysis is difficult as we lack good methods for characterizing the space of questions that we are interested in answering, and good methods for accounting for the impact of the knowledge engineer in such an investigation (eg. guarding against the knowledge engineer "fixing" the knowledge-base to answer just those questions). These are complex but important issues for further investigation.

Acknowledgements: Support for this research was provided by a grant from Digital Equipment Corporation and a contract from the Air Force Office of Scientific Research (F49620-93-1-0239). This work was conducted at the University of Texas at Austin.

References

1. James Crawford. Access-limited logic: A language for knowledge representation. Technical Report AI90-141, Dept CS, Univ Texas at Austin, Austin, TX, Oct 1990.
2. B. W. Porter, J. Lester, K. Murray, K. Pittman, A. Souther, L. Acker, and T. Jones. The botany knowledge base project. Tech Report AI-88-88, Dept CS, Univ Texas at Austin, Sept 1988.
3. Peter Clark and Bruce Porter. The dce help-desk project. (http://www.cs.utexas.edu/users/mfkb/dce.html), 1996.
4. Peter Clark. KM/KQL: Syntax and semantics. (Internal document, AI Lab, Univ Texas at Austin. http://www.cs.utexas.edu/users/mfkb/manuals/kql.ps), 1996.
5. J. F. Sowa. *Conceptual structures* Addison Wesley, 1984.
6. R. J. Brachman, D. L. McGuinness, P. F. Patel-Schneider, L. A. Resnick, and A. Borgida. Living with CLASSIC: When and how to use a KL-ONE like language. In J. Sowa, editor, *Principles of Semantic Networks*. Kaufmann, CA, 1991.
7. Adil Kabbaj. Prolog cg-object-oriented programming with PROLOG+CG. Master's thesis, Univ Montreal, Canada, May 1995.
8. Jon Doyle and Ramesh S. Patil. Two theses of knowledge representation: Language restrictions, taxonomic classification, and the utility of representation services. *Artificial Intelligence*, pages 261–297, 1991.
9. J. M. Crawford and B. J. Kuipers. Algernon – a tractable system for knowledge-representation. *SIGART Bulletin*, 2(3):35–44, June 1991.
10. R. MacGregor. Loom users guide (version 1.4). Tech report, ISI, CA, 1991.
11. Ronald J. Brachman, Richard E. Fikes, and Hector J. Levesque. KRYPTON: A functional approach to knowledge representation. In Ronald J. Brachman and Hector J. Levesque, editors, *Readings in Knowledge Representation*, pages 412–429. Kaufmann, CA, 1985.
12. William A. Woods. Understanding subsumption and taxonomy. In John F. Sowa, editor, *Principles of Semantic Networks*, pages 45–94. Kaufmann, CA, 1991.
13. Hassan Ait-Kaci and Andreas Podelski. Towards a meaning of LIFE. *Logic Programming*, 16:195–234, 1993.
14. Stefano Ceri, Georg Gottlob, and Letizia Tanca. What you always wanted to know about datalog (and never dared to ask). *IEEE Transactions on Knowledge and Data Engineering*, 1(1):146–166, Mar 1989.

Applying Conceptual Graph Theory to the User-Driven Specification of Network Information Systems

Aldo de Moor

Infolab, Tilburg University, P.O.Box 90153, 5000 LE Tilburg, The Netherlands,
e-mail: ademoor@kub.nl

Abstract. Users need to be strongly involved in the specification process of network information systems. Characteristics of user-driven specification are described, and process composition is proposed as a feasible approach. The knowledge representation framework used in the RENISYS specification method is introduced, using conceptual graph theory as its underlying formalism. The role of ontological and normative knowledge is explained. The presented theory is used to show how legitimate process definitions can be generated by the users. The facilitation of user-driven process composition is discussed.

1 Introduction

Research networks are goal-oriented networks of professionals that focus on supporting certain stages of the research process, such as the planning and conduct of research activities and the dissemination and implementation of the results.

One characteristic of such networks is that their activities are often highly complex as well as innovative in nature, implying that their work processes need to be continuously remodelled, and their supporting network information system redesigned. A second trait is that research networks, which are often Internet-based, make use of an ever increasing set of standard information tools, rather than custom-designed programs, to implement their information systems. These tools enable a fixed set of information and communication processes, which in turn allow participants to carry out their network activities. Another important feature is that for reasons such as the relative lack of hierarchical organization of the participants, and the participants themselves being the task-experts, user-driven network information system development is essential [1]. This means that the participants are responsible for determining the exact roles that a suite of tools plays in the enabling of their work processes.

In this paper, we aim to first delve in more detail into some key aspects of network information system development. More particularly we focus on how users should be involved in the system specification process through process composition. We introduce the knowledge representation framework used in a specification method for research network information systems currently under development (RENISYS), using conceptual graphs as the underlying formalism. The importance of ontologies as the conceptual foundation of specification knowledge is highlighted, and some operations to change ontological type definitions are presented. Based on the ontologies, sets of norms can be defined, which regulate user-behaviour both on the operational and the compositional level. By means of an example adapted from a concrete research network case we illustrate how legitimate process definitions can be generated. Subsequently, we briefly discuss how the previously developed theory can be applied to facilitate more active user-driven process composition.

2 Involving the Users in the Information System Development Process

2.1 What Is a Network Information System?

The network information system (NIS) can be described from an analysis as well as a (high-level) design perspective. From the analysis point of view, the NIS is seen as the set of meaningfully combined and configured information and communication processes which are necessary to support and coordinate the activities of the network participants in their various roles [1]. In the analysis view we thus focus on how meaningful process requirement definitions can be made. However, we do not take into account the roles that the available information tools play in the implementation of the required process functionality. These roles are determined during design, in which the set of required information and communication processes is mapped to the available enabling information tools. When looking at the information system in this way, we consider each tool to afford a number of generic information and communication processes, which can be used in a particular network to enable specific work processes.

2.2 Characteristics of User-Driven System Specification

User-driven NIS development is based on a paradigm very different from the one underlying traditional information systems development approaches, such as ISAC or SDM (Fig.1). Traditional methods are typically based on the waterfall paradigm. These methods are used in large scale projects, in which a group of external experts (represented in Fig.1 by the small pencils) analyzes the organization according to a pre-defined series of steps. Users play a rather passive role, which is often limited to being interviewed by the analysts. At a certain moment in time, the latter produce a blueprint of the information system, which is approved of by the users, and then implemented. However, this implementation is static and the functionality provided often rapidly becomes obsolete, due to quickly changing user requirements. When the mismatch between required and implemented functionality becomes too large, the whole process has to start all over again. An excellent analysis of the many problems encountered in this kind of large project-based software development is given by Brooks in his famous book [2].

Information system development for research networks takes place very differently. First of all, most of it is done by the users on their own. Network participants themselves discuss what activities they should carry out, and determine which information tools to use to accomplish their goals. Furthermore, as has been recognized for quite some time already, such user-driven system development should be based on an evolutionary approach [3]. In contrast with the waterfall-based approaches, the required and implemented functionality gradually evolves, rather than expands with leaps and bounds. Characteristically, a few researchers meet and decide to

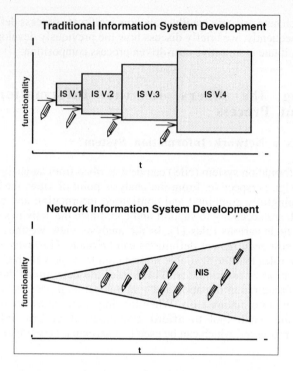

Fig. 1. Network Information System Development:
Evolution Instead of Revolution

form a small network, supported by, for example, a simple mailing list in order to freely exchange various kinds of information. Such a network can expand rapidly, however, both in organizational and task complexity, requiring ever more sophisticated information technological support. Another characteristic is that there is not a single, bird's eye view on the universe of discourse, which the external analysts of the waterfall methods (should) have. In user-driven development, there are rather multiple ants' eyes views, each user in general only being knowledgeable about, and interested in, the small part of the workflow she is involved in.

2.3 Process Composition

When classifying a NIS according to its purpose, it can be considered an ad hoc workflow management system. A workflow management system allows for the design, execution and management of work processes [4]. Ad hoc means that such a system supports creative knowledge activities. These activities are notoriously difficult to model; the workflow support provided can therefore at best provide some sort of control to ensure that tasks, responsibilities, etc. are delivered [5]. However, current workflow management modelling methods (input-process-output as well as speech act-based methods), and user-oriented development approaches (e.g. prototyping, radically tailorable tools), are not sufficiently capable of supporting such user-driven system specification [6,7].

A more promising approach seems to be the one taken by Fitzpatrick and Welsh to facilitate so-called process composition. A core concept they introduce is that of process space. This is `a semantically rich and relatively well defined space, both physically and conceptually, which constrains and bounds the very possibilities of work' [8]. Through process composition, participants can determine the (maximally) required information system functionality by describing their work processes, the total information system process space being equal to the sum of the individually composed processes.

Although some initial attempts have been made to formalise process composition support (especially by [8]), there is still a lot of research needed before it can be effectively used for network information system development.

The research agenda should include at least the following issues:
(1) How to adequately represent specification knowledge?
(2) How to ensure that only legitimate process definitions are made?
(3) How to foster more active user involvement in the specification process?
(4) How to match legitimate process definitions with the functionality enabled by the available suite of information tools?

2.4   The RENISYS Specification Method

The RENISYS method (**RE**search **N**etwork **I**nformation **SY**stem Specification method) is currently being developed to provide a concrete specification approach and tool for network participants to compose their own network information systems. The scope of the method and methodological design criteria were discussed in [1,6]. A model of the user-driven development process was described in [7]. In the model, we subdivide this process into three main subprocesses: the specification, implementation, and use of the network information system. Furthermore, we strictly separate specification and implementation, while on the other hand we strongly connect the specification and use process. For the method to be successful, it will at least need to incorporate answers to the points raised in the above-mentioned research agenda. For lack of space, in this paper we will only focus on providing an initial answer to the first three issues. It seems natural to address these questions first, since process requirements must be defined before they can be assigned to (enabling) tools in the system design, which the final issue is about.

2.5   Applying Conceptual Graph Theory

A user needs to be involved in a specification discourse which makes use of a restricted form of natural language. The language must be rich enough to allow the efficient expression of complex specifications. At the same time it must be sufficiently formal and constrained to allow meaningful specification inferences to be made for the method to adequately facilitate process composition. Conceptual graph theory is more than a syntactic variant of first-order logic because it can enforce conceptual definitions of concepts and relations in terms of natural language-related primitives [9]. The formal structures and operations of the theory have been shown to be useful for representing and processing terminological knowledge in concrete applications, see e.g. [10]. This natural language-focus is an important reason for choosing conceptual graph theory as the knowledge formalism of choice in RENISYS. We will illustrate its potential by giving some preliminary solutions to the research questions posed. In future research we hope to expand the application of CG theory in the method; the purpose of this paper is only to generate some ideas and discussion about the relevance of the theory to this particular kind of application.

3   Knowledge Representation in RENISYS

In RENISYS, the network information system is modelled from three different perspectives, resulting in different domains. The domains are combined in a *reference framework*. The framework is used to generate and store specification knowledge, of which two important kinds are ontological and normative knowledge.

3.1   The Reference Framework

The reference framework consists of three, interconnected domains: the problem domain, the human network, and the information system. In the problem domain, the universe of discourse is interpreted from a task perspective (what are the goals and activities?), in the human network it is described from the organizational point of view (what are the participant interaction processes allowed by the organizational positions?). Together, these domains form the usage context, which describes the determinants of the (generic) information and communication processes. These processes, having been assigned to the set of available information tools, constitute the information system. Directed mappings connect the entities from the different domains (for more details on the framework, we refer the interested reader to [7,11].

In RENISYS, we distinguish three types of knowledge. Ontologies contain descriptions of the terminology used in the different domains. Norms describe the desired behaviour of the various actors in the network. State knowledge can be used to describe the actual or potential behaviour of actors. In RENISYS, it plays an especially important role as a trigger of specification processes, when users use state knowledge to describe the actual situation they are in, and the workflow problems they experience. As the focus of this paper is on the specification process itself, rather than on how exactly it is triggered, and since the correct expression of state knowledge demands an ontological and normative knowledge basis, in the next sections we will study the format and use of ontological and normative knowledge only. In our treatment of these knowledge categories we have been influenced by the semiotic theory of Stamper [12], who takes a strong subjectivist instead of the more regular objectivist stance on the reality to be modelled. An extensive discussion on the pressing need for, and feasibility of such subjectivist information system development methods can be found in [13].

3.2   Ontologies

An ontology is an explicit specification of a conceptualization, which itself is an abstract, simplified view of the world that needs to be represented for some purpose [14]. In our case, this purpose is network information system specification. An ontology can practically be used to organize the storage of information and access to knowledge [15], which for us concerns specification knowledge. Thus, an ontology forms only part of a knowledge base, as it contains a vocabulary useful for describing a domain rather than knowledge about the state of the domain itself [14]. In this way, it is an important instrument in supporting the correct reuse and extension of already generated, complex knowledge structures.

The basis for the ontologies is the concept type hierarchy. Each concept used in any ontological, norm, or state definition must be in this type hierarchy. However, not all concepts that are included in the type hierarchy need to be defined by a differentia as well. For relation types we use a, slightly modified, fixed subset of the case relations used by Sowa [16]. For ontological purposes, type definitions as described in CG theory are useful. We use type definitions instead of schemas because we want ontological definitions only to contain *necessarily*, not just *possibly* existing knowledge. In this way, type definitions provide a clear canonical core for norm and state knowledge, and are useful in the enforcement of selectional constraints on what are considered to be meaningful specifications.

Sources of Type Definitions

In RENISYS, two main sources of type definitions exist: a stable set of theory-grounded definitions, and an evolving set of user-specified definitions.

Theories relevant to the analysis and design of research network information systems provide axiomatic primitives. Some of these primitives are useful for the natural expression of requirements by participants, some are relevant to the design of the network information system, a third category is used to describe the reference framework that provides the connection between these two sets of concepts. In our approach we draw from several such theories, notably language action theory (as introduced in [17] and concrete methodological approaches for organizational communication theory (e.g. the Dynamic Essential Modelling of Organizations method [18]. However, these provide only an initial set of specification entities, which needs to be adapted by the users to match their specific work processes. Our basic ontological goal therefore is not to come up with *the* definition of *the* determinants of the information system in a (research network) usage context. Rather, the basic, theory-grounded set of primitives forms a customizable conceptual foundation that users can tailor to their unique, evolving requirements.

The second, more interesting source of ontological definitions is the network participant herself. Users must be provided with the mechanisms to legitimately customize ontological constructs. Developing facilities which assist users in efficiently modelling their own worlds may furthermore help to considerably increase the limited intellectual capacity available for ontology construction. The pioneering, small groups of experts who currently try to do this cannot handle the sheer volume of the tasks involved [9]; user-defined ontologies can be used directly, or at least form an important input, in larger scale construction efforts.

Type definition operations form the basic tool set to implement such ontological construction mechanisms.

Type Definition Operations

We distinguish three kinds of type definition operations which prescribe how users can carry out such ontology customizations. These operations are: the *creation, modification, and termination* of type definitions.

• Type Creation

In the creation of a type definition, both the type label and its differentia (if needed) are generated. The type must be a subtype of an existing type, the differentia a specialization of the differentia of the supertype. This operation can be implemented using the standard form of type definition as described in [16].

Type creation can serve two uses: the specialization of a concept, or the generalization of a set of concepts. If a concept is to be specialized, no further action is required. If a set of concepts is generalized, all links to the original parent of each concept must be reassigned to the newly created supertype.

```
Example: The Creation of a Group_Report_Editor Type
```

Pre:
type EDITOR(x) **is**
   [ACTOR:*x] <- (agnt) <- [CONTROL] -> (obj) -> [EDIT].

Post:
type EDITOR(x) **is**
   [ACTOR:*x] <- (agnt) <- [CONTROL] -> (obj) -> [EDIT].

type GROUP_REPORT_EDITOR(x) is
   [EDITOR:*x] <- (agnt) <- [EXECUTE] -> (obj) -> [EDIT] -> (rslt) -> [GROUP_REPORT].

The example given is one of type creation for concept specialization purposes. Let us say a user intends to create a group report editor type. A group report editor is an editor who can perform the actual editing of a group report. An editor type has already been defined previously. In graph terms the operation could be represented as above (an execution is a subtype of a control process; items that have changed after the operation have been underlined).

• Type Modification

During the modification of a type, the type label is kept, only the differentia is changed. All (type, norm, and state) definitions including the modified type remain the same, although the role of this concept in these definitions changes due to the type modification.

```
Example: The Modification of a Group_Report_Editor Type
```

Pre:
type GROUP_REPORT_EDITOR(x) **is**
   [EDITOR:*x] <- (agnt) <- [EXECUTE] -> (obj) -> [EDIT] -> (rslt) -> [GROUP_REPORT].

type PUBLISH_GROUP_REPORT(x) **is**
   [TRANSFORMATION:*x] -
      (obj) <- [EXECUTE] -> (agnt) -> [GROUP_REPORT_EDITOR]
      (obj) <- [EVALUATE] -> (agnt) -> [CLIENT]
      (matr) -> [GROUP_REPORT]
      (rslt) -> [PUBLISHED_GROUP_REPORT].

Post:
type GROUP_REPORT_EDITOR(x) **is**
  [EDITOR:*x] <- (agnt) <- [EVALUATE] -> (obj) -> [EDIT] -> (rslt) -> [GROUP_REPORT].

type PUBLISH_GROUP_REPORT(x) **is**
  [TRANSFORMATION:*x] -
    (obj) <- [EXECUTE] -> (agnt) -> [GROUP_REPORT_EDITOR]
    (obj) <- [EVALUATE] -> (agnt) -> [CLIENT]
    (matr) -> [GROUP_REPORT]
    (rslt) -> [PUBLISHED_GROUP_REPORT].

In this example, the meaning of the concept type 'group report editor' is changed. For example, the actor who has the modification authority decides that no longer is the editor somebody who does the actual execution of the editing process, but the editor becomes the one responsible for the final assessment of the produced report. The actual execution can be distributed among a number of reviewers instead. This is a real-life situation often experienced in growing research networks. As the example shows, the differentia of 'group report editor' is changed, but the definition of 'publish group report', in which a group report editor plays the role of executor, remains identical. Changed type definitions thus have global implications, while at the same time responsibilities for these definitional changes are clearly divided among the network participants with the proper definitional authorities.

• Type Termination

When a type is terminated, both the type label and the type differentia are removed. In this case, all occurrences of the old type in any ontological, norm, or state definition must be replaced with for example the supertype or one of its subtypes, although other termination scenarios are conceivable as well.

| Example: The Termination of a Group_Report_Editor Type |

Pre:
type GROUP_REPORT_EDITOR(x) **is**
  [EDITOR:*x] <- (agnt) <- [EVALUATE] -> (obj) -> [EDIT] -> (rslt) -> [GROUP_REPORT].

type PUBLISH_GROUP_REPORT(x) **is**
  [TRANSFORMATION:*x] -
    (obj) <- [EXECUTE] -> (agnt) -> [GROUP_REPORT_EDITOR]
    (obj) <- [EVALUATE] -> (agnt) -> [CLIENT]
    (matr) -> [GROUP_REPORT]
    (rslt) -> [PUBLISHED_GROUP_REPORT].

Post:
type GROUP_REPORT_EDITOR(x): *removed*

type PUBLISH_GROUP_REPORT(x) **is**
  [TRANSFORMATION:*x] -
    (obj) <- [EXECUTE] -> (agnt) -> [EDITOR]
    (obj) <- [EVALUATE] -> (agnt) -> [CLIENT]

(matr) -> [GROUP_REPORT]
(rslt) -> [PUBLISHED_GROUP_REPORT].

In the example, the ontological definition of 'group report editor' is removed, for example because of a reorganization in the network. RENISYS must now find all definitions in which the concept to be removed occurs. Note that, like in the other type definition operations, the replacement of concepts in definitions with other concepts is not trivial. The owners of these concept definitions may need to be consulted, or at least notified that other actors plan to change the way in which 'their concept' is being used. Finding out what kind of protocols best suit user-driven specification is an important objective of our research. We have recently started modeling such protocols along the line of speech-act based discourse protocols, as described in [19].

3.3   Norms

Each network participant plays several actor roles. An actor is an interpreting entity capable of controlling processes. The actor states his requirements in terms of the operational actions he is involved in. An action is a combination of a control process (initiation, execution, evaluation) and a transformation (a process in which a domain object, such as a paper, is generated). To produce specifications, actors can also make compositions. These are combinations of control and definitional processes, such as the creation of a type definition.

The dual control/controlled process approach allows users on the one hand to define workflows and specification processes in terms of concrete deliverables, on the other hand to clearly define the responsibilities for these processes.

Responsibilities have to do with the rules that the network participants agree on, which specify their desired (non)behaviour, and have a normative character. However, to think of norms only as responsibilities produces specifications that are too coarse to properly model specification changes. Norms are therefore subdivided according to the role they play in restricting or affording behaviour. This results in the following categories: privileges, responsibilities, and prohibitions. Privileges comprise those actions and compositions that an actor is permitted to carry out. If an actor is obliged to carry out an action when appropriately triggered (e.g. by taking part in a workflow), the actor has a responsibility. Thus, responsibilities consist of mandatory actions and compositions. All responsibilities should be privileges as well, if not, respecification of norms is necessary. Prohibitions are actions and compositions that an actor is not permitted to carry out.

In order to define the required functionality of an information system, it is important to know what actions specific actors actually are - or are not - allowed to carry out. This knowledge we define in action norms, which of course must be expressed in terms of the concepts that have been constructed using the type definition operations described in the previous section. Besides action norms, in user-driven system specification we also have a need for compositional norms. These norms indicate which actors in the network can make what kind of knowledge definitions about the network.

In CG terms, the basic representation of action norms is:

[ACTOR] <- (agnt) <- [CONTROL] -> (obj) -> [TRANSFORMATION].

An action norm thus shows which control rights an actor has over which (operational) transformations. Compositional norms are similarly represented as:

[ACTOR] <- (agnt) <- [CONTROL] -> (obj) - [DEFINE] -> (rslt) -> [DEFINITION].

A definitional process is here either a creation, modification, or termination of a type, norm, or state definition.

4 Generating Legitimate Process Definitions

An important issue in process composition is how to determine what are legitimate knowledge definitions. Who must be involved in the evolution of norms and types? This is often very unclear in the non-hierarchical kind of professional networks which are our object of interest, as decision-making authority is role, instead of position-bound.

A legitimate definitional change is one of which (1) the canonicity, and (2) the authorization have been checked, thus combining meaningfulness with validity. This helps to ensure as much as possible a model of the network and its information system that represents the interests of, and is acceptable to all network participants.

4.1 Case: a Research Group Writing a Group Report

We will give a concrete illustration of the previously developed theory by producing a legitimate process definition. The example concerns the definition of a group report editor type, and is based on a real case. The B.C. Forests and Forestry Project Group (BCFOR) is an Internet-mediated group aiming to do 'public research' on deforestation in British Columbia, Canada[1]. Like most other such groups, it initially merely allowed unstructured discussion through a mailing list.

However, after a while, the group decided to create a structured group report on a specific discussion topic. More explicit workflows now needed to be defined, in order to allow for both an adequate division of tasks and more tailored information technological support. This specification process turned out to be very hard without a proper conceptual framework to manage the system evolution [7]. We expect an approach as described in this paper to allow participants in a network to better manage the complexities of network information system evolution.

4.2 The Problem: Creating a Group Report Editor Type

In this section, we will illustrate how legitimate (canonical and authorized) process definitions can be generated. In our current approach, the various knowledge categories are represented in simple graphs (only using complex referents to store definition graphs), as we expect this to considerably reduce the complexity of knowledge processing operations. In the implementation of the tool, the different kinds of

---

[1]BCFOR is part of the Global Research Network on Sustainable Development. More information is available at: http://infolabwww.kub.nl:2080/grnsd/proj/gp-bcfor/

knowledge could for example be distinguished by storing them in separate ontological and norm knowledge bases. In the example, the boxed comments indicate the status of the graphs (proposed type definition, accepted norm, etc.).

Example

At t=0, an editor(<actor), edit (<transformation), and paper (<object) type are distinguished in the type hierarchy, all in the problem domain. The editor and edit types also have a definition; the paper type (still) goes undefined:

Category: Type Definition, Modality: Accepted
---

type EDITOR(x) is
  [ACTOR:*x] <- (agnt) <- [CONTROL] -> (obj) -> [EDIT].

type EDIT(x) is
  [TRANSFORMATION:*x] -> (rslt) -> [PAPER].

When the group decides to write a group report, it is agreed that a special kind of editor, the group report editor, is required. The group does have the authority to create a subtype of editor, as the following compositional norm was already previously defined:

Category: Permitted Composition, Modality: Accepted

[GROUP] <- (agnt) <- [CONTROL] -> (obj) -> [CREATE] -> (rslt) -> [TYPE:[EDITOR]].

After some group discussion, based on the existing definition of editor, it is decided that a necessary condition for an editor to be a group report editor is that the edit process she is responsible for should result in the group report as an output. Thus, the proposed group report editor definition becomes:

Category: Type Definition, Modality: Proposed
---

type GROUP_REPORT_EDITOR(x) is
  [EDITOR:*x] <- (agnt) <- [CONTROL] -> (obj) -> [EDIT] -> (rslt) -> [GROUP_REPORT].

However, although the group has the authority to create this definition, the definition is not legitimate, because it is not canonical. The new concept type 'group report', which forms part of the definition, has not even been included in the type hierarchy. The method checks if there is an actor who has the default authority for doing this inclusion by (minimally) expanding the edit concept node in the editor definition (using the existing edit type definition), which results in the following graph:

Category: Type Definition, Modality: Derived (by method)
---

type GROUP_REPORT_EDITOR(x) is
  [EDITOR:*x] <- (agnt) <- [CONTROL] -> (obj) -> [EDIT] -> (rslt) -> [PAPER].

As (in RENISYS) a transformation can only result in a single type of output object, the method knows that a group report *must* be a subtype of paper for the proposed definition to be canonical. Subsequently, the method needs to identify the proper authority for a paper type creation operation. It does this by projecting the following query graph on its set of existing compositional norm graphs:

Category: Query, Modality: Derived (by method)

[ACTOR] <- (agnt) <- [CONTROL] -> (obj) -> [CREATE] -> (rslt) -> [TYPE:[PAPER]].

Let us assume that this projection results in just one permitted composition, namely:

Category: Permitted Composition, Modality: Accepted

[EDITOR] <- (agnt) <- [CONTROL] -> (obj) -> [CREATE] -> (rslt) -> [TYPE:[PAPER]].

The method can now propose the users who play the editor role to create the requested group editor type. If the editor does not agree, a negotiation discourse between editor and group should be initiated and supported by RENISYS.

5 Facilitating User-driven Process Composition

Knowing how to create legitimate knowledge definitions is a necessary, but not yet a sufficient condition for successful user-driven system specification. The dynamics of the specification process deserve special care, due to the fact that users have only a very limited interest in participating in this process. One possible solution to this problem is to present users with customized views on the total process space, views that make them better comprehend their privileges, responsibilities, and prohibitions in the network, and that increase the incentives for users to initiate and participate in more useful specification discourse. Therefore, we refine the concept of process space into action and process composition spaces, and facilitate discourse initiation by focusing on breakdowns.

5.1 Action and Process Composition Space

All permitted actions of an actor together form his action space. An example of a possible action space of an actor 'author' (author<participant) at a certain moment in time is the conceptual graph shown here:

[AUTHOR] -> (attr) -
    [PERMITTED_ACTION: [PARTICIPANT] <- (agnt) <- [EXECUTE] -> (obj) -> [WRITE] -
    (rslt) -> [CONTRIBUTION]]
    [PERMITTED_ACTION: [AUTHOR] <- (agnt) <- [INITIATE] -> (obj) -> [EDIT]].

The graph says that an author, being a kind of participant, can do the actual writing a contribution, as well as initiate an edit process. Depending on the exact mapping to the information and communication processes enabled by the available information tools, this would result in certain functionality specifications. As said before, we do not explain in this paper how this information system design should be done.

In order to *adapt* his set of possible actions, called making compositions in our terminology, an actor must have the authority to produce knowledge definitions, or otherwise negotiate with an actor who does have this authority. The total set of compositions that an actor is allowed to make we define as his (process) composition space. The composition space of an actor generally, but not necessarily, comprises some compositions required to update definitions that shape his own action space, plus other kinds of compositions he has been authorized to carry out. An illustration of a possible composition space for actor 'author' is given here:

[AUTHOR] -> (attr) -
    [MANDATORY_COMPOSITION: [AUTHOR] <- (agnt) <- [EXECUTE] -> (obj) -
        [CREATE] - > (rslt) -> [TYPE: CONTRIBUTION]]
    [PERMITTED_COMPOSITION: [AUTHOR] <- (agnt) <- [INITIATE] -> (obj) -
        [MODIFY] -> (rslt) -> [PERMITTED_ACTION: [EDITOR] <- [CONTROL] -
            (obj) -> [EDIT] - > (matr) -> [CONTRIBUTION]]]].

The graph should be interpreted as meaning that an author must decide on creating subtypes of the concept type 'contribution' (for example, 'abstract' or 'paragraph'), when requested. Furthermore, an author may initiate the modification process of the norm which says that an editor is permitted to control the editing of contributions.

5.2   Recognizing Breakdowns

A fundamental idea regarding the facilitation of process composition is that the method should not trigger users to define requirements according to the rigid analysis and design steps that are prescribed by waterfall methods. Instead, it should support them in resolving concrete functionality problems. These problems are experienced when they play the roles as defined in their action spaces.

According to Winograd and Flores, 'breakdown' plays a key role in the adequate design of artifacts. "A breakdown is not (necessarily) a negative situation to be avoided, but a situation of non-obviousness, in which the recognition that something is missing leads to unconcealing some aspect of the network of tools that we are engaged in using" [17, p.165]. Thus, the proper identification and handling of breakdowns, as soon as they occur, is crucial for users to become interested and actively involved in the (re)specification of their network information system.

The specification method, besides helping the user in becoming aware of his breakdowns, must also help formulate the breakdowns in network terminology, involve relevant other actors in the specification discourse, and support the process of making the actual changes in knowledge definitions.

How exactly this should be done is in our current research focus. The human computer interface now being constructed allows the method tool to have a pseudo-natural language dialogue with users who either explore their own breakdowns, or are involved in a specification discourse triggered by other users resolving their particular breakdowns. Users are prompted to participate by tool-generated e-mail, and once they log on to the RENISYS (web) server will be presented immediately with the appropriate dialogue screens.

It is important to realize that users are not forced to participate in specification discourse. They are allowed to delegate their authority. If they do not respond at all, while not having delegated their responsibilities, network-agreed upon decision rules should determine how the legitimacy of knowledge definitions should be established.

6 Conclusions

In this paper, we have attempted to indicate how conceptual graph theory can help to enable the complex process of user-driven specification of network information systems. Its intuitive semantics resulting from its roots in natural language, combined with its formal power, make the theory a 'logical' candidate as the underlying knowledge representation and reasoning formalism to be used in the RENISYS specification method. We do not claim that our current use of conceptual graph theory to represent the various specification constructs is the most appropriate way; rather, it is an initial attempt meant to generate discussion on better representations, and the algorithms needed to produce them.

Since the early days of conceptual graph theory, a lot of progress has been made in extending and refining its terminology and operations. However, only little attention has so far been paid to finding practical applications of the generic theoretical constructs [20]. We hope that the RENISYS method can serve as such an application, the research on the method both benefiting from the wealth of existing conceptual graph resources and contributing to their further development.

RENISYS is going to be implemented as a client/server system, the clients being standard web browsers. This will allow for maximum participation by the average user, who is the main source of specification knowledge. Our original intention was to develop the complete server in TCL, a powerful scripting language. However, it would be worthwhile to see to what extent one of the 'standard CG workbenches' [21] could be used as the heart of the server.

Acknowledgment

The author wishes to express his gratitude to Hans Weigand for his helpful suggestions for improvement of the original manuscript.

References

1. De Moor, A. Toward a More Structured Use of Information Technology in the Research Community. *The American Sociologist*, 27(1), 1996, pp.91-101.
2. Brooks, F. *The Mythical Man-Month: Essays on Software Engineering*. Addison Wesley, anniversary edition, 1995.
3. Knight, K., editor. *Participation in Systems Development*. Applied Information Technology Reports, Unicom, 1989.
4. Abbott, K., Sarin, S. Experiences with Workflow Management: Issues for the Next Generation. In Furuta, R., Neuwirth, C., editors, *Proceedings of the ACM Conference on Computer Supported Cooperative Work, Chapel Hill, October 22-26, 1994*. ACM, pp.113-120.
5. Khoshafian, S., Buckiewicz, M. *Introduction to Groupware, Workflow, and Workgroup Computing*. John Wiley & Sons, 1995.
6. De Moor, A. Coordinating the Specification Process of Information Systems for Research Networks: Methodological Design Principles. In Fidler, C., editor, *14th International Association of Management Conference, Toronto, August 2-6, 1996, Information Systems Proceedings*, pp.95-103.
7. De Moor, A., Van der Rijst, N. Fostering Active User Involvement in the Specification of Network Information Systems. In *Dutch Interdisciplinary Research*

Conference on Information Science, December 13, 1996, Delft University of Technology, pp.105-118.

8. Fitzpatrick, G., Welsh, J. Process Support: Inflexible Imposition or Chaotic Composition? *Interacting with Computers,* 7(2), 1995, pp.167-180.

9. Lehmann, F. CCAT: The Current Status of the Conceptual Catalogue (Ontology) Group, With Proposals. In Ellis, G., Levinson, R., editors, *Proceedings of the Third International Workshop on PEIRCE: A Conceptual Graphs Workbench, University of Maryland, August 19, 1994,* Lecture Notes in Artificial Intelligence, Vol. 835, Springer-Verlag, pp.18-28.

10. Angelova, G., Bontcheva, K. DB-MAT: Knowledge Acquisition, Processing, and NL Generation Using Conceptual Graphs. In Eklund, P., Ellis, G., Mann, G., editors, *Proceedings of the 4th International Conference on Conceptual Structures: Knowledge Representation as Interlingua, Sydney, August 19-23, 1996,* Lecture Notes in Artificial Intelligence, Vol.1115, Springer Verlag, pp.131-134.

11. Van der Rijst, N., De Moor, A. The Development of Reference Models for the RENISYS Specification Method. In Nunamaker Jr., J.F., Sprague Jr., R.H., editors, *Proceedings of the 29th Hawaii International Conference on System Sciences, January 3-6, 1996,* pp.455-464.

12. Stamper, R. A Semiotic Theory of Information and Information Systems / Applied Semiotics. In *Invited Papers for the ICL / University of Newcastle Seminar on "Information", September 6-10, 1993.*

13. Hirschheim, R., Klein, H., Lyytinen, K. *Information Systems Development and Data Modeling - Conceptual and Philosophical Foundations.* Cambridge University Press, 1996.

14. Gruber, T. Toward Principles for the Design of Ontologies Used for Knowledge Sharing. Technical Report KSL 93-04, Knowledge Systems Laboratory, Stanford University, 1993.

15. Mizoguchi, R. Knowledge Acquisition and Ontology. In *KB&KS '93, Tokyo,* 1993, pp.121-128.

16. Sowa, J. *Conceptual Structures: Information Processing in Mind and Machine.* Addison-Wesley, 1984.

17. Winograd, T., Flores, F. *Understanding Computers and Cognition - A New Foundation for Design.* Ablex Publishing Corporation, 1986.

18. Dietz, J. Modelling Business Processes for the Purpose of Redesign. In *Proceedings of the IFIP TC8 Open Conference on Business Process Redesign,* North-Holland, 1994, pp.249-258.

19. Chang, M., Woo, C. A Speech-Act Based Negotiation Protocol: Design, Implementation and Test Use. *ACM Transactions on Information Systems,* 12(4), 1994, pp.360-382.

20. Lukose, D., Mineau, G., Mugnier, M.-L., Möller, J.W., Martin, P., Kremer, R., Zarri, G. Conceptual Structures for Knowledge Engineering and Knowledge Modelling. In Ellis, G., Levinson, R., Rich, W., Sowa, J., editors, *Proceedings of the Third International Conference on Conceptual Structures - Conceptual Structures: Applications, Implementation and Theory, Santa Cruz, August 14-18, 1995,* Lecture Notes in Artificial Intelligence, Vol. 954, Springer-Verlag, pp.126-137.

21. Mann, G. What Conceptual Graph Workbenches Need for Natural Language Processing. In Ellis, G., Levinson, R., Rich, W., Sowa, J., editors, *Proceedings of the Third International Conference on Conceptual Structures - Conceptual Structures: Applications, Implementation and Theory, Santa Cruz, August 14-18, 1995,* Lecture Notes in Artificial Intelligence, Vol. 954, Springer-Verlag, pp.70-78.

Generic Trading Service in Telecommunication Platforms

A. Puder[1] and K. Römer[2]

[1] Deutsche Telekom AG
Research Centre
P.O. Box 100003
64276 Darmstadt, Germany
puder@tzd.telekom.de

[2] Department of Computer Science
University of Frankfurt
60054 Frankfurt, Germany
roemer@cs.uni-frankfurt.de

Abstract. Telecommunication companies are currently defining a *Telecommunications Information Networking Architecture* (TINA) to meet the future demands of electronic markets. As part of such standardization efforts generic services are being defined. As well we demonstrate how a knowledge–based approach using conceptual graphs helps to develop a generic service for the mediation of services. The prototype of a knowledge–based trader is built according to the client/server paradigm which allows access to this service in a heterogeneous environment. This paper describes service mediation in general and the details of matching service offers and requests.

Keywords: TINA, RM–ODP, type system, interface definition language, trading.

1 Introduction

The constant growth of digital networks in telecommunication systems has opened up new opportunities with respect to the provision of new services. The telecommunication companies have an interest not only to transport data but to provide user–oriented end services over these digital networks. This results in the migration of classical services like yellow pages towards computer–supported services in the digital network. Telecommunication companies place high demands on the properties of the underlying infrastructure which are generally not met by the current Internet technology. Aspects like reliability, broadband communication, accounting and security require new technologies and new software distribution platforms.

Such demands require standardization of the underlying infrastructure in order to support interoperability in a world–wide heterogeneous network. Several institutions work on technology–independent standards which define frameworks for open distributed systems. For example, the *International Organization for Standardization* (ISO) and the *International Telecommunication Union*,

Telecommunication Standardization Sector (ITU–T) have developed the *Reference Model for Open Distributed Processing* (RM–ODP) (see [2]). The RM–ODP describes an architecture which supports the distribution, interoperability and portability of software components. The telecommunication companies have adopted the RM–ODP and refined it according to their specific needs. The resulting standardization effort is led by the *Telecommunications Information Networking Architecture* (TINA) Consortium; an international collaboration of telecommunication and IT companies aiming at defining and validating an open architecture for information and telecommunication services, which meets the future market requirements (see [1]). The TINA architecture is based on distributed computing and object orientation.

Generic services are of particular interest as they can be customized easily for different environments. The Deutsche Telekom AG is developing in co–operation with the University of Frankfurt, a prototype of a particular service: a generic trading service. In this novel approach we make use of conceptual graphs as a service type specification language. The structure of this paper is as follows: In Section 2 we first provide the necessary background of service trading and then show how conceptual graphs can be used for type specifications and the matching of service offers and requests. In Section 3 we show in more detail how the matching of conceptual graphs works. The current prototype of our knowledge–based trader is described next in Section 4. In Section 5 we present a conclusion and an outlook for future work.

2 Service Mediation in Open Service Markets

2.1 The RM–ODP model of service trading

The RM–ODP as well as TINA require the support of a trading function which assists in the search for services (see [3]). All participating parties in a telecommunication system assume the roles of *service providers* and *service requesters*, which have no *a priori* knowledge about each other. In particular, they are not linked statically to each other, but rather dynamically when they decide to participate in the open service market. During the mediation of services another party assumes the role of a *trader*, whose task is to match service offers and requests (see figure 1). According to the RM–ODP, a mediation process is divided into several steps. First a service provider exports its service offer to the trader (1). At a later point in time a service requester tries to import a particular service (2). If the trader finds a matching service offer, which has previously been exported, it responds a reference to the service requester pointing to the appropriate provider (3). After a successful mediation process the service requester and provider are bound to each other and start to interact (4). The trading of services as described by the RM–ODP easily scales up as one trader can delegate import operations to another trader.

The notion of a *type system* is essential for the discussion of a mediation process (see [8]). For one thing, a type system defines a *type specification language*,

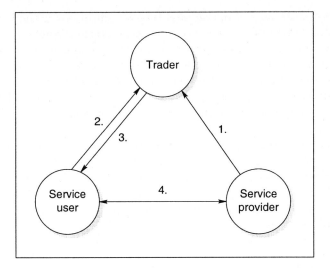

Fig. 1. Interactions during a mediation process.

which allows the specification of services in open distributed systems. A type specification abstracts from specific internal details as it decouples the service usage from its implementation. Furthermore, a type system defines *matching rules*, which allow the association of service offer and request, both represented as type specifications. A service type consists of an *intension* and an *extension*. The intension of a type is abstract in the sense that it serves as an abstract image of the properties of a service type. The intension is not to be confused with a physical representation, rather it is something we can think about. Any physical representation of an intension, such as a specification, is an extension of that intension. The extension provides a specific point of view of the intension of the service type. A service type specification (i.e., an extension) based on an intension is the only artefact which can be transported between users. Within an open service market, it is likely that two application users develop different extensions (or views) for the same intension.

Depending on the application domain, several different type systems have evolved. For example, the type specification language of the RM–ODP as well as those of the middleware platforms DCE (see [6]) and CORBA (see [5]) focus on operational aspects of a type (i.e., how a service is accessible from a programming point of view). The corresponding type specification languages are commonly called *interface definition languages* (IDL). The matching rules for type specifications* written in these languages are quite different in the RM–ODP, DCE and CORBA (this will be explained in more detail in one of the following sections).

---

* For the domain of operational interfaces the matching rules are generally called *syntactic subtyping rules*.

A general drawback of operational type specifications is the lack of semantics. The description of the semantics of a type specification is usually included in terms of informal comments which are ignored during service mediation. Although those comments are discarded by a middleware platform, they are nonetheless important during the mediation process where programmers browse IDL–repositories for appropriate interface specifications. It would therefore be helpful to make use of the information contained in a comment during a mediation process.

This implies that service type specifications have different levels of abstraction with respect to the level of detail contained in the specifications (i.e., operational interface specification vs. informal comments). Existing service mediators are tailored for one of these levels of abstraction. In the following we demonstrate how conceptual graphs (see [10]) can help to build a generic trader which embraces those levels of abstraction.

2.2   Type specifications through CGs

In this section we concentrate on the specification of service types at different levels of abstraction and postpone the discussion of the matching of those specifications until a later section. Consider an object–oriented decomposition of a bank application. There will most likely be a set of objects representing *accounts*. Conceptually, an account object keeps track of a balance which can be altered using designated operations. With this informal description one could write an operational interface specification using the notation according to the CORBA–IDL:

```
/* Operational interface for a bank account */
interface Account {
    /* Deposit a certain amount to an account */
    void deposit( in unsigned long amount );
    /* Withdraw a certain amount from an account */
    void withdraw( in unsigned long amount );
    /* Show the current balance of an accout */
    void balance( out long amount );
};
```

An object of type Account offers three operations at its interface. The operation deposit() takes one input parameter which denotes the amount to be deposited. Likewise the second operation withdraw() withdraws a specific amount. Finally the third operation balance() returns the current total of the account as an output parameter. Since the balance can be negative, the type of the output parameter is long and not unsigned long.

The COBRA–IDL is tailored for the specifiation of operational interfaces. Programmers require a semantic specification of the behaviour as well. These are usually attached in form of comments throughout the interface specification (denoted through /* and */ within the CORBA–IDL for example). The comments are ignored by a CORBA compliant platform. We propose a generic type specification language based on conceptual graphs which can represent the informal description as well as the formal specification of the operational interface using one notation.

For the following analysis we limit the discussion to the specific details of an operation as defined by CORBA. As can be seen in the interface Account, an *operation* consists of an *operation name* and an ordered *set of parameters*[**]. A parameter itself consists of a *directional attribute* (in, out or inout), a *parameter type* and an *identifier* denoting the name of the parameter. With this "anatomy" of an operation specification we can assemble a domain–specific conceptual and relational catalogue. We have developed such a catalogue which provides the "vocabulary" to express the information contained in an operational interface specification (see [4]).

The conceptual graph depicted in figure 2 shows how to translate the operation specification for deposit() (concept nodes are denoted by white rectangles and relation nodes by grey rectangles). The (ANNOTATION) relation connects the formal specification of this operation to an informal description. The informal description is derived from the comment associated with the operation deposit() in the interface Account. Note that the conceptual graph shown in figure 2 is itself embedded in a conceptual graph representing the complete interface Account. This example shows how formal and informal descriptions representing specifications at different levels of abstractions are mapped onto one conceptual graph.

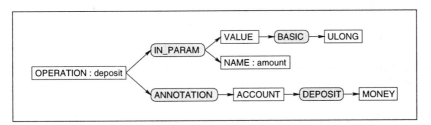

Fig. 2. Conceptual graph representing the specification of the operation deposit().

[**] There are more components to an operation specification like exceptions or contexts which are of no importance for the scope of this paper. See [5] for a more detailed description.

2.3 Syntactic subtyping

A conventional IDL allows the specification of a set of functions, which an object offers at its interface. Various IDLs have been standardized, such as the DCE–IDL or the CORBA–IDL. The RM–ODP standards do not prescribe any particular notation for an IDL. They rather define the characteristics that an IDL for open distributed systems should exhibit.

Each IDL, including RM–ODP's abstract characterization of an IDL, is complemented by a (syntactic) subtyping rule. Syntactic subtyping guarantees the substitutability of interfaces based on inclusion polymorphism. DCE, CORBA and the RM–ODP have different definitions of precisely what constitutes the subtype relation. In DCE an interface type T_1 is a subtype of an interface type T_2, iff all functions defined in T_2 are a subset of those defined in T_1 and appear in the same order. CORBA relaxes this rule by omitting the need for the concurring functions to be declared in the same order. The reason for this relaxation lies in CORBA's interface repository which stores runtime type information. Whereas DCE uses no runtime information except an UUID and a version number assigned to each interface, CORBA has access to type information at runtime via its interface repository. The RM–ODP extends CORBA's subtyping rule by additionally allowing parameter subtyping with contra–variance of input parameters and co–variance of output parameters as well as recursive subtyping.

Since operational interface specifications are translated to conceptual graphs during the mediation process as described in the previous section, there must exist different matching rules which are used by the trader during an import operation. The matching rule to be used depends on the context (i.e. the type systems the service user and provider employ). With our approach it is possible to match an operational interface written in DCE–IDL with another one written in CORBA–IDL using a matching rule according to the RM–ODP.

3 Matching Rules

As has been motivated above, the matching rules which are used to compare two conceptual graphs depend on the context. Thus, during a mediation process three parameters are required: the service provider's type specification (in the following called *type graph*), a service user's type specification (in the following called *query graph*) and a set of rules which determine whether type and query graph match. The service user has to choose a subset of rules as a parameter of the import operation. Our prototype of a knowledge–based trader has six matching rules so far:

1. DCE subtyping
2. CORBA subtyping
3. RM–ODP subtyping
4. Specialization
5. Quantity
6. Negation

The first three rules model syntactic subtyping rules of various middleware platforms which have already been described in [9]. The last three rules model linguistic matching rules. The specialization rule matches a type graph with a query graph if the specification contained in the former is more specific than the specification contained in the latter. The quantity rule matches a type graph [OBJECT] -> (QUANTITY) -> [UNIT:i_1] with a query graph [OBJECT] -> (COMPARATOR) -> [UNIT:i_2], if COMPARATOR (i_1, i_2) holds (where COMPARATOR is something like *equal*, *less* or *greater*). The negation rule is non–monotonic in the sense that it reduces the set of matching type graphs. This rule determines whether two subgraphs of a type and query graph have an opposite meaning. The opposite meaning is checked with the help of a lexicographical database which is part of the knowledge–based trader's architecture.

A matching rule has two input parameters — a type and a query graph — and one output parameter which reflects the *semantic distance* of those two graphs. The value of the output parameter is either -1 if the two graphs do not match or in the interval $[0, 1]$. A value of 0 means that the semantic distance is unknown and the value of 1 that there is a best possible match between type and query graph (i.e., minimal semantic distance). If more than one matching rule is selected to determine the semantic distance, then the overall semantic distance is determined by:

$$MATCH(Q,T) =_{df} \begin{cases} \max_{1 \leq i \leq n}\{MR_i(Q,T)\} & : \quad \forall i : MR_i(Q,T) \geq 0 \\ -1 & : \quad \text{otherwise} \end{cases}$$

where $MR_1, MR_2, \ldots, MR_k$ denote the matching rules which were selected. That is type and query graph do not match if at least one matching rule rejects them (i.e., returns -1), otherwise the overall result is the maximum match value determined by all the selected rules.

In the following sections we present formal specifications for the aforementioned linguistic matching rules. Although the matching rules return a semantic distance in terms of a numeric value, it should be noted that the individual nodes of the type and query graph are themselves not assigned a weight.

3.1   Specification of the Specialization Rule

The matching rule for specialization should obey the following design principles:

1. The more branches of a type and a query graph match modulo a lexicographical database, the better should the match be evaluated (i.e., the semantic distance should become less).
2. The influence on the semantic distance between type and query graph decreases for information deeply embedded within the type and query graphs. This guarantees that aspects of a service type specification located closer to the root node of a CG contributes more to the semantic distance than very specific details of a service deeply embedded in the CG.
3. If parts of a type graph are rejected (i.e., evaluated with -1), then this should not necessarily lead to a rejection of the type graph as a whole.

Now we present a formal specification of the specialization rule which follows the above design principles. The definition makes use of a 2–ary predicate $\sim_{R_1, R_2, \ldots, R_l} (W_1, W_2)$ which is true iff the words W_1 and W_2 are related via the lexicographical relations $R_1, R_2, \ldots, R_l$ in the lexicographical database. In the following definition the general structure of a type and a query graph is depicted in figure 3. Only two levels of successors of the nodes Q and T are shown. Each node Q_{ij} and T_{kl} can have further subgraphs.

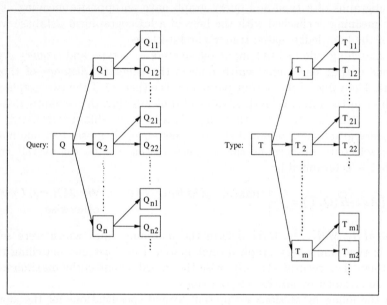

Fig. 3. General structure of a type and a query graph.

The formal definition of the matching rule for specialization is represented by a 2–ary function called *SPEC*. The value of *SPEC* is determined according to the following algorithm:

1. If not $\sim_{IS\_A, SYNONYM} (T, Q)$, then $SPEC(Q, T) =_{df} 0$, STOP.
2. If $n = 0$ (i.e., the query graph consists of only the root node), then $SPEC(Q, T) =_{df} 1$, STOP.
3. Calculate the semantic distances $w_1, \ldots, w_n$ of Q's subgraphs $Q_1, \ldots, Q_n$ as follows (note the recursive definition via the aforementioned function *MATCH*, which applies all selected matching rules to subgraphs):

$$w_i =_{df} \begin{cases} \max_{1 \leq j \leq m} \{MATCH(Q_i, T_j)\} & : \quad \forall j : MATCH(Q_i, T_j) \geq 0 \\ -1 & : \quad \text{otherwise} \end{cases}$$

4. If Q denotes a concept node, then *SPEC* is defined as:

$$SPEC(Q,T) =_{df} \begin{cases} \frac{1}{n}\sum_{i=1}^{n} w_i & : & \forall i : w_i \geq 0 \\ 0 & : & \text{otherwise} \end{cases}$$

Otherwise Q denotes a relation node and $SPEC$ is defined as:

$$SPEC(Q,T) =_{df} \begin{cases} \frac{1}{n}\sum_{i=1}^{n} w_i & : & \forall i : w_i > 0 \\ 0 & : & \text{otherwise} \end{cases}$$

STOP.

Example Let Q be a query graph and T_1 and T_2 be two type graphs as shown in figure 4. The semantic distances of $MATCH(Q,T_1)$ and $MATCH(Q,T_2)$ are to be computed where the set of selected matching rules is $\{SPEC\}$. Assume the lexicographical database contains entries such that $\sim_{IS\_A,SYNONYM}(X,X)$ is true for each $X \in \{a,b,\ldots i\}$.

During the execution of $MATCH(Q,T_1)$ $MATCH$ is recursively invoked on each node of the query graph. The numbers at the upper right corners of the nodes in figure 4 are the results of those recursive invocations of $MATCH$. That is 0.5 (the number at the root node) is the overall result of $MATCH(Q,T_1)$. Similarly 0.75 is the result of $MATCH(Q,T_2)$.

These examples hint that design goals (1) and (2) are met by the matching rule $SPEC$. Design goal (3) is met due to the fact that a reject (i.e., value of -1) is not propagated to the root but only one level up. $\square$

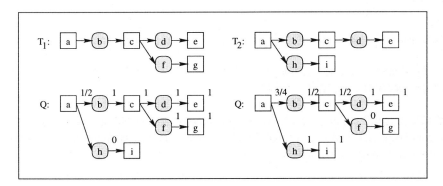

Fig. 4. Sample type and query graphs.

3.2 Specification of the Negation Rule

The design criteria for the negation rule are somewhat simpler than those for the specialization rule: reject (i.e., return -1) if type and query graph have an opposite meaning with respect to the lexicographical database.

In contrast to the specialization rule, both the negation and quantity rules are non–recursive in the sense that they do not apply themselves to subgraphs. But if used together with the specialization rule they are in fact applied recursively to all subgraphs due to the fact that the specialization rule calls $MATCH$ on subgraphs which in turn evaluates all selected matching rules on the subgraphs.

The formal definition of the matching rule for negation is represented by the 2-ary function NEG. Again the following definition refers to the general structure of a type and a query graph as depicted in figure 3. The value of NEG is determined according to the following algorithm:

1. If Q denotes a relation node then $NEG(Q, T) =_{df} 0$, STOP.
2. Calculate $NEG(Q, T)$ as follows:

$$NEG(Q, T) =_{df} \begin{cases} -1 &:& \exists i, j, k, l : isa(T, Q) \wedge antonym(T_k, Q_i) \wedge isa(T_{kl}, Q_{ij}) \\ 0 &:& \text{otherwise} \end{cases}$$

where

$$isa(T, Q) =_{df} \sim_{IS\_A, SYNONYM} (T, Q)$$
$$antonym(T, Q) =_{df} \sim_{ANTONYM, SYNONYM} (T, Q)$$

STOP.

Example Consider the following type graph T and query graph Q:

```
T: [CAT]->(HATE)->[DOG]
Q: [ANIMAL]->(LOVE)->[DOG]
```

$MATCH(Q, T)$ with $\{NEG\}$ as the set of selected matching rules evaluates to -1, assuming the lexicographical database contains entries that make all of the following predicates evaluate to true:

$$\sim_{IS\_A, SYNONYM} (CAT, ANIMAL)$$
$$\sim_{ANTONYM, SYNONYM} (HATE, LOVE)$$
$$\sim_{IS\_A, SYNONYM} (DOG, DOG)$$

□

3.3 Specification of the Quantity Rule

Using concept nodes of the form [UNIT:val], one can specify a quantity of val units of UNIT (e.g., [DAY:10] corresponds to ten days). The quantity rule should accept (i.e., return 1) if the type graph specifies a quantity i_1 of some unit and the query graph specifies a quantity i_2 of the same unit together with a comparator r (e.g., $<, =, >$) and the condition $r(i_1, i_2)$ is fullfilled.

The formal definition of the quantity matching rule is represented by the 2–ary function $QUANT$. The following definition assumes the existence of a unary function $value(C)$ that returns the value specified within concept node C. If C contains no value, 0 should be returned. Figure 3 shows the general structure of a type and a query graph. The value of $QUANT$ is determined according to the following algorithm:

1. If Q denotes a relation node then $QUANT(Q,T) =_{df} 0$, STOP.
2. Calculate $QUANT(Q,T)$ as follows:

$$QUANT(Q,T) =_{df} \begin{cases} 1 & : \quad \exists i,j,k,l: \begin{array}{c} isa(T,Q) \wedge \\ iscomp(Q_i) \wedge isquant(T_k) \wedge \\ syn(T_{kl}, Q_{ij}) \wedge \\ Q_i(value(T_{kl}), value(Q_{ij})) \end{array} \\ 0 & : \quad \text{otherwise} \end{cases}$$

where

$$isa(T,Q) =_{df} \sim_{IS\_A,SYNONYM} (T,Q)$$
$$syn(T,Q) =_{df} \sim_{SYNONYM} (T,Q)$$
$$isquant(Q) =_{df} isa(Q, "QUANTITY")$$
$$iscomp(Q) =_{df} isa(Q, "COMPARATOR")$$

STOP.

Example Consider the following type graph T and query graph Q:

```
T: [CAT]->(AGE)->[YEAR:5]
Q: [ANIMAL]->(LESS)->[YEAR:10]
```

$MATCH(Q,T)$ with $\{QUANT\}$ as the set of selected matching rules evaluates to 1, assuming the lexicographical database contains entries that make all of the following predicates evaluate to true:

$$\sim_{IS\_A,SYNONYM} (CAT, ANIMAL)$$
$$\sim_{IS\_A,SYNONYM} (AGE, QUANTITY)$$
$$\sim_{IS\_A,SYNONYM} (LESS, COMPARATOR)$$
$$\sim_{SYNONYM} (YEAR, YEAR)$$

□

4 Architecture of the Knowledge–based Trader

The knowledge–based trader consists of three components: a *service database*, a repository for *matching rules* and a *lexicographical database* (see figure 5). The service database stores service types based on conceptual graphs. A set of service providers is associated with every service type. The service database provides a mapping of types to instances of these types. The repository of matching rules is used to define different metrics to compute the semantic distance between two graphs. All six matching rules mentioned before are implemented within the PROLOG subsystem of the knowledge–based trader. The lexicographical database maintains a comprehensive semantical network with linguistic information that is used by the matching rules. Thus, the lexicographical database provides "background" knowledge needed to match two conceptual graphs.

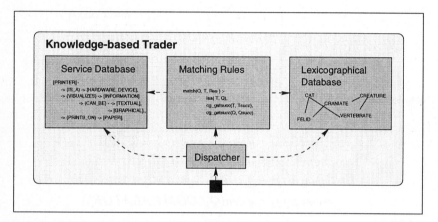

Fig. 5. Architecture of the knowledge–based trader.

The knowledge–based trader is generic in the sense that it works solely with conceptual graphs. Beside the service mediation specific translators allow the conversion between the DCE– and CORBA–IDL. As part of the trading system, we have built tools for translating DCE– and CORBA–IDL to and from CGs. This allows the trading of service types indepently of the IDL that was used to specify their operational interfaces. Besides the benefit of providing a homogeneous framework for trading IDLs, with such a translation capability the trader can be used to translate specifications written in different IDLs. We have performed timings of the three matching rules which handle subtyping of IDLs according to the conventions defined by DCE, CORBA and RM–ODP (see [9]).

The implementation of a matching rule within the knowledge–based trader has access to the details of the type and query graphs to be matched via built–in predicates. The following six predicates are available for the implementation of matching rules:

match(+*queryGraph*, +*typeGraph*, −*matchResult*)
> Entry point of a matching rule. Compares *queryGraph* with *typeGraph*. The result of the match (i.e., the semantic distance) has to be unified with *matchResult*.

cg_match(+*queryGraph*, +*typeGraph*, ?*matchResult*)
> This predicate applies all selected matching rules recursively with *queryGraph* und *typeGraph*. The semantic of this predicate is identical with match/3.

cg_getsucc(+*cgNode*, ?*successorList*)
> Unifies *successorList* with a list of sucessor nodes of *cgNode*.

cg_getname(+*cgNode*, ?*nodeName*)
> Unifies *nodeName* with a readable string associated with *cgNode*.

cg_gettype(+*cgNode*, ?*nodeType*)
> Determines whether *cgNode* is a concept or relation node. The variable *nodeType* is bound to concept or relation accordingly.

cg_matchword(+*word1*, +*relationList*, +*word2*)
> Corresponds to $\sim_{R_1, R_2, \ldots, R_l}$ (word1, word2) where *relationList* contains $R_1, \ldots, R_l$. This predicate fails if no path between the words using the given lexicographical relations can be constructed.

A matching rule which makes use of the above predicates has access the other components of the knowledge–based trader. The predicate cg_matchword/3 accesses information contained in the lexicographical database. The predicates cg_getsucc/2, cg_getname/2 and cg_gettype/2 access information in the service database. See Appendix A for an example.

5   Conclusion and Outlook

In this paper we have presented the architecture of a generic trading service for telecommunication systems, where conceptual graphs are used as a type specification language. As has been shown, this methodology allows service type specifications at different levels of abstraction. With the proper tool support, operational interface specification can be translated to conceptual graphs. The matching of import and export during a mediation process depends on the context which is modeled by different matching rules. Each rule is tailored for a specific context.

There remain a couple of refinements which need still to be investigated. The performance measurements carried out by us indicate that new approaches must be taken if the knowledge–based trader were to be exposed to a commercial environment. Especially promising seem to be techniques for computing hash values from CGs to support caches. Furthermore, the human computer interface to the knowledge–based trader would benefit from natural language to CG translators to facilitate the formulation of import operations. Finally we intend to extend the RM–ODP model of service mediation to relax the strict 1–to–1 mapping of request and offer. It seems feasible to support a 1–to–n mapping where several service providers are "composed" to achieve in a combined effort the needs of a service requester.

Appendix A: Implementation of the Specialization Rule

In the following we demonstrate the implementation of the matching rule $SPEC$ as described in Section 3.1. The predicates prefixed by cg_ are explained in Section 4. The code as shown below is taken without modification from the initialization files of our knowledge–based trader (see [7]). The code itself is located in the component for matching rules within the knowledge–based trader.

```
1  @define-matching-rule 'Specialization'
2
3   isa(Node1, Node2)  :-
4          cg_getname(Node1, Name1), cg_getname(Node2, Name2),
5          cg_matchword(Name1, ['IS_A', 'SYNONYM'], Name2).
6
7   match(Q, T, Res)  :-
8          cg_gettype(Q, concept), isa(T, Q),
9          cg_getsucc(T, Tsucc), cg_getsucc(Q, Qsucc),
10         match_lists_con(Qsucc, Tsucc, Res, 0, 0).
11
12  match(Q, T, Res)  :-
13         cg_gettype(Q, relation), isa(T, Q),
14         cg_getsucc(T, Tsucc), cg_getsucc(Q, Qsucc),
15         match_lists_rel(Qsucc, Tsucc, Res, 0, 0).
16
17  best_match(_, [], MaxVal, MaxVal).
18  best_match(Q, [T|Tlist], MaxVal, Res)  :-
19         (cg_match(Q, T, Val) ; Val is 0),
20         (Val < 0
21         ->      Res is Val
22         ;               best_match(Q, Tlist, max(Val, MaxVal), Res)).
23
24  match_lists_con([], _, 1000, _, 0)  :- !.
25  match_lists_con([], _, Res, Sum, NSucc)  :-
26         Res is round(Sum/NSucc), Res > 0.
27
28  match_lists_con([Q|Qlist], Tlist, Res, Sum, NSucc)  :-
29         best_match(Q, Tlist, 0, BestMatch),
30         (BestMatch > 0
31         ->      NewSum is Sum+BestMatch
32         ;               NewSum is Sum),
33         match_lists_con(Qlist, Tlist, Res, NewSum, NSucc+1).
34
35  match_lists_rel([], _, 1000, _, 0)  :- !.
36  match_lists_rel([], _, Res, Sum, NSucc)  :-
```

```
37          Res is round(Sum/NSucc), Res > 0.
38
39   match_lists_rel([Q|Qlist], Tlist, Res, Sum, NSucc) :-
40          best_match(Q, Tlist, 0, BestMatch), BestMatch > 0,
41          match_lists_rel(Qlist, Tlist, Res, Sum+BestMatch, NSucc+1).
42
43 @end
```

The predicate isa/2 in lines 3—5 implements $\sim_{IS\_A,SYNONYM} (Q,T)$. Predicate match/3 defined in lines 7—15 is the entrypoint of the matching rule, see Section 4 for details. To avoid using floating point numbers, the result is scaled by 1000, that is Res $\in \{-1, 0, 1, 2, \ldots, 1000\}$. If Q denotes a concept node, then the first branch of match/3 in lines 7—10 is taken, otherwise the second branch in lines 12—15. best_match/4 in lines 17—22 implements step 3 of the algorithm in Section 3.1. The predicates match_lists_con/5 and match_lists_rel/5 in lines 28—33 and 35—41 implement the two cases of step 4 of the algorithm.

References

1. F. Dupuy, G. Nilsson, and Y. Inoue. The TINA Consortium: Toward networking telecommunications information services. *IEEE Communications Magazine*, pages 78–83, November 1995.
2. ITU.TS Recommendation X.901 — ISO/IEC 10746–1: Basic Reference Model of Open Distributed Processing Part 1: Overview and Guide to the use of the Reference Model, July 1994.
3. ODP Trading Function, ITU/ISO Committee Draft Standard ISO/IEC DIS13235 Rec. X.9tr, May 1995.
4. C. Kaiser. Spezifikation operationaler Schnittstellen mittels kanonischer Konzeptgraphen–Templates. Diplomarbeit, Fachbereich Informatik, Goethe Universität, Frankfurt, August 1996.
5. Object Management Group (OMG), The Common Object Request Broker: Architecure and Specification, Revision 2.0, July 1995.
6. Open Software Foundation. *Introduction to DCE*. Open Software Foundation, Inc., 1992.
7. A. Puder. Introduction to the AI–Trader Project. http://www.vsb.informatik.uni–frankfurt.de/projects/aitrader/, Computer Science Department, University of Frankfurt, 1995.
8. A. Puder and C. Burger. New concepts for qualitative trader cooperation. In A. Schill et al., editors, *International Conference on Distributed Platforms*, pages 301–313. Chapman & Hall, February 1996.
9. A. Puder and K. Geihs. System support for knowledge–based trading in open service markets. In *7th ACM SIGOPS European Workshop*, Connemara, Ireland, September 1996.
10. J.F. Sowa. *Conceptual Structures, information processing mind and machine*. Addison–Wesley Publishing Company, 1984.

Assessing Sowa's Conceptual Graphs for Effective Strategic Management Decisions, Based on a Comparative Study with Eden's Cognitive Mapping

Simon Polovina
School of Computing, Information Systems and Mathematics
South Bank University, London, UK
Email: polovina@sbu.ac.uk

Abstract

This paper examines the potential for Sowa's conceptual graphs to improve practical strategic management decision making in business activity, by a detailed comparison between conceptual graphs and Eden's already established strategic business-level cognitive mapping technique. An appropriate, contemporary and realistic example of an office relocation problem is used to explore the comparison, which reveals that conceptual graphs do indeed significantly enhance Eden's technique. The paper therefore calls for this avenue to be exploited further for the benefit of strategic decision making, given its highly qualitative nature resulting in it being most difficult-to-model aspect of business activity.

1 Introduction

This paper examines the capability of conceptual graphs, as devised by Sowa and applied by Polovina, in practical business strategic management [1,2]. The paper attempts this by comparing conceptual graphs with the cognitive maps of Eden [3]. Eden's mapping technique, which is an established knowledge-based structured diagram technique for strategic planning, is based on the advanced personal constructs methodology begun by Kelly [4]. The cognitive mapping technique both a) employs a highly structured approach, and b) is designed as a practical human expert end-user support tool. Eden's cognitive maps thereby offer a valuable comparison with conceptual graphs. Should conceptual graphs sufficiently enrich cognitive maps then conceptual graphs can be implemented meaningfully to enhance business strategic decision making, especially given its highly qualitative nature resulting in it being most difficult-to-model aspect of business activity.

As its basis, the examination employs the realistic office location problem that Ackerman, Cropper, and Eden choose in highlighting the benefits of cognitive mapping [5]. An up to date discussion by Ackerman et al., employing the same example, can be found on the Web at http://www.scotnet.co.uk/banxia/depaper.html. An analysis of the same problem is performed using conceptual graphs.

2 The Example Problem

The example given by Ackerman et al. is as follows:

"We need to decide on our accommodation arrangements for the York and Humberside region. We could centralise our service at Leeds or open local offices in various parts of the region. The level of service we might be able to provide could well be improved by local representation but we guess that administration costs would be higher and, in this case, it seems likely that running costs will be the most important factor in our decision. The office purchase costs in Hull and Sheffield might however be lower than in Leeds. Additionally we need to ensure uniformity in the treatment of clients in the region and this might be impaired by too much decentralization. However we are not sure how great this risk is in this case; experience of local offices in Plymouth, Taunton and Bath in the South West may have something to teach us. Moreover current management initiatives point us in the direction of greater delegation of authority."

3 The Cognitive Map for the Example Problem

Ackerman et al. cognitively map the above problem resulting in **Figure 1**. This figure illustrates two essential elements underlying the cognitive mapping interpretation. Namely these elements are 'concepts' and 'links'. Each concept is represented as one emergent 'pole', which describes one side of the problem, and a 'contrasting pole' which is meant to focus the concept by a meaningful contrast to the first pole. Poles may lead to other poles by means of directed links.

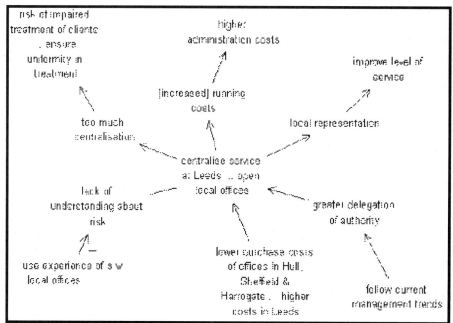

Figure 1: Ackerman et. al.'s cognitive map for the offices location problem

This can be clarified further by examining the text of the map for the computer software package *Decision Explorer* which, as shown by the diagram of **Figure 1** above, depicts these cognitive maps graphically (Decision Explorer runs on Windows 3.1, 95 or NT. More details about Decision Explorer can be found from the Web at http://www.scotnet.co.uk/banxia/demain.html, email info@banxia.co.uk or phone [+44] 0 141 552 3082). To begin with, the map's concepts can be represented by the following table:

```
1  open local offices...centralise services at Leeds
2  local representation...[not] local representation
3  increased running costs...[not] increased running costs
4  higher administration costs...[not] higher administration costs
5  improve level of service...[not] improve level of service
6  too much decentralisation...[not] too much decentralisation
7  risk of impaired treatment of clients...ensure uniformity of treatment
8  lack of understanding about risk...[not] lack of understanding about risk
9  use experience of s w local offices...[not] use experience of s w local offices
10 lower purchase costs of local offices...higher cost in Leeds
11 greater delegation of authority...[not] greater delegation of authority
12 follow current management initiatives...[not] follow current management initiatives
```

Note that a sequential number is usually added to signify each concept entered by the user. For any concept where a contrasting pole was not entered the term '[not]' is added to create a 'default' contrasting pole. The cognitive mapping methodology also happens to stress that it is important the emergent pole should always represent what the user can best identify with. However, this is likely to create confusion when it comes to making links as this consideration means an emergent pole may be required to lead to a contrasting pole and vice versa. The '-' symbol is thus added to the directional link to combat this problem.

Accordingly each directed link shows a pole to pole, and contrasting pole to contrasting pole, link. The added '-', or 'negative' link, shows a pole to contrasting pole, and contrasting pole to pole, link. The negative link occurs where 'use experience of s w local offices' leads to '[not] lack of understanding about risk', and '[not] use experience of s w local offices' to 'lack of understanding about risk'. Another type of link, the 'connotative' link, is employed when the user knows there is an insufficiently definable yet somehow valid connection between concepts. Such links would be shown as undirected links. The connotative link can be applied to the relationship between concepts '1' and '8' as overcoming 'lack of understanding about risk' may lead to either operating centralised services or opening local offices. Both the negative link and the connotative link are also reflected in **Figure 1**.

The above problem is now explored using conceptual graphs. The conceptual graphs representation of the above problem is based on the same cognitive map as identified above. This approach should ensure a common comparative basis, yet highlight vividly any distinguishing features between the two representations.

4 Modelling the Poles in Conceptual Graphs

Starting with the poles themselves, they appear to fall into two categories. The first category has user-defined contrasting poles whilst the second's contrasting poles remain undefined.

Concentrating on the defined concepts to begin with, these may be modelled initially by the conceptual graphs in **Figure 2**. In this figure the pair of poles become a conceptual graph by placing each pole into a separate conceptual graph concept and together surrounding them within a negative context. These negative contexts signify that whatever is contained within them, taken as a whole, is false. Therefore each graph provides contrast by stating that it is false that both poles can exist simultaneously. As elaborated below, if one of the concepts is true then the other becomes false.

Take the middle graph in **Figure 2**, which refers to the poles 'centralise services at Leeds.open local offices', as a representative instance. Lets say we decide to see what happened if 'centralise services at Leeds' was chosen.

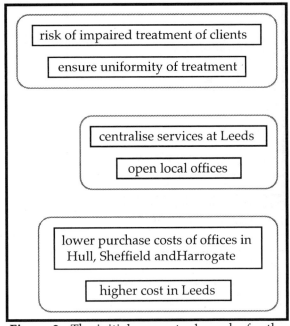

Figure 2: The initial conceptual graphs for the defined contrasting poles

As a conceptual graph this could be shown initially as in **Figure 3**. This true graph dominates its matching concept inside the middle graph in **Figure 2**, hence this inside concept can be removed, or deiterated, to yield **Figure 4**.

centralise services at Leeds

Figure 3

Figure 4 shows that 'open local offices' is false. This occurred because 'centralise services at Leeds' is true. Should the decision be 'open local offices'

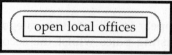

Figure 4

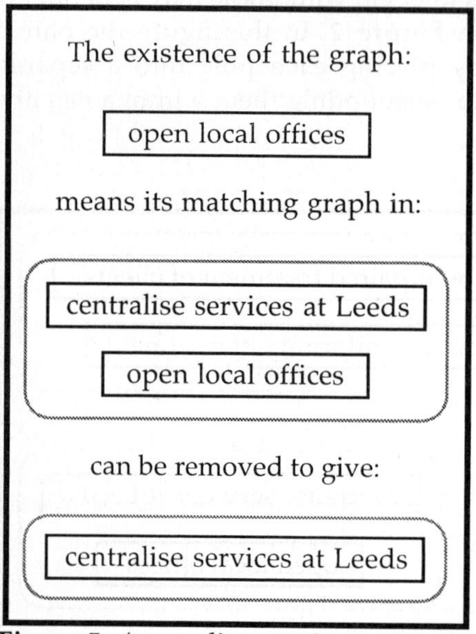

The existence of the graph:

open local offices

means its matching graph in:

centralise services at Leeds

open local offices

can be removed to give:

centralise services at Leeds

Figure 5: 'centralise services at Leeds' false when 'open local offices' is true

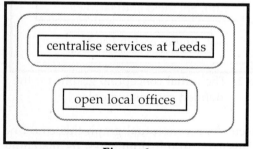

Figure 6

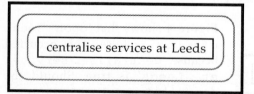

Figure 7

instead then 'centralise services at Leeds' would be false accordingly. The whole picture for this scenario is shown in **Figure 5**. Unlike the earlier cognitive map that only passively records the poles, a computer-based conceptual graph processor could make these new assertions automatically as the appropriate new graphs are added to its base of knowledge. The important repercussions of these inferences will become evident later.

Note that the above 'true-asserts-false' form does not assert one pole as true should the other be false. For example the poles 'centralise services at Leeds' or 'open local offices' cannot be asserted as true from their contrasting pole being false. To do this would require the additional 'false-asserts-true' graph shown in **Figure 6**. In this figure there are nested negative contexts. Remember that in conceptual graphs, a whole negative context and its contents can also be deiterated provided it matches an appropriately dominating negative context and its contents.

This removal can be illustrated from the graph in **Figure 4** ('open local offices' is false). This graph dominates the matching part in the false-asserts-true graph of **Figure 6** because it is surrounded by a *lesser* number of negative contexts. Hence its matching

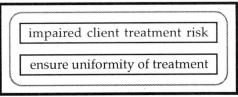

Figure 8(a)

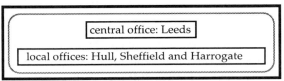

Figure 8(b)

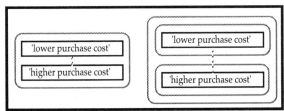

Figure 8(c)

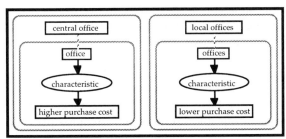

Figure 8(d)

Figures 8(a, b, c, d): Refined conceptual graphs

graph can be removed from the latter figure to yield the result shown in **Figure 7**.

This result has left two negative contexts around 'centralise services at Leeds'. These double negate to give the same graph as in **Figure 3** ('centralise services at Leeds' is true).

In the present cognitive mapping technique the 'false-asserts-true' aspect is insufficiently clear. The present approach may prefer the user to assume if one pole is false then the other is true, yet it is quite possible that the decision maker may for instance do nothing or decide to open mobile offices instead. In this case the above 'false-asserts-true' graph would be incorrect. The bipolar nature of the present method cannot cope with this scenario. Even worse, it could provide a too narrow framework which stifles originality of thought: The model does not lend itself to decision makers realising other alternatives, such as mobile offices. In view of this deficiency, the 'false-asserts-true' aspects cannot be transposed to the conceptual graph representation in a manner which guarantees validity.

5 Refining the Graphs

Moving on, it is possible to leave the conceptual graphs in this 'true-asserts-false' two concept form and manipulate them as elementary propositional logic statements (The details of propositional logic and predicate logic can be found through any seminal text on logic such as Kowalski [6]. Sowa, in Appendix A.5: "Symbolic Logic", pages 384–391 also discusses these matters [1]). Indeed Ackerman et al. stress that the sentences should remain as they are because the decision maker can identify with what he or she has stated directly. With conceptual graphs the above concepts could be refined nonetheless to the more powerful predicate logic level thereby capturing more about the problem, yet arguably remain human expert readable. The refinement is demonstrated by the graphs shown in **Figures 8(a) to 8(d)**, which refine the graphs in **Figure 2**.

The graphs in **Figure 8** now include relations, referents and coreferent links as well as essentially more proper hierarchical type labels. The left-hand graph inside the nested negative context of **Figure 8(d)** (or " **'8(d)** " for short) may be read as "The characteristic of an office is a higher purchase cost" for example. The referent 'Leeds' conforms to the type label 'central office' and 'Hull, Sheffield and Harrogate' conforms to 'local offices'. Part **'8(a)** is merely a shortening of one of the concept's phrases. This graph could easily be refined further, as indeed may all the graphs throughout the entire offices example, hence **'8(a)** may be viewed as an example of an intermediate step in model development.

The greater degrees of refinement are demonstrated by **'8(b)**, **'8(c)** and **'8(d)**. In **'8(b)**, 'Leeds' is an instance of a central office in that Leeds will have its own peculiarities but shares the same characteristics as any central office in general. This would permit inferences to be made about Leeds from both what is known about central offices in general and Leeds in particular.

6 Generalising the Model

The above shows that a knowledge-base can be built up based on the appropriate degree of generally applicable knowledge. This also prevents unnecessary duplication when the same knowledge applies to more than one particular concept. The degree can be appreciated by developing the Leeds example in a little more detail. It may be that certain things are applicable to Leeds in its own right, Leeds as a Yorkshire central office, as a northern central office, or an English central office as well as a central office. The same principles apply to the local offices. Taking the central office case as representative, the type hierarchy would then include (where *subtype < supertype*):

```
central office < office.
English central office < central office.
Northern central office < English central office.
Yorkshire central office < Northern central office.
```

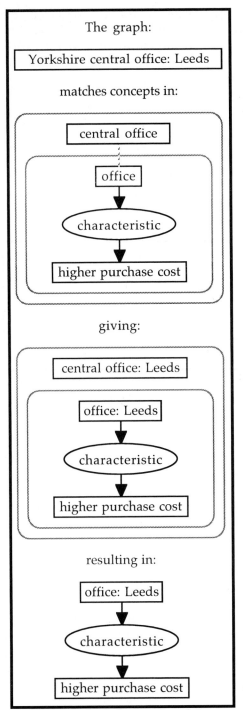

The graph:

Yorkshire central office: Leeds

matches concepts in:

central office

office

characteristic

higher purchase cost

giving:

central office: Leeds

office: Leeds

characteristic

higher purchase cost

resulting in:

office: Leeds

characteristic

higher purchase cost

Figure 9: An inference involving hierarchical relationships and referents

The most specialised conformity for 'Leeds' is 'Yorkshire central office'. This means Leeds conforms to *all* of the above central offices, but not to say 'Southern central office' (Southern central office < English central office). Thereby any inference in respect of Southern central offices would not apply to Leeds but any for Yorkshire, Northern, central office and office would.

The graphs in '8(c) and '8(d) concern the purchase costs of the offices. Examining '8(c), the left graph shows that if a purchase cost is higher then it cannot be lower and vice versa. The right graph shows that if one is false the other is true. The coreferent link in both cases establishes that they refer to the same cost. These graphs are therefore so general in nature that they can be used *beyond* the offices example.

Turning to '8(d), these graphs imply that a central office is an office which has a higher purchase cost whilst local offices are offices with a lower purchase cost. Should 'central office' or 'local offices' dominate these graphs respectively, the appropriate inference would be made accordingly. This is demonstrated in **Figure 9**.

Conceptual graphs thereby also raise the user's awareness through their inherent structure: As the user refined the graphs so they become more and more based on hierarchical type labels and specific instances within those labels, the user would have to think about the appropriate degree of relevance. The graphs as they currently stand apply to any local or general office. Alternatively they may be written to infer about Yorkshire offices only, in which case 'central office' and 'local offices' in the appropriate dominated graphs would instead read 'Yorkshire central office' and 'Yorkshire local offices' respectively.

7 Modelling the Undefined Poles in Conceptual Graphs

Continuing further, recall that where the contrasting pole is not defined on input, Decision Explorer creates it by prefixing the term 'not' to the input emergent pole as stated earlier. The concepts with undefined contrasting poles could thus be modelled initially in conceptual graphs as given by **Figure 10** (The concepts 'greater delegation of authority' and 'follow current management initiatives' however would not be modelled this way as explained later).

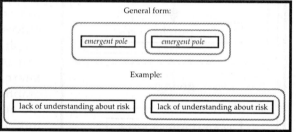

Figure 10: Conceptual graphs for concepts with undefined contrasting poles

On examining these graphs however, it rapidly emerges that there is no real need to include such poles as conceptual graphs at all. To illustrate, given 'lack of understanding about risk' was true or false, this would merely assert 'lack of understanding about risk' is true or false respectively. This tautology shows such concepts in fact turn out to be meaningless. Therefore they can be excluded from the conceptual graphs representation.

8 Modelling the Links in Conceptual Graphs

The cognitive map links may be modelled initially as implications in conceptual graphs as shown in **Figure 11**. The nature of these graphs is explained by **Figure 12**. As can be seen from these figures, without worrying about the graphs affected by double negation for the moment, the 'leads from' pole becomes a concept which is enclosed in a negative context. This context also encloses another negative context that encloses the concept of the 'leads to' pole.

Note that the negative link found in Decision Explorer becomes redundant because the order in which the poles are drawn are irrelevant in conceptual graphs. The user could still retain the visual order through arranging the shape of the graphs according as to what, say, that user would like to see at the top or bottom part of his or her graph drawings.

The concept 'use local office experience' has been refined to 'use local office experience:#256' as it describes a particular office experience identified by the serial number '#256'. This number may be a reference to the relevant documentation on this issue for example.

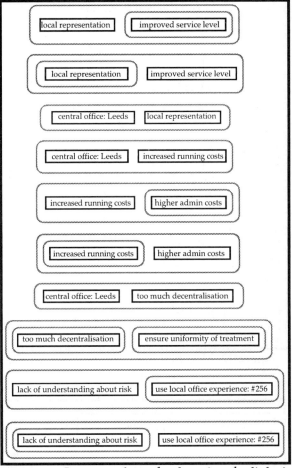

Figure 11: Conceptual graphs denoting the links in the cognitive map

As for the double negated graphs, the effect in the case of the graphs describing the false 'local representation', 'increased running costs', and 'too much decentralisation' implications of 'central office: Leeds' is they now appear to be like existing cognitive mapping concepts instead of its links. Hence these graphs show there are links that emerge to be *additional* contrasting poles. Conceptual graphs have yielded this fact explicitly and drawn it to the user's attention, whilst it remains unnecessarily implicit and thereby easily undetected in the existing cognitive map.

All the above of course highlights another question as to whether the present cognitive mapping technique should indicate that all default

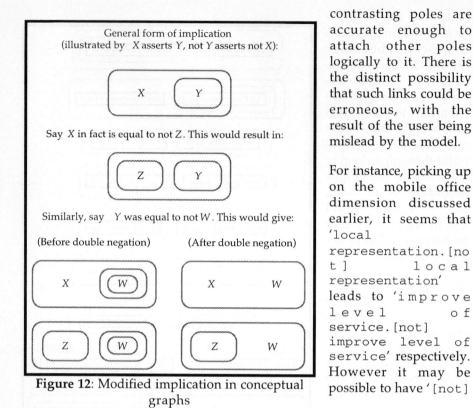

General form of implication
(illustrated by X asserts Y, not Y asserts not X):

Say X in fact is equal to not Z. This would result in:

Similarly, say Y was equal to not W. This would give:

(Before double negation)   (After double negation)

Figure 12: Modified implication in conceptual graphs

contrasting poles are accurate enough to attach other poles logically to it. There is the distinct possibility that such links could be erroneous, with the result of the user being mislead by the model.

For instance, picking up on the mobile office dimension discussed earlier, it seems that 'local representation.[not] local representation' leads to 'improve level of service.[not] improve level of service' respectively. However it may be possible to have '[not]

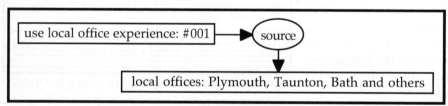

Figure 13

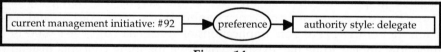

Figure 14

local representation' and 'improve level of service' through 'use mobile offices'. It would then be *false* that '[not] local representation' implies '[not] improve level of service'.

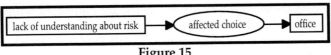

Figure 15

9 Modelling the Other Knowledge in the Example Problem

The concept 'use local office experience' has some background information relating to it about the source of that information from some actual offices in the South West. This may best be described by the graph in **Figure 13**, which can be added to the knowledge base and then called upon as necessary.

This leaves us with 'greater delegation of authority', 'follow current management initiatives' and the relationship between 'lack of understanding about risk' and the choice of office.

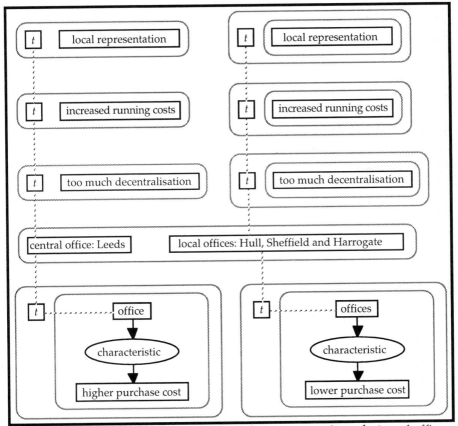

Figure 16: Directly linked interrelationships surrounding choice of office

The first two of these concepts are modelled by **Figure 14** whilst the latter relationship is modelled by **Figure 15**, which describes the 'lack of understanding about risk' relationship. Recall that, for Decision Explorer, the tacit 'lack of understanding about risk' relationship in the cognitive map was refined into a connotative link, which the graph is intended to reflect. Once again these graphs can be called upon as knowledge about the problem is elicited.

10 Allowing Inferencing

Now the conceptual graphs knowledge-base can sensibly start to infer new knowledge. An example, revealing the direct links arising from the choice of office, is shown by **Figure 16**. In this figure there are coreferent links to concepts labelled as '*t*', which may best be thought of as 'it' instead of the full concept's name, to aid readability by avoiding repetition. The '*t*' essentially equates to the conceptual graphs' universal supertype, but with an attached coreferent link that immediately specialises it.

To illustrate, let us state that 'central office: Leeds' is true. From **Figure 16** we can see that this statement implies that 'too much decentralisation' is false, thereby its contrasting pole, 'ensure uniformity of treatment' becomes true (See **Figure 11**). This in turn causes 'impaired client treatment risk' to be asserted as false (See **Figure 8(a)**).

11 Comments On Cognitive Mapping in Conceptual Graphs

The following general points have emerged from remodelling the offices problem as conceptual graphs:

a) Like the present cognitive mapping methodology, the concepts and relations in conceptual graphs can be based on a language that the decision maker identifies with. Conceptual graphs can also be arranged to retain the visual cues the end-user may require.

b) Unlike present cognitive maps, conceptual graphs allow the further refinement of the problem through, say, the interrelation of generalised and specialised knowledge. Initial conceptual graph models may start by being a literal paradigm of cognitive maps at the existing level. Subsequently they may be refined by graphs which, as illustrated by the office purchase costs, break down these phrases along increasingly greater expressive dimensions.

c) Through its bipolar limit, which conceptual graphs overcome, the current cognitive maps could stifle creative thought by the decision maker.

d) The '[not] emergent pole', which is the default contrasting pole in the present maps, turn out to be meaningless when modelled as a conceptual graph.

e) Conceptual graphs do not need any additional devices to show the 'negative' links unlike the present cognitive mapping technique. Any user visuality element therein need not be compromised by this link's absence.

f) By always implicitly linking concepts with default contrasting poles current cognitive maps obfuscate the distinction between legitimate and potentially damaging relationships. Conceptual graphs, on the other hand, remove this arbitrary situation by focusing the user's mind on what in fact are valid and invalid contrasting poles, including default ones.

12 Concluding Remarks

Clearly conceptual graphs can enrich cognitive mapping. Though it may successfully elicit knowledge through its contrasting poles and links, cognitive mapping cannot extract properly the genuine impact of these relationships nor put the user on enquiry to seek for further dimensions that may affect the problem. Moreover it can be wrong, as the references to mobile offices have revealed for example. A conceptual graphs processor could automatically recognise and deal with the contrasting aspects of the cognitive mapping technique. This would occur as a direct part of the negative contexts upon which conceptual graphs inference is based. As well as inference, the processor would also be able to check for any inconsistencies as they are entered into the knowledge-base. All this should free the user to declare merely what he or she believes and then review that mental model, or its computer paradigm, in the light of the processor's output.

Although critical of the current approach, this paper does not seek to dismiss it. As Eden states, the present cognitive maps can be drawn quickly and thereby get an immediate handle on the problem situation at hand, thus it remains a valuable initial modelling tool. However as a more permanent building block of knowledge, its limitations are simply too significant to ignore. Conceptual graphs supply a similarly visual but much more highly principled basis from which more meaningful knowledge can be eventually built. The conceptual graphs approach is therefore worthy of further exploitation for the benefit of strategic decision making, especially given its highly qualitative nature making it the most difficult-to-model aspect of business activity.

References

1. J. F. Sowa; *Conceptual Structures: Information Processing in Mind and Machine*, Addison-Wesley, 1984.

2. S. Polovina; *The Suitability of Conceptual Graphs in Strategic Management Accountancy*, PhD Thesis, Loughborough University of Technology, UK, 1993.

3. C. Eden; "Working on Problems Using Cognitive Mapping", *Operations Research in Management*, Littlechild, Stephen; Shutler, Maurice (eds.), Prentice-Hall, 1991.

4. G. Kelly; *The Psychology of Personal Constructs*, Norton, New York, 1955.

5. F.R. Ackerman, S.A. Cropper, C.L. Eden; "Cognitive Mapping for Community Operational Research–A User's Guide", Operational Research Tutorial Papers, A.G. Mumford, T.C. Bailey (eds.), Operational Research Society, 1991.

6. R. Kowalski; *Logic for Program Solving*, Amsterdam, North-Holland, 1979.

CGKAT: A Knowledge Acquisition and Retrieval Tool Using Structured Documents and Ontologies

Philippe MARTIN

University of Adelaide – Computer Sciences department, Australia
email: phmartin@gisca.adelaide.edu.au
This work was completed at the INRIA (ACACIA Project), France

1 Introduction

In Knowledge Acquisition (KA), the knowledge engineer must *model and represent* expertise into a knowledge base (KB). To do so s/he often *searches for information* in documents (e.g. interview retranscriptions and technical reports) and *structures* these documents in order to ease search and modelling. S/he also has to do *searches on the knowledge representations* to compare, *organize* and validate them.

In Information Retrieval (IR), the indexation of (parts of) documents by direct hypertext links, keywords or SGML–like tags do not allow the IR system to adequately answer queries expressed at different levels of generality or generate an organised view of the document contents. To allow this, an adequate *knowledge representation (KR)* language must be used. The more detailed the indexation is, the more precise the answers of the IR system will be. Like KA, precision–oriented IR implies the construction of an organised KB from documents and searches in documents via searches in this KB. It is also eased by a KA/IR system exploiting the *structure* of documents (i.e. the fact that the document elements are typed and may be linked by composition links or hypertext links, and that various presentation models may be associated and applied to them).

CGKAT [2,3,4] helps KA and IR in two ways.

1) CGKAT integrates a knowledge processor, the Conceptual Graph workbench CoGITo [1], with the structured document editor Thot [6], so that the user may use and *combine* a) an advanced *technique for representing, organizing, accessing and handling knowledge*, and b) an advanced *technique for displaying, organizing, accessing and handling document elements (DEs)*. More precisely, these two kinds of techniques may *be applied to conceptual graphs (CGs) and to any DE* since we allowed a) CGs to be edited and structured with and as any other DEs, and b) any DE (even a whole document) to be indexed by one or several CGs via hypertext links of types Representation and Annotation [3].

Since documents may store knowledge representations mixed with other DEs, there is no need to maintain a separate KB. Discrepancies between the KB and the KB documentation are thereby avoided. Moreover, Thot allows users to "include" a DE (e.g. a paragraph, a CG, a concept) into several other DEs and then enables hypertext navigation between the inclusions and their sources (in both directions), and automatically modifies the inclusions if the source is modified. This facility allows hypertext navigation between CGs via the concepts they share and eases the handling of modules or views.

A user may navigate from a DE to its indexations, then navigate between CGs according to the relations between their concepts or their context (described by the DEs which embed them), and then navigate to a DE indexed by a CG. Knowledge–based navigation is thus possible.

CGKAT can also merge words and their representations into an alphabetically sorted index table, and uses inclusions of this information for building the index table, thus providing a complementary way to compare and access knowledge representations, their authors, the viewpoints they use and the DEs they index.

We have designed *command language* for a) combination of queries in the KB or documents, b) the generation of virtual documents (i.e. views on parts of other documents) as answers to these queries, c) the storage of queries into scripts that may be associated to some DEs, thus allowing the use of virtual (dynamic) hypertext links as in some advanced knowledge–based hypertext systems, such as MacWeb. Using queries, the CGKAT user may generate documents that "include" the CGs satisfying conceptual constraints (e.g. the CGs specialising a given CG) and/or the DEs that are represented or annotated by these CGs. The use of inclusions allows users to combine searches by queries and searches by navigations. The scripts may combine commands on the KB with commands accessible from the Unix shell (and then with any tool behaving as a Unix filter). Such scripts may for example be used for testing the KB and generating explanations, (parts of) technical documents or new knowledge representations (e.g. with the provided maximal join command).

2) CGKAT *proposes libraries* for easing and guiding the structuration of documents, their representation and the reuse or extension of the representations [2]:

– *structure models and default presentation models* for various types of DEs, e.g. Article, Section, Paragraph, Image, Graphics and CG;

– an initial *concept type ontology* which merges various KA/KR *top–level concept type ontologies and the models associated to these types* (e.g. the KADS tasks models) and also the *natural language ontology* WordNet [5];

– an initial *relation type ontology* which merges various relation types ontologies, e.g. thematic, spatial, temporal and argumentative relation type ontologies.

Searches in such ontologies may be done by navigation, lexical queries (i.e. type name substrings) or conceptual queries (e.g. constraints on authors, domains and supertypes). The natural language ontology WordNet (90,000 types) is not proposed to be wholly included in the user ontology but the types retrieved in WordNet by queries or navigation may be included (with their supertypes) in the user ontology.

2 Architecture

CGKAT has a client/server architecture and also includes the above cited libraries.

The server is made up of the CG workbench CoGITo plus an additional functional interface to allow a) building and retrieval of CGs via Thot menus or textual commands callable inside Thot documents or from an Unix shell or script, and b) browsing ontologies (e.g. WordNet) and modifications on the user ontologies.

The client is the structured document editor Thot plus additional code to allow a) CGs to be edited, handled and stored inside structured documents using the Thot interface, b) the indexation of DEs by CGs, and c) the generation of virtual documents. When a Thot document including CGs is opened, CGKAT also automatically creates them in the base of CoGITo, and removes them when the document is closed. Thus, Thot documents may be used to load, display, browse, structure, document, edit and store selected parts of the KB (an editing operation on a CG via Thot is allowed only if it is accepted by CoGITo, i.e. if it does not violate conceptual constraints previously defined). Conceptual queries may also be done on these selected parts in order to retrieve some CGs or type definitions, or the DEs they index.

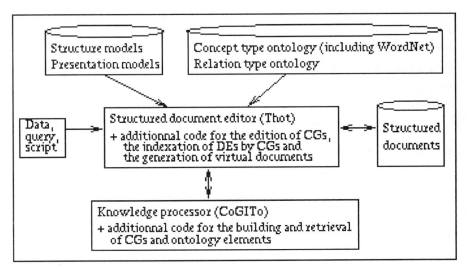

Fig. 1. The CGKAT architecture

3 Applications

CGKAT is a domain–independent KA tool and precision–oriented IR tool. Arbitrary precise representations are enabled by the CG formalism. However, the representations are done manually, therefore their precisions depends only on the users goals. CGKAT has already been used for modelling road accident expertises.

4 Limitations

The main limitations of CGKAT for IR, and to a lesser extent for KA, are the facts that: a) it does not help knowledge extraction (DE representation) by natural language processing techniques, b) no index on knowledge representations is exploited for accelerating their retrieval (except via their membership to documents), e.g. the search for the specialisations of a CG is done by projection of this CG on each CG loaded in main memory), and c) it does not allow the retrieval of paths of concepts and relations inside CGs (inside each CG or inside the CGs seen as a global semantic network). For these reasons, CGKAT is mainly interesting from a KA viewpoint.

5 Conclusion

CGKAT helps KA and IR by combining a CG workbench with a structured document editor, and by providing default general ontologies and functions to search and handle them. Thus, compared to other current KA or IR systems, it provides more ways or more *precise* ways to represent or index DEs and structure these DEs or the knowledge representations, and more *guidance* or freedom for representing or structuring information (KA or IR systems generally do not provide a default ontology or provide a non–extensible ontology). Articles related to CGKAT are accessible at *http://www.inria.fr/acacia/CGKAT/*.

CGKAT could be quickly extended by using extensions of CoGITo (e.g. with rules and CGs index), of Thot (e.g. Alliance for document cooperative edition and Amaya for Web–browsing) and of WordNet (e.g. EuroWordNet and International WordNet).

6   References

1. O. Haemmerlé, CoGITo: une plate–forme de développement de logiciels sur les graphes conceptuels. *Ph.D thesis, Montpellier II University, France, Jan. 1995.*

2. P. Martin, Using the WordNet Concept Catalog and a Relation Hierarchy for KA, in *Proceedings of Peirce'95, Santa Cruz, California, Aug. 18, 1995.*

3. P. Martin, and L. Alpay, Conceptual Structures and Structured Documents, in *Proceedings of ICCS'96, Sydney, Australia, Aug. 19–22, 1996.*

4. P. Martin, Exploitation de graphes conceptuels et de documents structurés et hypertextes pour l'acquisition de connaissances et la recherche d'informations. *Ph.D thesis, University of Nice – Sophia Antipolis, France, Oct. 14, 1996.*

5. G.A. Miller, WordNet: A Lexical database for English, in *Communications of the ACM. Nov. 1995.*

6. V. Quint, and I. Vatton, Combining Hypertext and Structured Documents in Grif, in *Proceedings of ECHT'92, D. Lucarella, ed., ACM Press, Milan, Dec. 1992.*

The WebKB Set of Tools: A Common Scheme for Shared WWW Annotations, Shared Knowledge Bases and Information Retrieval

Philippe MARTIN

University of Adelaide – Computer Sciences department, Australia
email: phmartin@gisca.adelaide.edu.au

1 Introduction

The World Wide Web (WWW) provides a simple means for users to make information available to others and to retrieve information by navigation. Additional information retrieval (IR) or collaborative facilities must be provided by servers. For example:
– *net search servers* for document retrieval by query, back–link capabilities, etc.;
– *annotation servers* for storing public/group annotations (e.g. comments, critics, indexations and relations) on (parts of) documents;
– *filtering servers* for controls on document content, form or sequence of desired document, e.g. for document aggregation, link indirection and selective access/display.

However, specific information is difficult to retrieve on the WWW. Net search servers only enable document retrieval and not *knowledge retrieval* because they do not represent the semantic content of documents. Similarly, hypertext links between documents or document annotations only support document retrieval. The smaller the indexed pieces of information are, and the more semantically organised they are, the more the IR system may provide precise answers, answers to queries expressed at different levels of generality, or organised views of documents content.

The WebKB set of tools is aimed to
– allow users to index any document element (DE) on the WWW (e.g. a document or one of its words) by arbitrary precise annotations, e.g. comments, knowledge representations or relations to other DEs,
– guide the building of precise DE representations,
– ease the access, display, modification and comparison of knowledge in a cooperatively built knowledge base (which may be a base of DE annotations),
– allow users or programs to do these operations remotely, e.g. via a WWW–browser,
– exploit such knowledge and indexation of DE by knowledge for generating documents answering precisely searches by query or navigation.

2 Architecture

Various modules or tools are necessary to obtain the previous functionalities, at least an HTML editor, a knowledge editor, an indexation link editor, knowledge/data bases, a data management system, a cooperatively built knowledge base handling system, a knowledge processor (inference tool), an IR processor, IR interfaces and a WWW–browser. A trader and a broker could also be added. A trader is meant to automatically find adequate knowledge/data bases to be exploited for answering information requests. A broker handles the exchange of information between the IR processor and knowledge/data bases having different information access protocols. The next figure shows dependencies between such modules or tools.

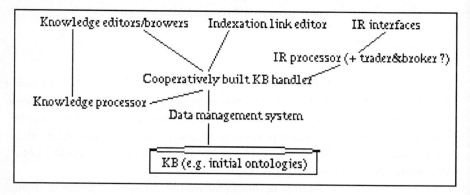

Fig. 1. The WekKB tools data flow

We have chosen to build separate, combinable and WWW–accessible tools. Thus, users may easily use them and exchange any of them with a tool they think more interesting (they only need to change the tool URL address; of course, if the first tool follows a special protocol, its substitute must follow the same protocol). They may save the outputs of a tool in files (which may then be used as inputs to other tools) or they may directly copy and paste the outputs into text entries of other tools.

Whenever possible, we have used HTML and Javascript in our tool implementation to allow users to easily customize their interface and behavior by modifying the HTML sources. As opposed to CGI servers (servers using the Common Gateway Interface), Javascript programs are directly executed on client machines (and at present more safely than Java programs). This implementation choice has the drawback that it restricts the usable WWW–browsers.

The WebKB set of tools will include at least the following components:
– a conceptual graph (CG) textual editor (already implemented in Javascript);
– an ontology editor for the edition of type definitions or relations between types and/or individuals, and a hierarchy browser for such relations;
– an indexation link editor (already implemented in Javascript);

– a CGI server and a Java server allowing the use a data management system;

– a cooperatively built knowledge base handling system (for that, we will implement the protocols we have described in the annexe 2 of [6]);

– the WordNet natural language ontology superseded by a top–level concept type ontology useful for knowledge representation and knowledge acquisition [4,6];

– a relation type ontology which merges various relation types ontologies, e.g. thematic, spatial, temporal and argumentative relation type ontologies [4,6].

– a CGI server allowing the use of the CG processor Peirce [1];

– a CGI server for the generation of documents (IR processor; this CGI server, as well as the previous one, might be reimplemented in Java for efficiency reasons);

– a javascript IR interface showing the results provided by the IR processor.

3 Applications

The WebKB set of tools have a broad range of applications: precision–oriented IR, knowledge acquisition (KA), computer supported cooperative work (CSCW).

We show in CGKAT [4,5,6] (also see our presentation of CGKAT in this volume) how ontologies and the combination of a structured document editor and a CG workbench can ease KA and IR. The WebKB set of tools reuses the underlying philosophy of CGKAT but is WWW–based, more easily extensible and customisable, and may exploit efficient tools such as databases. Similarly to CGKAT structured documents, HTML documents are used for organizing knowledge and data and displaying them (knowledge–based document generation). Although current WWW–browsers/ editors do not yet have all the facilities provided by the structured document editor used in CGKAT (e.g. graph editing, index generation, zoom, views handling), they include more and more similar facilities, e.g. DE presentation models (style sheets), and provide other ones (e.g. navigation history management and script languages).

The combination of facilities allowed by the WebKB set of tools (or similarly based set of tools) will probably allow users to develop applications that were previously too costly. The two most important facilities for KA and CSCW that the WebKB set of tools is aimed to provide are 1) some knowledge–based comparison and synthesis of information provided by different authors, and 2) the user control of generated document content, form or sequence via document descriptions using HTML, Javascript and conceptual queries.

4 Limitations

We do not intend to integrate in the WebKB set of tools facilities which are not based on knowledge handling or structured document handling, e.g. automatic indexation of documents, hypertext network visualisation and real–time interaction between users. However, the WebKB set of tools may easily integrate tools designed by others.

5 Conclusion

The WebKB set of tools is intended to combine various technologies for helping KA, IR and CSCW, notably the WWW–related technologies, the databases and the knowledge representation languages and processors. To do so, we rely on our previous experience in the development of CGKAT. The WebKB set of tools home page is at *http://www.gisca.adelaide.edu.au/~phmartin/WebKBtools/*.

The WebKB set of tools shares many goals and design principles with current WWW–based public annotations tools, e.g. ComMentor [8] and HyperNews [2], and WWW–based traders, e.g. AlephWeb[9], NetRepository [3] and AI–trader [7]. However, such annotation tools are not intended to index DEs by knowledge (they cannot exploit it for IR), and HyperNews, AlephWeb and AI–trader only allow the user to index documents, not arbitrary parts of them. AI–trader uses CGs for indexing documents, while NetRepository uses KIF for communications between knowledge servers. None of these tools can generate documents as answers to user queries.

6 References

1. G. Ellis, Managing Complex Objects. *Ph.D thesis, Queensland University (Computer Sciences Dept.), Australia, 1995.*

2. D. LaLiberte, Collaboration with HyperNews, in *Proceedings of Workshop on WWW and Collaboration, Cambridge, MA, September 11–12, 1995.*

3. C. Luigi Di Pace, P. Leo, and A. Maffione, NetRepository: A Networked Information Repository which Supplies Ontologies for Retrieving Information, in *Proceedings of ICCS'97, University of Washington, August 4 – 8, 1997*

4. P. Martin, Using the WordNet Concept Catalog and a Relation Hierarchy for KA, in *Proceedings of Peirce'95, Santa Cruz, California, August 18, 1995.*

5. P. Martin and L. Alpay, Conceptual Structures and Structured Documents, in *Proceedings of ICCS'96, Sydney, Australia, August 19–22, 1996.*

6. P. Martin, Exploitation de graphes conceptuels et de documents structurés et hypertextes pour l'acquisition de connaissances et la recherche d'informations. *Ph.D thesis, University of Nice – Sophia Antipolis, France, October 14, 1996.*

7. A. Puder, S. Markwitz, and F. Gudermann, Service Trading Using Conceptual Structures, in *Proceedings of ICCS'95, Santa Cruz, California, August 14–18, 1995.*

8. M. Röscheisen, C. Mogensen, and T. Winograd, Beyond Browsing: Shared Comments, SOAPs, Trails, and On–line Communities, in *Proceedings of the Third International World–Wide Web Conference in Darmstadt, Germany, April 1995.*

9. G. Rodríguez, and L. Navarro, AlephWeb: a CSCW Large Scale Trader. *http://www.pangea.org/alephweb.aleph/paper.html*

Deakin Toolset: Conceptual Graphs Based Knowledge Acquisition, Management, and Processing Tools

Brian Garner[1], Eric Tsui[2] and Dickson Lukose[3]

Department of Computing and Mathematics, Deakin University, Australia[1]
CSC Continuum and University of Sydney, Australia[2]
Department of Maths., Stats. and Comp. Sc., University of New England, Australia[3]
E-mail: brian@deakin.edu.au[1], eric@cs.su.oz.au[2], lukose@peirce.une.edu.au[3]

Abstract. In this paper, the authors will briefly describe the conceptual graphs based tools that was developed by various members of the Knowledge Engineering Group at Deakin University, Australia, in the mid-1980's. Using these tools, we have experimented in developing many different types of knowledge based systems.

1 Introduction

The Deakin Toolset is a comprehensive knowledge engineering environment offering rich conceptual structures and advanced graph processing capabilities, including (with restrictions) nested graph options! Knowledge engineering productivity benefits also stem from the domain-independent nature of the tools and the flexibility,through use of the interactive concept classifier,for exchange of conceptual structures derived from multiple sources, provided they conform to the basic Prolog data structure employed. The nature of the conceptual entities and primitives that can be constructed and processed is limited solely by the diadic constraint, formalized by John Sowa [21]. However, the variety of applications and the successful implementation of their associated conceptual structures fully vindicates this initial restriction. Complex prototypical knowledge structures (e.g., abstractions) and novel knowledge modeling regimes have been constructed to assist experts in knowledge acquisition, while on the other hand, a simple assertion-based input process enables relatively unskilled users to create simple graphs, including rule components, which may then be combined to form powerful inference nets. Recent work on advanced applications, in such areas as management intelligence systems (intelligent decision support) [2] and in the construction of intelligent agents for electronic trading, indicate the future promise of such technology and tools and complements the fruitful development of new theoretical ideas using conceptual graphs for logic programming (c.f. Peirce project).

2 Architecture of Deakin Toolset

There are three categories of tools (i.e., modules) The first is the knowledge acquisition tools, which allow the user to encode his/her own declarative knowledge base (in the form of conceptual graphs). The six knowledge acquisition modules that are available within the Deakin Toolset are listed below:

- **Knowledge Base Editor**: this tool enables the user to create conceptual structure, to construct concepts, conceptual relations, conceptual graphs, and abstractions like type definitions, schemata and prototypes. Details of this system are outlined in [13],[14] and [20].
- **Batch Processor**: this tool enables the user to create script files containing the knowledge structures to be built, and run the knowledge base construction as a batch process. Details of this system will be found in [8].
- **Interactive Concept Classifier**: this is a deductive question answering system to build and modify (i.e., classify) the concept type hierarchy. Detailed system is outlined in Garner et al. [4].
- **Rule Acquisition System for Conceptual Structures**: this tool enables the user to encode causal rules for inference programs to operate on. Detailed system is described in [10].
- **Goal Structure Processor**: this tool enables the user to construct a hierarchical goal structure, with each node in the structure being a conceptual graph. Details of this system are found in [11].
- **Rule Encoder**: this tool enables the user to encode Canonical Graph Model rules directly from input propositions. The Rule Encoder combined with the Semantic Interpreter and the General Purpose Inference Engine, provides an interactive environment for the novice user to create and execute rules in a rule base. Details of the Rule Encoder are found in [6].

The second category of tool which makes up the Deakin Toolset are the module utilized for efficient indexing, storage and retrieval (i.e., knowledge management activities). This function is performed by the following module:

- **Self-Organizing Conceptual Dictionary**: this dictionary is able to store (i.e., index) four kinds of conceptual patterns (i.e., concepts, graphs, rules and abstractions). Efficient indexing is facilitated by identifying the difference between the pattern to be indexed and the corresponding patterns in the dictionary. These 'differences' are then used as 'keys' for indexing and locating or matching these structures. Details of this dictionary are found in [5].

The third category of tools are for performing reasoning with these knowledge structures. This is facilitated by the following two modules:

- **Canonical Graph Processor**: this tool contains a set of high-level operators for transforming conceptual structures, and for performing both plausible and exact reasoning. Details of this module are outlined in [8], [13], and [20].

- **General Purpose Inference Engine**: this is the tool for performing reasoning with rules presented in Conceptual Graphs. The inference engine is the core part of the knowledge base system that performs reasoning. It implements a graph-based [18] inference strategy for implementing backward chaining, forward chaining, and mixed (backward and forward) reasoning. There are three sub-modules with this larger module. They are the Rule Consistency Checker, the Rule Compiler and the Rule Execution sub-module. Details of all these sub-modules are found in [20].
- **Executable Conceptual Structures**: this is a set of extensions to the conventional conceptual graphs based on the actor and object oriented principles. Two new forms of abstractions have been developed: Actor Graphs, and Problem Maps [15]. In recent years, it has been extended to a fully graphical executable conceptual modeling language called the MODEL-ECS [16]. MODEL-ECS consists of all the necessary forms of executable abstraction, and control structure to facilitate modeling of complex problem solving methods.

3   Applications: Canonical Graph Models

An overview of the Deakin Toolset with all the above modules are found in Garner, et al. [8] [9]. By utilizing the Deakin Toolset, a number of Canonical Graph Models have been implemented. The following table outlines a few of the Canonical Graph Models and the researchers involved in its development.

Canonical Graph Model	Researcher	Reference
Semantic Interpreter	D. Lukose	Lukose (1986)
Audit Planner	J.Koh	Koh (1986)
Personal Financial Planning Model	E. Tsui	Garner and Tsui (1988)
Goal Interpreter	D. Lui	Lui (1990)
Executable Conceptual Structures	D. Lukose	Lukose(1992)
Hypothesis Generator	F. Chen	Chen (1996)
Dynamic Knowledge Modeling	N.L. Smith	Smith (1996)
MODEL-ECS	D. Lukose	Lukose (1996)

Table 1 Canonical Graph Models

4   Limitations

The practical use of any CG environment requires the ability to restrict graphs to real world concepts and useful relations. Such graphs are called "canonical graphs", or more generally, "instances" of the entity (e.g., prototypes/abstractions of interest!). Schemata, for example, may also be defined as instances of a general schema! The original mechanism used at Deakin for restriction was the "Word Sense" dictionary which enabled us to define quickly only the useful entities of interest for the application in question! However, the general problem of access to a global ontology for business use has, not as yet, been resolved, although some progress has been made in defining taxonomies of re-useable objects, whether CG based or not! The heavy computational load on the computational system, if restriction is not imposed early in the processing stage, is a major impediment at present to wider use of tools, such as the Deakin toolkit, which was designed principally for research use. More recently, however, in work on creativity at Deakin using dynamic knowledge modeling [19], the use of temporal CG instances and the requisite algebra for manipulating such instances, has resulted in new, highly efficient graph processing options, believed to be of great potential in future extension of the toolkit. We are also attracted to the use of CG primitives, originally defined for temporal logic [1], but subsequently extended to application primitives, for obviating the problems in recursive processing of nested graphs.

References

1. Chan, M.C.,Garner, B.J. and Tsui, E. Recursive Modal Unification for Reasoning with Knowledge using a Graph Representation; Knowledge Based Systems, Volume 1, No.2,1988, pp94-104.
2. Chen, F. Hypothesis Generation for Management Inteligence. Ph.D. Thesis, School of Computing and Mathematics, Deakin University, Geelong, Victoria 3217, Australia, 1996.
3. Garner, B.J., Lukose, D., and Tsui, E. Parsing Natural Language through Pattern Correlation and Modification, Proceedings of the 7th International Workshop on Expert Systems and its Applications, May 13-15, 1987, Avignon, France, pp. 1285–1299.
4. Garner, B.J., Tsui, E., and Cheng, B., An Interactive Classifier for the Extendible Graph Processor, Technical Report 87/1, Xerox AI Laboratory, Department of Computing and Mathematics, Deakin University, Geelong, Australia, 3217, 1987.
5. Garner, B.J. and Tsui, E., A Self-Organising Dictionary for Conceptual Structures, in J.F. Gilmore (Ed.), Proceedings of Application of AI V, SPIE 784, 18-20 May, 1987, Orlando, U.S.A., pp. 356 - 363.
6. Garner, B.J., and Tsui, E., A Canonical Graph Model for Personal Financial Planning, in Proceedings of the International Computer Science Conference, December 19-21, 1988, Hong Kong.
7. Garner, B.J., and Tsui, E., Acquiring Lexical and Structural Knowledge for integrated Natural Language Understanding and Reasoning Systems, Proceedings of the IJCAI-89 Workshop on Lexical Knowledge Acquisition, August 21, 1989, Detroit, U.S.A.
8. Garner, B.J., Tsui, E., Lui, D., Lukose, D., and Koh, J., Progress on an Extendible Graph Processor for Knowledge Acquisition, Planning, and Reasoning, in Proceedings of Second Pan Pacific Computer Conference, 26-29 August, 1987, Singapore, pp. 150-165.
9. Garner, B.J., Tsui, E., Lui, D., Lukose, D., and Koh, J., Progress on an Extendible Graph Processor for Knowledge Acquisition, Planning, and Reasoning,in Tim Nagle, Jan Nagle, Laurie Gerholz and Peter Eklund (Eds.), Current Directions in Conceptual Structure Research, Ellis Horwood, 1992.

10. Lui, D., RASCS: Rule Acquisition System for Conceptual Structures, Internal Report, Department of Computing and Mathematics, Deakin University, Geelong, Australia, 3217, 1986.

11. Lui, D., Strategic Knowledge for the Identification and Construction of Goal States, MSc. Thesis, Department of Computing and Mathematics, Deakin University, Geelong, Australia, 3217, 1990.

12. Lukose, D., Canonical Graph Model for Discourse Understanding, BSc (Hons) Thesis, Division of Computing and Mathematics, School of The Sciences, Deakin University, Geelong, Victoria, Australia, 3217, 1986.

13. Lukose, D., Conceptual Graph Tutorial, Department of Computing and Mathematics, Deakin University, Geelong, Victoria, Australia, 3217, 1991.

14. Lukose, D., Goal Interpretation as a Knowledge Acquisition Mechanism, PhD Thesis, Department of Computing and Mathematics, Deakin University, Geelong, Victoria, 3217, Australia, 1992.

15. Lukose, D., Planning Knowledge Acquisition Techniques and Mechanisms, in *Proceedings of the ICCS'94 Workshop on Knowledge Acquisition using Conceptual Graphs*, M.L. Mugnier, M. Willims, B. Gaines and D. Lukose (Eds.), Maryland, USA, 1994, pp. 116-135.

16. Lukose, D., MODEL-ECS: Executable Conceptual Modelling Language, in Proceedings of the 10th Knowledge Acquisition for Knowledge Based Systems Workshop, Banff, Canada, November 1996.

17. Koh, J.E.K., Expert Planners/Audit Planner, BSc (Hons) Thesis, Division of Computing and Mathematics, School of The Sciences, Deakin University, Geelong, Victoria, Australia, 3217, 1986.

18. Neapolitan, R.E., Forward-chaining versus a graph approach as the inference engine in expert system, SPIE vol. 635, Applications of Artificial Intelligence III, 1986.

19. Smith, N.L., Dynamic Knowledge Modelling;Ph.D.Thesis; School of Computing and Mathematics; Deakin University, Victoria 3217, Australia, 1996.

20. Tsui, E., Canonical Graph Model, PhD Thesis, Department of Computing and Mathematics, Deakin University, Geelong, Victoria, Australia, 3217, 1989.

21. Sowa, J.F., Conceptual Structures: Information Processing in Mind and Machine, Addison-Wesley, 1994.

EGP: Extendible Graph Processor

Eric Tsui[1], Brian Garner[2] and Dickson Lukose[3]

CSC Continuum and University of Sydney, Australia[1]
Department of Computing and Mathematics, Deakin University, Australia[2]
Department of Maths., Stats. and Comp. Sc., University of New England, Australia[3]
E-mail: eric@cs.su.oz.au[1], brian@deakin.edu.au[2], lukose@peirce.une.edu.au[3]

Abstract. This paper describes a conceptual graph tool called the Extendible Graph Processor (EGP). We will review its architecture, application, limitations, and future developments.

1 Introduction

In the late 1980's Garner et al. [1] developed a large number of conceptual graphs tools. These tools are collectively known as Deakin Toolset. The Extendible Graph Processor (EGP) is a set of three basic Conceptual Graphs tools from this collection: Conceptual Graph Editor (CGE), Inference Net Editor (INE), and Conceptual Graph Processor (CGP). These three tools form the core of Deakin Toolset. The functions of each of these tools are as follows:

- **Conceptual Graph Editor**: this tool is used to construct conceptual graphs. It is made up of over thirty different operations to facilitate conceptual graph construction.
- **Inference Net Editor**: this tool is used to define abstractions like concept and relation type definitions, schema definition, schematic clusters, prototype definition, and composite individuals. It is made up of over a dozen operations to facilitate definitions of various abstractions.
- **Conceptual Graph Processor**: this tool is used to manipulate conceptual graphs. It consist of the four canonical formation rules: *Copy,Restrict, Simplify,* and *Join.* Further, it also has the following operations: *Maximal Join, Projection, Concept and Relation Type Expansion and Contraction,* and over thirty other operations to facilitate the use of this tool to develop sophisticated knowledge based systems.

For a detailed listing of all the commands available in each of these tools, interested readers are referred to [3].

2 Architecture of EGP

All the conceptual graphs tools that make up the Deakin Toolset is written on MUprolog [4]. Several PhD candidates have worked on these tools. The EGP forms the core part of the whole toolset. The architecture of EGP is depicted in Figure 1.

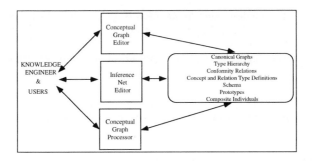

Fig. 1. Architecture of the Extendible Graph Processor

3 Application of EGP

EGP has been applied in the development of various types of knowledge based systems: Expert Systems, Natural Language Processor, Planning Systems, Knowledge Engineering Environments, Electronic Data Interchange, Auditing, etc. In this section, the author will provide a simple demonstration of how EGP can be used by a knowledge engineer to build knowledge bases. This example builds the differential graph for the concept KISS, define concept type KISS, and concept expansion and contraction on a data graph "workgr2".

```
7?- kbmsegp.
***************************************************************
*    E X T E N D I B L E    G R A P H   P R O C E S S O R    *
***************************************************************
[stuff deleted]
Enter the number of your choice: 2.
++++++++++++++++++++++++++++++++++
+ KNOWLEDGE BASE EDITOR (1.1) +
++++++++++++++++++++++++++++++++++
$INFERENCE NET EDITOR ( $INE> )
$INE> edit kissdef.
$$NEW GRAPH CREATED: kissdef
$$CGE(kissdef) > cr rel agnt.
$$CONCEPTUAL RELATION CREATED: -(AGNT)- ...(767651)
$$CGE(kissdef) > the agnt of touch is person.
$$CONCEPT CREATED: [TOUCH] ... (767653)
$$CONCEPT CREATED: [PERSON] ... (767654)
$$ASSERTION CREATED: The agnt (767651) of touch (767653) is person (767654).
$$SOWA'S NOTATION: [TOUCH]->(AGNT)->[PERSON]
$$CGE(kissdef) > cr rel manr.
$$CONCEPTUAL RELATION CREATED: -(MANR)- ...(767655)
$$CGE(kissdef) > the manr of touch is tender.
[stuff deleted]
$$CGE(kissdef) > lf gr.
[TOUCH:767653] -
                 ->(AGNT)->[PERSON:767654]->(PART)->[LIPS:767658]
```

```
                    ->(MANR)->[TENDER:767656]
                    ->(INST)->[LIPS:767658].
$$CGE(kissdef) > write gr.
$$WRITE TO FILE /home/profs/lukose/cgm/kb/rob/kissdef.pl COMPLETED
$$EXIT: CGE>
$ENTER: INE>
$INE> define type kiss as kissdef with head touch.
** INSERTED: Type Definition: type (kiss,touch,767653,767652)
[stuff deleted]
$$CGE(workgr2) > lf gr
[KISS:767667] -
                    ->(AGNT)->[PERSON:jane:767668]
                    ->(OBJ)->[PERSON:jack:767670].
$$CGE(workgr2) write gr.
[stuff deleted]
CGP 1> load workgr2.
 ... LOADING workgr2
 Total GIC = 1
CGP 2> expand kiss workgr2.
 ... LOADING kissdef
 Total GIC = 2
** Assumption: Aristotelian Type Hierarchy: kiss < touch
** Assumption: Aristotelian Type Hierarchy: kiss < touch
     [... DETECTED 2 CONCEPT(S) AND 1 CONCEPTUAL RELATION(S)
          FOR MAXIMAL JOIN ...]
[stuff deleted]
** <COPY>: Graph kissdef --> temp239 [DONE]
     GIC: - temp238
** <MAXIMAL JOIN>: [KISS] in workgr2 and
                    [TOUCH] in kissdef
                    gives temp239 [DONE]
** <TYPE EXPANSION>: Expanding [KISS] in workgr2 by type(KISS)
                    gives temp239 [DONE]
CGP 3> lf temp239.
[TOUCH:767677] -
                    ->(AGNT)->[PERSON:jane:767678]->(PART)->[LIPS:767680]
                    ->(MANR)->[TENDER:767679]
                    ->(INST)->[LIPS:767680]
                    ->(OBJ)->[PERSON:jack:767673].
CGP 4> contract temp239 kiss.
[stuff deleted]
** <CONTRACT>: temp239 by type(KISS) gives temp240 [DONE]
CGP 5> lf temp240.
[KISS:767687] -
                    ->(OBJ)->[PERSON:jack:767686]
                    ->(AGNT)->[PERSON:jane:767688].
```

4 Limitation of EGP

There are two major limitations in EGP. The first is due to its implementation language: MUprolog. This is a very old version of a prolog interpreter that does not have a garbage collector. Thus, the processing capabilities of EGP is limited by the memory size of the machine. We have successfully overcome this limitation by being very careful with our coding. The second limitation is due to the lack of a graphical user interface. This limitation has been one of the major difficulties in training new people to use these tools. The semi-structured English commands, and the linear form display of conceptual graphs are sometime quite difficult for novice users. Currently, the authors are investigating the options of building a graphical user interface to EGP by using the Constraint Graphs tool that was developed by Kremer [2]. The second extension to this tool is to move it onto the WEB.

Form the application development point of view, we identify the limitation of the EGP system to be the lack of industry-specific classes of knowledge structures (e.g., the definition of a policy, claim, rating etc. in the insurance domain) and a robust graphical system to display any kind of conceptual structures (i.e., concepts, relations, graphs, definitions etc.). For any significant commercial applications utilizing the EGP, it is difficult to imagine that all the fundamental concepts are, due to time and funding constraints, encoded from scratch. While the EGP is highly suited to function as a fraud detection and discovery engine (performing generalizations on certain known concepts and identifying any recurrent and abnormal patterns), the lack of an "off-the-shelf" library of conceptual structures has undoubtedly add more risk to a commercial project of this kind. Besides, users (i.e., fraud department officers) would generally prefer to "visualize" the relationships among various entities rather than dealing with the same information in a less iconic fashion. The EGP system is neither sharewere nor are they in public domain. For more details on EGP, please contact Professor Brian Garner at *Department of Computing and Mathematics, Deakin University, Geelong, Victoria, 3217, AUSTRALIA (Email: brian@deakin.edu.au)*.

References

1. Garner, B.J., Tsui, E., Lui, D., Lukose, D., and Koh, J., Progress on an Extendible Graph Processor for Knowledge Acquisition, Planning and Reasoning, *Current Directions in Conceptual Structure Research*, Tim Nagle, Jan Nagle, Laurie Gerholz and Peter Eklund (Eds.). Ellis Horwood, 1992.
2. Kremer, R., A Graphical Meta-Language for Knowledge Representation, (In Progress) PhD Dissertation, University of Calgary, Canada, 1997.
3. Lukose, D., Conceptual Graph Tutorial, Technical Report, Department of Computing and Mathematics, Deakin University, Geelong, 3217, Victoria, Australia, 1991.
4. Nash, L., MUprolog 3.1b Reference Manual, Melbourne University, September 1984.

CGKEE: Conceptual Graph Knowledge Engineering Environment

Dickson Lukose

Distributed Artificial Intelligence Centre (DAIC)
School of Mathematical and Information Sciences
The University of New England
Armidale, New South Wales, Australia
E-mail: lukose@peirce.une.edu.au

Abstract. This paper will describe briefly the Conceptual Graph Knowledge Engineering Environment (CGKEE), its application and limitations. Further, the author will also outline some of the work currently being carried out to overcome these limitations.

1 Introduction

The Conceptual Graph Knowledge Engineering Environment (CGKEE) is being developed at DAIC by the author and his research team. Munday [1] developed the first version of this tool, then extensively revised by other in t he development team [2]. Descriptions of this system is also found in [3] [4]. This systems is developed for the sole purpose of using it as a conceptual graphs knowledge base server. That is, CGKEE is designed to reside on a high speed machine, and for the users on the client machine to connect to this server for the purpose of knowledge retrieval, storage, and manipulation. CGKEE does not have any sophisticated inference mechanisms. It is purely a declarative knowledge engineering environment.

2 Architecture of CGKEE

CGKEE is made up of the following components: User Interface, Processor, Editor, Working Memory and Knowledge Base. Figure 1 depicts the architecture of CGKEE. Each of these components are described below:

- **User Interface** - enables the CGKEE Users to interact with the Processor. In addition to this, it also enables one to interrogate the Working Memory and the Knowledge Base;
- **Processor** - contains the Working Memory and has the following conceptual graph manipulation functions: *Join, Copy, Restrict, Simplify, Projection, and Maximal Join*; and
- **Editor** - enables the knowledge engineer to build conceptual graphs and other abstractions and store them in the Knowledge Base. There is no User Interface to the editor yet. So, at the moment, the knowledge objects are compiled with the source code.

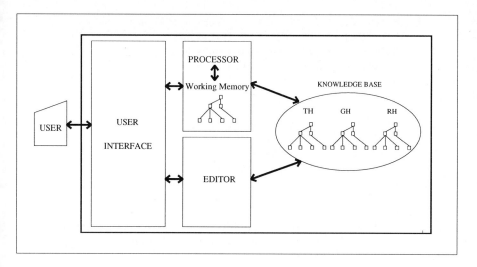

Fig. 1. Architecture of CGKEE

3 Application of CGKEE

CGKEE is used mainly for research and teaching purpose. The most extensive testing was done with a knowledge base of approximately 3 million conceptual graphs used in a data mining project. CGKEE is also used as a teaching tool in the "Knowledge Engineering" course that is offered the University of New England.

In this section, we will provide a sample run of CGKEE performing the following graph operations: *Copy, Restrict, Join, Simplify, Projection*, and *Maximal Join*.

- **Copy**: The copy operation is one of the Canonical Formation Rule. Before this operation can be carried out, the graph to be copied must exist in working memory.

```
CGKEE > load odieEatGarfield
Loaded the graph odieEatGarfield
CGKEE > copy odieEatGarfield
** MESSAGE: the graph 'temp258' has been created.
CGKEE >
```

- **Restrict**: There are two versions of the restrict command available in CG-KEE. The first version restricts some concept of type t to a subtype of t. The second version restricts a generic concept to an individual concept. Both versions of the restrict operation are shown below:

```
CGKEE > load personSueEatPie
Loaded the graph personSueEatPie
CGKEE > display personSueEatPie
graph personSueEatPie is
```

```
[EAT #256] -> (AGNT #254) -> [PERSON #25: Sue#12345]
[EAT #256] -> (OBJ #255) -> [PIE #257]
CGKEE > restrict 25 in personSueEatPie to GIRL
** MESSAGE: the graph 'temp24' has been created.
CGKEE > display temp24
graph temp24 is
[EAT #21] -> (AGNT #22) -> [GIRL #23: Sue#12345]
[EAT #21] -> (OBJ #36) -> [PIE #18]
CGKEE > load girlEatFast
Loaded the graph girlEatFast
CGKEE > display girlEatFast
graph girlEatFast is
[EAT #33] -> (AGNT #19) -> [GIRL #20]
[EAT #33] -> (MANR #37) -> [FAST #35]
CGKEE > restrict 20 in girlEatFast to individual 12345
** MESSAGE: the graph 'temp34' has been created.
CGKEE > display temp34
graph temp34 is
[EAT #260] -> (AGNT #262) -> [GIRL #259: Sue#12345]
[EAT #260] -> (MANR #263) -> [FAST #261]
CGKEE >
```

– **Join**: The join operation joins two graphs on identical concepts. The concepts must be identical both in type and referent. The operation could produce multiple results if there is more than one set of identical concepts in either of the graphs. The following sample run shows the join of two graphs on identical concepts:

```
CGKEE > join temp34 temp24
** MESSAGE: the graph 'temp273' has been created.
** MESSAGE: the graph 'temp283' has been created.
CGKEE > display temp273
graph temp273 is
[EAT #264] -> (AGNT #269) -> [GIRL #266: Sue#12345]
[EAT #264] -> (OBJ #272) -> [PIE #267]
[EAT #264] -> (AGNT #271) -> [GIRL #268: Sue#12345]
[EAT #264] -> (MANR #270) -> [FAST #265]
CGKEE > display temp283
graph temp283 is
[EAT #276] -> (AGNT #279) -> [GIRL #274: Sue#12345]
[EAT #278] -> (AGNT #281) -> [GIRL #274: Sue#12345]
[EAT #278] -> (OBJ #282) -> [PIE #277]
[EAT #276] -> (MANR #280) -> [FAST #275]
CGKEE >
```

– **Simplify**: The simplify operation removes duplicate relations from a graph. There are three versions of this operation. The first, simplifies a particular relation in a graph. This is done by specifying the ID number of the relation. The second version simplifies all duplicate relations of a given type. Finally, the last version of the operation produces the most simple graph possible.

– **Projection**: This operation projects one graph onto another. Multiple results can be produced if there is more than one projection.

```
CGKEE > load ACT_conCat
Loaded the graph ACT_conCat
CGKEE > load personSueEatPie
Loaded the graph personSueEatPie
CGKEE > display ACT_conCat
graph ACT_conCat is
[ACT #278] -> (AGNT #37) -> [ANIMATE #270]
CGKEE > display personSueEatPie
graph personSueEatPie is
[EAT #18] -> (AGNT #254) -> [PERSON #282: Sue#12345]
[EAT #18] -> (OBJ #255) -> [PIE #21]
CGKEE > project ACT_conCat onto personSueEatPie
** MESSAGE: the graph 'temp274' has been created.
CGKEE > display temp274
graph temp274 is
[EAT #257] -> (AGNT #279) -> [PERSON #280: Sue#12345]
CGKEE >
```

– **Maximal Join**: Two graphs can be joined on maximally extended common projections. Multiple graphs could result if there are more than one maximally extended common projection.

```
CGKEE > load girlEatFast
Loaded the graph girlEatFast
CGKEE > display girlEatFast
graph girlEatFast is
[EAT #283] -> (AGNT #285) -> [GIRL #284]
[EAT #283] -> (MANR #286) -> [FAST #276]
CGKEE > load girlEatFood
Loaded the graph girlEatFood
CGKEE > display girlEatFood
graph girlEatFood is
[EAT #280] -> (AGNT #265) -> [GIRL #266]
[EAT #280] -> (OBJ #267) -> [FOOD #281]
CGKEE > maxjoin girlEatFast girlEatFood
** MESSAGE: the graph 'temp287' has been created.
CGKEE > display temp287
graph temp287 is
[EAT #289] -> (AGNT #288) -> [GIRL #290]
[EAT #289] -> (OBJ #293) -> [FOOD #294]
[EAT #289] -> (MANR #291) -> [FAST #292]

CGKEE >
```

4   Limitations

The two major limitations of CGKEE are: (1) the lack of a graphical user interface; and (2) the lack of persistent knowledge base. That is, when the knowledge base is loaded from the file, all the objects have to be created each time. Our experiment with 3 million graphs demonstrated the difficulties in using this tool for very large conceptual graphs knowledge bases. These limitations are being addressed in the following manner. Firstly, the issue on graphical user interface is being address by incorporating a CG graphical interface developed by Kremer [5]. Further, since CGKEE is being used as a knowledge based server (its best application), the graphical user interface is designed as a plug-in to the Netscape web browsers. This will enable CGKEE to be available to everyone on the net. Also, a multi-user web interface is being developed in conjunction with the Sysphus 4 project that is being carried out at the Knowledge Science Institute (KSI), University of Calgary. Sysphus 4 is a world-wide project initiated by Professor Brian Gaines at KAW'96 (Banff). This project aims to develop knowledge acquisition and modeling tools and techniques for multi-user collaborative environment. The problem with persistent knowledge base is being addressed by developing a relational database to efficiently handle the storage and retrieval of conceptual graphs. The conceptual design of an appropriate relational database has been completed, but we have not implemented it yet. This is not an immediate issue, thus differed for later development.

There are many areas that need to be developed to make CGKEE much more complete and user friendly. Some of the future areas of development (including the above) are: Concept and Relation Expansion and Contraction Operation; representation of conjunctive set of conceptual graphs in both positive and negative context; and Actors. The author invites collaboration from members of the CG Community.

References

1. C. Munday. *The UNE Conceptual Graph Knowledge Engineering Environment: An Implementation of the Kernel.* Honours thesis, Department of Mathematics, Statistics and Computing Science, The University of New England, Australia, 1994.
2. C Munday, T Cross, J Daengdej, and D Lukose. CGKEE: Conceptual graph knowledge engineering environment user and system manual. Technical report, The University of New England, 1996. Technical Report No. 96-118.
3. C. Munday and D. Lukose. Object-oriented design of conceptual graph processor. In G. Ellis and R. Levinson, editors, *Proceedings of the Third International Workshop on Peirce: A Conceptual Graph Workbench*, pages 55–70, Maryland USA, August 1994.
4. C. Munday, F. Sobora, and D. Lukose. UNE-CG-KEE: Next generation knowledge engineering environment. In G. Ellis and P. Eklund, editors, *Proceedings of the First Australian Workshop on Conceptual Structures*, November 1994.
5. Robert Kremer. *A Graphical Meta-Language for Knowledge Representation.* PhD thesis, The University of Calgary, 1997. in preperation.

Menu-Based Interfaces to Conceptual Graphs: The CGLex Approach

Galia Angelova[1], Svetlana Damyanova[2], Kristina Toutanova[2], Kalina Bontcheva[3]

[1] Ling. Modelling Lab, Acad. G.Bonchev Str. 25A, 1113 Sofia, BULGARIA
`galja@lml.acad.bg`
[2] University of Sofia, Faculty of Mathematics and Computer Science, BULGARIA,
`{svetlana,kris}@lml.acad.bg`
[3] Dept. of Computer Science, Univ. of Sheffield, 211 Portobello Str., Sheffield S1 4DP
`K.Bontcheva@dcs.shef.ac.uk`

1 Introduction

Most of the existing CG tools aim at providing workbenches for Conceptual Graphs in (*i*) graphical representation (e.g., [5, 7]); (*ii*) linear form [4]; (*iii*) some internal representation (see [6], items identification under different operations).

Having in mind that a tool must support the creation, update and processing of thousands of type labels and graphs, the three approaches can be evaluated from an user-oriented perspective as follows:

- the most comfortable form should be the *graphical* one, when it is combined with functions that enable flexible browsing of the hierarchy and existing types, type definitions and graphs; search for type labels; easy updates; etc. Unfortunately, such a workbench is extremely complex and requires development efforts currently beyond academic capacity;
- *linear form* input requires a good command of the syntax; the resulting graphs are less readable by non-specialists. However, the linear notation is popular among CG specialists, especially in e-mail discussions;
- *direct encoding* using some internal representation is suitable for knowledge engineers only, who have no better choice. In addition to the usual difficulties of maintaining many type labels, they must be well aware of all internal data structures and often provide additional information – e.g., identifiers of types and graphs.

Alternatively, CGLex[4] offers a menu-based interface supporting the maintenance of a CG Knowledge Base (KB): type hierarchy, type definitions, concepts, individuals, contexts, and conceptual graphs. The two distinguishing features of CGLex are: (*i*) the underlying CG representation remains completely hidden and the internal identifiers are generated automatically; and more important, (*ii*) the menu based interface constraints the user to define only syntactically correct CG structures. Thus she can concentrate on the semantic consistency of the encoded knowledge.

---

[4] The project is funded by the Bulgarian National Science Fund under contract I-420/94, see `http://www.lml.acad.bg/projects/cglex.html`.

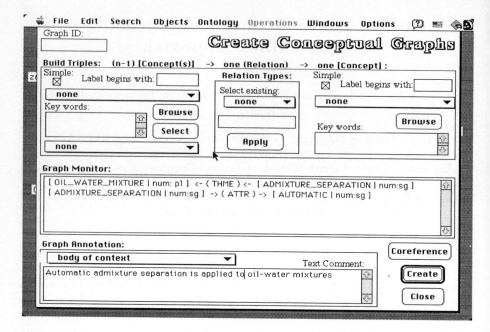

Fig. 1. Conceptual Graph Definition Window

2 CGLex Functionality

CGLex provides menus and windows for definition and revision of a conceptual
graphs KB. Figure 1 presents one of the most complicated windows – the conceptual graphs definition one. Initially, the window can display help texts in each
field, when this feature is selected in the **Options** menu. For instance, the help
text in the field for new relation types (the field between the pop-up menu of
existing relations and the **APPLY** button on Figure 1) was "Define new", which
was deleted after the choice of relations with the **APPLY** button.

Figure 1 illustrates an important CGLex feature: CGS are created element
by element, where each element (e.g., concept type, context) is already defined
using the CGLex interface. However, this requires a certain sequence of user
actions that create the necessary "building blocks". In CGLex we consider the
following design as most appropriate:

1. type labels are defined in the type hierarchy under the **Ontology** menu;
2. concept types should be fully specified with their referents using the window
   Create concepts under **Objects** menu. If no referent is specified, then a
   generic concept is created.

Only after these preparatory steps one can define contexts, graphs and type
definitions containing these items.

Now let us discuss briefly Figure 1, since it exemplifies the CGLex menu style
and design. At the top of the window there is an area for definition of the graph

n-tuples organised around conceptual relations, i.e. n-1[concept(s)] -> (relation) -> [concept]. The concepts (contexts) corresponding to the incoming arcs are to be chosen among existing simple concepts (the shown extraction is restricted by labels) and among existing contexts (which are browsed by keyword search on their NL comment, see [3]). The several selected items are displayed in a pop-up menu for revision and update. Then the concept or context corresponding to the outgoing arc is identified. If the same concept type is selected more than once, a special subitem of the Edit menu creates the concepts by generating indices, so they can be distinguished by the user. Finally, the user selects a conceptual relation and presses the APPLY button, which creates a n-tuple in the Graph Monitor field. All fields in the n-tuple selection area are deleted, so that the next n-tuple can be defined in the context of the graph fragments which are already stored in the Graph Monitor. The graph annotation contains a marker for graph kind (*context, graph, query graph, canonical, body of type definition*, etc. – all of them labels in the type hierarchy) and a text-edit field, where a free text comment about graph semantics can be entered. In the case of Figure 1, the CREATE button will create a new context with the attached text comment. New instances of the concept types [OIL-WATER-MIXTURE], [ADMIXTURE-SEPARATION] and [AUTOMATIC] will be automatically created as well. Later this context can be accessed either by its identifier (ID), or by search on the text comments until the context is found and displayed in a graph monitoring field.

An alternative approach to declaring new graphs is to focus on concepts instead of conceptual relations, i.e., to choose a concept and to enumerate all conceptual relations in which it participates. In fact, separate screens can be created for each strategy, so that the user can switch between them.

At present, CGLex offers the following functionality grouped in menus:

- **File** offers standard handling of knowledge base files;
- **Edit** manages CG structures: for instance, its items remove some elements selected on the front window, or duplicate concepts in case that more than one concept instances are to be included in a graph;
- **Search** also acts on CG elements, e.g., the user can check which type label participates in which graph;
- **Object** is the menu supporting definition of concept referents, graphs, contexts and individuals;
- **Ontology** provides a knowledge perspective to the encoded types and facts: the type hierarchy is defined and traversed here and it can also be drawn according to the depth selected in **Options**; the type definitions are managed as well as the hierarchy of graphs and individuals.

3 Application

CGLex was designed as a KA tool for a project in knowledge-based natural language processing [1, 2] and is used to manipulate larger knowledge bases of Conceptual Graphs. CGLex maintains several kinds of service information, e.g. participation of labels in graphs and arity of relation types, which could be used to enforce consistency when KB items are deleted and modified. The tool is currently implemented in LPA MacProlog and the graphical windows for the type hierarchy rely on its graphical facilities.

Current limitations: At present there are no CG operations integrated in CGLex and no linear notation input is supported.

4 Conclusion

To summarise, in CGLex (*i*) the system controls the user input to ensure consistent syntax and can provide context-sensitive help; (*ii*) all encoded facts can be tracked by the NL comments associated with each CG. Therefore, the user can think in terms of encoded facts as well as conceptual structures, i.e., can create a "library" of facts. Such an approach also offers a language-based exploration of the domain model and different models can be considered as collections of knowledge items regardless of their internal representation.

These features come at the cost of knowledge input spread across various screens and also providing NL comments. Most probably a skilled knowledge engineer will prefer a linear notation in case of more limited number of KB facts. However, browsing of NL texts instead of browsing graphs is very suitable for identification when the KB is processed.

We claim that the ideal CGTool would be a combination of graphics, NL labeling and flexible translation to/from linear form. The NL comment is extremely important, since it provides a higher level of abstraction from the CG structures.

References

1. G. Angelova and K. Bontcheva. DB-MAT: Knowledge Acquisition, Processing and NL Generation. In P. Eklund, G. Ellis, and G. Mann, editors, *Conceptual Structures: Knowledge Representation as Interlingua*, LNAI 1115. Springer Verlag, 1996.
2. G. Angelova and K. Bontcheva. Task-Dependent Aspects of Knowledge Acquisition: a Case Study in a Technical Domain. In *Proceedings of ICCS'97*, 1997. To appear.
3. G. Angelova, N. Boynov, K. Bontcheva, and S. Damjanova. CGLex: A Natural Language Based Tool for Conceptual Graphs. In *Proceedings of CGTools Workshop*, pages 9 – 10, Sydney, Australia, Aug. 1996.
4. G. Ellis and R. Levinson. The Birth of Peirce: A Conceptual Graphs Workbench. In T. N. H. Pfeiffer, editor, *Proceedings of the 7th Annual Workshop on Conceptual Structures (ICCS'92)*, LNAI 754, pages 219 – 228, 1992.
5. J. U. Moeller and D. Wiese. CGEditor: Editing Conceptual Graphs. In *Proceedings of CGTools Workshop*, pages 14 – 16, Sydney, Australia, Aug. 1996.
6. H. Petermann, R. Schirdewan, L. Euler, and K. Bontcheva. CGPro v. 1.0 – a Prolog Implementation of Conceptual Graphs. Memo 264, University of Hamburg, Computer Science Department, Oct. 1996.
7. M. Wermelinger and J. Lopes. An X-Windows Toolkit for Knowledge Acquisition and Representation Based on Conceptual Structures. In T. N. H. Pfeiffer, editor, *Proceedings of the 7th Annual Workshop on Conceptual Structures (ICCS'92)*, LNAI 754, pages 262 – 271, 1992.

Knowledge Extractor:
A Tool for Extracting Knowledge from Text

Walling R. Cyre

Associate Professor of Electrical Engineering
Virginia Polytechnic Institute and State University
Blacksburg, VA 24061-0111
cyre@vt.edu

1 Introduction

This paper describes a knowledge-base development tool being developed at Virginia Tech to assist domain experts (knowledge engineers) construct domain-specific knowledge bases. The methodology assumed here for knowledge-base development is to begin with a relatively small, basic ontology and extract conceptual graphs from text in the domain of interest. During this process, the ontology is augmented by the knowledge engineer as needed. The user scans the text, constructing conceptual graphs from sentences or other expressions, and joins the individual graphs into a knowledge-base.

A number of tasks in this process should be performed by a tool, so the engineer can focus on the difficult tasks of mapping expressions in the source text to concepts, linking concepts by relations, adding new concept types and relation definitions as needed. The readily mechanized tasks include displaying source text, reminding the engineer of the applicable concept and relation types, formatting the graphs, detecting similarities between concepts, and executing specified joins to integrate graphs into a knowledge-base.

The Knowledge Extractor is being implemented with Microsoft Visual C++.

Concept Type Hierarchy	*Relations Types*			
universal	attach	after	agent	attr
object	begins	before	contains	destination
device	duration	ends	enters	exits
value	facilitates	generates	if	in
behavior	initiates	instrument	manner	not
action	next	ordinal	operand	part
event	patient	purpose	quantity	result
state	source	temporal	terminates	triggers
attribute	when	or		

Figure 1. An Initial Ontology

2 Initial Ontology

Knowledge-base generation begins with a small, ontology constructed manually by an expert in the field. The initial ontology consists of a few high-level concept types (for example the Cyc upper hierarchy) and a few obvious relation definitions. In the present discussion, the example domain is digital computer systems. The top-level hierarchy and relations list are indicated in Figure 1. These have been derived from design representations including flowcharts, dataflow diagrams, Petri nets, schematic diagrams, and many other modeling notations for the domain [1].

3 Knowledge Capture

The knowledge engineer begins with the base ontology and a corpora of text from which the knowledge base will be extracted. In this scenario, the engineer selects a sentence of text and begins a new graph. To begin, the engineer will select a sentence word (token) which will generate the concept at which the graph is rooted. If this token has ever been processed, there may be an entry in the tool's dictionary for it, and the list of concept types into which the word has been mapped is displayed. The engineer selects one of the suggested concept types to instantiate the root concept for the graph. Specification of a concept type then invokes the list of relations which can have that concept type as its source or target, so the engineer can select the appropriate relation. If no relation is suitable, then the user must define a new relation using the ontology editor. If a suitable relation type is available, the user highlights it and a word in the text which is to satisfy the concept at the other end of the relation (the role filler or target of the relation). If the word and concept type are compatible, then the concept type is instantiated with the concept identified with that word in the dictionary. Otherwise, the user is invited to restrict the concept and cause a new entry to be added to the dictionary. When the user has completed the conceptual graph, by entering '],', then the graph is added to the knowledge base, and a new graph can be initiated. The text in a sentence may be translated into more than one graph, and some of the text can be ignored. If two distinct graphs are constructed using a common sentence word, then these graphs are automatically joined on the concepts into which the word is mapped. Thus, the engineer does not have to mentally parse very complex sentences, but can work on clauses and phrases. The tool will interconnect the resulting graphs.

Figure 2 illustrates the knowledge capture activity. Assume the engineer has rooted the graph on the main verb (awaits) and attached its subject ('the DLM interface') as the agent. Next, the word **activation** is selected in the source text while the terminate relation is selected in list of applicable relations. This adds the (terminates) <-[event] line. Note that **terminates'** is an inverse relation in the table, so the link direction is reversed. The subtypes of event are also displayed, but the word activation must not have occurred earlier since a suggested concept type has not been found in the dictionary. The engineer may select a subtype of event (*initiate* here) to see if it has a more suitable subtype. In this case, the *event* type in the graph is replaced by the *initiate* type. Assuming initiate is suitable, the engineer can use a dialog box to 1) add *activation* to the dictionary with *initiate* as its concept type, or 2) add *activate* as a subtype of *initiate*, thereby augmenting the ontology. If a suitable relation does not appear in the table, then a new relation has to be defined using a dialog box.

The various buttons below the source text window cause automatic formatting of the graph. For example, '[' begins a new graph, and ') - CR LF TAB [' starts a list of concepts for a relation. Since conceptual graphs can be hierarchical, it is possible that a graph must be installed in the referent field of a concept. this can be accomplished by selecting the ':' button to begin a referent field, and then the '[' button to initiate a new graph. This opens a new window for the nested graph.

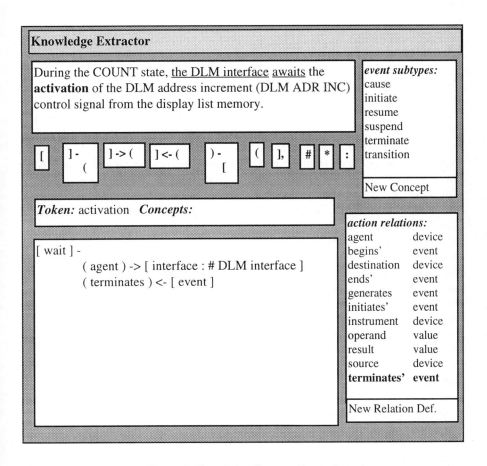

Figure 2. Knowledge Capture Example

4 Ontology Editor

The ontology can be modified during the knowledge capture activity as mentioned above. But, it is also desirable to be able to work on the ontology directly. This can be accomplished by invoking an ontology editor. This allows browsing of the concepts, the type hierarchy and the definitions of relations. Concept types and relation definitions can be added quite easily. Modification of the concept hierarchy, deletion of concept types or relations may be necessary. This raises new problems since a knowledge-base may then be inconsistent with the revised ontology. Clearly it will be necessary to evaluate the knowledge-base against the changes, and to update the knowledge base accordingly. Designs for facilities to support knowledge-base scanning and correction are currently being considered.

5 Coreferencer Feature

The knowledge capture process described above results in a sequence of disconnected conceptual graphs, each of relatively small size. For a useful knowledge base the graphs must be integrated by joining or linking concepts that have the same referents. In earlier research, a process for discovering coreferences in a running sequence of graphs extracted from text was investigated [2]. When a graph has been extracted from the text, its major concepts can be compared against similar concepts in earlier graphs. This is facilitated by maintaining a *definitions table* in which a concept plus its immediate roles (relations with immediately adjacent concepts) are maintained. In the current graph, a concept and its roles will compared against the entries of the table and positive comparisons can displayed to the knowledge engineer, who can decide if the concept is coreferent with a table entry. If so, a choice on how to link the concepts is available. A 'same' relation representing a coreference link can be established between the two concepts, or the concepts can be joined. The *same* relation allows the flexibility of retracting the links later, whereas a join cannot be identified and undone. If a concept in the current graph is not associated with an entry in the definitions table, the concept is classified as a definition and added to the table.

References

1. Walling Cyre "A Requirements Language for Automated Analysis," International Journal of Intelligent Systems, 10(7), 665-689, July, 1995.

2. S. Shankaranarayanan and W. R. Cyre, "Identification of Coreferences with Conceptual Graphs," Proc. 1994 International Conference on Conceptual Structures, College Park, MD, 45-60, August 16-20, 1994.

This work is being supported in part by the National Science Foundation.

The CG Mars Lander

Gil Fuchs[1] and Robert Levinson[1]

[1] Department of Computer Science, University of California, Santa Cruz, CA 95060
[2] in collaboration with TextWise, Syracuse, New York
E-mail:{gil,levinson}@cse.ucsc.edu
Phone: 408-459-2087

Abstract. The name of the system is "THE CG MARS LANDER", which stands for **T**ype **H**ierarchy **E**nhanced **CG** **M**atching **A**nd **R**etrieval **S**ystem **L**inear **A**ssociative **N**-tuple **D**eductive **E**mbedded **R**epresentation. The application is text retrieval. The technique involves translation of text fragments into an informationally ordered form (CGs) and retrieval on the induced subsumption ordering. This application is a linearized CG, NLP, query, matching and retrieving system. English discourse is converted into CGs which are stored in an associative database. The hierarchical database is guided by content morphology and associated ontological type hierarchy, to produce a partially ordered "more general than" hierarchy based on subgraph containment. Due to the nature of the ADB a query is answered by being inserted into the DB. The query's successors in the DB are the answers to the query, as well as the (more specific) retrieved relevancies.

1 Introduction

1.1 Background and motivation

CGs have a great promise and many applications, natural language processing and understanding, knowledge aquisition, modeling, data mining to name a few. Development and research in this domain, however, has suffered from the belief that this promising technology will not scale, even on multi processor computers. The initial timings of this project gave comfort to the sceptics. The first attempt (early 1996) at converting 100,000 English sentences to CGs and matching the CG database against a small set of queries took about 6.5 hours on a Sun Sparc Ultra enterprise 4000. Using a variety of techniques, including the progressive application of the Levinson associative database methods III through VI [8], we managed to reduce this to 19 seconds (of which 16 seconds is the overhead ontology loading and 3 seconds is the actual processing time) on the same Sparc station (early 1997). Our goal is to further speed the process up so that this technology can actually be useful on real world size problems.

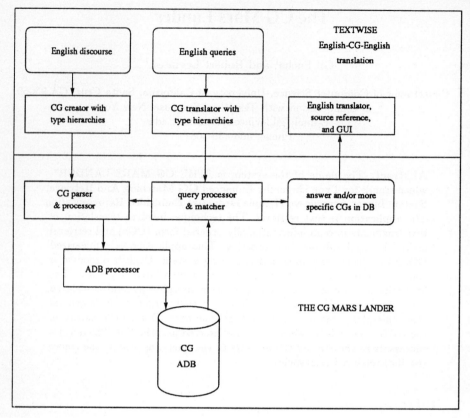

Fig. 1. The CG Mars Lander: Schematic View of Module Architecture

2 Architecture of the System

2.1 High level architecture

Figure 1 shows the overview of the CG tool. English discourse, coupled with ontology and queries are inputs to the system, and the result is an English answer with appropriate references to where in the English discourse the query gets answered.

2.2 Key features

Simple representation of CGs: By employing a linear tuple notation we provide for a simple and efficient implementation of CGs. A CG consists of n tuples, each of which, has any number of arguments, which, in turn, can either be a linguistic construct or a CG.

Expressiveness of CGs:

- **item negation:** Any CG, relation, or concept node arguments may be negated.
- **argument values:** Any concept node argument may have a value.
- **ranges:** Any place a single value is allowed, a range is also permissible.
- **dates:** These ranges, may be of **date** or **time** type.

Arbitrary nesting capability of CGs: An argument can, in turn, be a previously introduced CG tag. Hence, allowing nested contexts to any depth.

Methods III through VI of associative DB matching: Exploiting Levinson's algorithms for optimized graph retrieval.

Automatic maintenance of the type hierarchy: The type hierarchy is an **isa** partial order over some 100,000 English concepts.

Automatic cycle recognition for the type hierarchy: Should the above hierarchy contain cycles, a warning is issued.

DSPS codes: Each CG may contain an association to the document, section, paragraph and sentence from which it was derived, for bibliographic retrieval.

2.3 Levinson's UDS [7] design foundations slogans

- Every primitive data object, label or symbol should be stored only once with pointers used to denote the actual uses of the object.
- Every compound object should be stored with the minimum information required to represent the combination of its parts.
- Given no loss of accuracy, objects should be processed at the highest level of abstraction possible.
- If one were to implement a conceptual graph based on the diagrammatic representation, the costs associated with storage and matching would be much higher than they need to be.
- The same abstraction mechanism that goes from labels to graphs can be taken one step further to facilitate the storage and retrieval of nested context graphs.
- A graph is itself the best descriptor of its nodes.

3 Application of the System

3.1 Summary of results and timings

Our current program can accept a large set of CGs (tens of thousands), an ontology of some few hundred thousands words, and a set of queries. The program stores the CGs in a database which can be saved and restored, and answer queries, by returning relevant CGs from the previously saved DB. The timing statistics on a Sun Ultra Enterprise 4000 (with 4 UltraSPARC 167Mhz and 512KB External Cache CPU and 256BM of main memory) are as follows. Read, process, and store an 18,000 CG input file in 1 hour and 46 minutes. Reloading of above DB takes on the order of seconds. A 150,000 word ontology is processed in 16 seconds. Each query is handled in 5.5 seconds. For smaller databases (hundreds of CGs only), the time to handle a single query can be as low as 0.2

seconds. A typical CG consists of some ten tuples, each of which has two to five arguments. Some large CGs can, however, reach up to 30 tuples (with no effective limit in the program). CG processing includes the treatment of entity negation, item values and ranges (including dates), as well as nesting.

3.2   Sample session

Samples of elements in the package:

- An English sentence:

  The US government's privately managed Hungarian American Enterprise Fund has invested 1 million Dollars in Hungary's first business to establish automated teller machines nationwide.

- Tokenization and tagging using TextWise's parser and conversion to CGs. One sentence may give rise to several CGs, with embedding also possible.
- Standard CG notation:

  ```
  [[government]-->(AGNT) --> [be]] --> (AGNT) --> manage
      --> (OBJ)
  --->
  [[Hungarian_American_Enterprise_Fund] --> (AGNT)
      --> [invest] --> (OBJ) --> [dollars: 1000000]
      --> (IN) --> [First-Business]].
  ```

  Note that the sentence parsed above is not related to the question and answer which follow.
- Sample queries:

  ```
  /*   Q1: When did Rupert Murdoch own the New York Post?   */
  [[Rupert Murdoch] -->(AGNT) --> [own] --> (OBJ)
      ---> [New_York_Post]] --> (PTIM) --> [???].
  ```

- Answer:

  ```
  /*   A1: Rupert Murdoch owned the New York Post from
      1976 to 1988.
  */
  [[Rupert Murdoch] -->(AGNT) --> [own]
      --> (OBJ) --> [New_York_Post]]
      --> (PTIM) --> [year: (1976,1988)].
  ```

3.3   Cost benefit analysis

Suppose one only plans to do a small number of queries Q, over a database of N CGs. The question arises whether it is worth the overhead of creating a

Levinson Method III database, in which retrievals and insertions take approximately $\log_{10}^2(N)$ comparisons for a database of size N. The alternative is simply to compare each query to every CG.

Here we do the necessary math to aid in the decision: No precompilation of database: cost is $N * Q$ graph comparisons. (call this Method I)

Creating method III database: average cost per insertion is approximately $\log_{10}^2(\frac{N}{2})$, giving a cost to create the database of $N \log_{10}^2(\frac{N}{2})$ and a cost to answer the queries of $Q \log_{10}^2(N)$.

So the question is when NQ is $< (N \log_{10}^2(\frac{N}{2}) + Q \log_{10}^2(N))$.

Here is a table of results:

N	Q	Method I Cost	Method III Cost
10	1	10	5.00
10	10	100	14.90
10	100	1,000	104.80
100	1	100	296.60
100	10	1,000	328.64
100	100	10,000	688.60
1,000	1	1,000	7,293.43
1,000	10	10,000	7,374.44
1,000	100	100,000	8,184.40
1,000	1,000	1,000,000	16,284.44
10,000	1,000	10,000,000	152,823.78
10,000	10,000	100,000,000	296,823.78

Table 1. Comparison table between costs of respective methods

From this it can clearly be seen, that Method III is most cost effective except for very small ratios of queries to database size, and for very high ratios the savings grow exponentially.

Levinson's Methods IV-VI provide further order of magnitude improvements over these numbers.

4 Limitations of the System

4.1 Open questions and concerns

The translation from an English sentence to a CG is not, by all means, unique. One illustrative example might be the following sentence.

Rupert Murdoch owns Fox.

Which can be rendered either as a CG showing a characteristic (possession) or as a standard agent-verb-object relation.

```
[Rupert Murdoch] --> (POSS) --> [Fox].

[Rupert Murdoch] --> (AGNT) --> [own] --> (OBJ) --> [Fox].
```

Apart from the need for consistency, it isn't clear which of the above alternatives is better and why. Do we need to add graph transformation rules to exploit these equivalencies (in addition to the type hierarchy)? Should we create a mechanism for inferencing? At this time, these are still open questions.

5 Acknowledgements

The work on this project was done in collaboration with the research staff of TextWise, Syracuse, New York. TextWise and Manning and Napier Information Systems have provided the entire financial support for the project. Some of the work is part of a larger SBIR commercialization project at TextWise administered by Rome Labs. All commercial rights are the property of TextWise and Manning and Napier Information Systems.

Relevant papers on which this work is based [1, 2, 3, 4, 5, 6, 7, 8] are included in the list of references that follows:

References

1. C. Colin and R. Levinson, "Partial order maintenance," *Special Interest Group on Information Retrieval Forum*, vol. 23, no. 3,4, pp. 34–59, 1988.
2. G. Ellis, R. Levinson, and P. Robinson, "Managing complex objects in PEIRCE," *Special Issue on Object-Oriented Approaches in Artificial Intelligence and Human-Computer Interaction (IJMMS)*, 1994. To Appear.
3. R. Hughey, R. Levinson, and J. D. Roberts, eds., *Issues in Parallel Hardware for Graph Retrieval*, 1993.
4. R. Levinson, "A self-organizing retrieval system for graphs," in *AAAI-84*, pp. 203–206, Morgan Kaufman, 1984.
5. R. Levinson, "Pattern associativity and the retrieval of semantic networks," *Computers and Mathematics with Applications*, vol. 23, no. 6-9, pp. 573–600, 1992. Part 2 of Special Issue on Semantic Networks in Artificial Intelligence, Fritz Lehmann, editor. Also reprinted on pages 573–600 of the book, Semantic Networks in Artificial Intelligence, Fritz Lehmann, editor, Pergammon Press, 1992.
6. R. Levinson and G. Ellis, "Multilevel hierarchical retrieval," *Knowledge-Based Systems*, vol. 5, pp. 233–244, September 1992. Special Issue on Conceptual Graphs.
7. R. Levinson and G. Fuchs, "A pattern-weight formulation of search knowledge," Tech. Rep. UCSC-CRL-91-15, University of California Santa Cruz, 1994a. Revision to appear in Computational Intelligence.
8. R. A. Levinson, "Uds: A universal data structure," in *Proc. 2nd International Conference on Conceptual Structures*, (College Park, Maryland USA), pp. 230–250, 1991.

PCCG: An Operational Tracked Grid for Creating Conceptual Graphs

Randy P. Wolf and Harry S. Delugach

Computer Science Department
The University of Alabama in Huntsville
Huntsville, AL 35899 U.S.A.

Abstract. A method of acquiring knowledge from natural language (NL) and multiple experts into a standard knowledge representation is provided by the PCCG (Personal Constructs as Conceptual Graphs) operational tracked grid. A tracked grid is similar to a repertory grid with the difference that operational synthetic problems can be acquired naturally via an NL interface. The PCCG tracked grid provides an ability to guide a semi-automatic process of converting NL phrases to conceptual graphs. These phrases may be either stand-alone sentences or the elements and constructs of a tracked grid. If the phrases are from grids and from different experts then the PCCG operational grid will use graph similarity in close conjunction with case based (CB) reasoning to guide consistency checking and integration of the knowledge of the different experts.

1 Introduction

It is commonly accepted that having a method of transfering natural language to a computer would be beneficial. It is also very well known that it is much easier to desire this ability than it is to produce the ability. Numerous systems exist which exhibit this ability to some degree; more or less, mainly less. The PCCG system provides some aspects of this ability by concentrating on using two mutually supporting theoretical bases. The PCCG system does *not* presently use the traditional NLP theoretical base. PCCG does use the theoretical bases supplied by personal construct theory and conceptual graphs. An NLP system is; in general, charged with the task of speaking and understanding a natural language. This is a much more difficult task than the task performed by PCCG which is to inspect individual NL phrases and use a preset sequence of NL queries to create an approximate representation of the meaning of the NL phrase. This meaning is described using the notation of conceptual graphs. Once these conceptual graphs exist, then they may be compared. If the conceptual graphs are translations of constructs from repertory grids, then inconsistencies and commonalities may be detected. The theoretical basis for PCCG was described as the CCG method in [WolfDelugach96].

2 Implementation Method

The means of implementation chosen for PCCG is to use tracked grids. The theory of tracked grids are explained in [Wolf96] and [Wolf97]. Track grids are a form of repertory grid which are aware of their own nature. A repertory grid is based on asking the question "what constructs" about its elements. This question applies internally and implicitly within a single grid. Repertory grid systems also use questions to create new repertory grids and refer to this as 'laddering'. Tracked grids recogonize that both questions act as 'tracks' which are simultaneously questions and new grids. Recognition of the existance of the implicit, unstated question within single repertory grids and their similarity to laddering questions leads to the realization that other questions are possible. By ignoring the myriad of other types of questions, and concentrating on questions about operationaltiy and function, a particular set of tracked grids can be created which very naturally describe and provide a desired ability. In this case, the tracked grids provide the ability to inspect individual NL phrases and use a preset sequence of NL queries to create an approximate representation of the meaning of the NL phrase. The NL queries are naturally represented as track questions. The comparison process which detects inconsistencies and commonalities will be described as an operational tracked grid.

3   Method of Use

PCCG will be used to help convert a single NL sentence to a conceptual graph or to help integrate multiple grids created by multiple experts. There is a tutorial mode based on teaching the user to use PCCG to translate a single NL sentence. The basic interaction method for the PCCG operational tracked grid will be textual. Fig 1 describes state diagram of the system. Fig 2 provide architectural detail on the overall system.

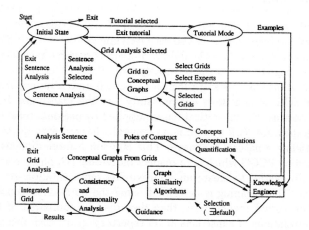

Fig. 1. State Diagram of PCCG Operation.

The first action of PCCG would be to determine via a NL question which form of use is desired. If the tutorial is selected, the user will be guided through the NL-phrase-to-CG process for a small set of phrases. Some phrases will be full sentences and some will be sentence fragments which refer to some element that will be described to the user by the tutorial. The intent of the tutorial will be to give a novice

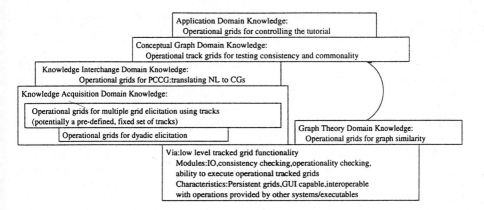

Fig. 2. Conceptual Design of PCCG and Supporting Architecture.

user training in the use of PCCG for both translating NL sentences and translating constructs. Comparison of conceptual graphs is handled by CB reasoning within PCCG and these cases must be built into the operational grid by the meta-KE.

If the function selected is to translate a NL sentence, then the user will be asked the questions as described in [WolfDelugach96] which elicit concepts, conceptual relations, and quantifications. Because the knowledge acquisition process is concentrating on acquiring the formal definition of the conceptual graph, neither a visual conceptual graph nor a textual conceptual graph is the result of the acquisition. The acquisition result is the predicate calculus statement which is the formal meaning of the conceptual graph. The translation result which is displayed will be a textual conceptual graph form of the predicate calculus formal semantic meaning.

If the primary function of PCCG is selected, then the user will be prompted to select the grid(s) which are to be analyzed. It is possible that multiple experts created their varying (or identical) opinions within a single grid. The system will be able to determine; by inspecting all existing tracks of the form "who created", the experts who are associated with the elements of the selected grid(s). In those cases where grids, elements, constructs, and/or poles share multiple creators, the system will inquire whether these shared items should be included in consistency and commonality checking. Other items within the grid(s) of interest are then filtered according to whether there presently exists a conceptual graph definition of the grid item. Any grid items which do not have a conceptual graph definition are passed along to the knowledge engineer (KE) for translation.

The KE translates as many of these items as desirable and then begins the comparison process. The assumption of this system is that the repertory grids which are being analyzed already exist before PCCG is activated. Another assumption which a different form of PCCG might make would be that PCCG should be applied during the original grid creation process. Since there already exist many repertory grids, there exists a need for a PCCG capability to analyze these pre-existing grids which is done by the present system. The integration of PCCG with the grid methodology for acquiring individual and multiple grids is future work. By choosing to first implement the present form of PCCG, it will be possible to concentrate on the issues relating solely to PCCG.

The results of comparison will be a description of related grid entities. One graph similarity algorithm will be the default similarity algorithm. The grid editing ability provided by the *via* prototype will provide the ability to modify existing conceptual graph translations of grid items. This is because the conceptual graphs will be defined using the structure of tracked grids. Modifying the conceptual graphs is equivalent to modifying the associated tracked grids.

The most appropriate graph similarity algorithm will be the default graph similarity algorithm and will define the (sub)cases used by the CB reasoning component of the system. The CG translations of poles which are similar in ways defined within the operational grid for PCCG are by definition cases as seen by CB reasoning. The tracked grids which define these cases will potentially have additional discriminating ability to determine subcases. Associated with these (sub)cases will be operational grids which define the necessary queries needed to compensate for detected inconsistency and commonality.

4 Summary

The PCCG system will by no means be a mature system of its type. It will be the first system of its general type. Despite the fact that CCG has two well-known and established theoretical progenitors, the combination of these two theoretical bases is only beginning to be explored. For example, it is reasonable to wonder about the relation of established ontologies to the results of PCCG analysis. Little work has been done to establish the relationship. This falls under the category of necessary future work. Tracked grids certainly have the ability to capture an ontology; the ability the capture the technique for capturing an ontology. The comparison process between such tracked grid versions of ontologies and of PCCG results is outside the scope of the present work.

References

[WolfDelugach96] Randy P. Wolf and Harry S. Delugach, "Knowledge Acquisition via the Integration of Repertory Grids and Conceptual Graphs", in *Auxiliarry Proceedings, 4th International Conference on Conceptual Structures*, (pp. 108-120), P.W. Eklund, G. Ellis and G. Mann, eds., 1996, University of New South Wales, Sydney, Australia, Aug. 19-23 1996. ISBN 0 7334 1387 0

[Wolf97] Randy P. Wolf and Harry S. Delugach, "An Example of Knowledge Acquisition Via Tracked Grids", Randy P. Wolf and Harry S. Delugach, in *10th European Workshop on Knowledge Acquisition, Modeling, and Management* (submitted) Catalonia, Spain, Oct. 15-18, 1997.

[Wolf96] Randy P. Wolf and Harry S. Delugach, "Knowledge Acquisition Via Tracked Repertory Grids", Technical Report No. TR-UAH-CS-1996-02, Computer Science Department, Univ. of Alabama in Huntsville, 1996, ftp://ftp.cs.uah.edu/pub/techreports/TR-UAH-CS-1996-02

Author Index

Springer
and the
environment

At Springer we firmly believe that an
international science publisher has a
special obligation to the environment,
and our corporate policies consistently
reflect this conviction.
We also expect our business partners –
paper mills, printers, packaging
manufacturers, etc. – to commit
themselves to using materials and
production processes that do not harm
the environment. The paper in this
book is made from low- or no-chlorine
pulp and is acid free, in conformance
with international standards for paper
permanency.

 Springer

Lecture Notes in Artificial Intelligence (LNAI)

Lecture Notes in Computer Science